C0-AUT-958

THE BANNING OF ANTI-PERSONNEL LANDMINES

The legal contribution of the International Committee of the Red Cross

The International Committee of the Red Cross (ICRC) has played a key role in the international efforts to ban anti-personnel landmines. This book provides an overview of the work of the ICRC concerning landmines from 1955 to 1999. It contains ICRC position papers, working papers, articles, and speeches made by its representatives to the international meetings convened to address the mines issue, including the 1995–1996 Review Conference of the 1980 Convention on Certain Conventional Weapons and the diplomatic meeting which adopted the Ottawa treaty banning anti-personnel mines. These documents provide critical insights into the development of international humanitarian law on this issue, and will form the basis for discussions on landmines and other conventional weapons for years to come.

Louis Maresca is Legal Advisor to the ICRC's Mines/Arms Unit.

Stuart Maslen is an independent consultant who advised the ICRC on landmines between 1997 and 1998.

THE BANNING OF ANTI-PERSONNEL LANDMINES

The legal contribution of the International Committee of the Red Cross

EDITED BY

LOUIS MARESCA

STUART MASLEN

PUBLISHED BY THE PRESS SYNDICATE OF THE UNIVERSITY OF CAMBRIDGE
The Pitt Building, Trumpington Street, Cambridge, United Kingdom

CAMBRIDGE UNIVERSITY PRESS
The Edinburgh Building, Cambridge CB2 2RU, UK http://www.cup.cam.ac.uk
40 West 20th Street, New York, NY 10011–4211, USA http://www.cup.org
10 Stamford Road, Oakleigh, Melbourne 3166, Australia
Ruiz de Alarcón 13, 28014 Madrid, Spain

First published 2000

Printed in the United Kingdom at the University Press, Cambridge

Typeface Minion 10½/14 pt *System* QuarkXPress™ [SE]

A catalogue record for this book is available from the British Library

ISBN 0 521 78317 8 hardback

The International Committee of the Red Cross (ICRC) is an impartial, neutral and independent organization whose exclusively humanitarian mission is to protect the lives and dignity of victims of war and internal violence and to provide them with assistance. It directs and coordinates the international relief activities conducted by the Movement in situations of conflict. It also endeavours to prevent suffering by promoting and strengthening humanitarian law and universal humanitarian principles. Established in 1863, the ICRC is at the origin of the International Red Cross and Red Crescent Movement.

CONTENTS

PART 3 THE OTTAWA PROCESS. FROM REGIONAL INITIATIVES TO AN INTERNATIONAL PROHIBITION OF ANTI-PERSONNEL MINES

FOREWORD BY CORNELIO SOMMARUGA

PRESIDENT, INTERNATIONAL COMMITTEE OF THE RED CROSS

The signing by 123 States of the Convention on the Prohibition of Anti-personnel Landmines (Ottawa treaty) in December 1997 was the culmination of lengthy efforts to lay down international rules against the use of anti-personnel mines. Only nineteen months earlier many in the international community had been disappointed by the failure of the 1995–1996 Conference reviewing the 1980 Convention on Certain Conventional Weapons to take decisive action against anti-personnel mines. Rather than consider the issue closed, governments and civil society continued to push for the comprehensive ban they felt was essential to halt the carnage caused by this weapon. By late 1997, some fifty governments had committed themselves to the treaty. When the signing ceremony was held this number had more than doubled and it took only ten months to attain the forty ratifications needed to bring the treaty into force. This was the fastest-ever entry into force of a multilateral arms-related agreement.

Bringing about a ban on anti-personnel mines was truly a remarkable achievement. Never before had such a diverse group of governments, organizations and UN agencies come together to put an end to a crisis of this type; never before had so many people around the world felt compelled to voice their outrage at the effects of a weapon designed to strike indiscriminately at soldiers and civilians alike. This had been a unique alliance between civil society and governments to bring into existence a treaty of international humanitarian law. The process involved was a true manifestation of what humanitarian law describes as 'the dictates of public conscience' and showed how that concept can change the world.

Like other humanitarian organizations, the International Committee of the Red Cross (ICRC) has been a direct witness to the horrific effects of landmines in war-torn societies. It was in the late 1980s and early 1990s that our medical staff began to sound the alarm, warning that the mines' impact on civilians had reached intolerable levels. The ICRC came to believe that a total prohibition on anti-personnel mines was the only truly effective solution to the crisis; existing restrictions were not effective or were not being followed. From the humanitarian point of view, a total ban was the only viable option.

Given the modest results of the 1995–1996 Review Conference, the ICRC and dozens of National Red Cross and Red Crescent Societies embarked on an unprecedented public campaign to raise awareness of the landmine problem, the need for a treaty banning them and the plight of mine victims themselves. The Red Cross / Red Crescent campaign was complemented by the extremely effective activities of the International Campaign to Ban Landmines, which was awarded the 1997 Nobel Peace Prize for its work. The campaign's goal was to stigmatize the weapon in the public mind and help generate the political will to outlaw it.

From the beginning, the ICRC stressed the need for a comprehensive ban based on an unambiguous definition of the anti-personnel mine. As the weapon was already in widespread use, its production, stockpiling and transfer also had to be prohibited. An essential step forward was to remove the ambiguity found in amended Protocol II as to what an anti-personnel mine actually was. A ban based on an imprecise definition might result in attempts to bypass the prohibition; and it risked having little impact in the field. Mine clearance, mine awareness and programmes to help the victims were also essential. Through the tireless efforts of many in the diplomatic and non-governmental-organization communities, all these elements came to be included in the Ottawa treaty.

The ban on anti-personnel mines bears witness to the power of humanity. Even States unable for the moment to adhere to the Ottawa treaty have nevertheless recognized the high price in human terms that is paid for these weapons. Many non-signing States have instituted a moratorium on their export, shown a willingness to reconsider their military doctrine, and are searching for alternatives that will allow them to comply with the ban at a future date.

Despite the speed with which a ban was achieved, the materials presented here are evidence of the huge investment in expertise from the medical, legal and political realms that was required by the process. Owing to its mandates in the fields of both humanitarian assistance and international humanitarian law, the ICRC was able to bring these resources together in a unique and credible manner. Yet this compilation of ICRC materials is only part of the broader mass of documents prepared by humanitarian agencies and non-governmental organizations. For its part, the ICRC is delighted to have been part of the effort to bring about the ban. Our organization remains committed to strengthening further and universalizing the law and to continuing its work to assist the victims of both mines and other effects of war.

FOREWORD BY AMBASSADOR JACOB S. SELEBI

SOUTH AFRICA

Throughout the turbulent times of our recent history, the International Committee of the Red Cross has persistently and unstintingly focused the attention of the world on the devastation caused by war on innocent people and sought to strengthen the rules and principles of international humanitarian law, in order to save lives and alleviate suffering caused during and after armed conflict.

The ICRC was amongst the first campaigners to address the horrifying civilian casualties caused by anti-personnel mines long after conflicts have ended. On the African continent, which is particularly affected by this landmine scourge, these mines remain hidden to prey on those who venture out to seek firewood or to fetch water for the family, or dare to hunt or to plough the fields. These deadly killers lay to waste economic infrastructure and stifle socio-economic development.

Governments have for too long argued that anti-personnel mines are a necessary instrument of war. Nations have acquired large quantities of these mines in the misguided belief that such weapons will give them security.

The sustained efforts by the ICRC, amongst others, to raise awareness of the effect such mines have on civilians, prompted States to recognize that the right of parties to an armed conflict to choose methods of warfare is not unlimited. From this belief was born the Convention on Prohibitions or Restrictions on the Use of Certain Conventional Weapons Which May be Deemed to be Excessively Injurious or to Have Indiscriminate Effects (CCW). As an instrument of international humanitarian law, the objective of this Convention is to save lives and alleviate suffering during armed conflict. However, realizing that the only lasting solution to the anti-personnel mine problem is the banning of such mines, and mindful of the limitations of the CCW to achieve this goal, the Ottawa process was initiated. This process built momentum towards the conclusion of a legally binding international agreement, complementary to the CCW, to ban anti-personnel mines.

The Ottawa process stemmed from the recognition that the extreme humanitarian and socio-economic costs associated with the use of these landmines required urgent action on the part of the international community to ban and eliminate this scourge to society.

As a result of this initiative we now have the Convention on the Prohibition of the Use, Production, Transfer and Stockpiling of Anti-Personnel Mines and on their Destruction, a treaty which sets a benchmark in the achievement of international disarmament, and establishes an international norm by also addressing humanitarian concerns.

The provisions for assistance to landmine victims and the requirements to clear emplaced landmines in support of humanitarian assistance and economic development, especially in the field of agriculture, are central to the comprehensiveness of the treaty. This Convention is not only about banning a particular type of weapon, but also about the restoration of communities which have literally been crippled by the presence of these mines in their midst. It provides us with the tool to move swiftly into action now to meet these challenges. The major issues requiring urgent attention are the task of coordination, of the removal of the millions of emplaced landmines which are causing thousands of casualties each year, and addressing the priority needs of mine victims, in terms of both adequate medical attention and rehabilitation, as well as social and economic reintegration. These elements are interdependent and need to be addressed in a comprehensive manner.

The conclusion of this Convention would not have been possible had it not been for the success of the ICRC and the International Campaign to Ban Landmines, along with many other non-governmental organizations involved with the landmine issue, in generating widespread public support for such a ban. Without the ground-swell of public opinion in favour of a ban, we would probably not have seen such a high level of political will to ban these mines on the part of governments around the world.

This publication, documenting the development of international humanitarian law on landmines, is a fitting tribute to the lobbying, campaigning, negotiating and commitment of our leaders, civil society, diplomats, military, landmine survivors and so many others who made the banning of anti-personnel mines a reality.

Ambassador Jacob S. Selebi was president of the Oslo Diplomatic Conference on an International Total Ban on Anti-Personnel Landmines. He is currently High Commissioner for the Police in the Republic of South Africa.

FOREWORD BY AMBASSADOR JOHAN MOLANDER

SWEDEN

Fourteen years after its adoption, preparations began for the First Review Conference of the 1980 Convention on Prohibitions or Restrictions on the Use of Certain Conventional Weapons Which May be Deemed to be Excessively Injurious or to Have Indiscriminate Effects (CCW) and, in particular, its Protocol II on Landmines. Governments approached the subject matter warily. At the very outset, few proposals for amendments were made – and they were modest at best. Generally, governments considered the Protocol a good treaty text. The problem was rather insufficient adherence and lack of implementation. Not until the third meeting of the preparatory Group of Governmental Experts did one country (Sweden) formally submit a proposal for a ban on the use of anti-personnel mines.

However, the cumbersome diplomatic process, based on universality and consensus, set in motion a chain reaction that was difficult to foresee. It created the ideal focal point for the international efforts to ban landmines. The haggling over seemingly unimportant details and procedure in comfortable Geneva, on one hand, and the nameless suffering of children, women and men torn to pieces by the hidden killers in the rice paddies of Cambodia, the valleys of Afghanistan or the fields of Angola, on the other – this contrast was too stark, too brutal not to bring home the message to millions around the globe that anti-personnel mines represent an evil that must be stopped.

The complications of the review process grew. Some mine-using and mine-producing countries stiffened their resolve to make only concessions that would be compatible with continued routine use of anti-personnel mines, at least in international conflicts. Changing instructions to other delegations, however, went in the opposite direction; the number of countries supporting a total ban was steadily growing. A fourth meeting of the preparatory Group had to be added in January 1995. Its final – and still heavily bracketed report – was adopted, thanks to exhaustion, only at 5 a.m. on a bleak Geneva Saturday morning.

When the Review Conference itself finally convened in Vienna, in September 1995, differences between delegations had again increased. So, however, had the expectations for substantive results among non-governmental organizations and the public at large.

Thus, the only success of the Vienna session, its adoption of Protocol IV banning blinding laser weapons, went unnoticed in the uproar created by the failure to agree on an amended Protocol II on landmines.

The new delay, until the Conference reconvened in Geneva in May 1996, was effectively exploited by international campaigners to galvanize public opinion. The President of the Review Conference, for his part, made good use of the time for intensive shuttle diplomacy, which took him to a number of capitals. He also arranged informal and private consultations in Geneva. Eventually, the necessary compromises were struck and the amended Protocol adopted. The political price for failure had become too high.

The outcome of the review was in substance far better than could have been expected originally. Let me just point out the extension of the scope of application to internal conflicts, the prohibition of non-detectable anti-personnel mines, the first transfer provisions in a humanitarian law treaty regarding a widely deployed weapon, the strict responsibility of mine-laying parties, the provisions aimed at protecting humanitarian missions, etc. Still, a total ban was unobtainable in a forum based on universality and consensus; that is where the Ottawa process took over. In a setting which excluded a number of major mine-using and mine-producing countries, some hundred States were able to translate their understanding of the 'dictates of public conscience', in the words of the Martens clause, into a treaty text banning anti-personnel landmines.

So where are we now?

The reality is that a number of countries, including major military powers of the North and the South, as well as key countries in conflict-prone regions, are ready to restrict in different ways the use and transfer of anti-personnel mines but are definitely not yet ready to renounce their use.

At the same time, a vast number of States, including significant military powers, have made the decision that the humanitarian cost of continued use of anti-personnel mines outweighs their military utility and have therefore banned them. It is also clear that some States, which currently depend on anti-personnel mines in their defence planning, would be willing to renounce their use, as and when alternative area-denial technologies develop.

It is my personal view that the two new international instruments are inadequate to respond to this situation, which is inherently dynamic.

Ideally, the two Conventions should complement each other, making the passage from the restrictions under Protocol II to the prohibition under the Ottawa Convention as easy as possible. As they now stand, there is a risk that they are seen as in some way antagonistic.

Protocol II of the CCW contains significant new restrictions and prohibitions, but they could be undermined if use is made of the unnecessarily long transition periods allowed. Furthermore, the text has become so complex that it could hardly survive another amendment, unless it were a radical and simple one – a total ban on anti-personnel mines.

The Ottawa Convention could have been formulated as a potential 'Protocol V' of the CCW permitting States Parties to add to their obligations the key provision not to use, transfer, produce or stockpile anti-personnel mines. It seems to me, however, that the Ottawa drafters have added elements to the treaty which make it unacceptable to a number of countries even if, at some point, these countries were ready to undertake to ban anti-personnel landmines. In plain words: the Ottawa Convention provides excuses not to join it.

I have two particular points in mind.

The first – which might be of minor importance – is the redefinition of an 'anti-personnel mine' under the Ottawa Convention. Admittedly, the new definition is marginally better, from the humanitarian point of view, than the one under Protocol II because it is more specific. The definition in the Protocol, however, was elaborated with the concurrence of all major military powers. If we want them to join, was it a clever move to change their definition?

The second point is more troublesome. It was evident from the proceedings of the CCW Review Conference that a number of important countries vehemently resist the idea of verification, in particular of production and stockpiling, for reasons ranging from principle to practicality. Still, the Ottawa Convention contains provisions on verification which make an unusual combination of total intrusiveness in principle and considerable weakness in practice.

The scope, in principle, for inspections is extremely far reaching, while the political filter for triggering it is so tightly knit that only a politically totally isolated State would run the risk of being the object of challenge inspections under the Ottawa Convention.

Thus, in my view, the treaty provides an excuse for countries resisting intrusive inspections regimes in principle not to join it. At the same time, it sets an unsatisfactory precedent for verification procedures in other disarmament negotiations, in particular if they deal with offensive capabilities.

In conclusion, we have two international instruments dealing with the same weapon, albeit on the basis of slightly different definitions.

Protocol II is universal in the sense that it was adopted by consensus in a global forum. In substance, however, it does not go far enough. It represents an unsatisfactory lowest common denominator.

The Ottawa Convention, in substance, attains what the 'dictates of public conscience' will eventually dictate to all. It, however, also contains provisions unrelated to its essential subject matter, which create obstacles – which I personally find unnecessary – to future adherence by those who did not participate in the process, and who currently use and produce anti-personnel landmines.

The landmine crisis is not over and will not be over for a long time. The situation calls for continued efforts in diplomacy including international humanitarian law, as well as in mine clearance, assistance, etc.

Efforts must continue to press for adherence to both international instruments. While it is clear that some countries will not be able to join the Ottawa Convention, they must be made to ratify the amended Protocol II, which they have adopted. Likewise, it is equally important that countries that ratify the Ottawa Convention do not forget to adhere to the CCW. While insufficient, Protocol II contains important advances in international humanitarian law. Furthermore the CCW as a whole is an international Convention of global importance to the continued development of international humanitarian law, as demonstrated by the adoption of Protocol IV on blinding laser weapons.

Finally, the CCW represents the only accepted global forum in which mine-users and those favouring a ban can continue to discuss landmine issues. It is therefore all the more important that all those that adhere to the Ottawa Convention also adhere to the CCW. The relationship between the two instruments must be discussed. It would be unfortunate if the present situation resulted in two different camps, each with its own Convention. It would bring the mine issue no further and the 1980 CCW, created as a dynamic instrument elaborated to provide legal responses to dangerous technological developments, would fall victim and risk becoming irrelevant.

There is thus a continued need for dialogue and creative diplomacy.

As is clear from this publication, the ICRC has already made an extremely valuable contribution to both the CCW and the Ottawa treaty. I am confident that it will continue its fruitful work on behalf of landmine victims and these important treaties in the years ahead.

Ambassador Molander was president of the 1995–1996 Review Conference of the States Parties to the 1980 CCW and is currently the Permanent Representative of Sweden to international organizations in Geneva, Switzerland.

The notes for each of the original documents reproduced in this publication are to be found at the end of the documents concerned.

INTRODUCTION

The International Committee of the Red Cross (ICRC) is perhaps best known for its work in the midst of armed conflict bringing aid and assistance to the injured and sick, prisoners of war and civilians affected by the fighting. Yet, the institution also has a long history of being closely involved in the development of international humanitarian law. It was the efforts of the founder of the ICRC, Henry Dunant, which led to the adoption of the first humanitarian law treaty, the 1864 Convention for the Amelioration of the Condition of the Wounded in Armies in the Field. Since that time the ICRC has continued to play an important role in the subsequent development of humanitarian law. It prepared the drafts which were the bases for the negotiations of the 1929 Geneva Convention on Prisoners of War, the four Geneva Conventions of 1949 and the Additional Protocols adopted in 1977.[1]

The formal basis for the ICRC's role in this area is found in the Statutes of the International Red Cross and Red Crescent Movement. This movement, which is comprised of the ICRC, National Red Cross and Red Crescent Societies and their International Federation, and which works

[1] The full titles and dates of signature or adoption are: Convention Relative to the Treatment of Prisoners of War, signed at Geneva 27 July 1949; Convention (I) for the Amelioration of the Condition of the Wounded and Sick in Armed Forces in the Field, signed at Geneva 12 August 1949; Convention (II) for the Amelioration of the Condition of the Wounded, Sick and Shipwrecked Members of the Armed Forces at Sea, signed at Geneva 12 August 1949; Convention (III) Relative to the Treatment of Prisoners of War, signed at Geneva 12 August 1949; Convention (IV) Relative to the Protection of Civilian Persons in Time of War, signed at Geneva 12 August 1949; Protocol Additional to the Geneva Conventions of 12 August 1949, and Relating to the Protection of the Victims of International Armed Conflicts (Protocol I), adopted at Geneva 1 June 1977; Protocol Additional to the Geneva Conventions of 12 August 1949, and Relating to the Protection of the Victims of Non-International Armed Conflicts (Protocol II), adopted at Geneva 1 June 1977.

closely with all States Parties to the Geneva Conventions of 1949,[2] has charged the ICRC with the task of working '*for the understanding and dissemination of knowledge of international humanitarian law applicable in armed conflicts and to prepare any development thereof*'[3] (emphasis added). Thus, the ICRC's involvement in the development of a ban on anti-personnel landmines was neither a unique nor an unexpected action. It was part of a long tradition.

What was unique, however, was the public campaign the ICRC embarked upon to raise awareness about the landmine problem and the plight facing mine victims and to help create the political will in favour of a treaty banning anti-personnel mines. While the ICRC has spoken out publicly in the past about the use of specific weapons, such as poison gas, its dissemination activities and advocacy on the landmine issue were of a scale never before undertaken by the institution. Its campaign focused on stigmatizing anti-personnel mines in the public conscience and supporting international, regional and local efforts to address the mines problem.

Thus, expert and non-expert literature was prepared and distributed throughout the world by National Red Cross and Red Crescent Societies and ICRC delegations. The ICRC organized scores of meetings and conferences around the world in which its medical and legal experts participated, and ICRC field delegations in mine-affected countries hosted visits of hundreds of journalists, film crews and officials, including the highly publicized visit of Princess Diana to Angola. Print and television advertisements were also produced and placed in international media on a *pro bono* basis. The value of the donated space by the media is estimated to be over 3 million US dollars, with a potential audience of over 700 million people. Never before had the ICRC undertaken such an initiative.

Long before the public campaign was launched in 1995, the ICRC was engaged in efforts to reduce the effects of the weapons and to help develop restrictions or prohibitions on their use. It consulted experts and convened several meetings and seminars to shed light on the landmine problem and to study how it could be addressed, both by legal means and in the field. Of particular relevance are the expert meetings convened in the 1970s that formed the basis of the 1980 Convention on Certain Conventional Weapons (1980 CCW), the Montreux Symposium convened by the ICRC

[2] As at 1 May 2000, there were 188 parties to the 1949 Geneva Conventions. The set of treaties remains one of the most universally ratified.

[3] Article 5 (g) of the Statutes of the International Red Cross and Red Crescent Movement.

in 1993, and the ICRC's statements and contributions during the 1995–1996 Review Conference of the 1980 CCW. This work continued through the Ottawa process, and the negotiation and conclusion of the Ottawa treaty at the Oslo Diplomatic Conference in 1997.

Throughout this period the ICRC was also at the forefront of efforts to bring aid and assistance to mine victims and communities in mine-affected areas. The ICRC developed surgical standards for treating mine victims and pioneered the production of low-cost, high-quality artificial limbs. It also operated mine awareness programmes to teach those living in mined areas how to recognize and avoid potentially dangerous places. As hundreds of thousands of people remain affected by the weapons, these efforts still continue today.

The ICRC was, of course, not the only organization working to bring a comprehensive ban into existence. Parallel efforts were undertaken by the International Campaign to Ban Landmines (ICBL) which was awarded the 1997 Nobel Peace Prize for its work on this issue, National Red Cross and Red Crescent Societies, United Nations agencies and governments. The ICRC worked closely with all those striving for a comprehensive ban treaty. Today it remains engaged with these partners to ensure that the Ottawa treaty is implemented and respected.

This book is a chronology of the major events and a compilation of key documents charting the ICRC's contribution to the 1980 CCW, to the CCW's Review Conference held in 1995–1996, and to the Ottawa process. It is published to facilitate research and reflection on the role of the ICRC in international efforts to respond to the landmines problem. The documentation includes statements and declarations made at major conferences and meetings, the texts of many ICRC publications on mine-related issues, ICRC reports and contributions submitted to negotiating sessions as well as other texts on legal issues and victim assistance. The first part of the book provides an overview of the humanitarian law principles applicable to all weapons and the development of the rules regulating the use of anti-personnel mines in Protocol II of the 1980 CCW. The second part covers the period leading up to and through the 1995–1996 CCW Review Conference. Finally, the third part covers the Ottawa process and the development and adoption of a comprehensive ban treaty.

PART 1

From principles to rules: regulating mines up to the 1980 Convention on Certain Conventional Weapons

1

Historical background: the international law governing weapons

International humanitarian law is the branch of international law concerned with the waging of warfare.[1] It regulates the conduct of hostilities and the treatment of those not actively participating in the conflict (namely, civilians, the wounded and sick, and prisoners of war). It seeks to minimize suffering and ensure that both combatants and civilians are treated humanely. Although international treaties on the subject are of fairly recent origin, practices regulating armed hostilities are evident throughout history. Even before there were States, battles fought between tribes, clans or other groups were often governed by rules to mitigate the effects of armed violence. The ancient texts of many civilizations show that in war, prisoners were not to be killed but taken and well treated; women, children and the elderly were not to be harmed; and warriors should not use barbarous weapons or methods of attack.[2] While such practices were often founded on grounds of religion, morality or honour, they are the forerunners of the legal regime States have developed to regulate armed conflict.

International humanitarian law is based on the precept that the sole objective of war is to overpower the armed forces of the opponent.[3] Men become the legitimate object of attack solely because of their relationship

[1] International humanitarian law was traditionally known as the 'law of war' and today is also commonly referred to as the 'law of armed conflict'.

[2] See Sumio Adachi, 'A Process to Reaffirmation of International Humanitarian Law – A Japanese View', *Proceedings of the National Defence Academy*, 48 (March 1994), 437–477, on the Japanese code of behaviour 'Bushido', and Nagendra Singh, 'Armed Conflicts and Humanitarian Laws of Ancient India', in Christophe Swinarski (ed.) *Studies and Essays of International Humanitarian Law and Red Cross Principles in Honour of Jean Pictet* (Geneva: Martinus Nijhoff, 1984), pp. 531–536.

[3] H. Lauterpacht (ed.), *Oppenheim's International Law*, 7th edn (London: Longmans, 1952) vol. II, pp. 226–227.

with the making of warfare. In his renowned work *The Social Contract* (1762) Jean-Jacques Rousseau formulated one of the law's philosophical footings:

> War is in no way a relationship of man with man but a relationship between States, in which individuals are only enemies by accident, not as men but as soldiers'.[4]

From this, States have concluded that, at all times, a distinction must be made between the fighting forces of an adversary and its civilian population. Civilians cannot be the object of attack and the lives of soldiers who are wounded or lay down their weapons must be spared. Like its early antecedents, international humanitarian law is founded upon the precept that the infliction of gratuitous violence offends certain human values.

As the waging of warfare became the province of States, governments sought to ensure that many of the early practices would become legally binding rules and, in the late nineteenth century, began to codify some practices in international treaties. The 'father' of the Red Cross and Red Crescent Movement, Henry Dunant, helped initiate this process by the publication of *A Memory of Solferino* in 1863[5] as did Professor Francis Lieber, author of a document on the rules of war for government troops in the American Civil War.[6] Dunant's book drew world attention to the realities of war and the dangers posed by the 'new and frightful weapons of destruction which are now at the disposal of the nations'. His efforts prompted the Swiss government to invite many of the world powers to a diplomatic conference to adopt the first international humanitarian law treaty, the 1864 Convention for the Amelioration of the Condition of the Wounded in Armies in the Field. This helped set in motion the process through which the international community came to ban the use of exploding bullets, poison gas and bacteriological warfare and, more recently, blinding laser weapons and anti-personnel landmines.

Early international humanitarian law treaties did not specifically address deployment of landmines even though ancestors of the devices were used

[4] Jean Jacques Rousseau, *A Treatise on the Social Contract*, Book I, Ch. IV.

[5] Henry Dunant, *A Memory of Solferino* (Geneva: International Committee of the Red Cross, 1986).

[6] Instructions for the Government of Armies of the United States in the Field (General Orders No. 100) commonly referred to as the 'Lieber Code'. In addition to being one of the factors inspiring the codification of the laws of war, it was also the impetus for the development of military manuals.

in the American Civil War. These treaties did, however, prohibit the use of certain types of weapons and established a number of fundamental principles generally applicable to all weapons. Over time these principles were confirmed as part of customary international law and as such apply to all States and every side in an armed conflict.[7] Of particular relevance to the use of landmines are the following principles:

- The right of the parties to a conflict to adopt means of injuring the enemy is not unlimited.
- It is forbidden to use weapons which 'cause superfluous injury or unnecessary suffering'.
- In the conduct of hostilities, parties to a conflict must always distinguish between civilians and combatants.

From these restrictions, it follows that weapons which inflict injury or suffering greater than what is required to render a soldier *hors de combat* are prohibited. Furthermore, it is forbidden to attack civilian and soldier without discrimination and, consequently, any weapon which is inherently indiscriminate must not be used. While the development of international treaties concerned with anti-personnel landmines is discussed throughout the remainder of this book, it was these principles which were most often at the forefront of the legal discussions about the banning of the weapons. They are recognized in the preamble of the Ottawa treaty as one of the bases for the instrument's prohibitions[8] and remain valid restrictions on the use of anti-tank and anti-vehicle mines. Below is a brief overview of the international instruments outlining the development of the above-mentioned principles and providing additional historical and legal background for the comprehensive ban which came to fruition in the Ottawa treaty.

[7] International law is not only found in international treaties. Customary international law is unwritten law and is comprised of the practices which States undertake believing that they are under a legal obligation to do so. It often allows the law to develop without the need for convening formal negotiations but rather through the consensus of action. While a treaty applies only to those States that have formally adhered to it, customary law applies to all States unless they have consistently objected to the practice involved.

[8] The 11th paragraph of the preamble reads as follows, 'Basing themselves on the principle of international humanitarian law that the right of the parties to an armed conflict to choose methods or means of warfare is not unlimited, on the principle that prohibits the employment in armed conflicts of weapons, projectiles and materials and methods of warfare of a nature to cause superfluous injury or unnecessary suffering and on the principle that a distinction must be made between civilians and combatants'.

The Declaration of St Petersburg 1868[9]

The Declaration of St Petersburg is the first formal international agreement banning the use of a particular weapon. In 1868 the czar of Russia, Alexander II, invited governments to St Petersburg to 'examine the expediency of forbidding the use of certain projectiles in the time of war between civilized nations'. The impetus behind this conference was the development of a bullet which exploded upon impact with 'soft' substances, including the human body. This was an advance on an earlier bullet developed by the Imperial Russian Army, which detonated solely on hard surfaces, the primary purpose of which was to destroy ammunition wagons. When used against humans the new projectile was no more effective than the ordinary bullet yet caused injuries and suffering beyond what was required to render a soldier *hors de combat.* Recognizing the danger that the new bullets posed to the troops of all States, the representatives of nineteen governments adopted the Declaration of St Petersburg.

The declaration prohibits the use of lightweight explosive projectiles, which are defined as bullets weighing less than 400 grams and either explosive or charged with fulminating or inflammable substances. While the declaration is exceptional because it is the first formal agreement prohibiting the use of a certain weapon in war, it is also significant because it established a number of fundamental principles concerned with the conduct of hostilities and which would come to play an important role in the future development of international humanitarian law. In banning these munitions, the participating governments concluded that:

> The only legitimate object which States should endeavour to accomplish during war is to weaken the military forces of the enemy;
>
> For this purpose it is sufficient to disable the greatest possible number of men;
>
> This object would be exceeded by the employment of arms which uselessly aggravate the sufferings of disabled men, or render their death inevitable;
>
> The employment of such arms would, therefore, be contrary to the laws of humanity.

[9] Declaration Renouncing the Use, in Time of War, of Explosive Projectiles under 400 Grams Weight. St Petersburg. Entered into force 11 December 1868.

These principles build upon the canon set forth by Rousseau and from them one can conclude that the parties to a conflict do not have unlimited choice in the way they wage war, and that weapons which cause gratuitous suffering or injury or certain death are not to be used.

In renouncing the use of lightweight explosive projectiles, therefore, governments balanced the military value of such a weapon against humanitarian considerations. This balance would also become an important formula in the future examination of weapons. As was provided in the final paragraph of the declaration:

> The Contracting or Acceding Parties reserve to themselves to come hereafter to an understanding whenever a precise proposition shall be drawn up in view of future improvements which science may effect in the armament of troops, in order to maintain the principles which they have established, and to conciliate the necessities of war with the laws of humanity.

The Brussels Declaration of 1874

Following the meeting in St Petersburg, Alexander II again took the initiative and convened a conference to discuss a possible agreement outlining the laws and customs of war. Fifteen European governments attended the conference in Brussels and considered a draft treaty proposed by the Russian government. While the conference participants adopted the document with minor alterations, it was never ratified by States, and thus, did not become a binding international instrument. Article 12 of the document is particularly notable for including, in addition to the ban on the use of projectiles established in the Declaration of St Petersburg, a prohibition on the use of poison or poisoned weapons and 'arms, projectiles or material calculated to cause unnecessary suffering'. Although the text never came into force, the conference and the draft document were important steps in the movement towards the codification of the laws of war and many subsequent developments can be traced back to them.

The Hague Conventions of 1899 and 1907

A prohibition on specific types of weapons was also one result of the Hague International Peace Conference of 1899. This conference brought together twenty-six States and sought, among other things, the most effective means of 'limiting the progressive development of existing armaments' and 'the revision of the declaration concerning the laws and customs of war elaborated in 1874 by the Conference of Brussels, and not yet ratified'.[10] The Hague conference resulted in the conclusion of three conventions and two declarations relevant to the conduct of warfare, all of which were eventually ratified and became international law.

Most relevant to this discussion is Convention II and its regulations[11] which cover land warfare. Importantly, the Convention confirms the norms outlined in the Declaration of St Petersburg and those considered at the Brussels Conference.[12] It also affirms an obligation to distinguish between those persons taking part in the hostilities and those who are *hors de combat.*[13] Two declarations attached to the Convention outlaw the use of specific kinds of weapons. The first (Declaration IV, 3) prohibits the use of projectiles which expand or flatten upon entering the human body.[14] These so-called 'dum-dum' bullets cause injuries similar to the horrific wounds inflicted by the lightweight projectiles proscribed in 1868. They were developed and manufactured by the British in India for use in colonial warfare and their development and use were the subject of intense debate within and outside the United Kingdom.[15]

The second declaration (Declaration IV, 2) bans the use of projectiles diffusing asphyxiating or deleterious gases.[16] This reflects an initial attempt to ban gas warfare and its scope was later broadened by the 1925 Geneva Protocol presented below.

The Hague Convention of 1899 is also noteworthy for introducing the so-called 'Martens clause'. This clause, found in the instrument's preamble and named after its author, the Russian delegate de Martens, provides:

10 Russian Circular note of 30 December 1898.

11 Convention (II) with Respect to the Laws and Customs of War on Land and its annex: Regulation concerning the Laws and Customs of War on Land. The Hague, 19 July 1899. Entered into force 4 September 1900.

12 See Article 22 and Article 23 (e).

13 The Convention requires that prisoners of war are to be treated humanely (Article 4) and prohibits a declaration that no quarter will be given (Article 23(d)). In Article 21 it also affirms the obligations upon belligerents under the Convention for the Amelioration of the Wounded in Armies in the Field. Geneva, adopted 22 August 1864. Entered into force 22 June 1965.

14 Declaration (IV, 3) concerning Expanding Bullets. The Hague, 29 July 1899. Entered into force 4 September 1900.

15 Edward M. Spiers, 'The Use of Dum-Dum Bullets in Colonial Warfare', *Journal of Imperial and Commonwealth History* 4 (1975), 3–14. The bullets were so called because they were manufactured at the cantonment of Dum-Dum, located several miles north-east of Calcutta.

16 Declaration (IV, 2) concerning Asphyxiating Gases. The Hague, 29 July 1899. Entered into force 4 September 1900.

> Until a more complete code of the laws of war is issued, the High Contracting Parties think it right to declare that in cases not included in the Regulations adopted by them, populations and belligerents remain under the protection and empire of the principles of international law, as they result from the usages established between civilized nations, from the laws of humanity, and the requirements of the public conscience.

The Martens clause establishes a legal safety net whereby soldiers and civilians alike remain protected by basic humanitarian principles in the event that the existing rules of international law are inadequate or non-existent. It makes clear that, in the absence of positive rules, the conduct of warfare shall not be left to the arbitrary judgement of military commanders.

In 1907, a second Hague Peace Conference was held to continue the work of its predecessor. At this meeting Convention II on land warfare was slightly revised and again adopted.[17] Yet, for the most part, the rules and principles discussed above remained unchanged. The declarations on expanding bullets and asphyxiating projectiles were not reconsidered at the 1907 conference and remained as adopted in 1899.

1925 Geneva Protocol on Poisonous and Asphyxiating Gases[18]

Declaration II of the Hague Convention of 1899 prohibited the use of projectiles diffusing asphyxiating or deleterious gases. Nonetheless, during the First World War various types of chemical agents were used in gas form and dispersed into the wind through canisters on the ground as opposed to projectiles. Thus, Declaration II was not deemed to have been violated, at least in purely technical terms. However, the suffering which such toxins produced on the ground among the troops of all sides provoked outrage in both public and governmental circles. Subsequently, an international conference convened by the League of Nations adopted the 1925 Geneva Protocol which broadened the prohibition on gas warfare. The Protocol banned the 'use in war of asphyxiating, poisonous or other gases and of all analogous liquids, materials or devices'. As the instrument recognizes that such weapons have 'been justly condemned by the general opinion of the

[17] Convention (IV) respecting the Laws and Customs of War on Land and its annex: Regulation concerning the Laws and Customs of War on Land. Entered into force 26 January 1910.

[18] Protocol for the Prohibition of the Use in War of Asphyxiating, Poisonous or Other Gases, and of Bacteriological Methods of Warfare, signed at Geneva 17 June 1925. Entered into force 8 February 1928.

civilized world' and that their use is prohibited 'in Treaties to which the majority of Powers of the world are Parties', it supports and develops the principles and rules of earlier instruments. Furthermore, and with some foresight, the Protocol also banned the use of bacteriological methods of warfare, a manner of warfare which had not been extensively developed at that time.

For the most part, the treaty law regulating the use of weapons and the conduct of hostilities remained unchanged until the Additional Protocols to the 1949 Geneva Conventions, which were concluded in 1977. Nonetheless, the rules and principles that had been established comprise some of the most fundamental norms of international humanitarian law and, as was recognized by the International Military Tribunal at Nuremberg, they are part of the customary international law applicable to all States.[19] They reflect the early developments of the international law regulating the use of weapons in armed conflict and are the framework within which States, through the Ottawa treaty, came to ban anti-personnel landmines.

[19] *Trial of the Major War Criminals Before the International Military Tribunal, Nuremberg*, Vol. XXII, p. 497. See also International Court of Justice, *Legality of the threat or use of nuclear weapons*, Advisory Opinion of 8 July 1996.

2

The ICRC's draft rules to protect civilian populations 1955–1956

As early as the 1950s the ICRC identified the landmine as a weapon of special concern. One of the first efforts to limit the consequences of landmines was the ICRC's *Draft Rules for the Limitation of the Dangers Incurred by the Civilian Population in Time of War* published in 1956[20] (referred to below as *1956 Draft Rules*). These rules were issued in response to developments in armaments and military doctrine, and the widespread injury and damage to civilian populations during the Second World War. As international humanitarian law was founded on the fundamental distinction between combatants and non-combatants, the ICRC was concerned that the proliferation of so-called 'blind' weapons and methods of warfare, which were unable or failed to take this distinction into account, would severely undermine the protection it afforded. The *1956 Draft Rules* sought to affirm and supplement many of the principles found in earlier international humanitarian law treaties.

Work on the draft rules began in April 1954 with a meeting of outside experts. Based on the results of this meeting, the ICRC prepared a first draft document entitled 'Draft Rules for the Protection of the Civilian Population from the Dangers of Indiscriminate Warfare' (referred to below as the '1955 draft') which it then submitted to National Red and Red Crescent Societies and other individuals for comment. While this draft contained regulations on means and methods of warfare, there was no specific rule on the use of landmines. The ICRC nevertheless recognized the danger that these weapons posed to civilian populations. While Article 10(3) of the 1955 draft covers the use of 'delay-action' projectiles, the ICRC in its commentary discussed the broader category of 'delay-action' weapons, including both

[20] *Draft Rules for the Limitation of the Dangers Incurred by the Civilian Population in Time of War.* Published by the International Committee of the Red Cross with commentary, Geneva, April 1958.

submarine mines and landmines. Article 10(3) and the relevant parts of the ICRC commentary are reproduced below.

Following the comments of National Red Cross and Red Crescent Societies, an Advisory Working Group was constituted with a view to finalizing the document in time for the forthcoming conference of the Red Cross Movement. In the Advisory Working Group, the dangers mines pose to civilian activity were raised by several Red Cross Societies, which voiced concern about the impact of newly developed submarine mines on civilian navigation. Subsequently, in the *1956 Draft Rules*, the ICRC introduced an article to lessen the potential effects of mines in a post-conflict environment. Article 15, covering both landmines and submarine mines, is notable because, like the draft rules as a whole, it was intended to apply not only to armed conflicts between States but also to conflicts 'not of an international character'. [21] This would have set an important precedent, had the draft rules been developed into a binding international instrument. While the *1956 Draft Rules* were presented to the 19th International Red Cross Conference and submitted to governments for consideration, no action was taken to turn them into an international treaty. The use of landmines nevertheless remained subject to the norms established in the Declaration of St Petersburg and The Hague Conventions and Declarations of 1899 and 1907, which the *1956 Draft Rules* sought to reaffirm and develop.

Draft Rules for the Protection of the Civilian Population from the Dangers of Indiscriminate Warfare

Geneva
1955

Article 10, sub-paragraph 3

The use of so-called delay-action projectiles is only authorized when their effects are limited to the objective itself.

Commentary

I – The ICRC consulted the Experts who met in 1954 about the problem of incendiary and "delay-action" bombs, as being among the weapons likely to cause unnecessary damage.

[...]

[21] *1956 Draft Rules*, Article 2.

On going into the subject in greater detail, the ICRC was led to consider three cases of so-called "delay-action" weapons. The first is that of submarine mines, the use of which is, as we know, expressly governed by the Eighth Hague Convention of 1907. Then there are the delay-action mines which are used by armies during land operations and are, for example, buried in the ground or hidden in houses, and explode after the enemy has taken possession of the terrain. Finally, there is a third category – bombs and mines which are dropped from the air and have a delayed action in the sense that they explode after a given lapse of time or when they are touched.

Since the question of submarine mines is already governed by an international Convention, there is no call to consider it here. A legal solution of the second case, that of land mines etc., raises great difficulties and the ICRC's study of the subject is not sufficiently advanced to enable it to propose a rule concerning it. The Committee proposes to continue its study of the question, however – and would greatly appreciate any opinions expressed with regard to it – in view of the danger to which this type of mine exposes the civilian population when they reoccupy their homes, even after peace has been re-established.

The ICRC has accordingly confined itself, for the time being, to drafting a rule referring solely to the third category, that is to say, to delay-action projectiles – having more particularly in mind the missiles of this type which would be used in "strategic" bombing. It appeared difficult, however, to prohibit the use of such weapons completely, for if they are confined to the military objectives themselves, their use is really equivalent to repeating the attack on the objective, and that being so, the justice of prohibiting them in particular might be questioned. [...]

Draft Rules for the Limitation of the Dangers Incurred by the Civilian Population in Time of War

Geneva
1956

Art. 15 – Safety measures and devices

Paragraph I.
"If the Parties to the conflict make use of mines, they are bound, without prejudice to the stipulations of the VIIIth Hague Convention of 1907, to chart minefields. The chart shall be handed over, at the close of hostilities, to the adverse Party, and also to the authorities responsible for the safety of the civilian population."

Commentary

In its Comments on the Draft Rules (1955), the ICRC had indicated that it was continuing to study the question of mines employed in land operations. These devices, whether buried in the ground or hidden in buildings, may constitute grave dangers for the civilian population. Many children have been killed as a result of mines which exploded while, all unwitting, they were playing with them.

For their part, several Red Cross Societies drew the ICRC's attention, in their Remarks and Suggestions on the Draft Rules (1955), to the question of submarine mines. They stressed the new developments in connection with these devices, which explode not only when the enemy comes into contact with them, but also as a result of other factors (such as pressure). These Societies considered that, in consequence, the VIIIth Hague Convention of 1907 on Automatic Submarine Contact Mines no longer regulates this problem satisfactorily. The result is an increase in the danger to civil navigation.

As noted in connection with Article 5, the ICRC could not in the present Draft and at this stage, settle this particular question which is one for international law relating to maritime war.

The ICRC decided to adhere to a very broad rule after studying the question of landmines, which appears difficult to circumscribe within strict limits, and in view of its attitude to the problem of submarine mines. The content of the present paragraph, in so far as it concerns landmines, conforms to the general practice of armed forces and in so far as it concerns submarine mines, to the spirit of the VIIIth Hague Convention, the provision of which, incidentally it reserves.

A rule of this kind, however limited, may afford valuable safeguards. Moreover it forms a starting point and it will be for the military and Government Experts to say to what extent they wish to go beyond it. Would it be possible, for example, in certain cases to hand the charts of minefields to the adverse Party and to the Authorities who are responsible for the safety of the population before the termination of active hostilities?

3

Expert contributions to the Diplomatic Conference on the Reaffirmation and Development of International Humanitarian Law Applicable in Armed Conflicts 1973–1977

In its role as depository for the Geneva Conventions, the Swiss government convened a diplomatic conference in the mid-1970s to supplement the existing international humanitarian law. The Diplomatic Conference on the Reaffirmation and Development of International Humanitarian Law Applicable in Armed Conflicts (1974–1977) sought to increase the protection afforded to the victims of armed conflict, particularly against the effects of hostilities, and resulted in the adoption of the two Additional Protocols to the Geneva Conventions of 1949.

In preparing the drafts which were the bases for the diplomatic negotiations, the ICRC did not include prohibitions or restrictions on the use of particular weapons. Instead, because of the potentially sensitive nature of such discussions, it chose to restate the fundamental rules governing the use of weapons in earlier international humanitarian law treaties, which by this time were considered to be part of customary international law.

Nonetheless, in spite of the absence of additional restrictions in the drafts, the experts of nineteen governments requested the ICRC to consult with specialists on the problem of conventional weapons which may cause unnecessary suffering or have indiscriminate effects. The purpose was to have detailed reports and information available should governments wish to address the regulation of these weapons at the Diplomatic Conference.

Three expert meetings were convened between 1973 and 1976:

- Expert Meeting on Weapons that may Cause Unnecessary Suffering or Have Indiscriminate Effects, held in Geneva, 26 February to 2 March and 12 to 15 June 1973;
- Conference of Government Experts on the Use of Certain Conventional Weapons, First Session, held in Lucerne, 24 September to 18 October 1974; and

- Conference of Government Experts on the Use of Certain Conventional Weapons, second session held in Lugano, 28 January to 26 February 1976.

The Diplomatic Conference created an *ad hoc* working group to examine the issue and the group was provided with the results of the expert meetings. In the end, however, the conference chose not to include additional prohibitions or restrictions. Instead, it called for the issue to be dealt with in the framework of the United Nations. The United Nations General Assembly endorsed this recommendation which led to the negotiation and adoption of the Convention on Prohibitions or Restrictions on the Use of Certain Conventional Weapons Which May be Deemed to be Excessively Injurious or to Have Indiscriminate Effects.

Each of the three expert meetings convened by the ICRC between 1973 and 1976 produced a report outlining the types of weapons which might be considered to cause unnecessary suffering or have indiscriminate effects. These included small-calibre projectiles, blast and fragmentation weapons, time-delay weapons and incendiary weapons. The reports are closely interrelated so it is helpful to present them together. For the most part, only those sections pertaining to landmines, which were treated under the heading of 'Time Delay Weapons', are reproduced in the following pages. The paragraph numbers indicated are from the original reports. As only selected chapters are reproduced here, the numbering of paragraphs is not necessarily consecutive.

Weapons that may Cause Unnecessary Suffering or Have Indiscriminate Effects

Report on the Work of the Experts

held in Geneva, Switzerland
26 February – 2 March and 12–15 June 1973

Extracts from the 1973 report produced by the ICRC are reproduced below. The report summarizes the work of the forty-three experts from various countries, international agencies and non-governmental organizations participating in the meeting. It gives a descriptive overview of the technical characteristics of mines, the general strategy behind their use and the medical characteristics of the injuries they cause. The report does not provide analysis or conclusions on the weapon's characteristics or effects in relation to humanitarian law rules. The chapter on time delay weapons has been reproduced in its entirety. Also reproduced are the chapters on the existing legal prohibitions at that time and on the principal categories of weapons and their effects. The final remarks found at the end of the report are also reproduced.

CHAPTER I

Existing legal prohibitions or limitations regarding the use of specific weapons

1. General Principles

18. At the outset of this report a brief discussion may be useful of the concepts upon which the employment of specific weapons has in the past been viewed as prohibited. The prohibition of use of specific weapons might, of course, be agreed upon between States regardless of the concepts and criteria that may be discerned behind prohibitions adopted in the past. Nevertheless these concepts and criteria retain their validity and still offer guidance.

19. The legal concepts discussed here are of direct relevance only to the questions of the use of particular weapons and of whether that use should be deemed permissible, or be subject to or be made the subject of prohibition. These concepts have no necessary bearing on measures of disarmament in the sense of elimination of development, production and stockpiling of particular weapons, though such measures may be deemed desirable as regards one or more of the weapons discussed in this report.

20. Any legal evaluation of problems related to the use of weapons in armed conflicts must proceed from the principle that the choice of means and methods of combat is not unlimited (cf. Hague Regulations concerning the Laws and Customs of War on Land, Art. 22). Rules which specify this general principle are those which explicitly or by implication seek to prevent the use of weapons causing unnecessary suffering, and indiscriminate weapons or methods of combat. In addition, the general maxim embodied in the Martens clause (Hague Convention No. IV, preamble para. 8) may be referred to here, namely that in cases not included in applicable conventions, civilians and combatants remain under the protection and the authority of the principles of international law, as they result from the principles of humanity and the dictates of public conscience. This underlines the applicability of international law even in such cases where prohibitions of specific weapons do not appear in existing international conventions.

(a) *Unnecessary suffering*

21. The principle that weapons causing unnecessary suffering must be avoided is set forth in the Hague Regulations, Art. 23, sub-para. (e). However, the authentic French version and the English translation of this Article differ in some respects: the French text provides that "il est . . . interdit . . . d'employer des armes, des projectiles ou des matières *propres à* causer des *maux* superflus", while the English version reads "it is forbidden . . . to employ arms, projectiles, or material *calculated to* cause *unnecessary suffering*" (emphases added). The English version would be the narrower if its contents were taken to add a subjective element to the original rule. In conformity with the authoritative French text, the principle must be stated to be that – irrespective of the belligerents' intentions – any means of combat are prohibited that are *apt* to cause unnecessary suffering or superfluous injury. While the authentic French text uses the term "superfluous injury" (maux superflus), the phrase "unnecessary suffering" used in the English translation has acquired a relevance of its own through the practice of States. Hence, both concepts are of importance for the assessment of whether particular weapons shall be deemed prohibited for use.

22. The principle referred to in the preceding paragraph is, indeed, found expressed already in the preamble of the St Petersburg Declaration to the Effect of Prohibiting the Use of certain Projectiles in Wartime (29 November/ 11 December 1868), which states:

"...

(2) that the only legitimate object which States should endeavour to accomplish during war is to weaken the military forces of the enemy;
(3) that for this purpose it is sufficient to disable the greatest possible number of men;
(4) that this object would be exceeded by the employment of arms which uselessly aggravate the sufferings of disabled men, or render their death inevitable;
(5) that the employment of such arms would therefore be contrary to the laws of humanity."

The St Petersburg Declaration imposes a ban on the employment of "any projectile of a weight below 400 grammes, which is either explosive or charged with fulminating or inflammable substances". The Hague Declaration concerning the Prohibition of Dum-Dum Bullets (1899), in pursuance of the same object of avoiding unnecessary suffering, prohibits the use of "bullets which expand or flatten easily in the human body, such as bullets with a hard envelope which does not entirely cover the core or is pierced with incisions".

23. What suffering must be deemed "unnecessary" or what injury must be deemed "superfluous" is not easy to define. Clearly the authors of the ban on dum-dum bullets felt that the hit of an ordinary rifle bullet was enough to put a man out of action and that infliction of a more severe wound by a bullet which flattened would be to cause "unnecessary suffering" or "superfluous injury". The circumstance that a more severe wound is likely to put a soldier out of action for a longer period was evidently not considered a justification for permitting the use of bullets achieving such results. The concepts discussed must be taken to cover at any rate all weapons that do not offer greater military advantages than other available weapons while causing greater suffering/injury. This interpretation is in line with the philosophy that if a combatant can be put out of action by taking him prisoner, he should not be injured; if he can be put out of action by injury, he should not be killed; and if he can be put out of action by light injury, grave injury should be avoided. In addition the concepts "unnecessary suffering" and "superfluous injury" would seem to call for weighing the military advantages of any given weapon against humanitarian considerations.

(b) *Indiscriminate effects*

24. The basic principle that the parties to an armed conflict shall confine their operations to defeating the military objectives of the adversary, and shall ensure that civilians and civilian objects are respected and protected, is embodied in international instruments, e.g., the St Petersburg Declaration, preamble, para. 2, and the Hague Regulations. It was often violated by methods of indiscriminate warfare, but is nevertheless firmly established in international law. The XXth International Conference of the Red Cross (Vienna 1965), in its Resolution XXVIII, confirmed "that distinction must be made at all times between persons taking part in the hostilities and members of the civilian population to the effect that the latter be spared as much as possible". The same text was included in Resolution 2444 (XXIII), of the UN General Assembly which was unanimously adopted on 19 December 1968.

25. It follows from this principle that weapons and other means of combat must never be directed against civilians or civilian objects. This principle has not been taken to mean a ban upon weapons and other means of combat which, though directed against military targets, entail the risk of incidental civilian casualties or damage to civilian objects in the vicinity of the targets. It does imply, however, that weapons which by their nature are incapable of being directed with any certainty to specific military targets, or which in their typical or normal use are not delivered with any certainty to such targets, are in violation of this principle.

26. Some international conventions prohibit the use of certain weapons because of their indiscriminate effects. The Hague Convention No. VIII concerning the

Laying of Automatic Submarine Contact Mines prohibits "1st – to lay unanchored automatic contact mines, except when they are so constructed as to become harmless one hour at most after the person who laid them ceases to control them; 2nd – to lay anchored automatic contact mines which do not become harmless as soon as they have broken loose from their moorings; 3rd – to use torpedoes which do not become harmless when they have missed their mark". Likewise, the most significant instrument prohibiting the use of specified weapons, i.e. the Geneva Protocol of 1925 for the Prohibition of the Use in War of Asphyxiating, Poisonous and other Gases, and of Bacteriological Methods of Warfare, is at least in part based on the concept of preventing indiscriminate effects of combat.

27. The prohibition of indiscriminate warfare relates more often to methods of warfare and methods of using weapons than to specific weapons *per se*. All weapons are capable of being used indiscriminately. This is not, of course, sufficient ground for prohibiting their use in armed conflicts, but a ground for prohibiting such types of use. However, in some cases weapons may be indiscriminate by their very nature. Moreover, in some other cases the normal or typical use of the weapons may be one which has indiscriminate effects.

2. Military Manuals and Regulations

28. It is of considerable value to know in which way international law in force is reflected in some military manuals and regulations. Only a limited number of national service manuals were available for evaluation for the purpose of this report, and no manuals concerning rules of engagement have been available.[1] The manuals taken into consideration deal expressly with prohibitions of the use of weapons in armed conflicts. This permits some detailed comments touching on the implementation of the various legal principles in State practice.

29. In the light of the development which has taken place in the practice of States, confirmed by stands taken in their military manuals, the *St Petersburg Declaration* continues to apply under present-day conditions to projectiles weighing considerably less than 400 grammes that are not capable of incapacitating persons other than those who have received a direct hit, thus causing unnecessary suffering.

30. The *Hague Declaration concerning the Prohibition of Dum-Dum Bullets,* which was modelled on the lines of the St Petersburg Declaration, did not raise much legal controversy, nor did it lead to difficulties in State practice. The British Government originally objected to an express prohibition of dum-dum bullets (named after a British arsenal near Calcutta) but ratified the Declaration in 1907. The United States was not a party to the Declaration; yet US Army Department

pamphlet No. 27–161–2 (p. 45) points out that the U.S. delegation at the 1899 Peace Conference had pressed for the adoption of more stringent limitations; "The use of bullets inflicting wounds of useless cruelty, such as explosive bullets, and in general all kinds of bullets which exceed the limit necessary for placing a man *hors de combat* should be forbidden".[2] During the First World War, belligerent Parties accused each other in several instances of having used dum-dum bullets. There is, however, no reason to believe that the Declaration had not been implemented on the whole. Difficulties might arise from the somewhat uncertain definition of "bullets which expand or flatten easily in the human body". The examples given in the Declaration: "bullets with a hard envelope which does not entirely cover the core or is pierced with incisions" do not appear to be exhaustive. Hence the German ZDv 15/10 (para. 75) states that projectiles which produce the same effects by other means, e.g. hollow-point projectiles, are covered by the prohibition of the Declaration.

31. The prohibition of means causing unnecessary suffering set forth in the *Hague Regulations,* Art. 23 sub-para. (e) is reflected in almost every military manual dealing with the subject of international law of war. The United States FM 27–10 (para. 34) states that the question as to what weapons cause "unnecessary injury" can only be determined in the light of the practice of States in refraining from the use of a given weapon because it is believed to have that effect. Like the British manual (para. 110), the US manual stresses that this prohibition certainly does not extend to the use of explosives contained in artillery projectiles, mines, rockets or hand grenades. Both manuals give the same examples of means to which the prohibition of the Hague Regulations, Art. 23 sub-para. (e), would apply: lances with barbed heads, irregular shaped bullets, and projectiles filled with broken glass, the use of any substance on bullets that would tend unnecessarily to inflame a wound inflicted by them, and the scoring of the surface or the filing off of the ends of the hard cases of the bullets. The US DA PAM 27–161–2 (p. 45) adds that this amounts to an official prohibition of bullets which tear an unnecessarily large hole.

32. The use of *shotguns* during the First World War gave rise to legal controversy. In September 1918, the German Government lodged a protest with the United States against the use of shotguns by the US Army. The US Secretary of State replied in a note stating that in the opinion of the US Government the Hague Regulations Art. 23 sub-para. (e) did not forbid the use of shotguns, that in view of the history of the shotgun as a weapon of warfare, the well-known effects of its use, and a comparison of it with other weapons approved in warfare, the shotgun then in use could not be made the subject of legitimate and reasonable protest, and that the United States would not abandon its use.[3] The US DA PAM 27–161–2 (p. 45) quotes an opinion of the Office of the Judge Advocate General

of 1961 stating that, while there is no conventional or customary rule of international law prohibiting the use of shotguns as such, international law does impose restrictions on the types of bullets that may be used in both smoothbore and rifled small arms. According to the author of DA PAM 27–161–2, the legality of the use of shotguns depends on the nature of the shot employed and its effect on a soft target: while the use of an unjacketed lead bullet is considered a violation of the laws of war, the use of shotgun projectiles sufficiently jacketed to prevent expansion or flattening upon penetration of a human body, and the employment of shot cartridges, with chilled shot regular in shape, is regarded as lawful. The German ZDv 15/10 (para. 76) regards shotguns as an illegal means of warfare which offers no real military advantage while causing unnecessary suffering.

33. The prohibition of *poison or poisoned weapons* contained in Art. 23 sub-para. (a) of the *Hague Regulations* has raised the question of whether the poisoning or contamination of water supplies from which the enemy may draw drinking water can be made lawful by posting up a notice informing the enemy that the water has thus been polluted. The adoption of that practice by a commander of German troops in South-West Africa in 1915 led to a British protest which was rejected by the German Government. A former official American opinion which regarded it as legal to contaminate water by placing dead animals therein or by other means, provided that such contamination is evident or the enemy is informed thereof, was squarely in conflict with the British position (British manual, para. 112). The American FM 27–10 (para. 37) now refers to the prohibition of poison without qualifications, and so does the German ZDv 15/10 (para. 77).

34. The use *of flame throwers* and *napalm* has been a matter of dispute. The British manual (para. 110) regards these means as lawful only when directed against military targets, and states expressly that their use against personnel is contrary to the law of war in so far as it is calculated to cause unnecessary suffering. The US FM 27–10 (para. 36) states that it is not violative of international law to use weapons which employ fire, such as tracer ammunition, flame throwers, napalm and other incendiary agents, "against targets requiring their use". The US DA PAM 27–161–2 (p. 42) points out that these words have been inserted in order to preclude practices such as the wanton use of tracer ammunition against personnel when such use is not called for by a military necessity.

35. An express reference to *indiscriminate weapons* is contained in the German manuals. The HDv 100/1 Anhang Teil III (para. 607) states that flying bombs (e.g. rockets) must be designed in such a way that they can be launched against military objectives with sufficient accuracy. Accordingly, the ZDV 15/10 (para. 90) confirms that the use of such weapons would be illegal if their inaccu-

racy makes it likely for the civilian population to be hit with full power. Yet the ZDv adds that even inaccurate weapons may lawfully be employed against military targets if, due to the situation or extension of the target, there is but little danger that the civilian population will suffer disproportionately.

36. The use of *mines* is dealt with in a specific manner by the German manuals. The HDv 100/1 Anhang Teil III (para. 58) states that mines as such cannot be considered to be treacherous means, provided they are used in places where the enemy might reasonably suspect them. The HDv adds that the use of mines is forbidden in places that are exclusively used for peaceful purposes. The same rule is repeated in the ZDv 15/10 (para. 89).

37. *Delayed action weapons* and, more specifically, *booby-traps* should also be mentioned in this context. While the problems posed by their use relate apparently in the first place to methods rather than means of warfare, some weapons of this category might be considered as illegal means. The German ZDv 15/10 (para. 70) describes as prohibited booby-traps that look like "peaceful objects", e.g. fountain pens, watches, or toys. The Austrian manual (para. 40), too, declares that booby-traps camouflaged as toys are illegal means of combat.

3. Implementation and International Co-operation

38. The continuous development of new weapons and weapon systems necessitates an equally continuous assessment of such development in the light of the guiding principles of international law of armed conflicts, viz. the prohibition of use of means that cause unnecessary suffering/superfluous injury and the prohibition of indiscriminate warfare. These assessments must evidently take place in the first instance at the national level.

39. While there can be little doubt about the desirability for States to carry out negotiations and consultations with a view to determining whether the use of new weapons or weapon systems will be compatible with the laws of armed conflict, contemporary international co-operation in this field still calls for improvements. In the interest of international cooperation, topical problems touching on the employment of new weapons and weapon systems should be the object of periodical consideration among Governments, as was envisaged already in the St Petersburg Declaration. The progressive development of international law in this field will enhance the protection of civilians against indiscriminate warfare and of combatants against means of combat that cause unnecessary suffering.

CHAPTER II

Principal categories of weapon, and the questions of indiscriminateness and degrees of suffering or injury

40. This chapter provides a brief introduction to the major types of weapon and their effects in order to place the subsequent discussion in perspective. It summarizes briefly the principal features, both of those weapons whose properties are described in detail in subsequent chapters, and of those other weapons forming part of present-day arsenals whose properties are not so described. The chapter goes on to discuss, in broad terms, some of the military applications of the different categories of weapon with particular reference to the concepts of indiscriminateness, unnecessary suffering and superfluous injury. The legal significance of these concepts has been described in Chapter I.

1. The Principal Categories of Weapons

41. The major categories of weapon and their effects are summarized in Table II.1. Since this report is primarily concerned with the effects of weapons on people, and less with their effects on material, the focus is on the casualty-producing properties of the different categories.

42. *Explosive and penetrating weapons* cause a variety of different physical injuries. These may be grouped into injuries due to blast and injuries due to penetration of the human body by one or more missiles, such as projectiles or fragments. Penetrating weapons cause the latter type of injury, whereas explosive weapons may cause either or both types. As is described in Chapter IV, explosive weapons can be designed to maximize one or other of these two casualty effects, in which event they may be classified either as blast weapons or as fragmentation weapons.

43. Blast injuries sustained from the shock-waves created by explosive weapons result from the transmission of the shock-wave through the human body and its internal cavities. They vary in nature according to the medium (e.g. air, water or solid materials) through which the shock wave is transmitted to the body. They may be compounded by a variety of secondary effects, such as penetration by fragments, crush injuries from falling debris, and so forth.

44. As regards penetrating weapons, projectiles and fragments are responsible for the majority of injuries in modern conventional warfare. In recent conflicts they have caused 70–80 per cent of all battle injuries. Of this percentage, about three quarters of the injuries have resulted from fragments released by explosive weapons and about one quarter from single projectiles such as rifle or machine-gun bullets.

45. The wounds from penetrating weapons result from the transfer of kinetic energy during penetration of the human body by the projectile or fragment. This is described in more details in Chapter III. Death is particularly likely where vital organs are penetrated. Death may also result from loss of body fluids and shock, or subsequently from infection, particularly where there is the massive tissue destruction that may be caused by large projectiles, exploding projectiles, projectiles which flatten, expand or tumble on impact, or which enter the body with high velocity.

46. *Incendiary weapons* depend for their effects upon the action of incendiary agents. The latter have been defined by the UN Secretary-General as "substances which affect their target primarily through the action of flame and/or heat derived from self-supporting and/or self-propagating exothermic chemical reactions".[4] Against man, their casualty effects may sometimes also include asphyxiation and poisoning, for example by carbon monoxide generated during combustion, or by certain components of incendiaries such as white phosphorus. These burns are frequently difficult to treat, and death, due to a number of causes, may sometimes ensue several weeks after the initial injury. This is discussed in more detail in Chapter VI.

47. *Nuclear weapons* are, in effect, incendiary explosive weapons of great power. They are also radiological weapons because of the ionizing radiation released by the nuclear explosion, both immediately, and subsequently in the form of radioactive "fallout".[5] The dominant casualty effects are those of blast, thermal radiation and ionizing radiation. For the last of these, there may be an extended period of days, weeks or even years before symptoms of ill-health are displayed. Ionizing radiation may also delay the healing of other injuries, and affect the progress of certain diseases.

48. *Biological weapons* depend for their effects upon biological-warfare agents. The latter have been defined by the UN Secretary-General as "living organisms, whatever their nature, or infective material derived from them, which are intended to cause disease or death in man, animals or plants, and which depend for their effects on their ability to multiply in the person, animal or plant attacked".[6] Against man, the weapons might be used, conceivably over very large areas, either to kill or to cause disablement lasting for days, weeks or months. One of the principal characteristics of the weapons is the incubation period that extends between the initial infection by the biological-warfare agent and the onset of disease; this may be of between a day and a month. Another characteristic is the poor degree of control, whether in space or in time, which the user of the weapons can exert over their effects. The agent may be carried far beyond the intended target area by natural processes of wind or drainage, or by living carriers of the

Table II.1

THE PRINCIPAL CATEGORIES OF WEAPONS AND THEIR EFFECTS

Category of weapon	*Principal casualty effects*
Explosive	blast, fragmentation, other secondary effects
Penetrating	penetration, high-velocity effects
Incendiary	burns, asphyxiation, toxic effects
Nuclear	burns, blast, ionizing radiation
Biological	disease
Chemic	

agent or of the disease. In some cases also, the disease may be directly transmissible from man to man, thereby creating the risk of a spreading and persistent epidemic. These factors militate against the military utility of biological weapons.

49. *Chemical weapons* depend for their effects upon chemical-warfare agents. The latter have been defined by the UN Secretary-General as "chemical substances, whether gaseous, liquid, or solid, which might be employed because of their direct toxic effects on man, animals, and plants".[7] Against man, the weapons may be used either to kill or to cause disablement for a period of between a few minutes and a few days. Chemical weapons have much in common with biological weapons, but a greater degree of control can be exerted over their effects, and the time-lag before their effects become manifest rarely exceeds a few hours. They are therefore considered to have greater military utility than biological weapons. The area of effectiveness of a single chemical weapon may range between a fraction of a hectare and several square kilometres.

2. Military Classifications of Weapons and the Question of Indiscriminateness

50. The weapons described in the preceding paragraphs may also be classified according to certain of their military characteristics. In one widely used

classification of this type, a distinction is made between "antipersonnel" and "antimateriel" weapons. Another convention is to distinguish between point targets and area targets and, as a corollary, between "point weapons" and "area weapons". Area weapons, whose effects are extended in space, have a further counterpart in "time-delay weapons", whose effects are extended in time. These distinctions are discussed further in the following paragraphs. They are of assistance in relating concepts of military necessity to such questions as the discriminateness, or otherwise, of particular weapon applications.

51. Antipersonnel weapons are those which are primarily directed towards killing or otherwise incapacitating people. The distinction between them and antimatériel weapons is not clear-cut, however, for in the interests of flexibility there is a military requirement for multifunctional weapons. Thus, many of the weapons that are commonly described as antipersonnel are also intended to be effective against light matériel, such as trucks. Likewise, antimatériel weapons for use against armoured targets may exert their principal effects by penetrating the armour and then killing or injuring the crewmen. A spectrum of target "hardnesses" may be envisaged that necessitates a range of weapons having different combinations of penetrating and destructive abilities. Antipersonnel weapons are those which lie at the "softer" end of the spectrum. Those lying at the "harder" end of the spectrum, the antimatériel weapons, are, generally speaking, considerably more powerful than antipersonnel weapons; their effects on the human body may therefore be correspondingly greater.

52. Chemical, biological and radiological weapons are examples of weapons that are primarily antipersonnel in their nature, for they have little or no effect upon inanimate objects. The human body is also vulnerable to thermal and mechanical stresses such as those created by incendiary or explosive weapons. Thus all the principal categories of weapon described in the previous section of this chapter may be used as antipersonnel weapons.

53. In distinguishing between point and area weapons, it is necessary also to distinguish between, on the one hand, the means of warfare, which are the weapons available to field commanders, and, on the other hand, the methods of warfare, which are the ways in which the weapons are actually used.

54. Point targets are, by definition, well defined and usually small in size. Weapons for use against them, namely point weapons, are matched to the scale of the target and depend for their effectiveness upon accurate delivery. Point targets may, however, also be attacked with weapons whose area of effectiveness is substantially greater than the area of the target. This may be considered militarily necessary in cases where the target is moving, or where only its general location, not its precise location, is known.

55. Area targets are, by definition, large in size and present no specific aiming point to an attacker. Examples include enemy troops deployed over a wide area, or targets comprising many buildings or other fixed installations. These types of target may be attacked with a multiplicity of weapons that individually have so small an area of effectiveness as to be classifiable as point weapons. Artillery bombardment, or the use of high rate-of-fire machine guns, are illustrations of this. Alternatively, the targets may be attacked with a smaller number of weapons that, individually, are effective over a broad area; examples include large fragmentation or incendiary weapons, or poison gas.

56. Weapons which are effective over a broad area, whether they are used against point targets or area targets, are known as area weapons. Information on the areas of effectiveness of representative weapons having, *inter alia,* antipersonnel applications is given in Table II.2.

57. The area of effectiveness of a particular weapon is obviously a dominant factor in determining the discriminateness with regard to combatants and noncombatants with which the weapon may or may not be used. The closer the proximity between combatants and noncombatants, the smaller must be the area of effectiveness of the weapon in order for discriminate use of it to become possible. However, since there may be circumstances in which there is no proximity between combatants and noncombatants, it is not possible to define the degree of discriminateness of a weapon solely in terms of its area of effectiveness. All that can be said is that, in circumstances where combatants and noncombatants are sufficiently close to one another, area weapons will inevitably be less discriminate than point weapons, regardless either of the accuracy with which they are delivered or of the diligence of the user in attempting to avoid injury to noncombatants.

58. Proximity between combatant and noncombatant may be specified in terms not only of distance but also of time. The foregoing remarks about the discriminateness of area weapons apply equally to time-delay weapons. These may either be of the type that are fitted with delayed-action fuses set to function after a predetermined or random time interval; or they may comprise target-activated devices such as sea-mines, landmines, or traps. Weapons of these types may remain active for hours, days or even years, and since the circumstances of war may change considerably during this period, the weapons may expose noncombatants to a grave and prolonged hazard.

3. Problems in Measuring Degrees of Injury and Suffering Experienced from War Wounds

59. Towards the end of the last century, a number of military and medical experts attempted to calculate lower limits for the amount of energy needed to kill a man

Table II.2

AREAS OF EFFECTIVENESS OF REPRESENTATIVE WEAPONS HAVING ANTIPERSONNEL USES

Type of weapon	*Category of area effectiveness (a) (hectares)*
Hand grenade, high-explosive or fragmentation 70 mm rocket, high-explosive 81 mm mortar projectile, high-explosive 105 mm shell, high-explosive	0.01–0.05
81 mm mortar projectile, fragmentation *(b)* 155 mm shell, high explosive Antipersonnel mine, Claymore type 81 mm mortar projectile, white phosphorus	0.05–0.2
350 kg firebomb, napalm 70 mm rocket, multiple fléchette	0.2–1
155 mm shell, multiple fléchette 155 mm shell, chemical (sarin nerve-gas) 250 kg bomb, general purpose, high-explosive 250 kg cluster-bomb, fuel-air explosive 105 mm battalion fire, high-explosive shell *(c)*	1–10
350 kg cluster bomb, fragmentation *(d)* 105 mm battalion fire, sarin shell *(c)*	10–100
7000 kg bomb, light case, high-explosive	100–500

Notes:

(a) The range of areas that includes the area of effectiveness of the weapon concerned, "area of effectiveness" meaning the area within which an unprotected man standing in the open has at least a 50 % probability of becoming a casualty. The figures are derived from a number of different military manuals and similar sources which refer specifically to the anti-personnel applications of the weapons (rather than antimatériel or smoke-screening applications.) Some of the figures are not strictly comparable, for, apart from operational considerations, there are differences in the severities of the casualties concerned.

(b) Notched-wire controlled fragmentation.

(c) A total of 72 rounds fired from six howitzers.

(d) Pre-fragmentation (pellet) bomblets.

or otherwise place him *hors de combat.* They were principally concerned with the design of new rifle bullets. By the beginning of the present century several different estimates were current in different countries. They ranged from 40 joules up to 240 joules.[8] While these figures still have some currency today, it is realized that attempts to set a precise value on the minimum energy needed to incapacitate a man require so many qualifications that the estimates have little practical value. Factors such as the velocity of the projectile, the position of the wound, and the strength and morale of the victim, strongly influence the degree of incapacitation caused, regardless of the energy of the projectile. These factors are discussed in subsequent chapters.

60. Similar difficulties attend the quantification of the degrees of suffering imposed by, or experienced from, the different types of weapon injury. This matter is discussed below in some detail from a medical point of view. There seem at first sight to be three main criteria with which levels of suffering might be assessed, namely degree of pain, degree of permanent disability or injury, and probability of death. As will be seen, however, none of these provide much more than the crudest guidance.

(a) *Degree of pain from wounds*

61. Pain may be conceived as the product of two components, one physiological and the other psychological. A man will experience pain only when two processes have occurred: the reception by the brain of a particular type of nerve signal, and the response of the brain to that signal. The signal will originate in the site of injury, and the response of the brain will be a consciousness of pain.

62. It follows at once from this that the degree of pain imposed by a particular weapon-injury will depend, not only on the amount of damage, but also on the psychology of the victim. It is this subjective element that is the principal obstacle to the quantification of pain.

63. Related to this matter is the commonly accepted view that certain types of physical injury are likely to cause particular psychological distress. Such injuries include those that might lead to permanent disability or deformity, especially facial disfigurement, loss of one or more of the senses, and impairment of reproductive capacity. This particular psychological component must obviously be included in any assessment of suffering, but, as it too is largely subjective, it cannot reasonably be quantified.

64. Setting aside these psychological questions, it is possible, and common practice, to speak of "physical pain". This is more accessible to quantification because it is determined in the first instance only by physical damage.

65. Pain arising at the time of injury may be severe, but it is frequently the case in war situations that little initial pain is in fact felt because of the state of excitement of the victim. This phenomenon has been observed many times (e.g. soldiers have walked long distances holding their intestines in their hands, or after having broken a leg have walked on the wounded leg without feeling pain). However, during the hours after wounding there is an increase of pain due to release of pain-producing substances and the onset of infection.

66. The intensity of the pain that ensues during the period after injury may become exceedingly great. It is determined by many factors, such as the situation and severity of the wound; the quality of first aid by dressing, splintage and drugs employed; the type, duration and circumstances of transport; and the time and availability of surgical treatment. Infection associated with complicated wounds necessitating prolonged treatment and many surgical interventions prolongs the duration of pain until healing has occurred. Scars are sensitive, and even painful, for many months. Residual infections and deformities may prolong pain for many years.

67. As a very general rule, the more tissue that is damaged the more painful will be the wound. (The factors described in paragraph 66 are of course also involved, to say nothing of the psychological element referred to in paragraph 62; according to circumstances, these may in fact be dominant.) Thus, extensive fractures combined with crush injuries of the soft tissue will be particularly painful. So will multiple wounds caused by many fragments. The wound caused by a high-velocity missile, being large, will be more painful than the smaller wound caused by the same missile at low velocity.

(b) *Probability of death*

68. As a criterion of the degree of suffering or injury imposed by a particular weapon, the probability of death is no easier to predict or specify than the pain criterion. This is because the probability depends upon several factors that are only in part determined by the weapon itself or the manner in which it is used. They include:

(1) the localization of the wound in the body;

(2) the time lag between injury and treatment; and

(3) the state of physical resistance of the wounded person.

69. Different parts or organs of the body are important for life to varying degrees. Thus it is obvious that if a particular bullet penetrates the head of a person, it is likely to kill him, whereas if it penetrates the foot it will probably only put him out of action for a relatively short time.

70. A hit by a multiplicity of projectiles or fragments is more likely to injure a vital organ than a hit by a single projectile. This is self-evident, but the matter is complicated since two wounds that in themselves are not lethal, may together put such a stress on the body that the probability of death increases. The more organs that are injured the higher is the probability of death. If more than five different organs, for instance, are injured in the abdomen, death is statistically inevitable even if the victim survives the primary injury.

71. The time between wounding and adequate care will influence the prognosis of most war casualties. Good first aid and adequate surgical treatment will reduce the mortality rate among those surviving the primary injury. If the delay between wounding and surgical intervention is more than six hours, the mortality rises sharply.

72. With improvements in military medicine and surgery over the past decades, the mortality rate among the battle casualties of well-equipped armies has declined markedly. Thus, in the case of US battle casualties, the rate has fallen from 17% in World War I to around 2% in the Korean and Vietnam wars. Increasingly rapid transportation of the wounded has also been a prominent factor in this. Similar statistics relating to civilian war victims are not available; their mortality has almost certainly been considerably higher. Civilian victims may be of all ages and both sexes. Children, old people, and pregnant women generally have a reduced resistance towards war injuries compared with the soldier. Resistance is seriously reduced among people exposed to famine, thirst, or cold climates. By way of illustration, it may be noted that mortality from abdominal injury among civilian people in Germany during World War II was 49% as compared with 35% among British soldiers; both groups received the same type of treatment.

(c) *Degree of disability after injury*

73. Generally speaking, the majority of weapon casualties brought in for medical or surgical care within a few hours will be cured, and their disability be only of short duration. It has usually been the case that more than 50 per cent of combatant battle casualties have been able to return to duty within a few weeks. When they occur, however, the disabilities succeeding weapons injuries may be of great variety. They may be due to the injury itself, or to complications developing after the injury.

74. Since so many different factors play an important role in determining the resulting disability, not even this criterion is at all precise for assessing the suffering imposed by a particular weapon. Typical disabilities include loss of one or more of the senses (e.g. sight or hearing); permanent or temporary damage to

other parts of the nervous system; or damage to nonvital organs or to the extremities.

75. Injuries to the central nervous system (the brain and the spinal cord) frequently lead to especially severe disability. Here the disability is usually permanent. It may comprise partial or complete loss of mental capacity or of physical function. There may be loss of one or more of the senses (e.g. sight or hearing). Damage to the peripheral nervous system may lead to paralyses of various types and muscular atrophy. Modern techniques of neurosurgery and rehabilitation may considerably ameliorate these disabilities. Treatments of this type take a much longer time, however, than the treatment of most other weapons injuries; thus, even if the disability proves to be temporary rather than permanent, it will nevertheless be of prolonged duration.

76. During the rehabilitation and resocialization of a patient suffering from wound disabilities, it is necessary to take into account the psychological trauma to which he has been, and continues to be, subjected. The attendant psychological disability is very difficult to gauge. It may, for example, only develop after a substantial lapse of time, and it may well prove to be permanent, even though the original physical injury was relatively small.

CHAPTER V

Time-delay weapons

1. Technical Characteristics of Time-delay Weapons

152. The time-delay weapons described in this chapter are time-fused or target-activated explosive devices. There is also a discussion of traps of the non-explosive type, for these may be regarded as target-activated time-delay weapons. There are various other categories of time-delay weapon, notably certain chemical and biological weapons, but they lie outside the scope of the present report.

153. *Landmines* are the most familiar example of time-delay weapons. They are primarily designed as counter-mobility devices, usually being implanted below the surface of the ground in patterns that restrict possible enemy movement. Landmines to counter armoured or other vehicles depend on blast, most frequently making use of "shaped" or channelled explosive force to disable targets. Anti-vehicle landmines cannot normally be detonated by dismounted troops because most such mines are activated by higher pressures than a man can exert, or by acoustic or magnetic-induction fuses that discriminate between humans and vehicles.

154. Antipersonnel landmines are also in wide use, and most depend on fragmentation to produce casualties. Usually, they are detonated by pressure--sensitive contact fuses, but may also be activated by vibration sensors, trip-wires or other such devices. As a rule, antivehicle and antipersonnel mines are used together in a minefield. Some antipersonnel mines, when triggered, pop up out of the ground before exploding, thus optimizing horizontal fragmentation effects.

155. Most mines use metal casings and fuse components so that under favourable conditions they can be located by electromagnetic sensors. In World War II and since, large quantities of wooden and plastic-cased antivehicle and antipersonnel mines have been in general use. Still other materials have been used, less commonly, for landmine construction: glass, clay and concrete. In respect to antivehicle mines, non-metallic casings are just as effective as steel ones since the shape and size of the explosive charge is decisive, not fragmentation.

156. Sonic and soil-disturbance sensors and like devices have been proposed for non-metallic mine detection, but the surest method remains the tedious probing of the ground with a sharp object. Even for metallic mines, rapid detection is not possible, and this is the main reason why minefields continue to serve a countermobility function. From a military point of view, however, battlefield use of nonmetallic mines may be as troublesome to friendly troops as it is to the enemy, especially under fluid combat conditions. For this reason, mines are often used that explode or render themselves harmless after a predetermined period of time.

157. *Aircraft, artillery and naval gun-delivered mines* are, like the landmine, employed for interfering with free movement in areas distant or close to the zone of combat. They are, however, not easily marked or charted, as is usual practice with extensive landmine fields. As a rule, air-delivered mines are modified aerial bombs, 250–500 kg in weight, that are time-fused or metal-activated. Such mines are generally used in conjunction with numerous antipersonnel bomblets that are target-activated and generally fitted with trip-wire detonators. The bomblets prevent troops from de-activating the anti-vehicle mines, and the anti-vehicle mines prevent tanks or trucks from clearing the bomblets. Air-dropped antipersonnel mines sometimes consist merely of small bags of pressure-sensitive explosive; several thousand may be scattered by a single aircraft.

158. Several nations have in service, or are developing, gun or rocket-propelled mine-laying systems. These can place on the surface strings or strips of mines to counter advancing enemy troops and vehicles.

159. Although *booby traps* sometimes have an anti-vehicle purpose, they are primarily used as antipersonnel devices, designed both to slow movement and to

cause casualties. Activation of explosive-type booby traps may be by pressure, pull, tension-release, pressure-release or electrical means, and the explosive charge can be any size blast or fragmentation device available. For example, hand grenades, aerial bombs, shot-gun cartridges or blocks of explosive can be rigged as mines for ambush or other purposes. Construction of explosive-trap mines does not require specialized military training or equipment, and for this reason the use of booby traps can be expected in any war.

160. Examples have been noted of wiring dead bodies, or even wounded, so that their movement will explode munitions, but this use is uncommon. Innocent-appearing objects, such as valuable items, doors of houses, floor-boards or furniture can be rigged to trip fuses. Further, trails can be randomly mined, or ambush positions planted in remote areas where troops might patrol. Yet other booby traps might serve to alert friendly troops of hostile soldiers in the vicinity.

161. Non-explosive booby traps, also serving to restrict enemy movement and cause casualties, are relatively simple devices. Armed forces have encountered them in all wars, and no recent developments have improved on old methods. Sometimes the trap is a pit in the ground along a trail in which pointed sticks are placed and the hole camouflaged. Also, trip wires can fire arrows or release spring-loaded weights or spiked devices. Infective or toxic substances have occasionally been smeared on the spikes.

162. All of the explosive devices noted above can be equipped with fuses that are set to explode the charge at a pre-set time. Some landmines, for example, are designed to self-activate hours or days after emplacement, allowing temporary safe passage of friendly and even some enemy forces. Landmines have been equipped with counters that permit safe passage of a predetermined number of vehicles before exploding.

163. More often than landmines, air-delivered weapons have time-delay fuses, sometimes short ones to allow escape of the aircraft before explosion occurs, and sometimes to impede the clearance of, for example, a bombed airstrip. Time-delay aerial bombs have also been mixed with incendiaries in raids on built-up areas to prevent fire-fighting crews from containing the flames. Such use can also prevent medical teams from aiding the wounded.

164. Generally, large time-delay aerial bombs used for mining are easily detected and avoided, and, since that is their main purpose, do not generally cause large numbers of casualties. In contrast, however, small bomblets fused to explode immediately may be mixed with others that are equipped with time-delay fuses and then spread over a wide area; casualties produced by the initial attack may,

because of the time-delay bomblets, be isolated for days from medical assistance, being exposed to further hazards during the interval.

165. Most air-delivered time-delay explosives, in particular bomblets, are usually set to self-destruct within a few days. This is an important feature if friendly forces are soon to cross the area, or if the area is a populated one, or if the enemy is to be denied the bombs for his own use. However, self-destruct mechanisms of this type are frequently unreliable.

2. MILITARY APPLICATIONS

166. Time-delay weapons used in land warfare are like sea mines in that their primary purpose is to counter enemy mobility.

167. Before hostilities commence, target-activated weapons can be used along border areas as barriers against the movement of potentially hostile forces. Barrier minefields may serve to channel attacks towards areas that can be better defended. Once hostilities have begun, such barriers perhaps have less usefulness, but they can be employed to isolate sections of the battlefield, to deny terrain, and otherwise to hinder enemy activities.

168. Similar to barrier minefields are other forms of tactical minefield intended to restrict movement of hostile forces in the area of friendly positions, and to protect fixed defensive points in the combat zone. Minefields used in such ways are generally marked or fenced to prevent injury to friendly forces, civilians and livestock.

169. Barrier, denial, tactical and protective minefields containing antivehicle and antipersonnel mines usually produce relatively few casualties among enemy troops. Since the minefields will usually be under friendly surveillance, explosion of one mine will alert the defenders. Attacking forces are likewise alerted to the presence of a minefield, are made more cautious, and may be forced to find alternative approach routes. The military purpose of time-delay weapons is to keep the enemy at tactical arm's length.

170. While most tactical uses of mines in forward combat areas are controlled, other antipersonnel or antivehicle minefields in the same areas might be more haphazardly placed. The use of aircraft, helicopters or artillery to scatter time-delay weapons, or booby-trapping a zone, is militarily beneficial, not to produce large numbers of casualties, but rather for nuisance purposes to slow movement. Uncontrolled combat-zone mining is frequently as much of an annoyance to friendly forces, who might enter areas earlier mined by themselves, as it is to the enemy, to say nothing of the local inhabitants.

171. Nuisance minefields can be employed in areas distant from combat zones along lines of communication such as railways, highways and inland waterways.

Poorly trafficable terrain raises potential mine effectiveness, and perhaps allows for effective route blocking. However in view of the haphazard accuracy with which time-delay weapons are usually scattered from aircraft, distant mining serves more to slow than to stop movement.

172. Distant blocking, nuisance, and communications attacks are collectively called "interdiction", a campaign against the enemy's logistical organization. In the case of interdiction at great distances from the combat zone, the distinction between tactical and strategic use becomes unclear. In any event, the military advantages of interdiction are difficult to measure, becoming more uncertain the further from the front lines it is applied.

173. A further application of air-delivered time-delay weapons is the use of bomblets to suppress anti-aircraft fire prior to a bombing attack. Time-delay bomblets tend to neutralize defences until the defenders consider that all the bomblets have exploded. The question is open as to how much delay is militarily necessary under normal circumstances, and how much is designed to continue harassment for an indefinite period.

174. Booby traps are useful solely as nuisance devices. Rapid mobility of combatants lessens their utility.

3. Medical Effects

175. In conventional warfare, casualties from mines and booby traps have normally been quite low in proportion to casualties from other weapons. This applies to both combatants and noncombatants.

176. With the exception of non-explosive booby traps and such things as chemical or incendiary mines, time-delay weapons always produce their effects by fragmentation or blast. Broadly speaking, therefore, the medical characteristics of the injuries which they cause do not differ greatly from those described in the preceding chapters.

177. Mines of the explosive type generally injure the lower half of the body. As a rule, the injury is mainly due to blast. The result is more or less pronounced tissue destruction and shattered skeleton parts. The extent of the injury is proportional to the weight of the charge. The amount of explosive needed to cause amputation of a foot and thereby incapacitate a soldier is around 30 g. An increase of the explosive charge may cause the following additional injuries: loss of limbs, damage to the anal-genital area and injuries to the lower half of the body including the abdomen. The traumatic amputation of a leg or a thigh was the typical wound produced by mines during World War II. The particularly extensive damage to tissue, the forcing of dirt, etc., into the stump by the explosion, the

serious shock condition sustained as a result of haemorrhage and later infection, are the cause of the relatively high mortality rate as compared with other injuries to the extremities.

178. Fragmentation-type mines very often cause injuries of the high-velocity type, since the victim is usually in direct contact with, or very close to, the point of burst.

4. Salient Features of the Chapter

179. The principal military use of time-delay weapons is to impede enemy mobility, and for this purpose they may be designed either as anti-vehicle weapons or as antipersonnel weapons. In the latter case, they may act either through blast or through fragmentation, so that their effects on the human body are broadly similar to those described in Chapter IV. In the case of explosive antipersonnel mines, however, the close proximity between the exploding weapon and its victim may lead to extremely severe injury. Thus, the wounds from fragmentation mines may both be multiple and be of the high-velocity type.

180. In the case of blast antipersonnel mines, it is possible to specify rather precisely the minimum size needed to put a man out of action, for it is known that less than 30 grams of explosive may suffice to destroy a man's foot. When larger amounts of explosive are used, there may be amputation of the legs, damage to the reproductive organs and grave abdominal injury.

181. In that proximity between combatants and noncombatants may be defined in terms of time as well as distance, time-delay weapons have a tendency towards indiscriminateness. This is discussed in Chapter II, paragraph 58. It is, however, possible to incorporate into many types of time-delay weapon special devices that can render the weapon harmless after a predetermined time interval. Devices of this type have been developed to meet such military requirements as the need to avoid obstructing friendly forces; but they could also serve to mitigate the indiscriminate tendencies of time-delay weapons.

Final Remarks

243. The purpose of the present report has not been to present proposals for the prohibition or restriction of the use of any of the weapons or weapon systems discussed. The purpose has rather been to compile facts – legal, military and medical – relevant to any discussion to that end, which governments, intergovernmental organizations and other international bodies may undertake. Some final remarks are nevertheless called for.

244. It is clear from the preceding chapters that several categories of weapon tend to cause excessive suffering and particularly severe injuries or may, either by their nature or because of the way in which they are commonly used, strike civilians and combatants indiscriminately.

245. One example is provided by high-velocity ammunition for small arms. Although the use of the presently known projectiles in this category appears to offer some military advantages, these must be weighed against the fact – of which there is insufficient awareness – that they typically cause injuries much in excess of what is needed to put a combatant out of action.

246. Another example is provided by the fragmentation weapons. Weapons of this category cause the greatest number of casualties in modern armed conflicts. Hence evolution in this area of weaponry is of especial importance. It is obvious that the trend towards weapons which fragment into vast numbers of small fragments, and are susceptible of covering large areas, increases the risk of multiple injuries and the possibility that civilians will be affected.

247. Mention should next be made of incendiary weapons with the suffering they entail and the massive destruction they have sometimes brought about. Certain uses of antipersonnel mines and other time-delayed action weapons likewise can lead to indiscriminate effects and injuries far in excess of what is required to put combatants out of action.

248. The facts compiled in the report in regard to these and other weapons speak for themselves and call for intergovernmental review and action. Such action might be justified particularly in respect of two types of weapon apart from incendiaries, namely, high-velocity small arms ammunition and certain fragmentation weapons. The risks involved in their rapid proliferation and use would seem to constitute good reasons for intergovernmental discussions concerning these weapons with a view to possible restrictions upon their operational use or even prohibition. It is appreciated that the technical difficulties involved in such discussion are considerable. Nevertheless, even in this regard, it would appear that several approaches to the solution of these problems may be open.

NOTES

For technical reasons, the notes originally appearing as footnotes to the individual documents are reproduced in this volume as endnotes; the numbering of the notes will also be different from that of the original text, which may not be printed in its entirety.

1 *Austria:* Bundesministerium für Landesverteidigung: Grundsätze des Kriegsvölkerrechts, Anhang B der Truppenführung (TF), 1965 (pp. 253, 254); *France:* Décret N° 66–749 du 1er octobre 1966 portant règlement de discipline générale

dans les armées (Art. 34); *Federal Republic of Germany:* Bundesministerium der Verteidigung ZDv 15/10, Kriegsvölkerrecht, Leitfaden für den Unterricht (Teil 7), Allgemeine Bestimmungen des Kriegsführungsrechts und Landkriegsrechts, März 1961 (paras. 64–90); HDv 100/1, Truppenführung, Oktober 1962, Anhang Teil III, Völkerrechtliche Grundsätze der Landkriegführung (paras. 44–60), new edition in preparation; *Sweden:* Jägerskiöld-Wulff, Handbok i folkrätt under neutralitet och Krig, 1971, paras. 99–104; *Switzerland:* Eidgenössisches Militärdepartement, Handbuch über die Gesetze und Gebräuche des Krieges, Handbuch für die schweizerische Armee, 1963 (paras. 18–24); *United States:* Department of the Army Field Manual (FM) 27–10, 1956, The Law of Land Warfare, paras. 33–38, Department of the Navy, Law of Naval Warfare (NWIP 10–2); Department of the Army, Pamphlet (DA PAM) 27–161–2, International Law, Vol. II, 1962, pp. 39–46; *United Kingdom:* War Office, Code No. 12 333, The Law of War on Land being Part III of the Manual of Military Law, London 1958 (Chapter V).

2 Scott (ed.), *The Proceedings of the Hague Peace Conference, The Conference of 1899* (1920), p. 80.

3 Cf. Hackworth, *Digest of International Law,* Vol. VI (1943) pp. 271–272.

4 *Napalm and other Incendiary Weapons and all Aspects of their Possible Use: Report of the Secretary-General.* United Nations: New York, 1973 (document A/8803/Rev. 1).

5 For further information on the mode of action and effects of nuclear weapons, the reader is referred, in particular, to the report of the UN Secretary-General, *Effects of the Possible Use of Nuclear Weapons and the Security and Economic Implications for States of the Acquisition and Further Development of these Weapons,* United Nations: New York, 1968 (document A/6858).

6 *Chemical and Bacteriological (Biological) Weapons and the Effects of their Possible Use: Report of the Secretary-General.* United Nations: New York, 1969 (document A/7575/Rev. l). This definition was derived from a report prepared by the World Health Organization, subsequently published in revised form as *Health Aspects of Chemical and Biological Weapons.* WHO: Geneva, 1970. A broader definition of biological warfare agents is implicit in Article I of the Convention on the Prohibition of the Development, Production and Stockpiling of Bacteriological (Biological) and Toxic Weapons and on their Destruction, which was opened for signature in April 1972. This Article refers to "Microbial or other biological agents, or toxins whatever their origin or method of production, of types and in quantities that have no justification for prophylactic, protective or other peaceful purposes."

7 UN document A/7575/Rev. 1, *supra* footnote 1. This definition also was derived from the WHO report. The Protocol for the Prohibition of the Use in War of Asphyxiating, Poisonous or other Gases, and of Bacteriological Methods of Warfare, signed at Geneva on 17 June 1925, refers to "asphyxiating, poisonous or other gases, and of all analogous liquids, materials or devices".

8 The units of measurement given in this report are those of the recent internationally agreed *Système International d'Unités* (SI units), now coming into use for all scientific purposes to replace both the c.g.s. and the f.p.s. systems. The joule (J) is the derived SI unit of work or energy; it corresponds to one newton-metre, which is to say about 0.1 kg-metre.

Conference of Government Experts on the Use of Certain Conventional Weapons

Report

First Session held in Lucerne, Switzerland
24 September–18 October 1974

The first session of the Conference of Government Experts, attended by the representatives of forty-nine States as well as representatives from national liberation movements, international organizations, non-governmental organizations and the International Red Cross and Red Crescent Movement, continued the work of the 1973 expert meeting. Using the 1973 ICRC report as one of its starting points, the Conference sought to provide additional information and to examine further the question of prohibiting or limiting the use of certain weapons that may cause unnecessary suffering or have indiscriminate effects. Discussion on the possibility of formulating prohibitions or restrictions on the use of certain types of weapons also took place. Of particular interest is the opinion of some experts that certain types of anti-personnel mines ought to be banned because of their inherently indiscriminate nature and effects. Reproduced here are two chapters of the report: chapter II on the legal criteria, and chapter VI on delayed-action and treacherous weapons.

CHAPTER II

Legal Criteria

16. The Conference started its work with a debate on legal criteria. The purpose of this debate was to clarify as far as possible the criteria, and this in the presence of the military and medical experts, who would thereby, so it was hoped, gain a better understanding of the factors determining the admissibility or non-admissibility of weapons falling within the scope of their expert knowledge. In this respect, some expert lawyers felt a need for legal parameters far more accurate than the criteria presently existing or being envisaged, so that their application to a given weapon could be performed almost as a mechanical task. Others, who were convinced that such precise parameters would be impossible to achieve, would be satisfied, more modestly, if the debate brought out at least some degree of clarification. Others, again, while accepting that such an attempt at clarification would serve a useful purpose, emphasized that there would always be room for an assessment of weapons and their effect regardless of pre-existing or pre-formulated legal criteria.

17. At the outset of the debate, a paper entitled "Legal criteria for the prohibition or restriction of use of categories of conventional weapons" by a British expert was distributed to the Conference. This paper, which was welcomed by many experts as an important contribution to the debate, received much support as well as criticism from various sides. Hereinafter, and although it did not constitute a document emanating from the British Government, it will be referred to as the British paper.

18. Essentially, the British paper discussed three criteria. unnecessary suffering (or superfluous injury), indiscriminateness, and treacherous (or perfidious) character. These were the criteria most extensively discussed in the debate as well, although some other criteria were also mentioned, such as ecological damage and the prohibition of the use of force.

19. The criterion of "unnecessary suffering" (or "superfluous injury") was recognized as an existing legal precept by most experts. A few experts hesitated, however, to accept the concept of "unnecessary" suffering as this implied that other suffering would be considered "necessary". In the view of these experts, all suffering caused by war was, in a sense, unnecessary. An expert, replying to this remark, explained that the distinction between unnecessary and other suffering was not meant to condone the infliction of suffering of any kind, but merely was aimed at precluding certain forms or degrees of suffering in a situation (armed conflict) where the infliction of suffering could never be wholly avoided.

20. To clarify the standard of "unnecessary suffering", experts examined the language by which it found expression in Article 23 *(e)* of the Hague Regulations respecting the Laws and Customs of War on Land, of 1899/1907. The official British translation of the authentic French text of 1907 reads as follows:

> "In addition to the prohibitions provided by special Conventions, it is particularly forbidden . . . to employ arms, projectiles or material calculated to cause unnecessary suffering".

21. One question was how closely this translation corresponded to the French text, which refers to arms (etc.) "propres à causer des maux superflus". While it was generally agreed that "maux superflus" should perhaps better be translated as "superfluous injury" and that "injury" was a more objective concept than "suffering", several experts pointed out that the latter concept had come to be accepted in its own right, as distinct from that of "injury". They thought it would be unthinkable for this Conference, by merely expressing its preference for another translation of the original French text, to remove the subjective element contained in the word "suffering" from the body of international law.

22. A similar difficulty was posed by the words "calculated to cause". It was generally admitted that these contained an element of calculation or design, which might not be present in the French expression "propres à causer". Several experts stated that, to the extent that the English text might be construed as more restrictive than the French, the latter should prevail. One expert did not want to see the element of calculation discarded from the text, because he felt that without it weapon designers might not always refrain from deliberately designing weapons that would cause unnecessary suffering.

23. The concept of "injury" or "suffering" evoked some further comment. It was generally considered that this comprised such factors as mortality rates, the painfulness or severeness of wounds, or the incidence of permanent damage or disfigurement. Some experts considered that not only bodily harm but also psychological damage should be taken into account. Another expert could not accept such a wide interpretation of the concept at issue, as all wartime wounds, no matter how slight, could entail severe psychological harm.

24. A question of particular importance was what injury should be considered as superfluous, what suffering as unnecessary. There was widespread agreement among the experts that this involved some sort of equation between, on the one hand, the degree of injury or suffering inflicted (the humanitarian aspect) and, on the other, the degree of necessity underlying the choice of a particular weapon (the military aspect). It was also widely agreed that the equation would often be a particularly difficult one, as neither side of the equation could easily be reduced to precise figures and as, moreover, the two sides were so different that they were hard to compare.

25. According to some experts, the element of military necessity consisted solely in the capacity of a weapon to put an enemy *hors de combat,* this in conformity with the preamble to the St Petersburg Declaration of 1868, where it is stated that "the only legitimate object which States should endeavour to accomplish during war is to weaken the military forces of the enemy" and that "for this purpose it is sufficient to disable the greatest possible number of men". An expert, elaborating this idea, felt that the subjective element it contained could be reduced, e.g. by a formulation which would require that, if two or more weapons would be available which would offer equal capacity to overcome (rather than "disable") an adversary, the weapon which could be expected to inflict the least injury ought to be employed. Other experts held, in contrast, that the element of military necessity in the choice of weapons included, besides their capacity to disable enemy combatants, such other requirements as the destruction or neutralization of enemy materiel, restriction of movement, interdiction of lines of communication, weakening of resources and, last but not least, enhancement of the security of friendly forces.

26. Even if the first interpretation of the concept of military necessity were accepted, this would leave open how much injury is required to disable an enemy soldier. According to some experts, it might be necessary, particularly at short range, to inflict a severe wound for this purpose, as a comparatively minor injury might enable him to continue fighting.

27. A remark made in this connection by some experts was that, whereas the ideal solution might perhaps be that the soldier be equipped with a range of weapons from which he could select the one that would, in the concrete situation, put his enemy out of action with the least possible injury, this solution was impracticable and that, hence, even much graver injury than the minimum strictly required in a given situation could not always be avoided.

28. In the British paper, the view was expressed that a weapon which in practice would be found inevitably to cause injury or suffering disproportionate to its military effectiveness could be held to contravene the existing prohibition. Some experts supported the correctness of this view, while others questioned the correctness of the word "inevitably" in this statement; in their view, it was not a true statement of the law that only those weapons were forbidden which caused, without exception, disproportionate injury or suffering. The true test, according to these experts, was whether a given weapon would normally, or typically, entail such disproportionate effects. Other experts however considered that concepts like "normally" or "typically" were too vague a guide because what was the normal or typical use of a given weapon would vary from campaign to campaign and from one party to a conflict to another. The British paper did, moreover, also point out that even if a weapon was otherwise lawful, its use on certain occasions might be such as to contravene the basic general rule.

29. Indiscriminateness, although not clearly and unequivocally expressed in any international legal instrument in force, was generally accepted as a valid standard of the law of armed conflict. Opinions differed markedly, however, as far as the scope of this concept was concerned. According to some experts, it included not only the prohibition (recognized as valid by all experts) of indiscriminate attacks but also a prohibition of the use of "indiscriminate weapons". Other experts denied that the latter prohibition had already acquired the status of a rule of positive international law.

30. Both these latter and some other experts felt that such a prohibition of the use of indiscriminate weapons would be exceptionally hard to apply. Except for the case of a weapon intentionally designed to follow a random course and at the end of its trajectory hit whatever object happened to be there, all conventional weapons could be used in circumstances where the risk of hitting civilians was virtually nonexistent. Conversely, all weapons could be used without discrimination. According

to these experts, the method of use of a weapon, rather than its properties, would in general be the decisive element in determining whether the requirement of discrimination had been violated. Another expert gave, as examples of weapons which are inherently indiscriminate, gas and bacteria.

31. In the British paper, the latter view was given a formulation in the statement that clarification of the concept of "indiscriminate weapons" could take the form of a generic prohibition on the use of any weapon which cannot be accurately directed against military targets, that such a prohibition should not be extended to other weapons merely on the ground that they might have been used indiscriminately in the past and that the remedy must be to attempt to formulate a sensible restriction on such a method of use. This statement was more restrictive than the position taken in the report on the work of experts regarding "Weapons that may Cause Unnecessary Suffering or have Indiscriminate Effects" (supra, para. 8 *(a)* of Chapter I), where both the weapons indiscriminate by their very nature and those whose normal or typical use would be one which had indiscriminate effects were brought under the scope of the prohibition (Report, paras. 27, 244). Some experts were inclined to favour the position taken in the latter report. Other experts, taking a middle position, thought that those weapons ought to be regarded as indiscriminate which, having regard to their effects in time or space, cannot be employed with sufficient or with predictable accuracy against the chosen target. Yet another expert, however, warned against a reliance on accuracy in this context, as accuracy was a relative concept in all cases and could never provide a clear guideline.

32. A somewhat different approach was taken by an expert who considered that, starting from the distinction he made between point weapons and area weapons, the question ought to be examined whether the latter are necessarily indiscriminate.

33. Some experts advocated that the criterion of treacherousness be given a separate place among the legal criteria governing the admissibility of specific weapons. These experts would then have a preference for the term "perfidy", a concept now being developed in the context of the Draft Additional Protocols to the Geneva Conventions, of 1949, rather than maintain the old term "treacherousness" contained in Article 23 *(b)* of the Hague Regulations of 1899/1907.

34. Several experts, in contrast, felt that the concept of treacherousness or perfidy did not deserve such a separate place. In their view, the perfidious nature of the use of a given weapon in certain specific conditions would rather constitute an element in the determination whether the weapon caused unnecessary suffering or was used without discrimination.

35. In the quest for further criteria governing the use of weapons, reference was repeatedly made to the Martens clause contained in the preamble to the Hague Convention respecting the Laws and Customs of War on Land, of 1899/1907, according to which "the inhabitants and the belligerents remain under the protection and the governance of the principles of the law of nations, derived from the usages established among civilized peoples, from the laws of humanity, and from the dictates of the public conscience". The reliance placed on this clause assumed two forms; in the eyes of some experts, what one had to look for was principles of international law derived, e.g., from public conscience; others were inclined to regard the public conscience as a standard by itself.

36. For the former group of experts, it would be necessary to show that the influence of public opinion had resulted in a new principle being incorporated into the body of international law. For the latter group, a strong demand on the part of the public conscience would be sufficient to formulate a new rule.

37. Criteria mentioned in this respect by some experts included the prohibition against doing irreparable damage to the environment (the criterion of ecological damage) and the abhorrence evidently provoked by the use of certain weapons, such as napalm and other incendiary weapons. The latter criterion of the abhorrent nature of certain weapons was also related by some experts to the "laws of humanity" mentioned in the Martens clause.

38. Other experts felt that the public conscience did not constitute a legal criterion by which the admissibility of the use of specific weapons could be measured. In the view of these experts, while public opinion should never be disregarded and could in fact represent a strong driving force, it constituted a political rather than a legal factor, which governments should take into account in examining and assessing the various questions involved in the deployment and use of modern conventional weapons.

39. An expert thought that the entire question of legal criteria should be tackled from a completely different angle, taking into account the present state of international law and international relations. Present-day international law included such basic principles as the non-use of force and equal security for States. In his view, the prohibition of the use of force, enshrined in the Charter of the United Nations and reaffirmed in many treaties to which his government was a party (and to whose conclusion it had often taken the initiative), implied a prohibition against the use of any and all weapons of war, the only exception to this rule being the case when a people would exercise its right of self-defence. This being so, he felt that disarmament was the primordial goal which governments should strive to achieve. In this context of disarmament, it would be necessary to consider all

weapons together, both nuclear and other weapons of mass destruction and so-called conventional weapons, all of which were capable of producing terrible suffering. Admittedly, governments could, in the course of disarmament negotiations conducted in the proper forum (such as the Conference of the Committee on Disarmament, or the proposed World Disarmament Conference), conclude agreements proscribing the use of specific weapons. He emphasized, however, that such agreements, to be effective, ought to be general. His views were supported by some other experts.

40. Another expert expressed his profound scepticism at any efforts to place a ban on the use of specific weapons. According to this expert, it was not the force of legal instruments but the fear of retaliation which kept States from using certain weapons. In a situation where this fear did not exist and where a belligerent party expected to secure a military advantage by the use of a prohibited weapon, this belligerent might well decide simply to set the prohibition aside. This expert felt, therefore, that agreements restricting the use of specific weapons or defining the targets against which they could be used would be incomplete and, indeed, of no avail, if they did not at the same time proscribe the production, stockpiling, etc., of those weapons.

41. Yet another expert expressed the view that the prime task of humanitarian law lay in protecting man against the aggressive war machine. To this end, the use of this aggressive war machine should be condemmed as a war crime. In this context, a set of complete and detailed regulations prohibiting criminal means and methods of combat ought to be brought about. Inadequate and dangerous concepts such as "superfluous injury" and "unnecessary suffering", as well as "proportionality" and "military necessity", ought to be discarded. These considerations had led this expert's government to propose a number of amendments to the Draft Additional Protocols envisaging a better protection of the civilian population and of man in general, and prohibiting such methods as genocide, biocide and ecocide.

42. In view of the great diversity of opinion on applicable legal criteria which had emerged from the debate, the suggestion was made that the Conference set up a working group of legal experts which could then, after a thorough examination of all the questions involved, present the Conference with a set of suitably defined criteria for the assessment of specific conventional weapons. The general feeling was, however, that this would be premature; criteria would have to emerge, or find clarification, as much as the result of discussions on military and medical aspects of the use of specific weapons as of the work of legal experts.

CHAPTER VI

Delayed-Action and Treacherous Weapons

VI.1 Definition and classification

219. With certain specific reservations noted below, Chapter V of the ICRC report was accepted by experts as a basis for discussion of the agenda item on delayed-action weapons and treacherous weapons. Several experts observed that the agenda item covered a highly diversified group of weapons, and that some degree of definition and classification was therefore advisable.

220. One expert, subsequently supported by others, spoke of the need to distinguish between the weapons *per se* and the uses to which they could be put. In order to illustrate this, he stated that, insofar as the weapons and devices under consideration had any feature in common, that feature was the element of surprise; yet the element of surprise was also something which many methods of using weapons were designed to exploit. Thus, in his opinion, there was no difference in principle between, for example, an unmarked mine, a booby-trap or an ambush.

221. Implicit, and in some cases explicit, in the interventions made by experts on this subject was the following classification. First, there was the general category of delayed-action weapons comprising a variety of time-fused devices, such as delayed-action bombs and target-activated devices, such as mines and booby-traps. Next there were those types of delayed-action weapon which, in use, are intentionally disguised or hidden; an emplaced mine, if it is left unmarked, would fall within this category. Finally there were those methods of using hidden or disguised weapons which were characterized by one expert as "perfidious", by which he stated that he meant the use of any weapon in such a way that it placed the intended victim under a moral, juridical or humanitarian obligation to act in such a way as to endanger his safety (see further hereafter, paras. 251 to 253). There was general agreement among experts, however, that a precise distinction between the second and third categories would not be easy to define, having regard to such devices as letter-bombs and exploding toys.

222. In addition to the foregoing sub-classification, based essentially on intent, experts also provided classifications based on design. This is reflected in the summary of the discussion on military aspects contained in paras. 224 to 243, where time-fused weapons, naval mines, emplaced landmines, scatterable mines and booby-traps are treated separately.

223. On the subject of time-fused weapons in particular, and having regard to questions of discriminateness, one expert recommended a differentiation based on time-delay. In his opinion there would be advantages in the Conference

concentrating only on those weapons having a time-delay longer than, say, a minute or so. But other experts perceived disadvantages in this, and noted that the military considerations which had led to the proposal of such a limit would, in some cases, be equally applicable to much longer time-delays.

VI.2 Military aspects
Time-fused weapons

224. Concerning time-fused weapons, experts addressed themselves only to delayed-action air-delivered weapons. It was explained that these might be either general-purpose bombs or small cluster-dropped bomblets fitted with fuses that could be set to detonate the munition after an interval of anything between seconds or days. One use for such weapons was said to be in low-altitude ground strikes where it was necessary to introduce a brief time-delay in order to ensure that the attacking aircraft did not suffer from the blast or fragmentation of its own weapons. Another use, for which one expert stated that there was an overriding operational need, was the neutralization of airfields; here a combination of impact-detonating, short- and long-delay weapons was needed in order to hamper and prolong such repair work as the filling-in of bomb craters on runways.

225. One expert was of the opinion that time-fused antipersonnel weapons were of particular value against the crews of anti-aircraft batteries. Since the aim here would be to suppress anti-aircraft fire during the period of an attack, a time-delay no longer than the duration of the attack would, in his opinion, suffice.

226. The same expert also stated that delayed-action bomblets could be used for interdiction or against unlocalized military targets, where, in certain combat circumstances, the military value might be high.

Naval mines

227. Experts did not address themselves in any great detail to naval mines. One expert explained that these mines might either be target-activated or activated by remote control. Use of the latter type was restricted to coastal waters, being controlled from shore stations via underwater cable systems. Target-activated naval mines might either rest on, or be moored to, the sea bed; they might be activated either by direct contact with sea vessels (in the case of moored mines) or by some less direct influence.

228. The same expert went on to observe that international law stipulated that minefields at sea were not to be laid without public warning; and that moored mines coming adrift were to render themselves harmless automatically. With regard to the latter stipulation, he observed that corrosion after prolonged sea exposure often fouled de-activating mechanisms.

Emplaced landmines

229. It was widely agreed that the primary function of emplaced landmines, as well as scatterable landmines and booby-traps, was to counter enemy force mobility and to keep the enemy at tactical arm's length until such time as other weapon systems could be brought to bear on him. It was observed that landmines were primarily, though not exclusively, defensive weapons: they were used to channel enemy forces into defensible areas, to deny terrain which could not be covered by combat troops, to hinder enemy activities generally, and as close-in protective weapons for defending troops. There was also general agreement that landmines were unlikely to be used in isolation but rather as part of an integrated battlefield system comprising surveillance sensors, anti-tank weapons, artillery, tactical air power, and the individual soldier.

230. The opinion was expressed that fields of landmines were at present the most economical and effective artificial obstacle system that could be provided with conventional means; although men with weapons could restrict enemy movement without the need for special obstacle systems, available forces were often inadequate, thus making use of landmines a necessity. Several experts were of the opinion that the value of minefields as obstacles was directly related to the number and types of weapon systems covering them; if landmines were unsupported, they might be of little value, and it therefore followed, in the opinion of those experts, that minefields would be employed mainly in combat zones. Other experts stated that, while this might generally be so, it was not invariably the case, and that minefields could also have applications outside the area of battlefield operations.

231. Several experts observed that, because the use of minefields would normally form part of an integrated plan of action or operation, there would be a compelling need to record their location and extent most carefully. In order not to endanger friendly personnel, it would also, in most cases, be advisable to mark the minefields, or even to fence them off; and one expert noted that, for minefields emplaced by the ground forces of his country, this was already the practice.

232. Experts stated that landmines might be designed either as antipersonnel weapons or specifically for use against vehicles, especially tanks. With regard to antipersonnel mines, one expert noted that there were two basic types in common use: blast mines and fragmentation mines. They could be detonated either through the direct pressure of a foot, or by means of sensors, such as tripwires, arranged so that several enemy personnel within the immediate area of the mine could be affected simultaneously. It was stated that, in the case of fragmentation antipersonnel mines, the concentration of fragments discharged over any

particular area would be considerably less than that resulting from normal artillery fire. The primary value of antipersonnel landmines was, in the opinion of several experts, to prevent the rapid clearing of anti-vehicle minefields; in this role they served to reduce the degree of surveillance and firepower necessary to defend the minefield.

233. It was suggested by one expert that, since it was known that about 30 grams of high explosive were sufficient to put a man out of action by damaging his feet and legs, there was no need for blast type antipersonnel mines to contain a larger quantity of explosive. Disputing the value of this suggestion, another expert stated that it was necessary to qualify the implied incapacitation criterion by reference to situational conditions.

Scatterable mines

234. Experts stated that scatterable mines, which are a comparatively recent development, serve much the same functions as emplaced mines, but with the additional utility that follows from the rapidity with which they can be deployed. It was stated that scatterable mines could be delivered both by tube and rocket artillery and, more commonly, by aircraft. The view was expressed that, perhaps to a greater extent than with emplaced mines, scatterable mines can be used offensively by providing flank security for advancing forces or by securing newly-gained positions against counter-attack.

235. Experts also noted that whereas ground emplacement of mines permitted some control over the forward movement of the enemy, techniques of scatter-mining allowed lateral and rearward control as well. Moreover, the rapidity with which scatter-minefields can be laid permits areas to be left mine-free, and therefore available for friendly manoeuvre, until the last moment.

236. It was stated that both antipersonnel and anti-vehicle scatterable mines were available. The antipersonnel type, it was explained, were frequently very small, sometimes consisting of no more than a few grams of explosive contained in a sachet, and dispersible by the thousand over a wide area. Other types of scatterable antipersonnel mine were stated to be larger, sometimes capable of sending out a cobweb of trip-wires, and acting through fragmentation rather than blast. In the opinion of one expert, scatterable antipersonnel mines would be of particular value in harassing enemy activities in areas far to the rear, in locations which the user had no intention of occupying.

237. It was noted that scatterable anti-vehicle mines, like those of the emplaced type, were commonly designed to detonate only under the influence of more pressure than can be exerted by a man's foot or by other action not released by a single soldier.

238. It was generally agreed that the tactical use of scatterable mines had a greater propensity for endangering friendly troops and non-combatants than the use of emplaced mines. It was also agreed, however, that for the selfsame reason it would normally be in the interests of the user to maintain particularly tight control over mine scattering, as by careful recording and registration of minefields. Although experts stated that it would generally be less feasible to mark scatter-minefields than ground-emplaced ones, they noted that compensating security was available, and commonly used, through the incorporation in scatter-mines of self-destruct mechanisms that functioned after a pre-set time interval. This was stated to be mandatory for the scatterable mines currently in issue to, and under development for, the armed services of at least one country. One expert made the suggestion that markers such as flags or pyrotechnic flares, possibly designed according to an internationally agreed specification, might be incorporated in scatter-minefields in order to indicate their general location. Measures such as this, and the use of self-destruct mechanisms, would counter the potential for indiscriminate employment of scatterable mines to which a number of experts had drawn attention.

239. There was much discussion of the precision with which scatterable mines could be delivered. The general point was made by one expert that there were very high costs associated with the laying of scatter-minefields, and that scatterable mines would therefore be employed only if there was a substantial probability of imminent enemy contact with them. In his view, this itself demanded high precision, quite apart from considerations of discriminateness. He went on to state that scatter-minefields could be laid by artillery or by aircraft to within an accuracy of 100 metres. Other experts doubted whether this was a figure of general validity: much would depend, in their view, on the availability of sophisticated navigational and weapon-guidance systems, on user competence, on terrain, and on the prevailing conditions of combat and weather. One expert mentioned the possibility to disperse simple scatter-type mines over large areas to deny the enemy access to those areas and to disturb and harass e.g. logistic supply lines. Such use, he felt, could easily become indiscriminate.

240. There was also much discussion, again with regard mainly to questions of discriminateness, of the efficiency of the self-destruct mechanisms that are built into scatterable mines. Criticism was directed towards the ICRC report where it stated, in para. 165, that those mechanisms were frequently unreliable. Experts reiterated the view recorded in para. 238 that efficient self-destruct mechanisms were militarily a necessity in order not to compromise friendly battlefield mobility. One expert stated that a reliability of better than 99.9% was available.

Booby-traps

241. Experts stated that booby-traps could serve the function of impeding enemy mobility just as minefields could. Indeed, as a number of experts observed, the difference between an unmarked mine and a booby-trap was largely a semantic one. Booby-traps were commonly used as nuisance and delay devices to hamper enemy minefield-breaching operations, as a means of alerting friendly troops to the presence of hostile soldiers, and to delay the enemy by posing a threat to his survival. In the opinion of a number of experts, the use of booby-traps for some of those purposes was militarily essential.

242. One point of difference between booby-traps and emplaced mines in their normal application is that the former are, by definition, unmarked and disguised. Several experts noted that booby-traps could display innumerable variations both in the method of disguise and in wounding action. The latter might depend on explosives; or it might be non-explosive, as in the cases noted by experts of concealed pits, contaminated spikes or falling weights. With regard to disguise, reference was made, not only to conventional methods of concealment, but also to the packaging of explosives to resemble innocuous or even attractive objects, such as children's toys or sweetmeats. One expert observed that booby-traps could be set up even with scarce military resources.

243. In discussing questions of discriminateness, experts observed that the problem of confining the effects of booby-traps to enemy combatants was a temporal rather than a spatial one; with scatter-sown minefields both problems arose. A number of experts spoke of a need for careful destruction, self-destruction or programmed neutralization of any booby-traps placed to delay passage of enemy troops in built-up areas. One expert stated that such programmed destruction/neutralization should be effected immediately following the maximum delay-period which the retreating friendly forces had decided to impose on the advancing enemy forces.

VI.3 Medical considerations

244. All experts who spoke on the subject agreed that, because explosive mines and booby-traps produced their effects by blast and/or fragmentation, the wounds which they caused did not, broadly speaking, differ greatly from those caused by the blast and fragmentation weapons described in Chapter V. There was general concurrence with the description of the medical aspects of delayed-action weapons given in the ICRC report (paras. 175–178).

245. With regard to para. 175 of the ICRC report, where it is stated that in conventional warfare casualties from mines and booby-traps have normally been quite low in proportion to casualties from other weapons, a number of experts

stated that the same was not true of guerilla warfare, where mine and booby-trap casualties were often considerably higher. One expert stated that, while working in a medical facility treating only civilian patients in a recent theatre of war that was characterized by guerilla operations, nearly 20 per cent of his patients, several of whom were children, had been injured by mines and booby-traps, both explosive and non-explosive.

246. Concerning non-explosive booby-traps, one expert said that, as there were many different types of such weapons, they could cause a variety of injuries in different parts of the body, including puncture wounds from sharp objects which might, at times, be coated with toxic or infective material.

VI.4 Evaluation

247. Two criteria entered into the discussion on assessment of delayed-action and treacherous weapons, viz., the concept of indiscrimination and that of perfidy. The discussion turned moreover around the question of whether one would pursue the prohibition of weapons as such or, rather, the prohibition or limitation of certain types of use of such weapons. Attention focussed on two categories of weapon: mines and booby-traps.

248. As for indiscrimination, while some experts felt that mines and booby-traps, or booby-traps alone, were indiscriminate by their very nature, other experts contested this. Where mines were concerned, several experts noted the possibilities for marking mine deposits and for the application of reliable self-destruct devices. It was suggested by some experts that those possibilities could and should be turned into mandatory rules. One expert suggested that maps indicating the location of mines ought to be handed to the civilian authorities at the close of hostilities.

249. For the rest, these experts were of the opinion that mines, as other weapons, could be used with or without discrimination. If the precautions mentioned in the previous paragraph were observed, there would be no risk of the mines hitting civilians indiscriminately. One expert pointed out that civilians in close proximity to military objectives would always run a certain risk. The questions of utilization of weapons, these experts remarked, were on the agenda of the Diplomatic Conference where prohibition of indiscriminate attacks upon the civilian population, as well as the rule of proportionality, were under consideration.

250. Like mines, booby-traps could in the eyes of some of these experts also be used with discrimination, and it was therefore not right to regard them as inherently indiscriminate. A booby-trap, one expert remarked, was really a method of using a weapon rather than a weapon *per se*.

251. Regarding perfidy, one expert put forward a proposal for the definition of a perfidiously used weapon, mentioned already in para. 221 and reading as follows:

> The use of any weapon in such a way that it places the intended victim under a moral, juridical or humanitarian obligation to act in such a way as to endanger his safety, is perfidious.

252. Commenting on this proposed text, an expert welcomed the reference to moral and humanitarian obligations as possible additional elements in the search for adequate legal criteria. Another expert felt, however, that moral and humanitarian obligations lacked the necessary precision upon which to base the definition of perfidious use. Until specific rules of international law, perhaps based upon such considerations, were incorporated into international law, references to such vague and variable concepts were not appropriate in a legal definition. This expert stated that to be perfidious an act had to involve a breach of an international confidence or right, and gave as an example the booby-trapping of a dead soldier. Other experts supported that view.

253. An expert thought that the language of the proposal, to be understood by all, ought to be made simpler. Another expert, too, felt that the language proposed would give rise to all kinds of interpretation problems.

254. Regarding possible prohibitions, some experts expressed as their considered opinion that at least certain types of antipersonnel mines, as well as booby-traps, ought to be banned. Arguments advanced included the reference to their inherently indiscriminate or perfidious nature, and to the fact that, as far as mines were concerned, their use was contrary to the necessity to protect war victims and to evacuate them as soon as possible, a principle expressed in the Geneva Conventions and the Draft Additional Protocols.

255. An expert recognized the possibility for constraints to be placed on the construction or utilization of such weapons rather than on the weapons themselves. Thus, one might consider a ban on airdropped mines while permitting their delivery by artillery. Yet taking into account the need for simple rules to ensure their faithful application, he declared himself in favour of a categorical ban rather than a prohibition or limitation of certain types of use.

256. Other experts were of a different view. They were convinced that a complete ban on such weapons was both impracticable and unjustified, as the weapons could be used in ways that would be neither indiscriminate nor perfidious.

257. At one stage of the debate, it was suggested by the President that there might perhaps exist a consensus on two points: that the use of explosive devices perfidious by their very nature would be prohibited, and that there ought to be a

ban on the use of booby-traps representing a grave danger to the civilian population. As examples for the first category he mentioned such objects as fountain pens and children's toys, which would generally be used by civilians rather than combatants.

258. This suggestion drew support from various sides, although certain criticisms were also formulated. As regards the first proposition, it was pointed out that, with the possible exception of toys, objects falling within the category indicated above were often not exclusively used by civilians. It was noted that, even though one might feel great sympathy for the President's suggestion, it needed to be examined rather carefully. As for the second suggestion, several experts felt that, as formulated, it was too vague to be acceptable and needed clarification; it was also stated that, once again, the point at issue here was the method of utilization rather than the inherent characteristics of a given weapon.

259. As for methods of use, it was widely felt that in further deliberations on the subject stress should be laid on use against the civilian population.

Conference of Government Experts on the Use of Certain Conventional Weapons

Report

Second Session held in Lugano, Switzerland
28 January–26 February 1976

The examination of certain conventional weapons continued with a second meeting of government experts in 1976. The purpose of the meeting was twofold: to study any new information, new facts and new arguments related to the various categories of weapons under consideration; and to examine in further detail the possibility, contents and form of any bans or restrictions on use. The conference was divided into working groups to examine the proposals submitted by government experts. Reproduced below are the relevant discussions on landmines from the Report on Plenary Meeting Proceedings, the general debate in the Summary Records of Plenary Meetings, the Report of the General Working Group and the Report of the Working Sub-Group of Military Experts on Mines and Booby-Traps. The various proposals submitted to the conference, which are attached to the conference report in an annex, are also reproduced.

I. Report on Plenary Meeting Proceedings

CHAPTER 3

Delayed-Action and Treacherous Weapons

27. At the outset of the debate on this item of the agenda, a working paper was introduced (COLU/203, see Annex A.2) which put forward detailed proposals concerning the regulation of the use of mines and booby-traps. It was explained that these proposals sought to achieve an improved protection of the civilian population against the dangers ensuing from the use of such means of warfare (which were becoming ever more sophisticated) while at the same time maintaining a correct balance between humanitarian ideals on the one hand and the realities of armed conflict upon the other. Essential in the proposals was: that recorded minefields should be made public upon the cessation of hostilities; that remotely delivered mines should be equipped with a neutralizing mechanism, or that the area in which they were delivered should be marked; that civilians going about their daily business in populated areas not immediately forming part of the combat zone should not be exposed to the risks posed by mines, booby-traps and suchlike devices; and that the use of certain especially perfidious booby-traps, whether explosive or not, should be specifically prohibited, as should non-explosive traps specifically designed to cause cruel or lingering death or injury.

28. The introduction of this working paper was widely welcomed as a valuable contribution to the discussion. The various proposals which it contained attracted both general and specific comments. A general comment was that the proposals rightly laid emphasis on the element of protection of the civilian population; another merit was seen in the treatment of mines and booby-traps in one single paper, thereby avoiding problems of demarcation. An expert suggested that a distinction should be drawn between defensive and offensive use of mines and minefields, and that the defensive use should not be made the object of a prohibition.

29. Specific comments were offered on many aspects of the proposals. While some of these were simply in support of particular proposals, others were in the nature of queries or suggested amendments. Comments of the latter type were discussed further in the General Working Group.

30. Several experts, in commenting upon the proposal concerning remote delivery of mines, referred to the proposal already contained in document CDDH/IV/201 (see Annex A.21) which sought to prohibit the laying of anti-personnel landmines by aircraft. They continued to prefer this proposal to the one put forward in COLU/203; delivery from the air of such mines, they felt, presented the greatest risks to the civilian population and, one expert added, was

most likely to take place in asymmetrical conflicts between a poor people fighting in self-defence against a technically superior enemy. Experts of this group nevertheless expressed their readiness to give close consideration to the proposal contained in COLU/203.

31. An expert, who thought that the detailed list of uses of booby-traps singled out for prohibition in COLU/203 was not, and never could be, complete and that the attempt to draw up such a list was misguided, stated as his view that a general formula was to be preferred. He introduced a proposal to that effect (COLU/206, see Annex A.5). This proposal received the support of some other experts. An expert commented, however, that the language chosen (providing that the camouflage of explosive devices in objects in general use among civilians be prohibited) posed difficult problems to the commander in the field; he also objected to the fact that only explosive devices had been singled out for prohibition. Another expert replied that the use of non-explosive booby-traps was already prohibited.

32. An expert pointed to an aspect of the problems posed by the use of mines which had not been dealt with in the proposals mentioned above, viz., the maximum charge of anti-personnel mines. This constituted an important aspect to which attention should be given. This expert also saw the need to distinguish between anti-personnel and anti-tank minefields, which had very different technical characteristics.

II. Summary Records of Plenary Meetings

Sixth Meeting

General Debate on Delayed-action and Treacherous Weapons

1. Introducing document COLU/203, one of the co-sponsors recalled that mines and booby-traps had been discussed at Lucerne and much useful, important and new information had been submitted. It was clear that some restrictions on the use of such devices could be acceptable to most experts.

Paragraph 259 of the report on the Lucerne Conference stated that "it was widely felt that in further deliberations on the subject stress should be laid on use against the civilian population". As a result, the experts of two countries, with the advice of those from certain other countries, had drafted proposals which had been submitted as a working paper. The speaker hoped that a correct balance had been struck and that restrictions, unrealistic to a soldier in battle – and hence impractical in application – had been avoided. The aim was to increase protection of civilians against some of the hazards of war.

The actual proposals referred to the use of mines and booby-traps in armed conflict on land only and did not apply to sea warfare. The word "landmine" had been avoided as it implicitly raised the question as to whether it referred to mines designed for use on land or mines in fact used on land, it being perfectly practicable for certain purposes to use sea mines on land or landmines under water. To avoid ambiguity, mines used under water had been excluded from the proposals. For the purposes of the present proposals, the word "mine" did not include those detonated by remote control. Such mines were not a danger for civilians as they were controlled by a person who would detonate them only on observing a proper military target.

The first proposal was that all minefields of more than 20 mines be recorded as was already done by most armies. A legal requirement to that effect was, however, desirable. It would be difficult to extend such a rule to isolated mines laid hastily during combat as a defensive technique to harass the enemy. It might be too much to expect the individual soldiers who laid such mines to record them but, since they were laid solely in the direct path of the enemy, they would be unlikely to remain *in situ* and undetected for long periods. The main point of the proposal was that it required the location of all recorded minefields situated in territory controlled by an adversary to be made public on cessation of hostilities.

The second proposal related to "remotely delivered mines" which were called "scatterable mines" at the Lucerne Conference. The essential feature of such mines was that they were delivered at long range by aircraft, gun, rocket or mortar. Mines so delivered were more likely to endanger civilians than were those emplaced by hand. A distinction nonetheless needed to be made between such long range delivery and delivery by devices operating at short range, as used, for example, in quickly laying a defensive minefield. The latter type of field was always used where combat was – or would shortly be – taking place and was akin to manually placed fields. "Remotely delivered mines" had accordingly been defined as those laid by aircraft and those laid by other means at a range exceeding 2,000 metres.

The proposal required that such remotely delivered mines be permitted only when the mines were fitted with a neutralizing mechanism or when the area into which they were delivered was distinctively marked.

Having covered remotely delivered mines, the proposal next referred to all other mines, being those emplaced by hand or by the short range devices already mentioned. The proposed restrictions referred equally to booby-traps, as there was no clear line between mines and booby-traps.

Such devices were of the greatest danger to civilians when used in populated areas where no combat was taking place and where many civilians were going about their daily business. The explosion of such a device could cause death or injury to unsuspecting people. In such circumstances, manually emplaced mines and booby-traps should be permitted only if due precautions were taken to protect civilians from their effects. The only exception to that rule would be when the device was placed on or near a military objective where the risk to civilians would have to be accepted in the same way as when the objective was attacked from the air or by long range artillery.

The proposals referred to a third type of manually emplaced explosive device dangerous to civilians – the remotely or automatically detonated delayed-action device. Experience had shown that as such bombs were a terrible risk to innocent civilians they, too, should be prohibited – or precautions should be taken to safeguard civilians from their effects.

Finally, the proposals called for the total prohibition of certain booby-traps and similar devices, referred to at the Lucerne Conference as being perfidious by their very nature. They included devices attached to or associated with internationally protected emblems, signs or signals, sick, wounded or dead persons, medical facilities, equipment, supplies and transport, and children's toys. Also included were those non-explosive booby-traps specifically designed to cause cruel or lingering death such as devices which stabbed, impaled, crushed, strangled, infected or poisoned the victim, and which had no military efficacy in modern warfare, their only purpose being to intimidate through the use of terror. Such devices were in most cases already unlawful under the Hague Regulations (para. 23(e)) but a further specific prohibition seemed desirable in the interests of humanitarian law.

The above-mentioned proposals, which were both realistic and humanitarian, had been introduced as they covered an area in which agreement seemed possible.

2. The other co-sponsor of document COLU/203 said that the 1973 Report on the work of Experts on Weapons that May Cause Unnecessary Suffering or have Indiscriminate Effects and the report on the Lucerne Conference gave little space to the subject of mines. That suggested a lack of interest quite out of proportion to the number and severity of military and civilian casualties caused by mines even after cessation of hostilities.

Regulation of mines and booby-traps would be relatively simple and worthwhile and his delegation had for that reason co-operated with another in wording the proposal COLU/203.

3. An expert, welcoming document COLU/203, reminded the meeting of document CDDH/IV/201 proposing, *inter alia,* the total ban of some methods of mine remote delivery. He questioned whether the "neutralizing mechanism" and "marking" proposed in document COLU/203 – in other respects an improvement on document CDDH/IV/201 – were adequate. He also wondered why the proposed definition of "remotely delivered mine" stipulated 2,000 metres as the minimum range.

He suggested that the words "use of remotely delivered mines" in paragraph C of the proposal contained in document COLU/203 should be changed to read "remote delivery of mines".

He urged also that the proposal should give greater protection to the civilian population by deleting the word "either" from the fourth line of paragraph D 2 and replacing "or", at the end of sub-paragraph (a), by "and". In addition, the use of one or more specific weapons for the defence of military objectives might be banned because of the indiscriminate effects of their use.

He also urged supplementing the list of forbidden booby-traps and devices by the addition, for example, of objects of worship and foodstuffs.

4. One expert said that his delegation gave general support to the joint working paper, which contained useful and realistic proposals of the type that were likely to lead to practical recommendations. A proposal made by his delegation at the Lucerne Conference was very effectively covered by the proposal to prohibit the use of certain booby-traps and other devices in paragraph E (b) of document COLU/203.

5. One expert said that his delegation was in agreement with the prohibitions and restrictions proposed in document COLU/203. It felt, however, that the attempt to make a list of prohibited uses of booby-traps, in paragraph E (b), was misguided. The list was not complete and could never be complete. His delegation proposed that the list should be replaced by a general formula to the effect that "The camouflage of explosive devices in objects in general use among civilians shall be prohibited" (see document COLU/206). It would welcome a discussion of that proposal.

6. One expert said that the proposals in the joint working paper provided a useful basis for discussions. He felt, however, that while the definition of "military objective" given in paragraph A of the proposals might have been suitable to the conditions of the First World War, it did not apply to the conditions prevailing after the Second World War; the Conference should try to work out a more meaningful definition of "military objective" giving clear guidance to soldiers in the field. He was doubtful whether underwater mines should be excluded from the provisions of the proposal, and agreed that the limit of 2,000 m in the definition of remotely delivered mines would need further discussion. With regard to the proposal that remotely delivered mines should be fitted with neutralizing mechanisms, he wondered whether the developing countries could afford to supply them. Lastly, he thought that the list in paragraph E (b) should be extended, since it was a matter directly related to humanitarian considerations.

7. One expert, welcoming the joint proposals, said that he thought they went to the limit of what many delegations would regard as desirable and possible. One

merit of the proposals was that they covered mines, booby-traps and other devices in a single paper, which would avoid difficulties of interpretation and queries concerning the demarcation lines between them. Another merit was the emphasis laid on the protection of the civilian population and on the criterion of indiscriminate use of such weapons. He wondered whether "underwater mine" covered mines used in internal waters, such as lakes and rivers; and he also queried the desirability of the 2,000 m limit for remotely delivered mines. The phrase "So far as is feasible" at the beginning of paragraph B was rather vague, but since it referred to battlefield conditions, it was difficult to be more precise.

8. One expert congratulated the sponsors of the proposals in document COLU/203, which deserved to be studied in greater detail.

They should be systematically discussed in conjunction with other possible solutions, such as those in document CDDH/IV/201, of which his delegation was a co-sponsor, which contained, for instance, a proposal concerning antipersonnel mines. The definition of "military objective" proposed in paragraph A was, he thought, not as precise as that in Article 47 of Additional Protocol I. He agreed with the previous speaker that the phrase "So far as is feasible" in paragraph B was too vague. He welcomed the proposal that minefields should be recorded but thought it desirable for the other party and, in particular, civilians to be informed of them before the "end of hostilities". They should be informed as soon as the troops which laid the mines had definitively left the area. With regard to paragraph C, he preferred the proposal in document CDDH/IV/201. He thought the addition of the words "(other than remotely delivered mines)" in paragraph D 1(a) was unnecessary, since the paragraph dealt with the *place* where mines, etc., were not to be used. With regard to E, there seemed to be a certain lack of balance between sub-paragraphs (a) and (b); moreover, the list in (b) was incomplete and somewhat arbitrary. A formula should be found which covered booby-traps in general.

9. One expert said that he had read the joint proposals with interest and his delegation would willingly participate in any discussion of the problems they raised. In view of what had been said at the Lucerne Conference (see paragraphs 220, 229 and 230) concerning minefields, he took the view that the defensive use of minefields for the purpose of paralyzing enemy movements were acceptable and should not be prohibited. On the other hand, mines scattered from aircraft over a large area presented a danger for civilians. The best solution would be to admit delayed-action weapons only as defensive weapons and to prohibit their use for offensive purposes.

With regard to the Canadian suggestion in document CDDH/IV/203 for an obligatory system of marking minefields, any such system could, for technical reasons, only be somewhat haphazard, since the perimeters of minefields were not accurately determined. In the case of air-delivered mines, accuracy depended on the efficiency of navigation instruments, the competence of air personnel, combat conditions and meteorological factors. A certain number of mines would unquestionably be scattered outside the area aimed at and would constitute a risk for civilians.

Even normally emplaced landmines might remain dangerous long after they had served their military purpose. They were usually very difficult to detect because one of the purposes aimed at was that they should not be detected and deactivated too easily. His delegation agreed with the sponsors of the joint proposals that means should be found for the accurate recording of the location of minefields so that they could be rendered harmless by the local authorities after the end of hostilities. It would be better if mines could be fitted with self-destruct devices; but that, unfortunately, would add to the weight and cost of the weapons.

In the case of booby-traps, the civilian population was most exposed to risk when they were placed in populated areas, i.e. outside combat zones. He therefore agreed with the proposal in document COLU/203 that such weapons should be banned outside combat areas, except in the immediate vicinity of military objectives, unless precautions had been taken to protect civilians. Furthermore, certain specific booby-traps which were manifestly treacherous should be prohibited, particularly when used in association with protective emblems or children's toys.

The weapons covered by agenda item 4 provided a suitable subject for the Conference's discussions because they had benefited from the latest developments of science and technology and because – as stated in paragraph 259 of the Lucerne Report – the Conference's deliberations should lay stress on use against the civilian population.

10. One expert congratulated the co-sponsors of the joint proposals and, in particular, supported the proposal that minefields should be recorded and that the records should be kept (paragraph B of document COLU/203). Mines delivered from aircraft or other remote delivery systems (paragraph C) were likely to be scattered over a wide area, so that the extent of the field was difficult to check or record. For that reason, his delegation preferred proposal V in document CDDH/IV/201 (page 3), which could be amended to read "Anti-personnel landmines must not be remotely delivered", instead of ". . . by aircraft". While agreeing in general with the proposals concerning booby-traps in paragraphs D and E of document COLU/203, he agreed with the proposal that the word "or" in D(2a) should be replaced by "and", and with the proposal to replace E(b) with the formula in document COLU/206.

11. One expert said that, while the proposals in the joint paper were generally acceptable to his delegation, he thought that certain improvements might be made to them.

In paragraph A, he wondered whether it was necessary to exclude underwater mines or whether only under-sea mines should be excluded and not those in inland waters. It had not been explained why the limit of 2,000 m for remotely delivered mines had been set; he was not sure whether there were any land-based weapons which could lay mines for more than that distance. He did not see why sub-paragraph D 2(b) had been included but, if it was maintained, the proposal that the word "or" should be replaced by "and" would be an improvement. With regard to paragraph E(b), he thought that the list should be expanded and that the general formula in document COLU/206 should be added at the end. Many recent conflicts had been highly asymmetrical: one side had possessed aircraft, while the other had not. He accordingly preferred the proposal in document CDDH/IV/201 that landmines laid from aircarft should be totally banned.

12. The author of document CDDH/IV/202, referring to the section dealing with the recording of minefields in the working paper under discussion said that he would be remiss if he did not draw attention to the reference to the proposal for the marking of minefields embodied in CDDH/IV/202, which foreshadowed the requirement for the marking of minefields. He laid stress on the fact that the proposals in CDDH/IV/201 failed to deal adequately with the delivery of mines by aircraft and the concern felt as to the dangers to the civilian population in inadequately marked minefields. He pointed out that mines had become more sophisticated and that there had been considerable technical advances in mine laying. He welcomed the initiative taken by the sponsors of the working paper and welcomed the comments made by other experts who had made realistic proposals. He had been glad to note that the previous discussions had dealt with specific details which proved the value of the working paper. He drew attention to paragraphs 238 and 248 of the Report of the Lucerne Conference on the desirability of marking mines and minefields and the opinion expressed by some experts on the need for mandatory rules for minefield marking. He said that the working paper had tried to satisfy those opinions and that it opened up a new area by dealing with the question of booby-traps. He also wished to record his satisfaction with the definition of "remotely delivered" mines. Concerning the comments expressed by a previous speaker, he pointed out that it would be unrealistic to try to seek prohibition of mines and minefields and that it would be more practical to try to ensure the protection of the innocent civilian population.

Concerning the camouflaging of booby-traps, mentioned in COLU/206, he pointed out that this created for commanders in the field the difficult problem of determining whether they could or could not resort to the camouflaging of, and to use as booby-traps, objects in general use by civilians, particularly in urban and populated areas. The proposal in COLU/206 was restricted to explosives, whereas the working paper had a wider scope. A good start would be made if an enumerated list could be drawn up of easily identified objects which were not to be booby-trapped. Such a list should be kept within practical and realistic limits.

13. The opinion that the Working Paper gave a balanced account of and a positive approach to the problems involved and provided concrete proposals that could well serve as a basis for constructive discussions was expressed by another expert. He said, however, that he would like further clarification of Section A, Definitions, since it left open several important questions such as the status of underwater mines, a minimum range of delivery of remotely delivered mines, and some others which had already been mentioned by several experts. The paper contained a sound proposal on the recording of minefields (Section B) and, in his view, should be retained with perhaps some additional improvements. As to the use of remotely delivered mines, his delegation still preferred, as more comprehensive and reliable, the proposal contained in Working Paper CDDH/IV/201, of which his delegation was a co-sponsor, but was also ready to give serious consideration to the provisions of Section C. He felt, however, that two points called for clarification. The first concerned the choice to be made between a neutralizing mechanism and marking; both had advantages and disadvantages, but his delegation would prefer the former, as providing better protection against the indiscriminate use of mined areas. But, here again, further examination of the question was needed. The second point concerned the period of time allowed to elapse before a neutralizing mechanism should be set to complete its task. Possibly the safest solution would be to have a predetermined time period to be observed in all situations, although a more acceptable one might be to define the period in terms of estimated military need, i.e. of the expected duration of military operations in the area. There were other possibilities and the whole problem merited careful examination.

His delegation considered that the question raised in Section D, Use of Mines, Booby-Traps and Other Devices in Populated Areas, had already been covered to a large extent by other provisions of international law. He asked for clarification regarding paragraph 2 and asked whether the stipulation that prohibition would apply in areas "in which combat between ground forces is not taking place or is not yet imminent" meant that the rules were not applicable in areas where combat was taking place. If so, he thought the lives of civilians would be seriously jeopardized.

His delegation considered that Section E, dealing with booby-traps, could well be expanded, particularly the sub-paragraph on children's toys, which should be supplemented by including all objects commonly used by civilians, such as fountain pens, telephones, etc. But this should in no way imply the prohibition of the use of ordinary mine devices.

14. One expert said that since mines were a very effective means of defence, he felt that they should not be discussed in the context of their prohibition. He thought that the Working Paper constituted a good attempt to try to reach a consensus. He supported its approach to the problems and welcomed the fact that it covered both humanitarian and defence aspects.

15. Another expert said that his delegation associated itself with the excellent Working Paper and that it combined two qualities that were rarely seen together: progress towards humanitarianism and a realistic attitude to the needs of defence. He stated that his delegation wished to be included as co-sponsor of document COLU/203.

16. An expert thought that any rules concerning mines should aim at avoiding indiscriminate effects, as was in fact recognized in the Working Paper before them. His delegation's first reaction was that the paper proposed useful definitions that could serve as background material, but he reserved the right to return to these definitions.

In principle, his delegation supported the proposal regarding the "recording of minefields", but thought that the phrase "so far as is feasible" left the choice to the commander in the field as to what was "feasible". That greatly reduced the value of the proposed rule which, by its very vagueness, might render it nugatory. The limitation of recording of minefields exceeding 20 mines might also lead to the possibility of circumvention.

He considered that the scattering of mines by aircraft constituted a severe threat to civilian populations in the area concerned, whether these mines were equipped with neutralizing mechanisms or not. He wondered whether it was possible to produce a 100% or even 80% reliable neutralizing mechanism. He was also doubtful whether the limits of a minefield could be so marked as to ensure that civilians would avoid them. He considered that the proposal embodied in CDDH/IV/201, of which his delegation was a co-sponsor, was not only simpler but more reliable.

He approved of the general approach of the Working Paper and considered as constructive the proposals made in Section D, Use of Mines, Booby-Traps and Other Devices in Populated Areas, but considered that the last part of the proposal concerning the prohibition of use of certain booby-traps and other devices were not sufficiently elaborated. He was unclear as to the reason for the discrimination in the proposals in Section E between non-explosive and explosive booby-traps. He said that he would prefer a note in a preamble stating that some uses were already unlawful, rather than having a long list of proscribed items as in Section E (b) (i-v), particularly as it was possible that some States might not adhere to the list.

He considered that the proposal regarding children's toys was a useful one. He thought that the question of the maximum charge of anti-personnel mines was an important one as was the possibility of introducing a rule prescribing that material used in landmines should be detectable by X-rays.

17. After expressing his appreciation of the Working Paper, one expert said that the requirement to record the location of small minefields should not be limited to minefields containing as many as 20 mines, since even smaller minefields were a danger to civilians.

He proposed for Section D, paragraph 2, a new sub-paragraph (c), to read "they are marked in a distinctive manner". He suggested that the last word of sub-paragraph (a) "or", be deleted and the word "and" be added at the end of sub-paragraph (b).

18. Another expert said that he had been struck by the limited interest that had been shown in mines and booby-traps at the Lucerne Conference; these weapons caused more civilian casualties than any other conventional weapons. He attached great importance to the question of the removal of minefields after hostilities, based on practical experience in Denmark after World War II. He considered that particular importance should be attached to the use of mines in areas where there was civilian movement. The Working Paper was a constructive document and constituted a practical contribution to the work of the Conference.

19. One expert said that the proposals in document COLU/203 were very useful and could help the Conference in its effort to arrive at common conclusions and recommendations. The Lucerne Conference had stressed the importance of mines and booby-traps which, as many experts had pointed out, were more dangerous to the civilian population than most other types of weapon, especially if used in densely populated areas.

In his country, the presence of unrecorded minefields still caused casualties among the civilian population thirty years after the end of the Second World War. Nevertheless, any proposal to record minefields must be one which it was really possible to implement during hostilities. The phrase "So far as this is feasible" gave rise to problems, but, in combat conditions, there were real difficulties in the recording of minefields, even those consisting of more than twenty mines. Efforts should be made to find a wording of paragraph B which was strong enough to create a real obligation to record minefields, but at the same time was capable of practical implementation. The phrase "cessation of active hostilities" in that paragraph was also unclear: did it mean a cease-fire agreement or the local cessation of hostilities in the area in question?

Since the laying of minefields was designed to protect military units or even civilian objectives against an advancing enemy, it was unrealistic to propose that they should be marked by flags or flares. The proposal that they should be equipped with self-destruct or neutralizing mechanisms was more useful and should be taken into account.

A distinction should be made between anti-personnel and anti-tank minefields, the technical characteristics of which were very different. Such a distinction had not been made in document COLU/203, but he was not convinced that it was unnecessary. But it was above all necessary to provide for the elimination of minefields after they had served their defensive military purpose, especially in the case of minefields laid from the air, which was the commonest method used in modern warfare. A way of avoiding the difficulties involved in the phrase "cessation of hostilities" would be to provide for them to be placed under the control of the civilian administration once they had served their defence purpose.

With regard to paragraph D, it should be recalled that the use of booby-traps against the civilian population was already prohibited; the problem was to protect all personnel against the dangers they represented. The experts should therefore propose, as their common conclusion, that the use of booby-traps should be prohibited as such, adding that their prohibition was particularly important in populated areas.

His delegation welcomed paragraph E. It understood that the purpose of the list in E (b) was to give commanders clear instructions concerning the devices prohibited; but the list was not complete and might give the impression that other uses were allowed. His delegation would not object to a formula such as that proposed in document COLU/206 but would prefer a general prohibition of the use of all types of booby-traps. Lastly, his delegation was not convinced of the necessity to distinguish between explosive and non-explosive booby-traps.

Seventh Meeting

General Debate on Delayed-action and Treacherous Weapons (continued)

1. An expert briefly commented on and expressed his approval of working document COLU/203 submitted by two delegations and concerning landmines and booby-traps and the regulation of their use.

He pointed out that several preliminary statements had been made at earlier meetings and, although the proposal submitted could doubtless be improved, it was clearly a well thought out paper. There was, however, little prospect of agreement being reached on the prohibition of remotely delivered mines.

2. Another expert welcomed the concrete proposal made in document COLU/203 which showed how necessary it was to protect civilians further. At the previous meeting, most experts had agreed on the constructive nature of the proposals and on the fact that it deserved careful consideration.

A technical debate within the General Working Party was obviously called for and his delegation would be making detailed proposals in due course.

3. A member of one of the delegations that had submitted COLU/203 said that his delegation was prepared to hold off-the-record discussions with any other delegation that might wish so to do. He thanked an expert from another delegation which wished to be considered a co-sponsor of COLU/203.

4. The CHAIRMAN, noting that the debate commenced during the previous meeting had drawn to a close, proposed that the meeting rise and reconvene as the General Working Party.

III. Report of the General Working Group

Introduction

1. The General Working Group of the Conference of Government Experts on the Use of Certain Conventional Weapons, second session, was set up by the Plenary Meeting of the Conference in accordance with Rule 5 of the Rules of Procedure. The General Working Group, hereinafter referred to as GWG, designated as its Chairman Dr. E. Kussbach (Austria) and as Vice-Chairmen Mr. A. de Icaza (Mexico), Mr. K. Saleem (Pakistan) and Mr. B. Wozniecki (Poland). Dr. F. Kalshoven (Netherlands) was designated to act as Rapporteur. Mr. Y. Sandoz and Mr. B. Zimmermann, legal experts of the ICRC, acted as Secretaries to the GWG.

2. The mandate of the GWG was to study in detail the various subjects of the work programme of the Conference after they had been introduced in plenary. The GWG was authorized to set up special working groups as necessary to deal with specific questions. In the course of its proceedings, the GWG set up the following special working groups:

(a) a working group of military experts on mines and booby-traps; *(b)* a working group of technical experts on small-calibre projectiles; *(c)* a working group on legal issues.

Reports of the special working groups are annexed to this report (see III. 9, 10 and 11).

3. This report does not attempt to set out in any detail the debate held in the General Working Group. Instead, and following the recommendations of experts, it is structured so as to give prominence to:

(a) areas of agreement or disagreement concerning the types of proposal which might be advanced for future consideration, and

(b) new factual information of direct relevance.

Delayed-Action Weapons and Treacherous Weapons

38. Although the GWG discussed the question of delayed-action and treacherous weapons in some detail, it was agreed that adequate treatment of the military considerations involved required that the subject be referred to a special working group. A working group of military experts on mines and booby-traps was duly convened; its report is reproduced in III.10.

Proposals

39. The proposals put forward concerned five distinct topics:

(a) the use of time-fused munitions;
(b) the recording of minefields;
(c) the use of remotely delivered mines (such as those referred to in the Lucerne report as "scatterable mines");
(d) the conduct of mine warfare within areas of civilian population; and
(e) the use of booby-traps.

Two general cautions were advanced during the discussion of these proposals: due consideration should be given to (1) the requirements of defensive military operations and (2) the risk of compelling resort to more objectionable means or methods of warfare.

40. On topic *(a)*, a proposal was put forward in document COLU/213 (see Annex A.12). It was explained that the proposal sought to eliminate as far as possible, by means of an express time limit, the effects of time-fused munitions which would escape, whether in time or in space, the sphere of military operations properly speaking and which would, hence, pose grave risks for the civilian population or for rescue operations. Opposition to the proposal was expressed on the grounds that the specification of any one particular time limit could place undue restrictions on important military operations.

41. On topics *(b)* – *(e)*, proposals were put forward in document COLU/203 (see Annex A.2). These attracted arguments both *pro* and *contra*, and gave rise to several alternative proposals. One of these, noted in para. 5, also addressed the same four topics. The others were more limited in this respect; they are noted in paras. 6–18, which are ordered according to topic.

42. Proposals on topics *(b)* – *(e)* were also put forward in document COLU/215 (see Annex A.14), being presented as amendments to the COLU/203 proposals. (The wording referred to for the amendment of the proposal on the recording of minefields is set out in para. 44 below.)

43. Concerning the recording of minefields, doubts were expressed about a number of different aspects of the proposal contained in COLU/203, which also provided for the publication of minefield locations upon the cessation of active hostilities. The Working Group of Military Experts (on which see para. 38) considered these matters, and its report advances a revised proposal.

44. It was suggested that any provisions for the recording and reporting of minefields should be supplemented by provisions for the disposal of minefields. This suggestion was put forward in the form of additional wording for part B of the COLU/203 proposal, as follows:

"B*bis*. Disposal of Mines

(*a*) No mine may be deployed until means exist to ensure that it can be safely located and disposed of at the close of hostilities or when the mine no longer serves the military purpose for which it was placed in position.

(*b*) The military authorities of the adversaries shall co-operate as necessary at the close of hostilities to ensure the disposal of all mines or other unexploded munitions."

45. Concerning the use of remotely-delivered mines, a proposal to prohibit the use of aircraft for laying anti-personnel mines had already been advanced in document CDDH/IV/201 (see Annex A.21).

46. The COLU/203 proposal on remotely-delivered mines provided for the prohibition of the use of such mines unless the mines were equipped with a neutralizing mechanism or the area in which they were delivered was distinctively marked. The COLU/215 proposal would, in addition, prohibit the use of remotely-delivered mines in populated areas. A number of difficulties were perceived in these approaches and in the manner in which "remotely-delivered mines" were defined; these, and a range of possible solutions are set out in the report of the Working Group of Military Experts.

47. Another suggestion concerning remotely-delivered mines was put forward in the following terms:

> "The use of remotely-delivered mines outside the battle zone is prohibited. Within the battle zone, remotely-delivered mines must be fitted with a neutralizing mechanism or the areas in which these are delivered must be marked in some distinctive manner."

48. Remotely-delivered ammunitions were explicitly included within the scope of the proposal on time-fused munitions contained in document COLU/213, noted in para. 40 above.

49. Concerning the conduct of mine warfare and related activities within areas of civilian population, the proposal contained in document COLU/203 sought to prohibit, with certain exemptions, the use of mines, booby-traps and suchlike devices in circumstances where they might create undue dangers for the civilian population. The proposal was criticized on the grounds that it would afford insufficient protection for the civilian population, both because the range of munitions whose use was to be prohibited was considered to be unduly narrow, being restricted primarily to manually emplaced munitions, and because the exemptions made from the prohibition were considered to be unduly broad. The COLU/215 proposal was intended, so it was explained, to offer more restrictive language in both these respects. Another criticism of the COLU/203 proposal was that it amounted to nothing more than a restatement of prohibitions derived from existing international law. The COLU/215 proposal was intended, so it was explained, as a response to this criticism also. These and related matters were discussed further in the Working Group of Military Experts, as is noted in the report from the group.

50. Concerning booby-traps, the proposal contained in document COLU/203 offered a prohibition of the use of a number of specific devices or techniques falling within this category. This approach attracted a number of criticisms and alternative proposals.

51. The proposal put forward in document COLU/217 (see Annex A.16) recommended a specific addition to the list of booby-traps contained in the COLU/203 proposal.

52. The proposal put forward in document COLU/206 (see Annex A.5) was confined to booby-traps of the explosive type, this on the ground that in the view of its authors the prohibition of the use of non-explosive booby-traps already was contained in Article 23 of the Hague Regulations. Some experts added that other international legal instruments in force also had a bearing on this topic. Other experts argued, on the other hand, that since non-explosive booby-traps might be of substantial military importance for poorly-endowed armed forces, a prohibition of their use would be inappropriate.

53. The proposal put forward in document COLU/214 (see Annex A.13) recommended the deletion of the passage specifying non-explosive booby-traps from the COLU/203 proposal.

54. It was argued that it was inappropriate to distinguish between explosive and non-explosive booby-traps, since both were equally perfidious. It was therefore proper to envisage only a comprehensive prohibition of the use of booby-traps, a consideration which also argued against the approach of specific prohibitions used in the COLU/203 proposal and its congeners.

55. A fully comprehensive approach would require agreement on what precisely was to be regarded as a booby-trap. A proposal for such a definition was put forward in document COLU/219. This, and other matters relating to booby-traps, were discussed in more detail in the Working Group of Military Experts.

New data

56. Some new information was presented on the reliability of the self-destruct mechanisms fitted to certain mines. For these munitions, the standard of reliability aimed at was that only one in a hundred thousand should remain hazardous after the self-destruct time had elapsed; and a reliability of one in a thousand was specified as the lowest acceptable. In the mine system referred to, test experience had shown that something in the order of one in 30,000–40,000 mines failed to neutralize.

Report of the Working Sub-Group of Military Experts on Mines and Booby-Traps

Chairman: Colonel K. Troughton (Canada)
Rapporteur: Mr. R. Akkerman (Netherlands)

Procedure

It was first agreed that there be two basic approaches to the subject, namely the French-Anglo-Dutch working paper (doc. COLU/203) and the proposal contained in document CDDH/IV/201, and that the former be taken as the basis for discussion in the Group. Discussion of the latter would arise automatically when discussing Article C of the former working paper. It was also agreed that discussion of definitions be left until substantive discussion had been concluded. Formal proposals on the subject of mines and booby-traps are listed in the table of contents.

Discussion on the merits of Proposals

1. *Recording of Minefields*
Discussion centred on three issues:

(*a*) whether the term "so far as is feasible" was sufficiently mandatory and whether it afforded the civilian population adequate protection. Some experts generally felt that this sentence should be rephrased in a more mandatory way since it left too much room for interpretation. Other experts were generally of the opinion that this provision should be maintained in order to exempt the soldier from the obligation to record minefields under circumstances preventing his compliance. In this connection, an expert raised the question of the accuracy of minefield recording;

(*b*) whether "20 mines" was acceptable as a threshold. Some experts were of the opinion that the number of 20 was rather high. Other experts were of a contrary opinion, finding that number rather low. Again, some other experts thought that the number should be replaced by the *density* as a criterion, that being opposed by other experts;

(*c*) whether the provision for a post-hostilities exchange of information had been drafted in a satisfactory way. Suggestions were made by several experts that records of minefields should be handed over to civilian authorities after cessation of hostilities. One expert thought that the word "cessation" in Article B of COLU/203 should be replaced by "close". However, no agreement was reached thereon. A proposal was made by one of the sponsors of doc. COLU/203, supported by another expert, amending Article B as follows: "The location of pre-planned defensive minefields shall always be recorded. So far as is feasible, the location of all other minefields containing more than 20 mines shall be recorded. Such records shall be retained until after the cessation of active hostilities, at which time the location of all recorded minefields situated in territory controlled by an adversary party shall be made public." Widespread agreement was reached on Article B as quoted above.

2. Disposal of Mines

It was furthermore discussed whether a provision would have to be added to the paper as suggested by the representative of SIPRI and implied in doc. COLU/215, submitted by the experts of Spain and regarding the disposal of mines. The submitting experts suggested that the word "safely" be deleted in the SIPRI document. However, no agreement could be reached as to the insertion of such a provision.

3. Use of Remotely-Delivered Mines

The following issues were discussed by the experts:

(*a*) the concept of "remotely-delivered mine"
(*b*) the definition of "neutralizing mechanism"
(*c*) "marking" of minefields
(*d*) the extent of protection of the civilian population arising from this provision.

(a) *The concept of "Remotely-Delivered Mine"*

An expert of a delegation co-sponsoring doc. COLU/203 explained that one had here to deal with a weapon that, to a great extent, belonged to the future weapons category and that this feature had been a factor when drafting the relevant provisions in the above document.

(b) *The definition of "Neutralizing Mechanism"*

A number of experts were of the opinion, along the lines of doc. COLU/213, that this definition should contain a time-limit of 24 hours. Another expert thought that such a limit would not be realistic for many delegations and that it would indeed not be acceptable for the same delegations to accept any limit expressed in hours, days, etc.

One expert thought that the definition contained in doc. COLU/203, Article A, was a good one since the distinction that could be made with regard to neutralizing mechanisms, namely in devices that could be destroyed by remote control, those that were self-destroying and those that were self-sterilizing, was covered thereby.

(c) *Marking Devices*

Many experts thought that the marking of mines would reduce the military advantage thereof to a considerable extent. Several other experts thought that marking would establish the only real protection for the civilian population.

(d) *Protection of Civilians*

Some experts thought that the only way of protecting civilians in a satisfactory way was to provide all remotely delivered mines with a neutralizing mechanism. Experts from other delegations were of the opinion that marking remotely-delivered mines was the only means of full protection of the civilian population. One expert thought that remotely-delivered mines should, at the same time, be provided with marking devices and a neutralizing mechanism. Other experts thought that all mines delivered outside the battle zone should contain neutralizing mechanisms.

Experts from delegations co-sponsoring doc. COLU/203 defended the wording thereof (offering an alternative choice between the two devices) by stressing the future character of this application of mines and the relative novelty of the weapon-type itself. Many experts from delegations co-sponsoring doc. CDDH/IV/201 thought that the proposal contained therein and forbidding the use of air-delivered anti-personnel land mines was more protective of the civilian population.

An expert stressed that air-delivered minefields, when they have served their defensive purpose, ought to be eliminated or placed under control of civilian authorities at the close of hostilities.

Some experts were of the opinion that the use of devices to which this proposal applied should be expressly prohibited in any civilian population area, only allowing for some specific exceptions to that main rule in combat situations and if the civilian population would be duly protected.

A similar proposal was contained in COLU/215. Some experts expressed themselves in favour of the amendment proposed in that document.

One expert advocated the prohibition of anti-lifting devices in all mines.

There was broad agreement in the group that the proposed text of COLU/203, Article C, was a significant advance over current regulations and that it could serve as a meaningful basis for future elaboration and refinement. There was consensus that the proposed text was sufficiently complete to satisfy all concerned.

4. *Use of Mines, Booby-Traps and Similar Devices*

After considerable preliminary discussion on the meaning and implications of the original proposals, it was agreed that a revised version of Articles D and E would serve as a basis for discussion, reading as follows:

"D. *USE OF MANUALLY EMPLACED MINES AND OTHER MANUALLY EMPLACED DEVICES IN POPULATED AREAS*

1. This proposal applies to manually emplaced mines and all other manually emplaced devices (explosives and non-explosive) which are designed to kill, injure or damage and for that purpose actuated:
(a) by the presence or proximity of a person or vehicle;
(b) when a person disturbs or approaches an apparently harmless object or performs an apparently safe act;
(c) by remote control; or
(d) automatically after a lapse of time.

2. In any city, town, village or other area containing a concentration of civilians in which combat between ground forces is not taking place or does not appear to be imminent, devices to which this proposal applies may not be used unless either:

(a) they are placed on or in the close vicinity of a military objective; or *(b)* due precautions are taken to protect civilians from their effects.

IV. Annexes

A. PROPOSALS SUBMITTED TO THE CONFERENCE

1. COLU/203[1] (Original: English)

Working Paper
submitted by the Experts of France,
the Netherlands and the United Kingdom

LAND MINES AND BOOBY-TRAPS
AND PROPOSALS FOR THE REGULATION OF THEIR USE

1. Land mines, booby-traps and similar devices were discussed by the first session of the Conference of Government Experts on the Use of Certain Conventional Weapons held at Lucerne between 24 September and 18 October 1974 (hereinafter called "The Experts' Conference"). It was noted that these weapons might be used in a manner which could be characterised as "perfidious" and, further, that their use in certain circumstances might involve a degree of indiscrimination between military and civilian targets. The Experts' Conference reported that "it was widely felt that in further deliberations on the subject stress should be laid on use against the civilian population".

Recording of Minefields

2. Many armed forces today have a sophisticated system for recording the location of minefields, whether remotely delivered or manually emplaced. Such recording is primarily for the benefit of the armed forces themselves, but it does provide a ready means of locating such minefields in order that they may be removed after the cessation of active hostilities. Although it may not be possible to provide for the recording of the location of isolated mines hastily laid during combat, it would seem desirable to make a formal requirement for the recording of the location of even small minefields (exceeding 20 mines) and for public disclosure after the cessation of active hostilities of the location of all such minefields in territory controlled by an adverse party.

"Scatterable" (or "Remotely Delivered") Mines

3. One category of mines which was discussed in detail during the Experts' Conference was there referred to as "Scatterable mines". This category embraces mines delivered by tube and rocket artillery and, more commonly, by aircraft. The Experts' Conference recognised that these mines, which are a comparatively recent development, serve much the same function as emplaced mines but with the additional utility that follows from the rapidity with which they can be deployed. This category of mines can perhaps be better described as "Remotely

Delivered Mines" since they will normally be delivered from comparatively long ranges.

4. It was generally agreed at the Experts' Conference that the tactical use of remotely delivered mines had a greater propensity for endangering friendly troops and civilians than the use of emplaced mines. It was also agreed, however, that for the selfsame reason it would normally be in the interests of the user to maintain particularly tight control over mine scattering.

5. It was suggested at the Experts' Conference, and again subsequently during the deliberations of the *ad hoc* Committee at the Geneva Diplomatic Conference, that it would be possible to mark the location of remotely delivered mines through the use of markers such as flags or pyrotechnic flares, delivered simultaneously with the mines. An alternative suggestion was to incorporate in these mines a self-destruct or neutralization mechanism; it was noted that some such mechanisms were already in existence and were extremely reliable.

6. The adoption of either of the alternatives mentioned in the last paragraph would counter the potential for indiscriminate employment of remotely delivered mines and it is suggested that a proposal to this effect should be considered.

Booby-Traps

7. A booby-trap *stricto sensu* is an apparently harmless device concealing an explosive charge designed to go off when tampered with. However, the expression has come to include certain other devices; examples include concealed pits, sharpened wooden stakes under vegetation, and explosive devices connected to, although not concealed by, a harmless object. Additionally, as was recognized by the Experts' Conference, there may be little or no difference between an unmarked, isolated mine and a booby-trap; similar considerations may also apply to anti-personnel munitions which detonate automatically after a lapse of time or which are remotely detonated. The feature common to these devices is, of course, the inability to control the type of target against which they operate. There is an obvious danger that such a target may be civilian.

8. The greatest danger to civilians from the use of booby-traps and similar devices will occur in populated areas. It may well be difficult to place realistic prohibitions on the use of these weapons in such areas when combat between ground forces is taking place or is imminent, but in such circumstances civilians are normally either no longer present or have taken shelter. It is outside the battle zone, where the life of the community may well be proceeding relatively normally, that the use of the devices in question may often lead to the death or injury of civilians. Evidence of the grave danger to civilians in these circumstances was given at the Experts' Conference and it was suggested by the President that the

Conference might reach a consensus on prohibiting the use of booby-traps which represent such a danger. Although there was considerable support for the President's suggestion, the Conference felt that some clarification was needed. It is therefore now suggested that in populated areas outside the battle zones, the use of booby-traps and similar devices should be forbidden except on or in the close vicinity of a military objective or unless due precautions are taken to protect civilians from their effects.

9. The President of the Experts' Conference also suggested that there might perhaps exist a consensus on the prohibition of explosive devices perfidious by their very nature. Again, there was support in the Conference for the President's suggestion but it was felt that further careful examination was needed. It is now suggested that certain specific and obviously perfidious booby-traps could be forbidden, such as devices used in connection with recognized protective emblems, the sick, wounded or dead, graves, medical facilities or supplies and children's toys.

10. From time to time use has been made of non-explosive booby-traps designed to kill or injure by such means as stabbing, impalement, crushing or strangulation. In many cases these have been associated with matter likely to infect or poison the victim. Devices of this nature are likely to cause a cruel death or, at the least, very considerable suffering and from the military point of view are ineffective except for the purpose of intimidation. The use of such devices is in most cases already prohibited by Article 23(e) of the Hague Regulations as being calculated to cause unnecessary suffering but it would seem desirable to reinforce this ban by a further express prohibition.

Formulation of Prohibitions – scope of application

11. An attempt to formulate the prohibitions and restrictions in this paper is at Annex A.

12. The work of this second session of the Conference of Government Experts on Conventional Weapons, like that of the first session and that of the Diplomatic Conference on Humanitarian Law, has application only to conventional weapons as indeed is clear from this Conference's title itself. Accordingly, the prohibitions and restrictions on use suggested below must be taken as having no application to atomic, bacteriological or chemical weapons. The use of such weapons is of course already in part controlled by existing instruments of international law, such as the Geneva Protocol of 1925 and the Biological Weapons Convention of 1972.

PROPOSALS ON THE USE OF MINES AND BOOBY-TRAPS

A. Definitions

For the purpose of these proposals:

- "booby-trap" means an explosive or non-explosive device or other material deliberately placed to kill or injure when a person disturbs or approaches an apparently harmless object, or performs an apparently safe act;
- "military objective" means, so far as objects are concerned, any object which by its own nature, location, purpose or use makes an effective contribution to military action and whose total or partial destruction, capture or neutralization in the circumstances ruling at the time, offers a definite military advantage;
- "mine" means an explosive or incendiary munition placed under, on or near the ground or other surface area and designed to be detonated by the presence or proximity of a person or vehicle but does not include an underwater mine;
- "neutralizing mechanism" means a self-actuating or remotely controlled mechanism which is designed to render a mine harmless or cause it to destroy itself when it is anticipated that the mine will no longer serve the military purpose for which it was placed in position;
- "remotely-delivered mine" means any mine delivered by artillery, rocket, mortar or similar means at a range of over 2,000 metres or dropped from an aircraft.

B. Recording of Minefields

So far as is feasible, the location of all minefields containing more than twenty mines shall be recorded. Such records shall be retained until after the cessation of active hostilities, at which time the location of all recorded minefields situated in territory controlled by an adversary party shall be made public.

C. Use of Remotely-Delivered Mines

The use of remotely delivered mines is forbidden unless either each such mine is fitted with a neutralizing mechanism or the area in which they are delivered is marked in some distinctive manner.

D. Use of Mines, Booby-Traps and other Devices in Populated Areas

1. This proposal applies to:

(a) mines (other than remotely-delivered mines);

(b) booby-traps; and

(c) all other manually emplaced munitions designed to kill, injure or damage and for that purpose to detonate automatically after a lapse of time or to be remotely detonated.

2. In any city, town, village or other area containing a concentration of civilians in which combat between ground forces is not taking place or is not yet imminent, devices to which this proposal applies may not be used unless either:

(a) they are placed on or in the close vicinity of a military objective; or *(b)* due precautions are taken to protect civilians from their effects.

E. Prohibition of Use of Certain Booby-Traps and other Devices

It is forbidden in any circumstances to use any booby-traps or similar device which:

(a) is designed to kill or injure by a non-explosive means which stabs, impales, crushes, strangles, infects or poisons the victim; or
(b) is in any way attached to or associated with:
  (i) internationally recognized protective emblems, signs or signals;
  (ii) sick, wounded or dead persons;
  (iii) burial sites or graves;
  (iv) medical facilities, medical equipment, medical supplies and medical transport;
  (v) children's toys.

2. COLU/214 (Original: English)

Working Paper
submitted by the Experts of the Philippines

USE OF MINES AND BOOBY- TRAPS (COLU/203)

Delete paragraph E(a) of the "Proposals on the Use of Mines and Booby-Traps" attached to Working Paper COLU/203 (submitted by the Experts of France, Netherlands and United Kingdom).

3. COLU/215 (Original: Spanis h)

Working Paper
submitted by the Experts of Spain

LAND MINES AND BOOBY-TRAPS (COLU/203)

The following amendments are proposed to document COLU/203, submitted by the experts of France, the Netherlands and the United Kingdom:

to paragraph B. *Recording of minefields*

Replace with the wording proposed by the expert of SIPRI, which is fuller and more accurate.

to paragraph C. *Use of remotely delivered mines*

Replace the words after "delivered" (third line) with the words: ". . . is easily identifiable". Then add a separate paragraph, reading: "In every case, their use in populated areas is prohibited."

to paragraph D. *Use of mines, booby-traps and other devices in populated areas*

We propose the following text:

"1. This proposal applies to:
(a) mines; and
(b) any other device emplaced with the intent of causing death, injury or damage by delayed action or remotely controlled detonation.

2. The use of the above-mentioned devices is prohibited in any city, town, village or other populated area in which combat between ground forces is not taking place or is not imminent.

3. Remotely detonated devices may be used in populated areas only when they are placed in or near military objectives and when due precautions can be taken to protect civilians from their effects.

Note: This text requires the deletion of the word "booby-traps" from the title of paragraph D.

to paragraph E. *Prohibition of use of certain booby-traps and other devices*

Insert in the first line, after the word "circumstances", the words: ", in accordance with the regulations already established,".

4. COLU/217 (Original: English)

Working Paper
submitted by the Experts of Israel

USE OF MINES AND BOOBY-TRAPS (COLU/203)

Add a sub-paragraph:

(vi) "objects in general use among civilians"
to paragraph E of the "Proposals on the Use of Mines and Booby-traps" attached to Working Paper COLU/203 (submitted by the Experts of France, Netherlands and United Kingdom).

5. CDDH/IV/201

Working Paper
submitted to the CDDH by Algeria, Austria, Egypt, Lebanon, Mali, Mauritania, Mexico, Norway, Sudan, Sweden, Switzerland, Venezuela, Yugoslavia

Incendiary Weapons, Anti-Personnel Fragmentation Weapons, Fléchettes, especially Injurious Small-Calibre Projectiles, Anti-Personnel Land Mines

Explanatory Memorandum

Document CDDH/DT/2 of 21 February 1974, submitted by Egypt, Mexico, Norway, Sudan, Sweden, Switzerland and Yugoslavia contained a working paper with draft proposals for the prohibition of the use of specific conventional weapons. The proposals are now revised somewhat in the light of the discussions at the Lucerne conference of government experts on the use of certain conventional weapons (24 September–18 October 1974) and resubmitted. The comments which were submitted together with the draft proposals in the working paper at the 1974 session of the diplomatic conference are now presented in revised form and embodied in an accompanying explanatory memorandum.

V. Anti-personnel land mines

Anti-personnel land mines must not be laid by aircraft.

ANNEX

Explanatory Memorandum

The two draft Protocols contain some essential rules regarding methods and means of warfare. In one area, however, that of weapons, the draft rules hardly amount to more than a reaffirmation of existing law. Thus, in Article 33 of the first Draft Protocol the following rule is proposed:

"Prohibition of unnecessary suffering

1. The right of parties to the conflict and of members of their armed forces to adopt methods and means of combat is not unlimited.

2. It is forbidden to employ weapons, projectiles, substances, methods and means which uselessly aggravate the sufferings of disabled adversaries or render their death inevitable in all circumstances."

These fundamental rules largely reflect the contents of articles 22 and 23(e) of the Regulations respecting the laws and customs of war on land found in the Hague Conventions of 1899 (II) and of 1907 (IV) and the preamble of the 1868 St Petersburg Declaration. This reaffirmation of the general prohibition of use of one kind of weapon is in itself welcome.

A similar general prohibition of use of another category of weapon, namely, those which are by their nature or normal use indiscriminate in their effects, would be of interest *inter alia* in response to the expressed desire for rules against weapons which may cause ecological damage. However, such a rule is perhaps redundant in view of the even broader general rule proposed in article 46: 3 to reaffirm the customary rule prohibiting indiscriminate warfare:

> "The employment of means of combat, and any methods which strike or affect indiscriminately the civilian population and combatants, or civilian objects and military objectives, are prohibited . . ."

These rules of a general scope also express the philosophy behind the prohibitions of use which in the past have been adopted regarding specific types of weapons, e.g. the dum-dum bullet (1899) and the automatic unanchored contact mine (1907).

It is submitted that the general prohibitions of weapons apt to cause unnecessary suffering and of means and methods of warfare which are by their nature or normal use indiscriminate – proposed for reaffirmation – should now, as in the past, be supplemented with prohibitions of use of specific weapons which are deemed to fall within the general categories prohibited. Time would seem to be ripe for an examination of specific conventional weapons the use of which currently may be questioned from the viewpoint of compatibility with the general prohibitory rules which are to be reaffirmed. In the present working paper proposals are advanced for the prohibition or restriction of use of a number of such conventional weapons.

Proposals for the prohibition or restriction of use of other conventional weapons than those covered by this working paper could easily be added on the grounds that they risk having indiscriminate effect or causing excessive suffering. The list of proposed bans is thus not exhaustive, but may well be supplemented by other proposals.

In connection with the adoption of prohibitions of use, such as those contained in the working paper, consideration would have to be given to some related matters, viz. the question whether the rules should be absolute in character or binding only as between adversaries which have assumed the obligation to abide by the rules. Moreover, the question should be examined how, in the future, surveys can be made with a view to identifying weapons the use of which should be prohibited or subjected to restrictions for humanitarian reasons. A mechanism should be devised to facilitate such surveys to recur without too long intervals in order to ensure that weapon developments are always assessed in the light of humanitarian principles. Only in this way can there be some assurance that the broad prohibitory rules relating to the use of weapons will in fact be applied to specific weapons. But for such periodic review the technological development could lead to the production of ever-more cost-effective – but inhumane – weapons and weapon systems. Should the efforts fail to prohibit the use of specific weapons and to create mechanisms for review, the temptation to produce such new and cost-effective – but inhumane – weapons would be strong *inter alia* for the purpose of deterrence. The introduction now of prohibitions of use of specific weapons and agreement of regular reviews could discourage development of new particularly inhumane weapons.

Comments to the Proposed Prohibitions

V. Anti-personnel land mines

The use of anti-personnel mines is a generally accepted means of hampering enemy advance and of putting combatants out of action. However, certain ways of employing anti-personnel land mines may easily lead to injuries indiscriminately being inflicted upon combatants and civilians alike. The risks of such results are especially high if such mines are laid, perhaps in very large numbers, by aircraft. The limits of the mined area will often be very uncertain with this method. The results are apt to be particularly cruel if the mines are not equipped with self-destruction devices which will function reliably after a relatively short time. The risk of indiscriminate effects may be reduced also through marking of minefields – this is not possible, however, when the mines are scattered over a vast area.

NOTE

1 Takes account of document COLU/203/Add.1.

4

The United Nations Convention on Prohibitions or Restrictions on the Use of Certain Conventional Weapons Which May be Deemed to be Excessively Injurious or to Have Indiscriminate Effects

adopted 10 October 1980

The 1977 Additional Protocols to the Geneva Conventions of 1949 do not include prohibitions or restrictions on specific conventional weapons. None the less, the conference which negotiated and adopted the treaties, the Diplomatic Conference on the Reaffirmation and Development of International Humanitarian Law 1974–1977, recommended that a separate meeting be held to consider restrictions on conventional weapons which might be considered to be excessively injurious or have indiscriminate effects. This recommendation was subsequently endorsed by the United Nations General Assembly. Following two preparatory meetings, a United Nations Conference was convened in Geneva from 10 to 28 September 1979 and from 15 September to 10 October 1980 and adopted the United Nations Convention on Prohibitions or Restrictions on the Use of Certain Conventional Weapons Which May be Deemed to be Excessively Injurious or to Have Indiscriminate Effects. This convention, which was adopted 10 October 1980, is also known as the Convention on Certain Conventional Weapons (CCW or 1980 CCW).

The 1980 CCW contains a framework convention and three annexed protocols each regulating or prohibiting specific types of weapons. Protocol I prohibits the use of weapons which injure by means of fragments not detectable by X-rays. Protocol II governs the use of mines, booby-traps and other similar devices. Protocol III limits the use of incendiary weapons.

Protocol II regulates both anti-personnel and anti-vehicle mines. It prohibits their indiscriminate use and their manual emplacement in populated areas unless combat is taking place or is imminent or the mined areas are clearly marked. Remotely delivered mines are prohibited unless they are used in the vicinity of a military objective and their location can be accurately recorded or they possess an effective neutralizing mechanism. The Protocol also stipulates that all pre-planned minefields must be recorded and that, with regard to all other minefields, the parties to the conflict are to endeavour to record their location.

The ICRC participated as an observer in the negotiations of the CCW. Although there were few formal interventions by the ICRC during the negotiating process, the results of the three expert meetings convened by the ICRC on the subject of conventional weapons in relation to the 1974–1977 Diplomatic Conference provided much of the background material for the preparatory meetings and diplomatic negotiations. Several subsequent articles were published which specifically spoke about the treaty's obligations relating to landmines.

As at 1 May 2000, the 1980 CCW had been ratified by seventy-eight States, and Protocol II had been ratified by seventy States. The provisions of Protocol II would later be strengthened by the adoption of amendments during the 1995–1996 Review Conference of the 1980 CCW which would also adopt a protocol banning blinding laser weapons.

A New Step Forward in International Law: Prohibitions or Restrictions on the Use of Certain Conventional Weapons

by Yves Sandoz

Published in the *International Review of the Red Cross,* 220
January–February 1981

I. INTRODUCTION

On 10 October 1980, the "United Nations Conference on Prohibitions or Restrictions of Use of Certain Conventional Weapons Which May be Deemed to be Excessively Injurious or to Have Indiscriminate Effects" ended with the adoption by consensus of the following instruments:

- Convention on Prohibitions or Restrictions on the Use of Certain Conventional Weapons,
- Protocol on Non-Detectable Fragments (Protocol I),
- Protocol on Prohibitions or Restrictions on the Use of Mines, Booby-Traps and Other Devices (Protocol II),
- Protocol on Prohibitions or Restrictions on the Use of Incendiary Weapons (Protocol III).

In addition, at its first session, the Conference had adopted a Resolution on small-calibre weapon systems. All these texts are reproduced in this issue of the

Review. It should also be pointed out that the Conference took note of six draft resolutions and one proposition which it will submit in its report to the UN General Assembly.

We propose here to give an account of the stages which led up to the successful outcome of the Conference; to indicate the place of the Convention and the three Protocols in international law; to analyse briefly the contents of the instruments and the Resolution adopted by the Conference, and of the different motions and propositions; and finally, to attempt to assess the influence of this accord in humanitarian terms.

II. BACKGROUND

The Second World War clearly showed the necessity of ensuring better protection for the civilian population during armed conflicts. The Fourth Geneva Convention of 1949 represents a great advance in this respect, but is essentially concerned with the population in the hands of an enemy Power. The general protection of civilians against the effects of hostilities is still inadequately covered by this Convention. The ICRC soon realized this, and as early as September 1956 it drew up a set of "Draft Rules for the Limitation of the Dangers Incurred by the Civilian Population in Time of War". These rules included a chapter on weapons, entitled "Weapons with Uncontrollable Effects", proposing, in particular, that weapons whose harmful effects might escape the control of those using them and delayed-action weapons should be banned, and that belligerent parties making use of mines should be obliged to chart minefields and, at the cessation of active hostilities, to hand over the charts to the authorities responsible for the safety of the population. This proposal was presented in 1957 to the Nineteenth International Red Cross Conference which requested the ICRC to submit it to governments.

This move towards a further development of international humanitarian law was premature, however, since many States were still not parties to the Geneva Conventions.

The matter was taken up again in 1965, at the 20th International Red Cross Conference which in its Resolution XXVIII pointed out that "indiscriminate war constitutes a danger to the civilian population and the future of civilization" and that "the right of parties to a conflict to adopt means of injuring the enemy is not unlimited". The International Conference on Human Rights, held in Teheran in 1968, voiced similar anxieties, and the United Nations General Assembly, in resolution 2444, adopted the principles which these Conferences established on the subject.

In the report on the reaffirmation and development of the laws and customs applicable in armed conflicts presented to the Twenty-first International Red

Cross Conference in 1969, the ICRC set forth as its principal conclusions that the belligerents should abstain from using weapons:
- likely to cause unnecessary suffering;
- which, because of their lack of precision or their effects, affect civilians and combatants without distinction;
- whose harmful effects were beyond the control, in time or space, of those employing them.

The Conference requested the ICRC to continue its efforts in this field.

In the same period, studies on the subject were published by the UN Secretariat and again by the Stockholm International Peace Research Institute.

In 1971 and 1972, the ICRC organized a Conference of Government Experts on the Reaffirmation and Development of International Humanitarian Law applicable in Armed Conflicts.

The documentation presented to the Conference dealt with protection of the civilian population in time of armed conflicts and, in particular, with protection against certain types of bombing and against the effects of certain weapons. While not inviting the experts to discuss "prohibitions of specific weapons", so as not to overlap the work of bodies concerned with disarmament, the ICRC thought it possible for them to examine, in addition to general principles, the principles relating to weapons which in any case, owing to their effects or their lack of precision, might affect the civilian population indiscriminately. The experts' opinions fell into three categories. According to the first, the problem of weapons ought not to be dealt with by such a body. The second felt that, without dealing directly with the question of weapons of mass destruction (nuclear, biological, chemical), the necessity of banning them should be affirmed, since greater protection for the civilian population largely depended on such a ban. The third current of opinion held that the Conference should not consider weapons of mass destruction – under discussion by the Conference of the Disarmament Committee – but other particularly cruel weapons which were not being studied anywhere else.

This third tendency won the day, and at the second session of the Conference, in 1972, the experts of nineteen States asked the ICRC to organize a special meeting to consult legal, military and medical experts on the question of the explicit prohibition or restriction of conventional weapons likely to cause unnecessary suffering or to have indiscriminate effects. This consultation took place in Geneva in 1973. A purely documentary report was produced, without formulating any specific proposals. Its role was to stimulate further studies on the subject, and it was distributed to all National Red Cross Societies, all the governments of

States Parties to the Geneva Conventions and all the relevant non-governmental organizations.

The draft of the Protocols additional to the Geneva Conventions, as presented to the Diplomatic Conference which met in Geneva in 1974, contained general principles applying to weapons but no provisions on the use of any specific weapon. The Conference nevertheless set up an *ad hoc* Committee to deal with the problem. Again, the prevailing view was that the Committee's work should be restricted to conventional weapons.[1] With the encouragement of the Diplomatic Conference, the ICRC organized a Conference of Government Experts, which held two sessions, one at Lucerne in September–October 1974, the other in Lugano in January–February 1976.

Like the *ad hoc* Committee of the Diplomatic Conference, the experts discussed various conventional weapons; but in the end no article on the subject of a specific weapon was included in the Protocols. An article envisaging the creation of a committee on the prohibition or restriction of certain conventional weapons, whose task would have been to examine definite proposals on the matter and to prepare agreements, was dropped as it failed, by a few votes, to obtain the required two-thirds majority.

However, a resolution was adopted by the Diplomatic Conference (Resolution 22) recommending, *inter alia,* "that a Conference of Governments should be convened not later than 1979 with a view to reaching agreements on prohibitions or restrictions on the use of specific conventional weapons" and "agreement on a mechanism for the review of any such agreements and for the consideration of proposals for further such agreements".

The UN General Assembly supported this recommendation (see resolutions 32/152 of 19 December 1977, 33/70 of 28 September 1978 and 34/82 of 11 December 1979), and the proposed Conference, the subject of the present article, after a preparatory Conference which met in August–September 1978 and March–April 1979, took place in Geneva from 10 to 28 September 1979 and from 15 September to 10 October 1980.

III. CONTEXT

The specific prohibition of certain weapons belongs to two branches of international law, disarmament law and international humanitarian law applicable in armed conflicts. This dual relationship is not unimportant, since each of these laws approaches problems differently.

In matters of disarmament, stress is laid on problems of security. The aim is to proceed steadily toward general and total disarmament, without any sudden disruption in the balance of forces at some stage in the proceedings jeopardizing the security of the various States. Moreover, agreements on disarmament should cover not only prohibitions or restrictions on the use of any weapon, but also on its manufacture, storage and sale or purchase. In short, it should deal not merely with the use of a weapon but with its possession.

Problems of security are not entirely disregarded by international humanitarian law, but in that context they do not have the vital interest which they possess in relation to disarmament. The aim of international humanitarian law is, in fact, a modest one: "humanize" as far as possible those armed conflicts which cannot be avoided. Since it is by its nature subsidiary, operating only when the law prohibiting the use of force has failed to fulfil its role, international humanitarian law cannot claim to be a substitute for the other. It would be unrealistic to think that conflicts could be prevented by laying down such severe limits on means of combat that conflict would be made impossible. There is no reason whatever why such an obstacle should prove any stronger than that formed by the law prohibiting the use of force.

It is therefore imperative for international humanitarian law to confine itself to modest objectives. True, it has had its failures; but there have also been undeniable successes, and these have been due essentially to the fact that its provisions are of humanitarian interest to everybody while harming the military interests of nobody.

The considerations outlined above also apply in connection with weapons. It is highly unlikely that States will accept, as part of international humanitarian law, the prohibition of weapons of strategic importance which bedevil all discussions on disarmament. On the other hand, there are some weapons the possession of which does not materially affect the balance of forces in the world, and which are not essential from the military viewpoint, but whose effects are particularly cruel or cause extensive damage without military justification. Hence some people have remarked, understandably, that international humanitarian law should be satisfied with prohibiting useless weapons. Yet in the long run this is not as ironic as it seems. Obviously, if the only effect of international humanitarian law on armed conflicts were to prevent any use of force not strictly justified by military necessity, it would still save a great many lives and much suffering. However, the urgent need to improve the protection of the civilian population led the States, in the 1977 Protocols, to go further and agree to take humanitarian factors into account even at the sacrifice of some military advantage. The same could be said of the Conference on conventional weapons. But it should never be forgotten that it is not in the interests of international humanitarian law to venture too far in this direction. To force the pace might well lead to catastrophe.

Yet such considerations should not be understood as a suggestion to give up all efforts in this sphere. Nor should it be thought, as is sometimes the case, that military necessity is used as a pretext to reject any new humanitarian measure.

IV. CONTENT

As the Convention and its three Protocols are appended to this article, we will not go into their contents in detail. A few items, however, seem to be worthy of close study.

1. The Convention

The scope of the Convention was established by reference to the Geneva Conventions and to Protocol I additional to them. This means international conflicts, with the understanding that it includes "armed conflicts in which peoples are fighting against colonial domination and alien occupation and against racist regimes in the exercise of their right of self-determination".

The date on which the Convention comes into force will be six months after the date of receipt of the twentieth instrument of ratification, acceptance, approval or accession. It will be noted that there is a disparity between these instruments and the Geneva Conventions and their Additional Protocols, for which only two instruments of ratification or accession are sufficient. This disparity is explained by the fact that some States placed the debate in the sphere of disarmament. Any agreement in this sphere aimed at reducing the level of armaments and thus diminishing the military power of States can obviously be envisaged only if it is applied by all States or, at the least, by all the military Powers.

Observance of the Conventions belonging to international humanitarian law, on the other hand, should have no effect on the military efficiency of the States involved. But even if these instruments on conventional weapons are applicable only between parties to a conflict which have accepted them, it may be assumed that States which decide to ratify or accede to these instruments will forgo possession of the weapons they prohibit. These, while not of vital importance from the strategic viewpoint, still have implications for security, according to some States, who consequently demanded substantial support for the Convention and the Protocols before their entry into force. The figure of twenty is therefore a compromise between the States which held this view and those which unreservedly associated these instruments with international humanitarian law.

Another point to be noted is that a State cannot become party to the Convention alone; this is logical, since the Convention merely provides the legal framework within which the prohibitions contained in the Protocols are applicable. But the conditions fixed go further: a State becoming a party to the Convention must accept at least *two* of the Protocols. This requirement was aimed mainly at preventing any State from becoming a party only to Protocol I, which is at present of little practical significance (see below).

An interesting aspect is the system of relationships established when the Convention takes effect; the same flexible system as for the Geneva Conventions. A State which is party to the Convention is obliged not only to observe it with respect to another State also party to the Convention and having an ally not bound by it – in contrast to the rigid system adopted at The Hague Conferences of 1899 and 1907 – but must also apply the Convention to the ally if the latter accepts and applies the Convention (and the relevant Protocol or Protocols) and notifies the depositary State of this fact. It will be noted, however, that the formality of notifying the depositary is not required in the Geneva Conventions.

Concerning wars of liberation (in the sense of Article 1, paragraph 4, of Protocol I of 1977), the authority representing a liberation movement may undertake to apply this new Convention and the associated Protocols with respect to a State which is party to these instruments and likewise bound by the 1977 Protocol I. The Convention and its Protocol or Protocols then become applicable between that State and the liberation movement, as does the 1977 Protocol I.

But the real innovation lies in the fact that the authority representing a liberation movement may act in the same way toward a State party to the present Convention and to two or more of its Protocols, even if the movement is not bound by the 1977 Protocol I. Moreover, such commitment will result in the application, not only of the present Convention and its Protocols, but also of the Geneva Conventions as a whole. This means that the present Convention provides access to the whole body of the Geneva Conventions, something which was not envisaged by the Conventions.

The provision making this access possible calls for four comments.

1. It demonstrates clearly that recognition of the international character of wars of liberation, in the sense of Article 1, paragraph 4, of Protocol I of 1977, is not linked in international humanitarian law to this Protocol alone. The international character of such wars, already affirmed by numerous resolutions of the UN General Assembly,[2] here obtains additional confirmation and, above all, direct involvement in the applicability of the Geneva Conventions.

2. Logically, the hypothesis presented by this provision should not occur. It would seem inconsistent for a State to agree to the present Convention without also accepting the 1977 Protocol I, which reaffirms or develops the principles applied in this Convention and its Protocols. But the possibility cannot be excluded, since refusal to accede to the 1977 Protocol I might be due to provisions unrelated to the question of weapons.

3. While this new step may be seen as encouraging the wider application of international humanitarian law in wars of liberation, it should be emphasized that the principle of equality of rights and obligations of the parties to a conflict – a vital element in international humanitarian law – has not been disputed: in fact it has been clearly reaffirmed.

4. The unlikely hypothesis of a State's becoming a party to the Convention without being a party to the Geneva Conventions was not even envisaged. This demonstrates the recognized universal character of those Conventions and should encourage the few States not yet officially bound by them to accede to them without delay.

The procedure for revising the Convention was one of the crucial points in the negotiations. Agreement was finally reached on an ad hoc system although the opinion was also expressed that the matter should be entrusted to the disarmament Committee. A conference is to be convened at the request of the majority of the States parties to the Convention (but at least eighteen States, as some felt it to be unacceptable for only eleven States to possess such powers).

Revision of the Convention and its Protocols is to be decided solely by the States which are parties to them, while the addition of further Protocols may be decided by all those States attending such a Conference. Though there is no explicit mention of the fact, the Conference would probably reach its decisions by consensus, as did the Conference which produced the Convention, and this should make it impossible, even for revision of the existing instruments, for decisions to be made on the basis of a majority of the moment.

A Conference will probably be held at least once every ten years, since, if a period of this duration has elapsed without a Conference, a request from only one of the High Contracting Parties is sufficient for the depositary to be obliged to convene a Conference.

Establishment of this procedure was imperative, as it gives lasting value to the Convention by leaving the door open for the introduction of other restrictions and by urging all States to practise constant vigilance to ensure that conventional weapons conform with the principles laid down in the Protocol I of 1977. The revision method also represents a valuable addition to Article 36 of Protocol I, which binds all Contracting Parties to examine all new weapons to make sure their use is not prohibited by international law.

2. The Protocols

a) *Protocol on Non-detectable Fragments (Protocol I)*

This Protocol has little immediate importance, since the weapons concerned have not been used – or in any case not widely – up to now. But it constitutes a ban for the future and should prevent undesirable developments. The prohibition is an expression of the principle that the purpose of a weapon should not be to hinder the healing of wounds it causes, and this principle is certainly one of the basic elements for determining whether a weapon produces "unnecessary suffering".

b) *Protocol on Prohibitions or Restrictions on the Use of Mines, Booby-Traps and Other Devices (Protocol II)*

The purpose of Protocol II is to prevent or at any rate to reduce as far as possible loss and damage to civilians by the devices it covers, during hostilities and afterward, when those devices no longer have any military usefulness. The Protocol deals with a very definite problem: even today, many civilians are still being injured by mines, long after the events which led to the sowing of the minefield.

The Protocol does not tackle the awkward problem of mines laid during war at sea, a problem still covered by Conventions adopted at the beginning of the century. It might be thought, incidentally, that it is high time to consider updating international humanitarian law applicable to armed conflicts at sea.

In Article 3, Protocol II applies to the devices with which it is concerned the general principles which prohibit attacks on civilians and their property and indiscriminate attacks. This means that the use of certain booby-traps specially designed to attract civilians, or even children, is totally prohibited (see Article 6), while restrictions are laid on the use of mines, booby-traps and "other devices" defined in Article 2 (see Articles 4 and 5). A distinction is made between devices put in place from nearby and those delivered from a distance, i.e., "delivered by artillery, rocket, mortar or similar means or dropped from an aircraft". Those dropped from the air, especially, are very difficult to neutralize when they have ceased to fulfil their military function. The problem was solved by requiring either that they be supplied with a mechanism which makes them inactive after a certain lapse of time or that they be launched or dropped with sufficient precision for their positions to be recorded with accuracy. However, there was no agreement on more precise rules which might have determined, in particular, the height from which it was admissible for an aircraft to drop such mines.

Another aspect of this Protocol which should be stressed is the "international cooperation in the removal of minefields, mines and booby-traps" (Article 9). It is essential, if civilians are to be properly protected, for the parties to the conflict to collaborate, once active hostilities are over, by at least providing information concerning the mines they have laid. The text adopted does not go as far as was initially envisaged. In particular, it does not include the obligation to hand over, immediately after the cessation of active hostilities, charts showing the location of mines, even, to an occupying Power. Such an obligation was intended solely to give adequate protection for the civilian population, including those within occupied territory. Some delegations, however, found it impossible to envisage any cooperation whatever with an occupying force, even for humanitarian purposes.

Several of the rules in this long Protocol are consequently not very rigorous. For example, we may note that Article 3, paragraph 4, requests the parties to take "all feasible precautions" to protect civilians, that is, "those precautions which are practicable or practically possible taking into account all circumstances ruling at the time, including humanitarian and military considerations"; that advance warning is obligatory before remotely delivered mines are launched or dropped "unless circumstances do not permit" (Article 5, paragraph 2); that the parties "shall endeavour to ensure" the recording of the location of minefields which were not pre-planned (Article 7, paragraph 2).

The indecisive and complex character of the rules finally adopted indicates how acute were the problems encountered. Mines undeniably play an important part in military activities, but their indiscriminate use gives rise to inadmissible loss and damage to civilians. Protocol II is a typical offspring of the arranged marriage between military necessity and humanitarian imperatives, a union which has produced the whole of international humanitarian law. The legal protection of civilians against the effects of mines, booby-traps and other devices is far from perfect, but it is definitely better than it was.

c) *Protocol on the Prohibitions or Restrictions on the Use of Incendiary Weapons (Protocol III)*

Incendiary weapons are probably the conventional weapons which have the greatest impact on public opinion. Many felt that any agreement on conventional weapons which did not include a Protocol on incendiary weapons would have the distressing appearance of a fire brigade which had forgotten to bring the hose-pipe. If nothing had been achieved on this subject, it is likely that all the work of the Conference would have been wasted. This is a further reason to welcome the agreement finally obtained, during the last few days of the Conference, on this category of weapons.

Protocol III applies to incendiary weapons the general principle, reaffirmed in the 1977 Protocol I, that civilians should not be subject to attack. But it takes a big step further by placing severe restrictions on attacks on military objectives located within a concentration of civilians and particularly by prohibiting completely any attacks by air on such objectives. This provision is intended to prevent the terrible danger of huge concentrations of civilians being wiped out by fire.

Forests and plant cover are civilian property unless used for military purposes, and their protection is therefore included in the general rule prohibiting attacks on civilian property. Nevertheless, it was considered desirable to give prominence to this protection by mentioning it specifically, in view of the potentially disastrous nature of forest fires.

It will be noted that there is no provision to protect combatants against incendiary weapons. This is because emphasis was placed on the indiscriminate nature of these weapons and on the danger they present for civilians, rather than on their cruelty, an aspect which would have justified restriction of their use against combatants also.

Of course it may be argued that combatants are generally better equipped and can therefore deal more efficiently with the use of incendiary weapons. But among combatants too there have been extremely cruel burns, and several delegates regretted that no agreement could be reached after all on protecting combatants. A resolution on the subject was drafted by the Conference and sent to the UN General Assembly. This text "invites all governments to continue the consideration of the question of protection of combatants against incendiary weapons with a view to taking up the matter at the conference that may be convened in accordance with the provisions of Article 3 of the Convention adopted."

3. Resolution on Small-calibre Weapon Systems

The Conference did not produce a Protocol on small-calibre projectiles which tumble upon impact and transfer considerable energy into the victim's body, thus causing extremely cruel wounds. But at its first session it adopted a resolution inviting governments to carry out further research and appealing to all governments "to exercise the utmost care in the development of small-calibre weapon systems, so as to avoid an unnecessary escalation of the injurious effects of such systems".

One of the major working documents distributed at the Conference emphasizes that research at present is being carried on in two directions: one, to find a medium capable of being used to simulate living tissue, the other, to evolve a simple test to determine the energy-transfer characteristics of a projectile.

V. SCOPE

The attempt to place the Convention of 10 October 1980 and its three Protocols in their context indirectly raises the question of their scope.

Obviously, the Convention, like the rest of international humanitarian law, does not claim to resolve any political problems. At most it could be argued that the moderation which it introduces into conflict is a factor favouring settlement.

The significance of a Convention of this kind, therefore, is purely humanitarian. Its relation is solely to men, women and children who would otherwise have been blown to pieces by mines, had their faces mutilated by booby-traps or their bodies burned by napalm. Those who have been saved from these weapons will remain unknown, unlike those who, in spite of all efforts, will become victims. It is a peculiarity of such prohibitions that their merit truly is known through being breached.

Yet the potential victims who are spared because of the new law do exist. This is the firm belief and sole guiding motive of those who work for the development of international humanitarian law.

The link between the instruments adopted on 10 October 1980 and Protocol I of 8 June 1977 additional to the Geneva Conventions has not been settled categorically. It seems logical, however, to consider these restrictions and prohibitions as rules intended to put into concrete terms some of the principles laid down in the 1977 Protocol I, particularly in its Articles 35 and 51. Moreover, several points of the Convention's preamble give a clear indication in this direction. Yet it cannot be claimed that the prohibitions follow so naturally from the principles reaffirmed by the 1977 Protocol that an obligation concerning them existed before they were explicitly formulated. The protracted negotiations which were necessary to achieve these instruments plainly demonstrate that their content was by no means an obvious matter. So the Convention and its Protocols should be considered as a development of law and any condemnation of action taken previous to their enactment, by retroactive application of their underlying philosophy, would be, juridically, as sterile as it would be inadmissible.

We have already noted the conditions necessary for the Convention and its Protocols to be formally applicable. In particular, we have seen that they are to be applied only in international conflicts. Nonetheless, it seems undeniable that texts of this kind also carry great weight outside their official legal context.

The method of consensus, used very frequently in international conferences nowadays, undoubtedly confers a certain weight, in international circles, on the agreements reached at such conferences. The Vienna Convention on treaty law, very often cited well before it came into force, is a good example of this. But such a situation is true even more of humanitarian instruments. If States are agreed on the specially cruel character of certain weapons or certain combat methods and on the necessity of prohibiting them, can they decently fail to take such agreement into account even before they are legally bound to do so? In this connection, it is interesting to note that a draft resolution which the Conference sent to the UN General Assembly "calls upon all States which are not bound by the present Convention and which are engaged in an armed conflict, to notify the Secretary-General of the United Nations that they will apply the Convention and one or more of the annexed Protocols in relation to that conflict, with respect to any other party to the conflict which accepts and abides by the same obligations.

But although the Convention is applicable in principle only in international armed conflicts, it is improbable that governments will feel free to use against their own population, in conflicts not of an international nature or in internal unrest, weapons and combat methods which they have agreed to forgo against an alien enemy.

In international humanitarian law, more than in any other sphere, public opinion would demur at any recourse to purely legal arguments for refusing to observe principles whose value had been widely acknowledged. An interesting fact reported by various technical experts is that the discussions and trials carried on by experts in relation to small-calibre weapon systems, although they have not yet resulted in binding prohibitions or restrictions, have nevertheless had a beneficial influence on several States when renewing their stock of weapons of this kind. (See also the resolution on the subject adopted by the Conference, the text of which is given below.)

VI. CONCLUSIONS

The adoption on 10 October 1980 of a Convention and three Protocols marks the completion of a significant phase in the evolution of international humanitarian law, a phase whose prime purpose has been to provide better legal protection for the civilian population against the effects of hostilities. In order to accomplish this, it was felt essential to reintroduce into international humanitarian law, without ambiguity, principles concerning the conduct of hostilities which had been laid down at the beginning of this century, at The Hague Conferences in 1899 and 1907, and to develop those principles. This was done in the 1977 Protocols additional to the Geneva Conventions. But the principles alone, without precise rules to buttress them, were in danger of remaining mere words, and the merit of the Convention of 10 October 1980 and its three Protocols is that they have tackled the problem directly and specifically. In this sense, the instruments are valuable, or rather indispensable, supplements to the 1977 Protocols.

While the reaffirmation in international humanitarian law of principles concerning the conduct of hostilities was intended chiefly to give better protection for the civilian population, it must be acknowledged that these principles were originally formulated, above all, to alleviate the suffering of combatants. Simplifying the matter, it may be said that methods or means of combat having indiscriminate effects are prohibited because there is too great a risk of their harming the civilian population, while the ban or restriction on excessively cruel weapons takes into account combatants as well as civilians. Mines may be placed in the first category, non-detectable fragments in the second. Even so, there are weapons, such as incendiary weapons, which may be classified, depending on which aspect is considered, in one or other of these categories. The restrictions placed on the use of these weapons in Protocol III are motivated by the indiscriminate character of such weapons and the risk that they may injure civilians. Yet the reason that several delegations expressed the wish to continue work on the subject was that they considered these weapons – or some of their uses, at any rate – to be excessively cruel and for this reason wanted combatants also to be granted protection.

We have seen that some international value must undeniably be attributed to the instruments which have just been adopted, regardless of when they enter into force. Yet it is plain that formal accession to such instruments gives them much more weight and that lack of interest by the States might well lead to their being forgotten. It is to be earnestly hoped that the States will sign and then ratify these instruments rapidly and in very large numbers. Incidentally, many States refused to ratify the 1977 Protocols until or unless they were supplemented by an instrument concerning weapons. For those States, as for the great majority of others, the adoption of the Convention of 10 October 1980 and its Protocols should be the occasion of acceding to the whole of the corpus of modern international humanitarian law. The phase just completed was essential to maintain the credibility of this law. The States which have patiently worked together to produce the Convention should now, by acceding to it, indicate their determination to respect its humanitarian principles and rules.

The texts adopted in 1980, like those of 1977, indicate that the world is horrified by the massacre and mutilation of millions of civilians during the conflicts of our century. These texts are the result of patient efforts and we should welcome their adoption. But progress made in international humanitarian law is never completely satisfactory: there is always the question whether it could not have been taken a step further, whether more lives could have been saved, more suffering avoided. Alongside the advances made, however substantial, there is the shadow of those which have perhaps failed to come into being for lack of perseverance or persuasion.

The mixed feelings which greet any advance in international humanitarian law, however, are due to deeper causes, to be found in the nature of that law, able only to relieve and not eliminate the absurd suffering engendered by armed conflicts. In our time, as never before, the necessity of attacking the causes of evil and not merely its effects is obvious to everyone. The extent of the probable consequences of any large-scale conflict makes any efforts to attenuate them appear derisory. Those engaged in such efforts, therefore, even though convinced of the nobility of their task, must regard it as a contribution to peace and an urgent appeal to those capable of achieving it.

NOTES

1 This expression covers all weapons not included in the category of 'nuclear, biological or chemical' weapons.
2 In particular, resolutions 2105 (XX), 2621 (XXV) and 3103 (XXVIII).

Turning Principles into Practice: The Challenge for International Conventions and Institutions

by Yves Sandoz

Published in *Clearing the Fields: Solutions to the Global Land Mines Crisis*
Kevin M. Cahill, ed., 1995

The use of anti-personnel mines has reached proportions that take the issue beyond purely humanitarian concerns. The problem's economic and social implications, as well as its impact on the environment and its effects on population movements in particular, give it a universal dimension.

Different actors must address the problem at different levels, pursuing a variety of aims, with local or worldwide objectives, in the short or the long term. But the key word for all these efforts is *complementarity:* they should not compete with each other, but be designed to be mutually supportive.

To overcome the current crisis, the emphasis must be placed on preventive action. This will involve compliance with clear, effective, and universally accepted rules, which must be drawn up without delay. Two main levels of preventive action may be identified: the prevention of wars (*jus ad bellum*) and the prevention of abuses during wars (*jus in bello*). The reasoning applied to each is different, but the two must not be regarded as contradictory. A third level comes between these two and has a bearing on both of them: disarmament and arms control.

Prevention of Wars

One cannot disregard the fact that, whatever the circumstances, wars cause immeasurable suffering, and that the need for an international mechanism capable of imposing a ban on the use of force between States and of preventing, as far as possible, recourse to force within states is becoming ever more pressing.

The proliferation of mines is certainly a further argument in favour of such a mechanism. This type of solution is, however, a long-term prospect with little bearing on the current massive use of land mines.

Arms Control and Disarmament

Even in the short term, the problem of mines cannot be seriously addressed without being viewed as part of arms control and disarmament, fields that extend beyond the scope of international humanitarian law.

Given the reality of the proliferation and transfer of weapons, it is evident that a ban on the use of a certain weapon will not be completely effective as long as the weapon continues to be manufactured and stockpiled. Arms control and disarmament measures are therefore indispensable.

Such measures might include the prohibition or the restriction of exports, destruction of existing stocks that do not comply with possible new manufacturing standards, prohibition of the manufacture of certain types of mines, and establishment of verification mechanisms to accompany all these measures.

States should therefore seriously examine the possibility of a multilateral agreement to ban the development, manufacture, transfer, and use of at least certain types of land mines and the destruction of all existing stocks of those types. To this end, they should introduce the subject of mines in the Conference on Disarmament and could look to the process that led to the adoption of the Chemical Weapons Convention as a source of inspiration.

As an immediate and unilateral measure, an export moratorium on mines, as instituted by the United States and some other countries, should be encouraged.

Preventing the Use of Mines During Armed Conflicts: The General Framework of International Humanitarian Law

International humanitarian law originally limited the damage caused by weapons by altogether prohibiting the use of weapons that were perceived as excessively cruel or "barbaric." The customary prohibition of the use of poison was based on the perception of its treacherous nature and the fact that poisoned weapons inevitably caused death.[1] The prohibition of the use of explosive bullets by the St Petersburg Declaration of 1868[2] was similarly based on the wish to outlaw weapons that inflicted excessively cruel injuries or that usually killed the victim. Subsequently, humanitarian law prohibited the use of expanding (dum-dum) bullets.[3] For the same reason, but also because of their indiscriminate nature, the use of chemical weapons, mentioned in the Treaty of Versailles in

1919 as contrary to international law,[4] was specifically prohibited, together with the use of bacteriological weapons, by the Geneva Protocol of 1925.[5]

Between 1925 and the adoption of the 1980 Conventional Weapons Convention, however, international humanitarian law did not make any significant progress in prohibiting the use of specific weapons, despite the numerous discussions on incendiary weapons[6] that took place between the two world wars and the reaction to the use of nuclear weapons on Hiroshima and Nagasaki.[7]

As for the general rules of international humanitarian law governing the conduct of hostilities, it is primarily in the 1977 Additional Protocol I to the Geneva Conventions of 1949 that they were reaffirmed and developed. It is evident that rules limiting the use of weapons and methods of warfare "of a nature to cause superfluous injury or unnecessary suffering"[8] have little chance of being effective unless they are supplemented by rules specifying which weapons or methods of warfare are prohibited.

The importance of the reaffirmation in 1977 Protocol I of the ban on the use of "a method or means of combat which cannot be directed at a specific military objective"[9] and of the other rules that protect civilians cannot be denied. They reflect the very essence of international humanitarian law – that is, the distinction between combatants and non-combatants.

But the key question is the relationship between the specific rules introduced in the 1980 Convention, in particular those governing the use of land mines, and the general rules of international humanitarian law. There is no doubt that the rules governing the use of land mines are based on the principles and rules laid down in 1977 Protocol I. Furthermore, Protocol I reaffirmed and developed international humanitarian law. Whereas certain provisions, such as those specifically concerning the environment, are obviously developments, others are a reaffirmation of pre-existing rules, often considered as pre-emptory norms of general international law (*jus cogens*).[10]

Thus it can be claimed that the indiscriminate use of anti-personnel mines is a violation not only of the 1980 Convention, but also of Additional Protocol I, and even of a rule of *jus cogens.*

Review of the 1980 Convention

The military view on mines is expounded elsewhere in this volume, and we shall not dwell on it here. It should just be recalled that a meeting of experts on the subject convened by the International Committee of the Red Cross (ICRC) came to the conclusion that "no alternative fulfils the military requirement in the way that anti-personnel mines do" and that "the anti-personnel mine is the most cost-effective system available to the military."[11]

This is a very important point because, as experience has shown, international humanitarian law has no chance of imposing exigencies that would have a decisive effect on the outcome of a conflict. If the law is to be applied, it must be seen as useful, or at least not detrimental, to all concerned. Military commanders will not take the risk of giving up an effective weapon unless it is guaranteed that their potential adversaries have also done so.

Consequently, among the numerous conditions that have to be met to ensure that the review of the 1980 Convention brings real progress, the first, apart from parallel action in the field of disarmament, is universal acceptance of the Convention's rules.

UNIVERSAL ACCEPTANCE OF THE CONVENTION'S RULES

Everything possible must be done to ensure that the 1980 Convention is universally accepted. The fact that only forty-one States are party to the Convention makes its revision almost futile. The review process should, therefore, prompt a large-scale mobilization of the international community to incite States to ratify or accede to the Convention, especially in regions, notably in Asia and Africa, where mines have taken and are still taking such a heavy toll.

Furthermore, since the 1980 Convention and its three Protocols reflect the rules reaffirmed or developed in Additional Protocol I of 1977, they also concern the States party to 1977 Protocol I, or even all States in some respects. It is therefore important that all States ratify or accede to the Additional Protocols of 1977, so as to give their provisions as broad a base as possible. In view of the US role in the international scene, it seems evident that ratification by the United States of the 1977 Additional Protocols could give new impetus to the indispensable progress of international humanitarian law toward universal acceptance. This objective can be achieved in the medium term, and its achievement would certainly have a favourable effect on the global land mine problem. Meanwhile, we need to consider interim measures.

REDEFINING THE SITUATION OF THE CONVENTION AND THE REVIEW PROCEDURE

This review system provided for in the 1980 Convention is flexible,[12] allowing for rapid adaptation to developments in the weapons field. Indeed, the procedure can be seen as an intelligent manner of implementing Article 36 of 1977 Protocol I, which stipulates: "In the study, acquisition or adoption of a new weapon, means or method of warfare, a High Contracting Party is under an obligation to determine whether its employment would, in some or all circumstances, be prohibited by this Protocol or by any other rule of international law applicable to the High Contracting Party."

Anti-personnel laser devices that could be used to blind combatants are a prime example of a newly developed weapon that could, and indeed should, be the subject of a further protocol.[13]

A major difficulty, however, is that the High Contracting Parties are bound by the Convention, six months after ratification, only in relation to other States party. This is also the case for the annexed Protocols. This aspect is far from satisfactory for two reasons: First, it gives the impression that the prohibitions or restrictions in question arise entirely from the Convention itself. In fact, for the most part they are only an illustration of the more general rules contained in the 1977 Additional Protocols, or of rules that form part of *jus cogens.* Who, then, is to determine which rules are binding only on States party to the 1980 Convention and the Protocol concerned, which ones are also binding on States party to 1977 Additional Protocol I, and which ones are binding on all States? No one; hence the deplorable situation of uncertainty in a domain in which States, and their armed forces in particular, have to know exactly where they stand. Second, since certain restrictions have obvious military implications, how will a State's armed forces go about imposing rules on their soldiers in certain circumstances (for instance, when they are in conflict with another State party) and not in others? Should the troops nevertheless be trained in the use of weapons prohibited by the Convention?

These considerations raise the question of whether it might be necessary to rethink the entire philosophy of the 1980 Convention and make it an instrument that is linked much more closely with the Additional Protocols of 1977.

In reality, everything would be much easier and clearer if participation in the Additional Protocols of 1977 automatically meant participation in the 1980 Convention, which would become a sort of technical annex that could be adapted to keep pace with technical developments. Additional Protocol I of 1977 already has one such technical annex, in the form of regulations concerning identification, which has the purpose of allowing procedures for the identification of medical personnel, units, and transports to be adapted in accordance with technical advances. Indeed, this annex has its own review procedure.[14]

This review procedure should also have a bearing on the reflection on the 1980 Convention. Under this procedure, any amendment proposed by a conference of States Party is communicated to the parties to the Convention by the depositary, and "shall be considered to have been accepted at the end of a period of one year after it has been so communicated, unless within that period a declaration of non-acceptance of the amendment has been communicated to the depositary by not less than one-third of the High Contracting Parties." The amendment then "shall enter into force three months after its acceptance for all High Contracting Parties other than those which have made a declaration of non-acceptance."[15]

Such a procedure has enormous advantages. First, it is very speedy. The 132 States party to 1977 Protocol I are today bound by the revised technical annex to the Protocol,[16] only four years after the start of the review process. Second, it obliges States to take a position, since they have to declare their *non-acceptance* of proposed amendments rather than their agreement to be bound by them. This is a vital point when one considers the heavy workload and, occasionally, the apathy of government administrations. Indeed, it is probable that a great majority of the States that have not ratified the 1980 Convention have failed to do so not because they reject its provisions, but simply because they have not found the time to examine them.

This type of procedure has other precedents, as well. One can be found in the Convention on International Civil Aviation, signed in Chicago on 7 December 1944. That Convention provides that international standards and recommended parties, designated as annexes to the Convention, "shall require the vote of two-thirds of the Council at a meeting called for that purpose and shall then be submitted by the Council to each contracting State. Any such Annex or any amendment of an Annex shall become effective within three months after its submission to the contracting States or at the end of such longer period of time as the Council may prescribe, unless in the meantime a majority of the contracting States register their disapproval with the Council."[17] Here, the procedure goes even further than the one provided for in the technical annex to 1977 Protocol I, since the new rules become binding even on minority States that have refused to accept them. This is quite understandable in the case of rules governing international civil aviation, which obviously have to be standard.

This further step, or some similar measure, could perhaps be taken in regard to prohibitions or restrictions on weapons that will be fully complied with only if those bound by them are convinced that their potential adversaries will do likewise. In any event, all these matters should provide us with food for thought. Although such a proposal should prevent neither an immediate mobilization in favour of the 1980 Convention nor efforts to improve its efficiency during the ongoing review procedure, it should be submitted to States without delay for examination as a medium-term objective.

EXTENSION OF APPLICABILITY[18]

The 1980 Convention formally applies only to international armed conflicts. However, in recent times, most conflicts have been internal, and it is these conflicts that are largely responsible for the human suffering caused by land mines. It is therefore of the utmost urgency to formally extend the scope of application of the Convention to non-international armed conflicts through an amendment specifying this new scope.

Nevertheless, we cannot disregard that some States would undoubtedly object to such an extension of the applicability of the Convention, invoking national sovereignty. This foreseeable reaction probably arises from a misunderstanding. Several States, particularly in the Southern Hemisphere, oppose the tendency toward intervention on humanitarian grounds to bring assistance to people in need. They fear that such operations might have a destabilizing effect on governments or might even serve as a pretext for seeking to overthrow them. They are therefore extremely reluctant to see any development of rules pertaining to non-international armed conflicts.

Such views can perhaps be understood in regard to large-scale relief operations. They are totally unwarranted, however, when it comes to the weapons issue: the application of rules banning the use of weapons does not imply any outside intervention. From a moral standpoint, moreover, it is inconceivable that a government would consider using against its own population weapons it has agreed not to employ against an enemy State because of their cruel nature.

But there again, the military advantages of a weapon might make a government hesitate if it fears that insurgents might easily be able to lay their hands on it. Hence the importance of universal rules and parallel efforts in the sphere of disarmament. The Chemical Weapons Convention demonstrates that if the manufacture, possession, and transfer of a weapon are prohibited, the question of its use is automatically resolved, and controversies on possible different approaches to international and non-international armed conflicts become pointless. In the framework of international humanitarian law, the formal link between general rules and specific weapons envisaged above should relate to both Additional Protocols of 1977, and not only to Protocol I.

It should be noted, finally, that progress toward harmonization of the rules governing hostilities in all types of armed conflict might be achieved by establishing cooperation among the military, particularly in terms of drawing up military training manuals. In practice, military training generally makes no distinction between preparations to use weapons and means of warfare in international or in internal conflicts.

INTRODUCTION OF IMPLEMENTATION MECHANISMS

Although the 1980 Convention contains no implementation provision, it reaffirms the rules of international humanitarian law found in other treaties. Implementation measures under those treaties are therefore also relevant to the

1980 Convention. Nevertheless, it may be desirable to include specific measures in the 1980 Convention; these could be drawn mainly from the humanitarian law treaties, or from other instruments of international law. It bears emphasis, however, that this possibility should be considered as secondary to the more radical reform suggested above, whereby the 1980 Convention would become an annex to the 1977 Additional Protocols.

Provision of Legal Advisers

This is now required under Article 82 of Additional Protocol I to the Geneva Conventions. A similar provision in the 1980 Convention could provide for legal advisers to give guidance on matters relating to the use of weapons, to be incorporated at all levels down to brigade or equivalent level, and to be included in planning staff.

Requirements for Training in Humanitarian Law

The requirement to instruct the armed forces in the law is embodied in the Hague Convention IV of 1907, the four Geneva Conventions of 1949, and their Additional Protocols of 1977. Such a requirement ought to be contained also in the 1980 Convention, including such provisions as the following:

- Training in the use of weapons in accordance with humanitarian law should be given in cadet academies and during all command and staff training courses;
- Manuals on weapon systems should include the law specifying their correct use, in the languages of the user countries;
- The packaging of weapons should bear a warning as to the legal limitations on their use; and
- All military training of foreign nationals should include training in humanitarian law.

Incorporation in Domestic Law

The 1980 Convention should be translated into local languages, and appropriate national laws and regulations should be adopted. This suggestion is similar to the provision set out in Article 84 of 1977 Protocol I.

International Fact-Finding Commission

The International Fact-Finding Commission provided for in Article 90 of Additional Protocol I to the Geneva Conventions could also be used to investigate possible violations of the 1980 Convention. In the context of the 1977 Protocol, the Commission's competence is based on consent that can be given either in advance, in the form of a declaration, or on an ad hoc basis. It would have to be decided whether the same formula would be appropriate for the 1980 Convention and whether it should also be based on confidentiality, as in the 1977 Protocol.

Creation of a Supervisory Body

A number of international treaties create supervisory bodies designed to help implement the treaty. These bodies typically receive periodic reports submitted by States parties on the measures they have taken to implement the treaty. They also receive complaints about alleged violations, undertake investigations, and discuss their findings with the States concerned. They often undertake promotional activities to improve compliance with the law.

The review conference could consider whether it might be appropriate to create an analogous body for the 1980 Convention, or whether the role of the international fact-finding commission could be extended to cover this Convention.

NECESSARY CLARIFICATIONS OF PROTOCOL II[19]

Article 3, enumerating general restrictions on the use of mines, booby-traps, and other devices, is based on the generally accepted distinction between military objectives and civilian objects. Such a distinction, however, is difficult to maintain once a military target has moved away from a mined area, leaving behind anti-personnel mines. Moreover, the duty to protect civilians from the effects of those weapons is couched in very weak terms, as para. 4 of this Article makes reference to all "feasible" precautions. The term *feasible* allows for considerable flexibility in interpretation. Furthermore, the provision is weak because feasible measures would include installation of fences or signposts, although these tend to be removed by members of the local population, either out of ignorance or for the sake of the profit to be derived from such items. This Article might be the right place to introduce a ban on the use of anti-personnel mines without a self-destruct mechanism.

Article 4, restricting other types of mines, has the same shortcomings as Article 3.

Article 5 prohibits the placement of remotely delivered mines outside areas that are military objectives unless one of two conditions is fulfilled: that their location can be recorded or that they are fitted with self-neutralizing mechanisms. It is difficult to record accurately the location of mines delivered by fixed-wing aircraft, artillery, and rockets. The recording requirements in the absence of a neutralizing mechanism, set out in para. 1(a), are therefore not applicable when these methods of delivery are used. Problems remain with the implementation of para. 1(b), regarding self-destruct and self-neutralization devices, because these devices are not reliable enough to guarantee the safety of a mined area; moreover, there is no maximum time limit for the active life of these mines. The wording of para. 1(b) also creates confusion between self-destruct and self-neutralizing mechanisms. Furthermore, para. 2 of this Article requires but does not define "effective advance warning."

In fact, Article 5 embodies a striking contradiction between an approach based on the military utility of the weapon and a view of the problem that takes the economic and social cost of its use into account. As pointed out above, mines have an acknowledged "military utility," especially in countries that do not have the technological capacity to develop alternatives. Moreover, from the military standpoint, mines would be slightly less cost-effective if there was a general obligation to fit them with self-destruct or self-neutralizing mechanisms. This requirement, too, would have a greater impact on technologically less advanced countries, where the production of such devices might be problematic.

It is thus the poorer countries that put up the most vigorous opposition to a total ban on anti-personnel mines or to a general obligation to fit all mines with self-destruct or self-neutralization mechanisms. Yet it is these same States that suffer most from the effects of mines, and in which the concept of the "military utility" of these weapons becomes absurd in view of the economic and social costs to which they give rise.

Article 7, on the recording and publication of the location of minefields, mines, and booby- traps, contains a major flaw in that it gives no definition of a "pre-planned" minefield, which is the only type that requires recording. With regard to all other minefields, parties are required only to "endeavour" to record them, which is a rather weak provision. In practice, there are still other difficulties. For instance, regular armies have followed strict procedures with respect to mine laying, and there are clear rules for marking and recording minefields. However, such records are properly created and kept by only very few armies. They also are quite frequently lost.

Even with such records, successful minefield clearance rarely can be guaranteed for a number of reasons. Mines tend to move, sometimes long distances, over a period of time, owing to the effects of the weather, soil erosion, and on occasion the action of animals. This is especially true in the case of scatterable mines. Furthermore, even the most conscientiously maintained minefield record can be subject to human error by soldiers who may be tired or under stress.

Article 8, addressing the protection of UN forces and missions from the effects of mines and other such devices, should be expanded to cover organizations other than the UN, such as CSCE missions and private mine clearance agencies. In fact, the UN itself has acknowledged the need to coordinate mine clearance activities, and for that purpose, the Department of Peace-keeping Operations, which includes the Mine Clearance Center, has established a database to which any contributions are welcome. Expert reports from other mine clearance organizations point out that, more often than not, mine clearance is an extremely hazardous exercise, principally because records are not properly kept and there are often no maps or signposts. As of April 1993, in Afghanistan, for instance, mine clearance activities had resulted in twenty-four deaths and twenty-eight amputations, and no fewer than nineteen operatives had been blinded.[20]

Mine clearance is expensive given the high cost of experts' fees, personnel insurance premiums, and such support expenses as medical and casualty evacuation costs. Article 8 should therefore be extended to afford protection to third-party missions and, logically, to humanitarian organizations working in regions affected by mines.

Article 9, on international cooperation in the removal of minefields, mines, and booby-traps, does not impose an obligation to remove mines, as the words used therein are "shall endeavour to reach agreement." Moreover, this agreement relates only to "the provision" of such information and assistance as "necessary to" remove or render ineffective mines and minefields, and thus in no way imposes a specific obligation to actually do so. This is a major shortcoming of the law. Further, this Article does not deal with other issues of crucial importance in a mine-devastated country after the cessation of active hostilities, such as repatriation and land reclamation requirements.

PROHIBITION OF UNDETECTABLE MINES

A ban on undetectable mines should be viewed as secondary or complementary to other proposals. It must be mentioned, however, because the difficulty in detecting mines makes mine clearance – so vitally needed today – an almost insurmountable task.

If a provision on the subject is introduced, it should stipulate that the detectable element in the mine be not easily removable. Provisions should also be made for verification that mines not conforming to these specifications are not manufactured.

SPECIFICATIONS FOR ANTI-TANK MINES

The mines considered here are essentially anti-personnel mines. Antitank mines frequently have self-neutralizing devices because they are expensive and often need to be reused. It would therefore be useful to require that all antitank mines be fitted with neutralizing mechanisms, and it would have to be verifiable that only this type is manufactured.

Reparation for Damage

Although the principle of reparation is reaffirmed in international humanitarian law,[21] its application has proved uncertain. Cases in which an arrangement is reached between belligerent parties are generally settled after the armed conflict is over, on terms imposed by the victor. At this stage, considerations associated with *jus ad bellum* apply, and not the humanitarian exigencies of *jus in bello.* Moreover, requests that the responsible State make reparation are submitted through the State of origin of the injured persons (individuals or bodies corporate).

The ambiguous nature of many current conflicts and the frequent lack of a clear passage from war to peace make the problem doubly difficult to resolve. The question of reparation for damage resulting from acts contrary to international humanitarian law is, however, a very topical issue, owing in particular to the work of the International Law Commission and the reports of the special rapporteur of the sub-commission on the prevention of discrimination and protection of minorities on questions relating to "the right to restitution, compensation and rehabilitation for victims of gross violations of human rights and fundamental freedoms."[22]

In the framework of this chapter, the key point is that no progress in the field of reparation will be possible unless responsibility for violations of international humanitarian law can be clearly determined. That can be achieved only if the rules are explicit, and this is an additional argument in favour of formulating prohibitions that are precise and cannot be circumvented.

Repression

Responsibility for punishing those who commit grave breaches of international humanitarian law lies first and foremost with governments in regard to their own population, and in particular to members of their armed forces. At the international level, two events have created a new impetus for fulfilment of the obligation incumbent on each State party to the Geneva Conventions to punish or extradite any war criminal on its territory: the establishment of an international tribunal for the prosecution of persons suspected of perpetrating grave violations of international humanitarian law committed on the territory of the former Yugoslavia since 1991,[23] and the examination of a draft code of crimes against the peace and security of mankind and of a draft statute for an international criminal tribunal.[24]

The preventive role that such tribunals can play, insofar as they meet the challenges before them (stringency, consistency, independence, cooperation of States, and so forth) cannot be underestimated. But the rules pertaining to mines must be clear so that acts committed in violation of international humanitarian law, war crimes in particular, can be determined unequivocally and action can be taken to prosecute the perpetrators.

Role of Humanitarian Institutions: The Example of the ICRC

International and non-governmental organizations that are directly or indirectly concerned with the problem of mines can certainly play an important role in this collective effort. The ICRC feels that it has a special mission in this respect, in view of the mandate entrusted to it by the international community to work for the application and development of international humanitarian law.[25]

The ICRC's operational work in the field, carried out with the support of National Red Cross and Red Crescent Societies, gives it, first of all, the role of a witness: the surgeons it sends to the theatres of war are only too well placed to see the devastating effects of mines on the civilian population, and on children in particular; the orthopaedic technicians it sends to set up and manage orthopaedic workshops in many countries can but acknowledge that the needs greatly exceed their capacity. The thousands of ICRC delegates and local employees who work in conflicts worldwide are faced every day with the destruction wrought by mines. The ICRC therefore feels duty-bound to alert the general public and governments to the magnitude of the problem, and to this end has published a number of brochures that have been widely distributed.[26] Through its regional delegations and its cooperation with National Red Cross and Red Crescent Societies and their International Federation, the ICRC seeks to spread awareness in all parts of the world of the mines problem, and indeed of all issues concerning international humanitarian law.

Second, the ICRC has tried to grasp all the aspects of the problem and to make them clear to others. In particular, it organized a multidisciplinary symposium in Montreux, Switzerland, in April 1993. The general objective of the symposium was to collect the facts and ideas necessary to coordinate future action by bodies interested in improving the situation of mine victims and in taking preventive action. Its more specific aims were to gain as accurate a picture as possible of actual mine use and the consequences thereof; to analyse available methods designed to limit such use or alleviate the suffering of the victims, and to identify the inadequacies of such methods; to decide on the best remedial action; to establish strategies for coordinating the work of the different bodies involved in such action; and to draft a report that could be used as a reference for future efforts.[27] The symposium took place in a positive atmosphere. Following intensive work by the different working groups, the participants stressed the necessary complementarity of the various courses of action envisaged. These courses were set out in a highly detailed report.[28]

Third, the ICRC feels that it has a moral duty to make its position public. The organization's president has done this, coming out in favour of a total ban on the use of anti-personnel mines.

With regard to international humanitarian law, the ICRC takes action on two levels. The first involves consolidation of the basic rules of international humanitarian law. To this end, the ICRC has published commentaries on the Geneva Conventions and their Additional Protocols[29] and engages in diplomatic approaches to encourage States to ratify or to accede to these treaties. It also organizes numerous seminars throughout the world, in cooperation with the local National Red Cross or Red Crescent Societies, to make the law more widely known and to invite States to adopt national legislation and other measures necessary for its implementation.

In parallel, the ICRC has always helped prepare for the development of the instruments of international humanitarian law. As regards conventional weapons, it organized two meetings of experts in the mid-1970s, whose results were published and served as a basis for the conference that drafted and adopted the 1980 Convention.[30]

In preparation for the review of the 1980 Convention, the ICRC has organized four experts meetings and has published a detailed report on a recent and terrifying invention, namely, blinding weapons.[31] It has also held a conference of army experts to examine the military utility of mines and the possibility of finding alternatives. Finally, it has drafted a report for the review conference containing many suggestions for improvement of the Convention itself, and more particularly of the Landmines Protocol.[32]

Finally, the entire International Red Cross and Red Crescent Movement is engaged in ongoing efforts to bring help to the victims of mines. Numerous other organizations are also involved in the mobilization against the use of mines, especially in relation to mine clearance and in raising public awareness.[33] All these efforts are important and necessary, but considerable room for development remains.

Conclusion

In view of its social, economic, and ecological implications, the massive use of anti-personnel mines in present-day conflicts is a problem that concerns us all. Greater awareness and large-scale mobilization among the public are necessary if governments are to be spurred to make a genuine commitment to address the issue. But the problem is a complex one and should be tackled simultaneously on different fronts, with the accent on cooperation rather than confrontation. Strengthening the means of preventing war, seeking a total ban on the use of anti-personnel mines, or seeking the prohibition of the use of certain mines in wartime are valid objectives that stem from the conviction that the current situation is unacceptable. They are just different levels of the same endeavour.

The goal of a world organized in such a way as to eliminate war, considered Utopian by some, is at any rate a distant prospect. The military utility of mines leads one to believe that a total prohibition can be contemplated only within the framework of a disarmament agreement comprising strict verification procedures, and this is bound to take time. Immediate efforts to broaden, strengthen, and clarify existing agreements are therefore also necessary.

The following practical proposals, while not constituting a plan of action as such, are useful steps toward alleviating the land mine crisis.

1) Efforts to mobilize public opinion must be increased, especially in countries that suffer the most from mines. Everything possible must be done to alert the public to the scale of the phenomenon, its pernicious nature, and its dramatic consequences in the long term. Land mines cause horrendous damage in human, social, economic, and ecological terms, and there is a critical shortage of resources to cope with that damage, particularly in the countries worst affected. All these considerations call for vigorous and determined action on the part of the entire international community.

2) There is little hope of serious progress unless the issues of the production and transfer of land mines are addressed. Prohibition of *certain uses* of mines is not enough, as effective control of illicit use is practically impossible for this type of weapon, and such a measure would thus have only marginal effects. It is therefore essential to set standards that will make it possible to tackle the matter at the levels of production and trade, in the framework of disarmament negotiations. Such standards will have to be accepted by a wide consensus and be accompanied by verification procedures. As in the case of a contagious disease, measures taken to control the use of mines will be effective only if they are universal. Like microbes, mines do not stop at borders.

3) A total ban on anti-personnel mines would be by far the best solution. If this objective is not immediately attainable, the standards chosen must be capable of containing the problem. The active life of a large proportion of the mines produced today is almost unlimited, and they are extremely difficult to detect. At the very least, every mine should be fitted with a self-destruct mechanism, or a self-neutralization mechanism in the case of the larger models, and include an element making it easier to detect. The cost of self-destruct mechanisms or elements facilitating detection is negligible in

comparison with the extra cost of clearing mines that are not fitted with such devices.

4) Rules requiring extra expenditure and technological expertise must not, however, be seen as favouring rich countries. Such requirements must be accompanied by a dialogue that takes this aspect of the problem into consideration and envisages the transfer of technology and financial assistance.

5) Universal recognition of the rules governing the use of certain weapons during armed conflict must be secured to avoid any ambiguity: members of the armed forces need clear directives and have to know the rights and duties of their adversaries. Whereas the Geneva Conventions have achieved almost universal acceptance, only about two-thirds of States are party to their Additional Protocols of 1977.[34] Three of the permanent members of the UN Security Council, which in that capacity have a special responsibility in regard to the use of force, are not yet party to the Protocols: France (party to Protocol II only), the United Kingdom, and the United States. It is essential that ratification, examination, or re-examination under way in those States be speeded up and completed, so as to give a decisive boost to universal acceptance of the rules of international humanitarian law.

6) Moreover, a clearer, more formal, and more automatic link between the general instruments of international humanitarian law and the rules that prohibit or restrict the use of certain types of weapons would be highly desirable. In addition, the review procedure for those rules must be flexible, allowing for rapid adaptation to new developments and for the universal acceptance, within a very short period, of new or amended rules. The possibility of making them an annex to the 1977 Additional Protocols, along the lines of Annex I to Protocol I, merits consideration.

7) At the same time, a considerable effort of mobilization must be made to persuade States to ratify or accede to the 1980 Convention.

8) It is morally and militarily unjustifiable for weapons banned during international armed conflicts not to be prohibited during internal conflicts. Moreover, disarmament measures aimed at prohibiting the production of such weapons imply a total ban on their use. Any prohibition on the use of anti-personnel mines or of certain types of mines must therefore also apply to non-international armed conflicts.

9) Preparations for the future cannot be made without regard to the past; the entire international community must make a concerted effort to clear the vast tracts of land rendered unusable by mines and to provide mine victims with the care and assistance they need and deserve. Those who use mines,

but also those who produce them and States that tolerate the mines trade, must assume their responsibilities.

10) These objectives call for an effort on the part of all who are aware of the scale of the problem and who refuse to see this evil that is afflicting our planet spread further.

The measures taken may be directed toward a whole range of short-term or long-term goals. Care must be taken, however, to ensure that the different approaches do not appear contradictory, and that they are seen as a coherent attempt to achieve a common objective. In a wider perspective, the effort to control mines is also an effort for peace.

NOTES

1 See, in particular, Yves Sandoz, *Des armes interdites en droit de la guerre* (Geneva: Grounauer, 1975), 84.
2 "Declaration Renouncing the Use, in Time of War, of Explosive Projectiles under 400 Grammes Weight, St Petersburg, 29 November / 11 December 1868," in D. Schindler and J. Toman (eds.), *The Laws of Armed Conflicts: A Collection of Conventions, Resolutions and Other Documents* (Dordrecht: Martinus Nijhoff Publishers / Geneva: Henry Dunant Institute, 1988), p. 101.
3 "Declaration (IV, 3) Concerning Expanding Bullets, The Hague, 29 July 1899", in *Laws of Armed Conflicts*, 109.
4 Article 171 of the Treaty of Versailles of 29 June 1919, states: "L'emploi des gaz asphyxiants, toxiques ou similaires, ainsi que de tous liquides, matières ou procédés analogues, étant prohibé . . ." For more details about this article, see Sandoz, *Des armes interdites*, 31–32.
5 "Protocol for the Prohibition of the Use in War of Asphyxiating, Poisonous or Other Gases, and of Bacteriological Methods of Warfare, Geneva, 17 June 1925", in *Laws of Armed Conflicts*, 115.
6 See Sandoz, *Des armes interdites*, 106.
7 Ibid.
8 See Article 35, para. 2, of 9 June 1977, Additional Protocol I to the Geneva Conventions of 12 August 1949.
9 See Article 51, para. 4(b), of Protocol I.
10 See Article 53 of the Vienna Convention on the Law of Treaties. On this question, see also, in particular, the "Declaration on the Rules of International Humanitarian Law Governing the Conduct of Hostilities in Non-international Armed Conflicts", *International Review of the Red Cross* 278 (September–October 1990): 404.
11 See *Report of the International Committee of the Red Cross for the Review Conference of the 1980 United Nations Convention on Prohibitions or Restrictions on the Use of Certain Conventional Weapons Which May Be Deemed to Be Excessively Injurious or to Have Indiscriminate Effects* (Geneva: International Committee of the Red Cross, February 1994), p. 57.

12 See Article 8, para. 1(a) of the United Nations Convention on Prohibitions or Restrictions on the Use of Certain Conventional Weapons Which May be Deemed to be Excessively Injurious or to Have Indiscriminate Effects, 10 October 1980.

13 On this subject, see L. Doswald-Beck (ed.), *Blinding Weapons: Reports of the Meetings of Experts Convened by the International Committee of the Red Cross on Battlefield Laser Weapons 1989–1991* (Geneva: International Committee of the Red Cross, 1993), p. 371.

14 See Article 98 (Revision of Annex I) of 1977 Additional Protocol I.

15 See Article 98, paras. 4 and 5, of 1977 Additional Protocol I.

16 Of the 132 States, only 2 expressed minor reservations on the amended Annex 1.

17 Article 90 of the Convention on International Civil Aviation, Chicago, 7 December 1944.

18 On this subject, see *Report of the International Committee of the Red Cross* (1994), 21–24.

19 Ibid., 43–44.

20 *Symposium on Anti-personnel Mines, Montreux, 21–23 April 1993, Report* (Geneva: International Committee of the Red Cross, 1993), p. 154.

21 See Article 91 of 1977 Additional Protocol I.

22 *Study Concerning the Right to Restitution, Compensation and Rehabilitation for Victims of Gross Violations of Human Rights and Fundamental Freedoms.* Final report submitted by Mr Theo van Boven, Special Rapporteur, Doc. E/CN.4/Sub.2/1993/8, 2 July 1993.

23 See UN Security Council resolution 808 of 22 February 1993.

24 See UN General Assembly resolution 48/31 of 9 December 1993.

25 See Statutes of the International Red Cross and Red Crescent Movement, adopted by the 25th International Conference of the Red Cross, 1986, Article 5, para. 2.

26 See, in particular, *Mines: A Perverse Use of Technology* (Geneva: International Committee of the Red Cross, 1992), p. 19.

27 To ensure a multidisciplinary approach, the Symposium brought together eminent experts in different fields related to the issue of anti-personnel mines use and its effects. They included military strategists; mine specialists and manufacturers; experts in international humanitarian law and disarmament; surgeons and orthopaedists; and representatives of mine clearance agencies, non-governmental organizations and the media.

28 *Symposium on Anti-personnel Mines*, 321.

29 Jean S. Pictet (ed.), *Commentary on the Geneva Conventions of 12 August 1949*, 4 vols. (Geneva: International Committee of the Red Cross, 1952–1959); Y. Sandoz, C. Swinarski and B. Zimmermann (eds.), *Commentary on the Additional Protocols of 8 June 1977 to the Geneva Conventions of 12 August 1949* (Geneva: International Committee of the Red Cross, 1987).

30 See *Conference of Government Experts on the Use of Certain Conventional Weapons (Lucerne, 24 September – 18 October 1974), Report* (Geneva: International Committee of the Red Cross, 1975), and *Conference of Government Experts on the Use of Certain Conventional Weapons (Second Session: Lugano, 28 January – 26*

February 1976), Report (Geneva: International Committee of the Red Cross, 1976).

31 See Doswald-Beck, *Blinding Weapons.*

32 See *Report of the International Committee of the Red Cross* (1994).

33 One product of these efforts that deserves particular mention is the volume *Landmines: A Deadly Legacy* (New York: Human Rights Watch, Physicians for Human Rights, 1993), p. 510.

34 As of 31 December 1993, 185 States were party to the 1949 Geneva Conventions; 130 States were party to Additional Protocol I of 1977; and 120 States were party to Additional Protocol II.

PART 2

The Review Conference of the 1980 Convention on Certain Conventional Weapons

An Initial Response to the Landmine Crisis

1

Introduction

During the negotiation of the 1980 Convention on Certain Conventional Weapons, relatively little attention was paid to the question of mines, as the international community concentrated their efforts on trying to control incendiary weapons. By the beginning of the 1990s, however, it was clear that the use of landmines, especially anti-personnel mines, was escalating dramatically and was causing a major humanitarian crisis. The alarm was first sounded by surgeons of the ICRC and the staff of non-governmental organizations working in war-torn areas. Their call for action to end the increasing number of civilian casualties led to the convening of the Montreux Symposium in 1993. Later that same year, France called formally for the first Review Conference of the 1980 Convention, in particular to study the possibility of strengthening the provisions of international humanitarian law governing landmines.

In accordance with a request from States Parties, a series of three meetings of a group of governmental experts to prepare proposals for the Review Conference was scheduled to take place in Geneva from early 1994. In February 1994, just before the first of these meetings, the President of the ICRC, Cornelio Sommaruga, declared that a total prohibition on anti- personnel mines was the only effective solution to the humanitarian emergency created by landmines. At that time, few believed that such a measure was feasible.

States Parties also invited the ICRC to take part as an expert observer, both in meetings of the group of experts and in the Review Conference itself, and to prepare documentation and proposals for these meetings on the basis of its field experience and expertise in the sphere of international humanitarian law. The ICRC submitted the reports included below and played an active role in the negotiating process, commenting both formally and informally on the issues and proposals which emerged. The ICRC was also permitted to submit its own proposals.

Once the meetings of the group of governmental experts got under way it swiftly became clear that any substantial improvement in the law restricting landmines would be extremely difficult to achieve in the prevailing climate. A further obstacle was the system of consensus which governed negotiations in the Review Conference. Although in August 1994 Sweden became the first State formally to propose a total ban on anti-personnel mines, such a far-reaching measure was not seriously envisaged by the group.

In view of the difficulty of reaching consensus on an acceptable text to be forwarded to the Review Conference, a fourth meeting of experts in January 1995 became necessary. As Ambassador Molander mentions in his foreword to this work, agreement was reached only in the early hours of the final Saturday morning. Thereafter final preparations began for the Review Conference, which was scheduled to take place in Vienna from 25 September to 13 October 1995.

Once the Conference got under way, however, the expert group's fairly modest package of measures which had appeared to be broadly acceptable to all States Parties soon began to unravel. Two days before the Conference was due to end, Ambassador Molander, the President of the Conference, announced to the assembled delegates that they had run out of time and that no consensus was within reach. The reaction both inside and outside the Conference hall was one of extreme disappointment, mitigated only by the success in adopting a new Protocol IV prohibiting the use of blinding laser weapons.

Two further sessions of the Review Conference were hastily arranged back in Geneva in January and April to May 1996 to try to break the deadlock. At the end of the third session of the Conference, States Parties finally adopted by consensus a revised version of Protocol II governing mines, booby-traps and other similar devices. So many compromises had been made and so many potential loopholes introduced into the text during the three tortuous years of negotiations that there was little enthusiasm or optimism as to its potential effect on the ground. The ICRC described the restrictions on use as 'woefully inadequate'. By the end of the Conference, however, more than thirty States had become convinced that the only effective solution to the landmine crisis was the total prohibition of anti-personnel mines. A critical mass was forming that would become the core of the 'Ottawa process'. The world was taking the first steps down the path towards a global ban on anti-personnel landmines.

2

ICRC Symposium on Anti-Personnel Mines

(Montreux Symposium) Montreux, Switzerland

21–23 April 1993

The Montreux Symposium was the first meeting organized by the ICRC specifically to address the issue of anti-personnel mines. It was convened in response to the increasing magnitude of suffering caused by the weapons, as witnessed in the field by ICRC delegates and other humanitarian workers. The meeting brought together experts from various disciplines with the purpose of obtaining an accurate overview of the scope of the problem and the measures and mechanisms available to limit anti-personnel mine use and to alleviate the suffering of mine victims. Participants included military strategists, mines specialists and manufacturers, legal experts, surgeons, rehabilitation specialists, and representatives of demining organizations and concerned non-governmental organizations.

The report on the Montreux Symposium became an important source of reference for the ICRC, non-governmental organizations and governments in their future activities in pursuit of a ban treaty. A copy was sent to all governments in August 1993. Reproduced below are selected chapters dealing with the humanitarian aspects of the landmine problem, as well as the table of contents of the full report.

Montreux Symposium Report

(selected chapters)

INTRODUCTION

Every year, thousands of men, women and children are victims of anti-personnel mines. The use of these often extremely pernicious weapons has resulted in a tragedy that is all the more acute in humanitarian terms as, apart from the appalling number of victims they cause, anti-personnel mines not only kill but mutilate horrendously, strike blindly at all human beings alike, and continue to spread terror for years or even decades after the hostilities have ended. Moreover, massive and indiscriminate sowing of mines renders whole regions useless for human habitation and activity, thereby resulting in substantial population movements and consequent economic destabilization in other neighbouring regions. As a method of warfare, such weapons are not in conformity with certain fundamental rules of international humanitarian law governing the conduct of hostilities, which call upon parties to distinguish between civilians and combatants, prohibit attacks against the former and therefore also the use of indiscriminate weapons, and do not permit the use of such weapons as are liable to cause excessive suffering.

In the seventies already, the ICRC, concerned by the need to deal more effectively with certain weapons' abuses, organized two important symposia of government experts at Lucerne in 1974 and Lugano in 1976. These meetings provided valuable support for the United Nations Conference that followed in 1979–80. More recently, the magnitude of the suffering caused by anti-personnel landmines as witnessed by its delegates in the field pushed the ICRC into organizing the Montreux Symposium.

The general objective of this Symposium was to collect the necessary facts and ideas to coordinate future action by bodies that are interested in improving the fate of mine victims and in undertaking preventive action. More specifically, the aims of the Symposium were to gain as accurate a picture as possible of the actual use of mines and its consequences; to analyse the mechanisms and methods that presently exist to limit this use or alleviate the suffering of victims, as well as to identify the lacunae in such methods; to decide on the best remedial action; to establish a strategy on how to coordinate the actions of different bodies involved in such action; and, to write a report on the conference which could be used as a reference for future actions and which would equally serve to mobilize governments as well as the public in general.

In order to ensure a pluridisciplinary approach, the participants consisted of established experts from different disciplines related to the whole issue of the use of anti-personnel mines and their effects, and included military strategists, mines specialists and manufacturers, legal experts in the international humanitarian law and disarmament fields, surgeons and orthopaedists, representatives of demining organizations, concerned non-governmental organizations, and the media. The gathering was characterized by an open-minded and constructive approach. Prior to the Symposium, preparatory reports, prepared by certain participants, were sent to all participants. These reports dealt with seven themes, namely, a picture of the present use of mines in reality, the trade in mines, the humanitarian consequences, technical characteristics of anti-personnel mines, demining and mine detection, the professional military perception of the use of mines, and the legal situation.

The Symposium began with an introductory statement followed by a brief presentation of the aforementioned reports by their authors, who equally answered some questions in connection with them. The second day was devoted to discussions held within six working groups into which the participants were divided. Each group dealt with one of the six following subjects: rehabilitation, demining, the 1980 United Nations Convention, proposals for further humanitarian law rules, possible restrictions on the method of manufacture of mines, and, possible arms control measures relating to the commerce in mines and their stockpiling. The purpose of the working groups was to conduct an in-depth examination of the advantages and disadvantages of various possibilities and their viability in terms of putting them into practice, as well as of the extent to which they would actually solve the problems presently created by mines. At the end of the day, each group produced a report. These six reports, together with the twenty-five reports previously distributed and introduced on the first day, have been incorporated into the present report.

The third and final day was spent in a thorough discussion of each of the six reports of the working groups, held in a plenary session of the conference. This session ended with a set of conclusions that emerged from the reports and the discussions, and which bear principally upon a coordinated strategy for future action. A summary of the discussions on the working group reports, as well as the general conclusions of the Symposium, that consist of recommendations for future action, have equally been included in the present report.

ITEM (1)

MINES AND HUMANITARIAN ACTIVITIES

Report by: Jean-Michel Monod
General Delegate for Asia and PaciWc
International Committee of the Red Cross

"Mines may be described as fighters that never miss, strike blindly, do not carry weapons openly, and go on killing long after hostilities are ended. In short, mines are the greatest violators of international humanitarian law, practising blind terrorism."

Jean de Preux

In the execution of its humanitarian mandate, the International Committee of the Red Cross (ICRC) is confronted daily with the harrowing consequences of the use of landmines.

The use of landmines, mostly anti-personnel mines, to terrorize and intimidate civilians by inflicting upon them atrocious suffering and irreversible after-effects is a blatant disavowal of the basic principles of international humanitarian law, which stipulates that persons not participating or no longer participating in hostilities must be protected and assisted.

During the discussions, specialists in this field will explain how mines, at first essentially defensive weapons, have been turned into vicious devices used to spread terror and suffering among civilians, to create additional difficulties for the parties to the conflict by forcing them to care for a growing number of amputees, to paralyse the economy or even to contribute to causing a famine.

In such circumstances, it follows that mines can also be used to hinder, or even prevent, humanitarian assistance.

The use of mines has become common practice in virtually every conflict situation where the ICRC is present. From Angola to Cambodia, from the former Yugoslavia to Afghanistan, the long list of regions affected by this scourge is only too well known.

Let us first consider the obstacles, or in any case the additional difficulties, that the use of mines causes for humanitarian assistance operations.

For obvious reasons, the security of persons providing humanitarian assistance in a conflict is of utmost importance. One of the safer ways to move about in the field is to use air transport, although the danger is always there during take-off and landing. However, while minimizing security risks, this solution maximizes costs. For example, in Angola in 1988, the cost of delivering one tonne of relief supplies from Lobito to Huambo overland, by rail, would have been 56 Swiss francs. By Hercules, the cost was approximately Sfr 1,000. From Huambo to Municipio by truck cost an additional Sfr 50. – whereas by Twin Otter, the cost was Sfr 1,600/tonne. The final price was therefore Sfr 2,600/tonne, whereas it could have been Sfr 106/tonne, or 25 times less. (As it was, this did not include subsidiary costs such as insuring the planes against war risk, etc.)

Security concerns aside, this is often the only way to break the isolation of the civilian population affected by a conflict. Indeed, if no assistance reaches them, they themselves have no means to go and get it.

In the same vein, let us consider all the security measures that must be taken by a delegation working in such conditions, the delays that may arise while waiting for sufficient security guarantees to be given by all the parties concerned, without forgetting the anxiety of the delegate at the head of a relief convoy, the stress and accelerated wear and tear on the staff involved, at times making it necessary to speed up staff turnover and hence giving rise to administrative difficulties and additional costs.

And still, despite all these precautions, owing to the number of journeys made, physical weariness, bad luck or a stroke of fate, accidents do happen and in certain circumstances can even call into question the very principle of relief activities.

ICRC communication to the press, 9 February 1993

Quote

ICRC condemns indiscriminate use of mines

The International Committee of the Red Cross is deeply concerned by the widespread and indiscriminate use of land mines in various parts of the world, and condemns this practice in the strongest terms.

In the last few days, there have been three incidents in which staff of humanitarian organizations working in Africa have been killed or wounded by mines.

On 25 January, seven first-aid workers of the Senegalese Red Cross were killed and four others injured when their vehicle hit a mine in Casamance.

On 5 February, in Zambezia Province, Mozambique, a mine blast killed two Mozambicans, one of whom was an OXFAM employee, during a mission in which the ICRC was taking part. Three other people were wounded.

One the same day in Lasanod in north-western Somalia, a third mine incident cost five people their lives, three of them local ICRC staff. Three other Somali employees were injured and taken to hospital. . . .

Unquote

Apart from delivering relief supplies to those who need them, setting up surgical hospitals or improving existing facilities to save lives, even if all too often the arms and legs of the victims are too shattered to be saved, apart from setting up orthopaedic workshops, making artificial limbs suitable for conditions in the country and fitting the largest possible number of amputees, what can the ICRC do to limit the effects of this scourge?

According to the Statutes of the International Red Cross and Red Crescent Movement, the role of the ICRC is in particular:

– *to undertake the tasks incumbent upon it under the Geneva Conventions, to work for the faithful application of international humanitarian law applicable in armed conflicts and to take cognizance of any complaints based on alleged breaches of that law,*

and

- *to endeavour at all times – as a neutral institution whose humanitarian work is carried out particularly in time of international and other armed conflicts or internal strife – to ensure the protection of and assistance to military and civilian victims of such events and of their direct results.*

The latter category obviously includes mine victims.

While the ICRC does not, as a rule, make any statements of principle on the use of certain weapons or methods of warfare, it does not rule out taking steps if it considers that use of a specific weapon or method causes a situation of exceptional gravity, which is certainly the case with the use of mines, as we have already seen.

On this basis, the ICRC may choose several complementary approaches to try to limit the effect of the use of mines on civilian population.

Where there is reason to believe that abuses have been committed strictly with respect to mine-laying, owing either to ignorance of existing legal texts or to wilful disregard of such texts, the ICRC may set up a dissemination programme for military officers and those who actually lay the mines. Another possibility is to conduct a preventive mine-awareness programme for civilians, for example in the event of an organized, mass return of refugees.

First-aid assistance could also be provided to save lives if the first part of the programme (dissemination) proved insufficient to prevent accidents.

- The ICRC, which is often active in areas where few other organizations are present at this stage, may draw the attention of the government authorities concerned, specialized agencies, NGOs, etc., to the specific needs of certain regions, particularly in terms of mine-clearing. In 1945, France mobilized some 55,000 mine-clearers. Many of them were prisoners of war who had not necessarily been given a choice as to whether or not to participate in the operation, and they paid dearly, since 3,000 died and 6,000 others were injured in action over an 18-month period. But this effort made it possible to deactivate or destroy some 13 million mines, most of which were metallic and perhaps easier to spot. By comparison, the action taken by the international community in present-day theatres of hostilities is not nearly as intensive, and could certainly be stepped up.
- In its capacity as a neutral intermediary, the ICRC could play a role in the mutual notification by the parties of the location of mine fields in a conventional conflict, should the need arise.

- During a conflict, the ICRC must make direct representations to the parties involved whenever flagrant violations including the utilization of mines, are committed against civilians or against humanitarian activities. It can also remind parties that prisoners of war or other categories of protected persons are not used in mine-clearing operations.

The ICRC's professional expertise, both in the law and in war surgery, to mention only those two areas, together with its neutrality and impartiality, which have stood the test of time, have earned the institution the confidence of Governments as well as of numerous armed movements of all kinds, all parties to the conflicts that are ravaging far too many regions of the world.

This fund of credibility should be used sparingly to defend and advance the cause of victims forced to endure intolerable situations. The violations committed every day through the use of mines, and particularly anti-personnel mines, are an eloquently cruel example of one such situation.

ITEM (1)

THE REALITY OF THE PRESENT USE OF MINES BY MILITARY FORCES

Report presented by: Rae McGrath
Director
Mines Advisory Group

1. Introduction

Probably the first person to fully recognize the insidious nature of landmines was the British war correspondent, Christopher Buckley, who, as early as 1943, professed to be worried by mines as a weapon

> "... *because human qualities were not directly involved ... that was the danger of mines, buried and invisible*".[1]

Buckley may have become a powerful lobbyist against the proliferation of mines had not his fear been cruelly justified when, in August 1950, he was one of three passengers in a jeep that hit a mine in Korea – all were killed.

Yet what Christopher Buckley identified in 1943 was the crux of the issue that faces more than twenty mine-infested countries today – the persistent and uncontrolled nature of mines. It is also worth recognizing that Buckley was making an observation relating to *easily detectable* mines laid in *pre-planned* minefields and, very probably, minefields which were mapped, marked and recorded. As this paper will seek to illustrate, such controlled strategies are an exception in modern military terms, where the mine is neither designed nor disseminated with any thought given to its long-term impact nor to the eventual removal of the mine as a threat to the civilian community in the post-combat period.

This paper will also raise a critical question – are mines designed to meet the planned or stated needs of the military? Or is the development led by manufacturers?

References to mine-related strategies in this paper are largely based on practical experience of the situation that exists in Afghanistan, Cambodia, Angola, Mozambique, Iraq and Somalia, either from the author's own observations or those of Mines Advisory Group specialists or both.

This report is designed to stimulate debate rather than to present a comprehensive picture of military mines strategy, it should not be seen as limiting the discussion to the issues raised but should be perceived as a starting point. It is deliberately focused on accepted military structures rather than smaller groups involved in mine-laying since mines are manufactured, in the first instance, for sale to such financially viable customers.

2. Targeting mines

Most military spokesmen would argue that mines have several clearly-defined and acceptable roles in combat. They are:

To protect military bases and key installations;
To channel or divert the enemy forces;
To deny routes and strategic positions to the enemy.

These uses of mines have commonality in as much as that they should allow for mapping, recording and marking of mined locations by a responsible command structure. In fact, it could be argued that the marking itself of such minefields would often achieve the desired purpose, thus the common usage of dummy minefields.[2]

However, military strategy demands more of the mine as a weapon in the modern theatre of war than these basic deployment strategies. And here it is necessary to subdivide military use of mines into two sub-categories of deployment philosophy – the *conventional war scenario* and the *counter insurgency campaign.* This terminology has no bearing on the type, scale or scope of the military operation, rather, it relates to the *perception* of the enemy forces. For instance, the second Gulf War would fall into the former category, as would the Falklands/Malvinas conflict, while the Vietnam War and the Soviet/Afghan War fall into the latter category of conflict.

In both scenarios mines are deployed using strategies that target enemy forces in such a manner as to ensure that a long-term humanitarian problem will exist in the post-combat period.

a) The Conventional War Scenario

The enemy forces are perceived as *an organized army, with rank structure, uniforms etc.*[3] and, in addition to the three, generally accepted and admitted, strategies for mines dissemination they are subject to the following, less openly discussed, strategies:

Deep strike: deployment of mines into the enemy force's rear areas.[4]

Cut off: deployment of mines behind a retreating, or in front of an advancing, force.

Both the above strategies involve the remote[5] dissemination of scatterable mines by fixed-wing aircraft, helicopters, artillery, rocket or mortar. The former strategy would normally be aimed at key junctions on enemy main supply routes (MSR's), supply dumps, loading areas, workshops and headquarter elements. The purpose of the latter, *cut off* usage of scatterable mines, is self-explanatory.

What both strategies have in common is the lack of any reliable methods of recording, mapping or, immediately or subsequently, marking such concentrations of mines.

b) The Counter Insurgency Campaign

The terms *insurgent, guerrilla* or *terrorists* have been usefully employed by conventional military forces to describe an enemy force which does not conform to the *accepted* norms – in many cases this may simply mean that they have less resources (although possibly more popular support) than the *real* armies involved in the conflict. It may also, as in the case of the US campaigns against Laos and Cambodia, the Soviet invasion of Afghanistan and the Iraqi campaign against the Kurds, be seen as a more generally acceptable way to describe an illegal prosecution of war against an indigenous population.

It is these category of actions that have resulted in the most inhuman and persistent mine disseminations strategies:

Random and widescale mining of agricultural and community land

Deliberate use of mines as an anti-morale or terror weapon against civilians

Mining of villages, water sources, religious shrines, etc.

The military use the fact that the enemy are *hiding in the community* to justify many of these actions and, thus, do not feel restricted to any specifically defined military targets since, under such operational definitions, virtually *everywhere* becomes a justifiable target. The use of mines by Soviet forces and by the Kabul Regime against the *Mujahideen* in Afghanistan is a classic example of this genocidal tactic – in some areas virtually all mountain grazing land was remotely mined and the whole agricultural infrastructure brought to a halt by the widescale mining of fields, karez[6] and surface irrigation systems.

It is apparent that the mining of agricultural land in order to restrict the supply of food to the enemy (and also, by design or otherwise, the civilian population at large) is now accepted by military strategists and field commanders as *a normal* strategy.

In addition it is clear that the use of mines as weapons of terror against the civilian population is also increasingly accepted as a military tactic. It should come as no surprise that the Iraqi government used such strategies against the Kurds and is now using the same tactic against the Shia population in the southern marshes of Iraq, nor that the MPLA used a similar strategy in Mozambique, but this strategic thinking is not confined to extremist groups and the armies of dictators. Both Chinese and British training of Khmer forces opposed to the Vietnamese-backed government in Phnom Penh stressed the use of the mine (and improvised booby traps) as an anti-morale weapon targeted against the civilian infrastructure.

To summarize:

In the modern conflict mines cannot be said to be *targeted* at military formations in any reasonable definition of the word, since the very design of a great proportion of the mines in use and their method of dissemination is such that civilian casualties and long-term infestation of the land are inevitable rather than coincidental.

Military philosophy regarding the use of mines is such that, even where devices *are* targeted, their impact on civilian populations is unacceptable by any humanitarian definition.

3. Design-led strategy?

There is a strong argument that the military have less to do with the development of mines than the manufacturers – in other words that minelaying strategy has become manufacturer/design-led. Obviously such a suggestion may be expected to meet with strong rebuttal from both the military and the manufacturers, but the facts support the hypothesis.

a) The development of the remotely delivered scatterable mine should really have little attraction to the military strategist. Rather like the use of chemical and biological weapons it is a dangerous strategy to embark upon; mines deployed can offer as great a threat to one's own advancing forces as to the enemy in retreat.

   Using unrecorded mines to deny the enemy key areas will inevitably involve the use of valuable manpower and probably lives of one's own troops when those areas are overrun.

b) It is inconceivable that the military would themselves call for the design of mines which were virtually undetectable, to the enemy and to themselves. But this has been the direction of development over the past thirty years and the advancement of the low-metallic content mine has far outstripped the development of effective detection technology.

   Most manufacturers offer metallic detection inserts as an *option*, a fact that indicates that there is little or no demand for such an addition from their customers.[7] But why is such an option not popular with the military? It is the knowledge imparted to the enemy that a minefield has been laid in a particular area that achieves the main military objective – there are many more effective methods of killing and maiming enemy troops.

   The conclusion must be that:

   either the military *aim* to deny land to civilians in the post-combat period and have a *policy* of killing and maiming civilians long after the cessation of hostilities, or

   international military strategic thinking is so inhuman and short-sighted that they are unaware or uncaring of such considerations, or

   they use what is designed and made available by manufacturers and develop strategy based on available, and not necessarily desirable, technology.

4. Conclusion

Accepting the factual nature of the points raised in this paper must lead to one frightening conclusion – the military, by perpetuating and refining mine-laying strategies which will inevitably kill and maim non-combatants during the period of combat and continue to kill and maim civilians following a ceasefire, and lead to the long-term denial of land to rural communities, have shown a total disregard for humanitarian law, principles and simple human decency.

Presentation and discussion

Mr McGrath said that an important task of the Symposium was to inform the general public throughout the world of the reasons why mines existed and of the ways in which they were used. In addition to the military use of mines to protect the armed forces and to cut off the enemy's advance and retreat, there was their more insidious use to damage the enemy's morale.

There was a tendency to distinguish between conventional and irregular forces with respect to responsibility, but the random scattering of mines by artillery or from the air by conventional forces, with no possibility of recording where the mines were dropped, constituted irresponsible use of those weapons by forces normally regarded as responsible.

Mine laying in such countries as Afghanistan and Laos, where tens of thousands of people were being killed or maimed by mines every year, was justified by the alleged need to take defensive action against insurgents hiding among the civilian population, and the strategy received an impetus from the mine manufacturing industry. It should be stressed that mine clearance took a very long time, and that meanwhile hundreds of thousands more mines were being scattered, in an exercise of mindless violence against innocent victims.

In reply to questions, **Mr McGrath** said that there had been no discernible change in the use of mines since the adoption of the 1980 Convention, since the existence of that instrument was unknown to most users. The prime mine producers often could not be identified, since the mines were manufactured under licence in other countries.

It was pointed out that it was difficult to design a weapons system unless there was a military demand for it, and that even self-destruct and self-neutralizing mines had to be cleared. In conclusion, it was considered necessary to focus on the attitudes of mine users and manufacturers.

ITEM (1)

OVERVIEW OF THE PROBLEM OF ANTI-PERSONNEL MINES

Report presented by: Kenneth Anderson
Director
Arms Project of Human Rights Watch, New York

The Arms Project of Human Rights Watch ("HRW"), like all the groups represented in this Symposium, is appalled and dismayed at the spreading abuse of anti-personnel landmines in armed conflicts throughout the world. The purpose of this memorandum is briefly to set forth a series of general propositions concerning why, how, and the extent to which the abuse of anti-personnel landmines has spread so widely and in so relatively short a period of time, in order to allow us to analyse possible solutions.

Most of these propositions will be discussed at greater length during other parts of the Symposium, and so they are here presented briefly, without exhaustive discussion. In addition, while some of these propositions are uncontroversial, others may provoke sharp disagreement. They are offered here simply as a starting point for discussion.

The propositions concerning anti-personnel landmines are these:

1) Anti-personnel landmines, which in conventional military doctrine seek to deny ground to the enemy on the battlefield, overcoming the problem of insufficient force to space, have gradually shifted to become a strategic, as well as tactical, weapon. That is, they have shifted from being a weapon whose importance is on a particular battleground to being a weapon whose importance is control of a whole theatre, a country, and its inhabitants.

2) Collateral to becoming a strategic as well as tactical weapon, anti-personnel landmines have become an offensive, as well as defensive, weapon in virtue of the development of remotely deliverable mines.

3) Also collateral to becoming a strategic as well as tactical weapon, anti-personnel landmines have become a weapon often aimed directly against civilian populations, i.e., a weapon used to deny ground to civilians, as though they were enemy combatants, for such purposes as starvation, emptying a zone to prevent insurgent operations, creating a refugee problem, and ethnic cleansing. This is not to ignore that in many cases, civilian casualties are simply unjustified collateral damage by armed forces that are indifferent to civilians, for better or worse.

4) Anti-personnel landmines have become a key offensive and defensive weapon in insurgency and counterinsurgency armed conflicts around the world, and their use in these types of conflicts is central to the problem of landmines today. In addition, landmines have become central to several ethnic civil wars, in which neither side worries about devastating the geographic zones of the other ethnic group; this type of internal war differs from counterinsurgency war in that it is not a struggle between central government and guerrillas, both of which may use mines, albeit for very different military reasons.

5) Anti-personnel landmines devastate civilian populations within a given country, but they are not a “destabilizing” weapon in traditional international

security terms. International wars have rarely (if ever) been caused by landmines, and so anti-personnel landmines have not received attention from the "security," as distinguished from "humanitarian," perspective. The fact that the international community is now showing itself willing to intervene in places like Somalia, where no small part of the breakdown of society and starvation is due to such "non-destabilizing" weapons as small arms and landmines should cause the international security community to rethink weapons and stability.

6) The reasons for the above shifts in the military use of landmines are complex, but among them are, first, the rise in internal armed conflicts where the outcome is truly at issue, so that neither side necessarily has the immediate expectation of winning and thus having to worry about the consequences of landmines in the long term and, second, the rise of remotely deliverable mines that allow them to be laid in immense quantities and as an offensive weapon.

7) The problem of anti-personnel mines is increasing worldwide, rather than decreasing. Nor is the problem of anti-personnel mines limited to one or two infested countries, as is sometimes thought in the press, such as Afghanistan and Cambodia, but is spread far and wide in Africa and parts of Asia. While not seeking to list all the major problem countries, certainly such a list would include Cambodia, Afghanistan, Iraq, Angola, Mozambique, Somalia, Ethiopia, former Yugoslavia, El Salvador, and Nicaragua.

8) Anti-personnel mines are cheap and easily manufactured by the weapons plants of a wide range of countries. The economics of landmines production and trade are such that production can easily be shifted from place to place; mines are readily available on the open market now, and attempts to shut down particular suppliers would result in a shift to others. In addition, large quantities of mines are available from stocks left over from various Cold War conflicts.

9) The Landmines Protocol has been a nearly complete failure with respect to controlling the use of mines, recording their emplacement, and removal; indeed, there is no known internal conflict where maps of any kind have been maintained in accordance with Protocol procedures or any more rudimentary way.

10) The failure of the Landmines Protocol coincides with the rise of new forms of use, such as widespread use in civil wars and insurgency-counterinsurency, and new forms of technology, such as remotely deliverable mines, that go beyond the situations contemplated by the Protocol.

11) The Landmines Protocol is arguably weaker than the application of more general principles of customary law of war, such as rules against indiscriminate use, etc., and arguably, the hand of those seeking to control landmines use in the field would be freer under general principles than under the specific rules of the Protocol.

12) Attempts to control field use of landmines by means of humanitarian law governing use alone have failed utterly, and this argues for shifting the forum for control from use to production and transfer, along the same lines as the prohibition on certain types of ammunition such as dum-dum bullets.

13) The economics of production and trade in landmines suggest that actual control of the field use of mines by means of control of production and trade is also immensely difficult, given the ease with which mines are produced and their easy availability from a wide variety of sellers in the arms markets.

14) Shifting the focus from controlling field use to production and transfer has the advantage of greater verifiability, and it is also true that many producers, or at least their home countries, are more easily "stigmatizable" by the charge of abuse than the ultimate groups that use the mines. There are always likely to be "rogue" suppliers, even in the best-willed world, but the supply issue would become more manageable in a world where production and transfer were outright prohibited. It is for this reason, among others, that HRW has joined various NGOs in calling for a complete ban on production, transfer, and use; it is fully aware of the difficulties both with the Landmines Protocol and any new standard based on production and export.

15) The US one year moratorium on landmines exports is, at this moment, unique, and similar legislation deserves support in other countries.

16) Although conventional armies would certainly oppose a ban on production, transfer, and use, in some cases, opposition might be less than otherwise, given that, for example, US forces contemplate a mobile insertion role in which landmines are largely a risk rather than a favourable weapon.

17) The degree of landmine infestation in several conflicts would arguably be much less than it is today had the super-powers not made mines available in large quantities, e.g., Angola; those responsible for supplying such weapons in pursuit of their geopolitical goals have an obligation for their use and, by extension, landmines cleanup.

18) The human costs of anti-personnel landmines will continue to rise, in the form of deaths and mutilations and, in addition, in lost development

opportunities, inability to resettle refugees, restart agriculture following conflicts, etc. New and cheaper technology is needed in prosthetic limbs.

19) Demining can never take the place of preventing mine emplacement in the first place, although the existing scope of the landmines problem makes demining an urgent necessity. To hope to make a dent in the existing landmines problem, however, demining must be expanded drastically, funded to a far larger extent, and new technology must be developed to make it quicker, more extensive, and cheaper.

20) The value of demining must be considered carefully in situations where the conflict is on-going; in Cambodia, for example, new mines are being laid faster than that old ones are being taken out, and the same is true in former Yugoslavia.

21) The proper international responsibility for demining must be made clear; within the United Nations system, responsibility for demining has never been established, and the broader international moral responsibility of the superpowers to clean up mines laid in "their" past conflicts must be agreed upon.

22) The future of landmine technology runs in three directions: first, the distinction between anti-tank mines and anti-personnel mines is gradually being blurred by new technology that "packages" them together in a single system of area control; second, mines are becoming ever smaller and less detectable and, third, high technology mines are being developed that are able to select targets and also shut themselves off.

23) The possibility of self-shut-off mines ought not to deter adoption of a complete ban on production, transfer, and use of mines; the most significant use of mines is in low-tech conflicts where the combatants will not have advanced mines and may have no incentive to use them if their military purpose is to attack civilians. The development of expensive, high tech-mines is a red herring for the real sources of abuse, which would almost certainly continue.

24) The real technological changes that should concern the world are less-detectable mines, ones that will cause ever greater injuries and are harder than ever to eradicate. The fact of these technological changes is a still stronger argument for a ban on production and export as well as use.

*****

These propositions are offered for the purpose of laying out the major issues for debate in the Symposium. Some of them are elementary, while others are controversial and likely will be discussed in subsequent papers at the Symposium.

These propositions have been structured as an argument in favour of an international ban on production, transfer, and use, which is the position HRW has adopted. The counterargument against HRW's position that it is pointless to adopt a regime more utopian than one the world ignores now is hardly, on the face of it, unpersuasive. HRW's position is frankly strategic and instrumental rather than categorical; it is based on an assessment of what parties can be embarrassed or stigmatized by international scandal into compliance.

But HRW's position is also based on the recognition that the nature of mine warfare has changed for good, and the assumptions underlying the old landmines regime no longer hold. In this situation, in the view of HRW, only an attempt to tie the hands of producers and exporters can reduce the supply available to users who will, in HRW's view, always break the rules.

Presentation and discussion

Mr Anderson said that over the past 20 years there had been an increase in the different kinds of armed conflict, with the result that mines had become strategic as well as tactical weapons and also offensive as well as defensive weapons; moreover, they were now often aimed directly at civilians.

From the practical point of view, Human Rights Watch regarded anti-personnel mines as irresistible weapons, which would be used by any side in a conflict that possessed them; it therefore considered that the focus shift to include not just restrictions on the use of mines but also to restrictions on production and trade in them. Mr Anderson stated Human Rights Watch's desire for a ban on production, transfer and use.

In the ensuing discussion, it was suggested that restrictions on the use of mines might be strengthened by encouraging ratification of Additional Protocol I of 1977, Article 51 of which included a prohibition of the use of weapons having indiscriminate effects.

ITEM (3)

HUMANITARIAN CONSEQUENCES OF MINE USAGE

Report by: Dr. Robin Gray
Surgical Coordinator
International Committee of the Red Cross

Introduction

Mines are weapons which when used legitimately by military forces are intended to frustrate and hinder the enemy; they are used in the pursuit of military goals and objectives. For civilian populations they are singularly pernicious when employed haphazardly, without methodical control by marking and documentation, when used indiscriminately in large numbers to block routes of supply or population movements and when used to terrorize populations with the goal of denying access to land. The special characteristics of anti-personnel mines are that they cannot be directed towards a specific target, they cannot distinguish between friend, foe or non-combatant and they remain an active menace until inactivated or destroyed.

There has scarcely been a modern conflict in which mines have not been used; wherever they are laid they cause problems unrelated to their intended military use. Demining carries risk of injury and death, land recognized to be mined but considered too dangerous or too expensive to be cleared is unusable and haphazardly placed mines are a danger to all.

Loss of life is a human tragedy, injury and physical handicap in a civilian population adds to the load of the medical facilities providing care for victims; the loss of land available for agriculture adds to the socio-economic burden of a country.

Epidemiology

Accurate data on mine injury is not easily available because of the organizational and social disruption existing in the countries currently afflicted. Global statistics given for a country are at best estimates based on partial information, intelligent guesswork, assumptions and extrapolations.

That is not to say that the magnitude of the problems have been under-estimated; huge problems clearly exist.

For Cambodia amputation rates due to mines of 1 per 236 members of the population have been quoted using the estimate of 36,000 amputees and assuming a population of 8.5 million. Figures of 1:470 for Angola, 1:650 for Somalia, 1:1,100 for Uganda, 1:1,862 for Mozambique and 1:2,500 for Vietnam are quoted and give indications of the scale of the problems (ref. 1).

In a detailed study in North West Cambodia incidence rates of mine injury in the adult population ranged in different districts from 3.4 to 339.5 per 100,000 of the population per year. The incidence rates for adult males in the same districts ranged from 20.2 to 1,803 per adult males per year (1 injured per 4,950 adult males per year to 1 injured per 55 adult males per year). The same study showed that 87% of mine injured were males of whom just under half were soldiers; 58% of those injured by mines were therefore non-combatants. Military activities accounted for only 26% of injuries in soldiers while collecting firewood or fruit in the forest was the major cause (45%) of civilian casualties. Thirty-seven per cent of the women were injured during cultivation in fields. It was noted that the biggest detrimental effect of amputation on social and economic status was experienced by those yet unmarried, particularly women (ref. 2). What can be added is that unknown numbers of mine victims die without official registration or documentation.

Different anti-personnel mines have variable capacities to kill and maim. Those triggered by foot pressure principally damage the legs while bounding and directional mines will kill all within a certain radius or distance. Statements that there are two deaths for every landmine injury must therefore be taken with caution (ref. 1). A study of 25 cases of injury from bounding mines reported 15 deaths and 10 survivors. Survivors were from incidents in which mines malfunctioned (ref. 3). Of 305 mine incidents in Northern Cambodia, 65.5% involved one person, in 15.7% of incidents another individual was injured or killed and 18.7% of incidents affected more than two people. There were 27 deaths (5.4%) reported from the 305 incidents which gives a figure of 11 wounded for every death (ref. 2).

ICRC hospital mortality from a sample of 4,396 mine-wounded in the Cambodian and Afghan conflicts is 2.9%; for those admitted within 6 hours of injury, 4.5% and for those admitted after a delay greater than 72 hours, 1.5%. Rapid evacuation brings those whose wounds are fatal in spite of medical assistance. The lower hospital mortality for those having slow evacuation results from a proportion of those more severely injured dying in transit.

ICRC has hospitals which have served the wounded from the conflicts in Cambodia and Afghanistan. It is not known what proportion of all wounded have been evacuated to these hospitals. Khao i Dang Hospital in Thailand was the hospital serving wounded Khmer principally coming from the border regions which were known to be heavily mined. Opened in 1979 the hospital saw a constant flow of mine injured in its 12 years of existence; no month went by without the hospital receiving mine-injured. Seasonal fluctuations are noted in all hospitals. The counterpart hospital in Mongo Borei on the opposite side of the border inside Cambodia has a similar experience. Khao i Dang treated 6,137 war-wounded in its last nine years; 3,452 were mine-injured (56%).

The ICRC hospitals in Peshawar and Quetta serve wounded coming from inside Afghanistan and ICRC had an independent hospital in Kabul from 1988 until 1992. Of the 11,067 wounded treated in Kabul 1,719 were due to mine injury (15%), of which nearly one quarter were children of 14 years or less. This high percentage is due to the fact that children are employed to look after the family cattle and are therefore at particular risk.

Just under half of all casualties seen in Kabul were women and children, certainly innocent victims of war. Children make up 23% of all wounded treated by ICRC in the Afghan and Cambodian conflicts.

In comparison the hospitals of Peshawar and Quetta see a higher proportion of mine injured. In 1992 44% (1,530) of the wounded admitted (3,461) were due to mines. From April to July 1992, the proportion of mine-wound victims increased substantially on the Afghanistan border due to the partial return of refugees to Afghanistan encouraged by the fall of Kabul to the Mujahideen.

A study conducted in Peshawar between June and December 1992 showed that 85% of the 528 mine-wounded were engaged, when wounded, in non-military activity such as farming, travelling between villages or tending cattle; 77% were returnee refugees.

Surgical aspects

The surgery of mine injury follows the basic principles of war wound management: that is, the excision of all dead tissue and the removal of foreign material followed by delayed closure of the wound (ref. 4). The author is sad to report that the military medical services of one European country supporting the United Nations operations in the former Yugoslavia and in Cambodia are failing to practise these fundamental surgical principles.

Three clinical patterns of mine injury are recognized (ref. 5). Mines which are ignited by foot pressure damage by blast resulting in traumatic amputation of part of the lower limb (pattern 1). The tissues of the foot are simply exploded when the charge is small; or when the charge is larger the lower leg or legs are blown away. Damage is rarely confined to one leg; the contra-lateral limb is usually damaged to a greater or lesser extent and the genitals, upper limbs, chest and face may also be involved. Even those patients with lesser injury cannot escape amputation as the foot can never be saved. The wearing of boots is said to channel the explosive energy and so produce more extensive damage than would occur without such footwear.

The surgeon must recognize that dirt, debris and footwear may all be pushed up into the lower leg and this material may extend considerably further up between the muscles than is immediately apparent. Failure to remove all the dead tissue and to leave debris lodged between muscles will result in infection and protracted recovery, usually involving multiple operations.

From a purely theoretical point of view a patient undergoing an amputation for mine injury needs only two operations if managed properly. In ICRC hospitals this only happens to one- third of patients undergoing major amputations of the lower limbs; the remainder require three or more operations. Delay in evacuation increases risk of infective complications and surgical difficulty. Only 28% of ICRC mine injured arrive in hospital within 6 hours of injury when the wound is considered to be contaminated only. Wounds older than 6 hours have established infection and are therefore more difficult to treat.

Stake, bounding and directional mines damage principally by the projection of fragments (pattern 2). Those very near the weapon will be killed by a combination of direct blast and the effects of receiving multiple fragment wounds. Those more peripheral to the lethal radius of stake and bounding mines or beyond the lethal distance of directional mines will receive fragment wounds. The amount of damage will depend on the energy available at entry and less damage will be produced the further away the individual is from the detonation. The anatomical site or sites involved will clearly be the key factor determining the potential lethality or otherwise of the wound.

The third pattern to be recognized are those wounded while handling. They sustain variable injury to the hands, arms, chest and face. The extent of damage will depend on the munitions involved; the PFM-1 butterfly mine can blow away the hand of a child.

The ICRC experience of 4,396 injuries produced by mines is that 27% undergo a major amputation of a lower limb and that 84% of lower limb amputations are performed for mine injury. Only 2% of mine-injured avoid injury to the lower limbs and in 18% the injury is confined to the legs.

Medical services for mine-injured

The level of health services in the majority of countries afflicted by mines is at best poor; they are ill-equipped to deal with the injuries produced. Much unnecessary loss of limbs and life results from poor medical facilities and inexpert surgical skill. First aid is often rudimentary and ICRC surgeons not infrequently face the consequences of inexpertly applied tourniquets requiring higher levels of amputations than strictly necessary. The surgery of mine victims is particularly demanding and time consuming when done properly and places considerable strains on blood bank services where they exist. An ICRC study on the use of blood in its hospitals found that for every 100 wounded 44.9 units of blood were required while 103.2 units were required for every 100 mine injuries (ref. 6). Inadequate or absent blood transfusion services add to the difficulties of providing for mine-injured in poor countries. The provision of hospitals and surgical services in the developing world has until recently been neglected in the interest of providing primary health care. This has now been recognized. Surgery is one of the most expensive services for a health care system.

References

1. Indiscriminate Weapons: Landmines, The Medical Educational Trust, London, June 1992.
2. Fiona King, Landmine Injury in Cambodia: a case study, London School of Hygiene and Tropical Medicine, M.Sc. thesis, September 1992.
3. Adams D.B., Schwab C.W., Twenty-one-years' experience with land mine injuries, J. Trauma 1988; 28 (suppl): S 159–62.
4. Amputation for war wounds, International Committee of the Red Cross, Geneva, 1992.
5. Coupland R.M., Korver A., Injuries from anti-personnel mines: the experience of the International Committee of the Red Cross. B.M.J., 1991; 303: 1509–12.
6. Eshaya Chauvin B., Coupland R.M., Transfusion requirements for the management of war-wounded: the experience of the International Committee of the Red Cross B.J. Anaes., 1992; 68: 221–23.

Presentation and discussion

Dr Coupland (ICRC) illustrated his presentation of Dr Gray's report with a series of vivid slides. Graphs of data from five separate hospitals and covering some 17,500 patients showed that, overall, around a quarter were suffering from mine injuries.

There was, however, some variation and the proportion was as high as 60% in one of the hospitals. It was not a simple matter to discover how many of those injured by mines were non-combatants, since patients were often reluctant to admit to being combatants.

Making the broad assumption that all young men were combatants, and that children, women and old men were not, it appeared that over 30% of the mine-injured were non-combatants. Of all mine-injured, 31% had one limb amputated and 3% had both limbs amputated. Following the fall of Kabul to the Mujahideen, there had been a huge increase in the number of mine-injured; according to data from Peshawar, 77% of them claimed to be returnees. Special surgical techniques had been developed to deal with mine injuries, and a video was available demonstrating those techniques.

In reply to various comments, he said that secondary amputations generally resulted from an inadequate primary amputation, when there was a failure to excise all dead tissue or when foreign matter remained.

Other complications included a new growth of bone or a benign tumour of the nerve end (neuroma) which made it impossible to use a prosthesis. In his opinion, if a wound became infected, the fault lay with the surgeon.

Infection could arise if particles of foreign matter had been left in the wound, or if an inappropriate material (such as silk) had been used for sewing up. **There was thus an enormous need to train surgeons. To support that end, the video was being widely distributed through National Societies.** In particular, military medical services were being informed of its availability. Where amputations were performed by ill-trained surgeons and where proper care (especially nursing care) was unavailable, the prospects for patients were grim.

A member of the ICRC medical staff stressed that surgeons had to be supported by an adequate infrastructure, including nursing care and the availability of hospital drugs and medicaments.

One participant, commenting on the percentage of non-combatants injured, pointed out that the ICRC figures had been derived in war time. After the end of hostilities, 100% of those injured by mines were non-combatants.

Another participant asked whether the use of plastic in mines created any particular problems.

Dr Coupland recognized that the question had legal and technical ramifications; from a purely surgical point of view, however, plastic posed no greater problems than the bits of foot, mud and grass blasted into the wound by the mine. None of these materials was detectable by X-ray and all had to be removed. Surgical techniques remained the same.

A question was asked whether a country case study had ever been done to provide data on acute medical needs, including training needs.

Dr Coupland said that, although such data would be most useful, it was likely to prove difficult to obtain in a country where the population was not even known.

ITEM (3)

ICRC REHABILITATION PROGRAMMES ON BEHALF OF WAR-DISABLED

Report presented by: Alain Garachon
Rehabilitation Service
International Committee of the Red Cross

Faced with the realization that war-disabled in many countries receive little or no attention, the ICRC decided in 1978 to create a special department within the Medical Division to bring assistance to these victims. The majority are amputees injured by mines; paraplegics represent a lesser problem in terms of numbers.

In the last 14 years, the ICRC has implemented 33 orthopaedic programmes for amputees and two for paraplegics in 17 countries: Angola, Ethiopia, Mozambique, Chad, Zimbabwe, Uganda, Kenya, Sudan, Colombia, Nicaragua, Lebanon, Syria, Afghanistan, Pakistan, Burma, Cambodia and Vietnam.

These programmes have been run or are still run in collaboration with local partner organizations, mostly ministries of health and National Red Cross Societies from European countries.

The aims of these actions are:

a. To manufacture artificial limbs locally so as to avoid purchasing imported and expensive orthopaedic spare parts.

b. To organize training programmes for local prosthetists so that continuity can be ensured after the ICRC withdraws.

   Amputees and the paraplegics are permanently affected and in consequence the services to assist them should be permanent.

   - The first step in tackling this problem is to set up rehabilitation programmes which take into consideration the socio-economic problem of the country concerned. Setting up a prostheses production unit is a relatively easy task which requires money and skill (both available abroad) but to maintain these activities is another story. Prostheses are expensive items and the price of technology used in rich countries is totally out of reach of the victims in Cambodia or Afghanistan for example. Furthermore, a prosthesis for a child should be replaced every 6 months and for an adult every 3 to 5 years. A child injured at 10 years of age with a life expectancy of another 40 or 50 years will need 25 prostheses which at US$ 125 each amounts to US$ 3,125. In countries where average incomes are of the order of US$ 10 to 15 a month, one can easily understand that crutches are all that are available to the population. The manufacture of a prosthesis is individual and it takes many hours to equip an amputee. Orthopaedic workshops can rarely meet the demand and patients may have to wait years before they receive an artificial leg.

Statistics:

1. Since 1979, the ICRC programmes have produced: 59,015 prostheses for 42,104 amputees and 19,070 orthoses.

2. A study on 3,262 mine injury cases in ICRC war surgery hospitals showed:

   – 45% need prosthesis = 15% upper limb prosthesis
   30% lower limb prosthesis

   – 28% above knee
   81% below knee

   – 7.1% Female
   93% Male

   – 19.3% below age of 14
   6.5 % above age of 60

   – average cost of 1 prosthesis = US$ 125

   – training period for 1 technician = 3 years

3. Lower limb amputations by weapon, study on 1,253 lower limb amputations (Cambodia and Afghanistan):

   – mines : 84%

   – fragments : 11%

   – bullets : 3%

   – other : 2%

Presentation and discussion

Mr Garachon, presenting his paper, said that the rehabilitation of amputees did not provide a satisfactory solution to the permanent problem of mine injuries. Even the best orthopaedic devices were not suitable for long-term use. Many of those injured by mines were to be found in destitute countries, and organizations such as the United Nations could not be expected to remain in those countries forever.

The thousands of artificial limbs produced through ICRC programmes – some 60,000 in 13 years – were not enough to meet the needs of the huge and growing numbers of amputees. Neither was the training of local prosthetists an adequate response, as most of the countries in which the ICRC and other non-governmental organizations operated did not have the financial means to carry on programmes with counterpart staff. Rehabilitation could not, therefore, be considered to be an answer to the injuries caused by anti-personnel mines.

In reply to a question on the proportion of those fitted with artificial limbs who were able to resume their normal activities, he said that the young who had suffered the amputation of a lower limb – a common mine-injury victim – could usually go back to work. It was more difficult for the old.

The ICRC was particularly concerned with medical treatment and the fitting of artificial limbs, rather than vocational rehabilitation and training. The latter were, however, very important.

But it should be borne in mind that, in countries where unemployment was widespread, it was difficult to create jobs for the handicapped; for example, a project in Lebanon for that purpose had not succeeded.

On the other hand, self-employed agricultural workers, once fitted with artificial legs, are often able to take up their former work again.

One participant emphasized the importance of drawing attention to the need for rehabilitation, citing the cases of the occupation by militant amputees of Red Cross premises in Angola and the refusal of the Government of Mozambique to admit the existence of military handicapped.

Another added that most countries recognized the existence of amputees, and a protest by a group of the disabled in Ethiopia had influenced Government policy.

ITEM (3)

SOCIAL CONSEQUENCES OF WIDESPREAD USE OF LANDMINES

Report presented by: Jody Williams
Coordinator
Landmines Campaign,
Vietnam Veterans of America
Foundation

This paper consists of general narrative to give an overview of the problem, followed by country-specific statistics which then begin to convey a sense of the global nature of the devastation caused by the largely indiscriminate use of landmines.

* * * * *

Most of the countries contaminated today by landmines are countries in the developing world with the least resources available to respond to the social consequences of the widespread, generally indiscriminate use of the weapon. Afghanistan, Angola, Cambodia, El Salvador, Mozambique, Nicaragua . . . and the list goes on to include more than twenty countries around the world.

It is fair to call the landmine aftermath of armed conflict “contamination.” For decades after conflicts have ended, landmines are capable of killing and maiming. Large tracts of land cannot be safely used because of the lethal nature of the seeds they now carry. Perhaps once sown for military purpose, the landmine cannot determine when the battle is over. Once sown, it lies in wait indefinitely – and as has been pointed out by others familiar with the weapon, it is capable of killing or maiming the grandchildren of those who laid it.

The countries most severely affected are also for the most part rural and agricultural societies. And within those societies, it is the subsistence farmer, the nomad and the herd, and fleeing refugees and the displaced who most often are affected – those sectors of the society who most must rely on their physical fitness for the most basic subsistence and who least can afford the care necessary to treat landmine injuries.

For the affected countries, this translates into the devastation by landmines being felt at all levels: individual, family and community, and societal. Each individual bears the personal and psychological trauma felt by survivors; the burden of their care often falls upon family members and/or the local community of which that individual is a part. Inevitably the accumulated effects of all casualties are felt throughout the society as a whole.

Not only must the countries plagued by landmines respond in some way or another to the immediate and rehabilitation needs of landmine victims, but they must also respond to the problem in its entirety: How can a country rebuild when its national territory is sown with landmines? How can it move from peace-keeping to peace-building with the post-conflict legacy posed by hundreds of thousands, often millions of landmines? How can refugees and the displaced be returned to places of origin along routes long since mined? And if and when they do get home, how can they begin to rebuild a life surrounded by landmines? Where will they plant their crops? Where will they graze their livestock? Where will they gather food? Where will they gather firewood?

Immediate Needs of the Individual

When an individual steps on a landmine, if the initial impact of the blast is survived s/he must receive medical treatment. If that individual is alone, herding grazing animals for example, s/he may bleed to death before being found. If not alone, transportation to medical facilities, which themselves are too frequently inadequate, is difficult. Mine victims are often carried to treatment or transported over rough terrain in carts. Once first aid has been administered, it can be days before hospital treatment is received.

If the individual manages to get to a hospital, what s/he often finds is overcrowding and lack of personnel (and often the personnel that is there is seriously lacking in skills) and supplies. The lack of competent care often results in the most rudimentary treatment which can produce infected wounds and a need for secondary and even tertiary amputation. The overall lack of resources can mean the patient receives the most minimal treatment in the shortest amount of time possible with no chance of rehabilitative care. Facilities which can barely provide for the survival needs of the patient cannot even begin to consider the longer-term psychological implications of an amputee returning to a society which often has no place for the disabled.

Family/Community Needs

As mentioned above, all too often the landmine survivor must return to society without either the physical or psychological rehabilitative services available to help his/her return to community. In agricultural societies, generally the only thing an individual has to offer is his/her labour and this requires an intact body. Amputees who can no longer either vie for themselves or provide for the sustenance of their families become a burden not only to themselves but also on the family.

Cultural and/or religious stigma can also be attached to individuals who are "no longer whole." If the individual is unmarried prior to amputation, whether male or female, chances of participating fully in the community as a married adult with children are minimal. And if the individual is alone, the burden of care can fall upon the community.

Society

When needs as outlined above are multiplied by 30,000 as in the case of Cambodia, or 20,000 as in Angola, the strains on society become staggering. (And these numbers do not tell the entire story. Precise numbers of victims are virtually impossible to tally. All too often injuries occur in remote areas where the individual dies alone. Even when not alone, lack of immediate care results in death and often quick, unrecorded burials.)

The developing countries most plagued by landmines simply do not have the resources to provide the complexity of care required by landmine victims: usually, there are no medical transport vehicles to get casualties to hospitals; if they reach the hospital, care is minimal; once released, there is little rehabilitative care; the need for prostheses far outstrips the ability to provide them; there are few, if any, services designed to reintegrate the victim into the larger community; psychological trauma goes untreated.

In some countries, when these inadequately cared for landmine survivors have felt even more victimized by the society for which they often feel they have made the supreme sacrifice with part of their body, deep resentment and schisms within the society are created. In some cases, amputees have organized to fight back and demand the care that they feel is their right.

Beyond the question of how a society deals (or does not deal) with the immediate and life-long needs of the landmine victim, lies the issue of how a society deals with a countryside contaminated with mines. Not only must the society attempt to respond to the needs of casualties, it must consider reconstruction of the society as a whole in countrysides that are safe for no one.

Refugees and displaced have often fled large areas of their countries to escape armed conflict. Once these conflicts are over, return must be planned. All too often, the areas depopulated by war have also been severely littered by landmines making easy return impossible. Additionally, the mining of large tracts of land critical to sustaining a returning population can threaten them with at the very least malnutrition, and sometimes starvation, because land is simply not available for their crops and animals.

Often roads that would lead the refugees home are mined. Those who survive that journey must attempt to live in areas strewn with minefields, usually unmapped and often unmarked. How can community life be rebuilt when fields, wells, grazing land are full of mines? When the problem is multiplied beyond one community to many, how can the society respond? Peace-building and reconstruction has to concern itself with reintegrating returned refugees and displaced into the larger socio-economic life of the country. If they are cut off from productive life by minefields, it can only have a larger effect on the society as a whole.

If mines cannot be cleared from fields, pastures and villages nor from roads, railroads, and bridges, the complete economic reconstruction of the country is imperilled. Landmine contamination thus limits the recovery from war and prolongs the suffering caused by conflicts whose battles are now long silent. If a country cannot have complete access to its territory to rebuild, enough strain can be put on the society to threaten real peace.

*****

The information, "facts," which follow do not pretend to be exhaustive – nor even necessarily definitive. In fact, the often conflicting and contradictory information available regarding the scope of the problem posed by landmines only further affirms the need for detailed investigation into the problem.

The information outlined below is intended to give a picture of the nature and extent of the problem caused by widespread landmine contamination. This picture graphically illustrates the social consequences of the widespread, indiscriminate use of landmines in conflicts that have taken placed primarily in the developing world.

The information is taken from a number of sources, including Africa Watch, Asia Watch, International Committee of the Red Cross, Middle East Watch, Mines Advisory Group, United Nations reports and news articles and discussions.

Some Basic Statistics on the Effects of the Use of Landmines

Globally

It is estimated that there are more than 100 million uncleared mines in the world; CMS, a Florida-based manufacturer of mines and mine clearance equipment, says 120 million; British and US militaries estimate up to 200 million.

Countries and regions around the world are affected by landmines. This list, which is not exhaustive, includes: Afghanistan, Angola, Armenia, Azerbaijan, Burma, Cambodia, Colombia, El Salvador, Ethiopia, the Falklands-Malvinas, Guatemala, the Golan Heights, Honduras, Iran, Iraq, Kurdistan, Kuwait, Laos, Mozambique, Nicaragua, Rwanda, Somalia, Sudan, Uganda, the former Yugoslavia, Vietnam . . .

Out of 14,221 war-wounded treated in ICRC hospitals in Asia between January 1991 and June 1992, 23% were injured by anti-personnel mines; out of 3,262 mine victims, 21% were women and children.

In 1991, 7,876 amputees – 26% women and children – were fitted by ICRC with prostheses.

As only a small fraction of all mine victims actually reach treatment sites, it must be assumed that the total number injured is much higher and that many die of their wounds.

For every mine victim surviving a blast, two die. In some countries, approximately 75% of survivors require amputation.

Mine clearance and disposal requires massive effort and resources. It is also hazardous: on an average, one disposal expert is killed and two injured for every 5,000 mines cleared.

Mine clearance is a lengthy process. The experience from Second World War shows that even in Western Europe, where resources were made available for mine and munitions clearance, decades later the affected countries were still engaged in clearance. In those countries affected where clearance was not a priority concern, as in northern Africa, landmines are as much a problem today as they were at the close of the war.

Afghanistan

There are approximately 10 million mines in Afghanistan. One-fifth of the one million deaths in the Afghan civil war were caused by landmines; 400,000 have been wounded by them.

One district suffered a 1.95% population loss in two years (1989–90) due to landmines; 3.5% suffered landmine injuries.

With the return of refugees to Afghanistan after the change of government in April of 1991, there has been a dramatic increase in the number of mine victims. Between April and September of 1992, 1,400 mine victims were treated in ICRC hospitals – and increase of 130% over figures for the same six-month period in 1989, 1990 and 1991.

In Afghanistan, over 70% of landmine injuries required amputations, compared with only 18% from other fragmentation weapons and only 2% from firearm injuries.

In late 1989, UNOCHA launched a mine-clearance program. By the end of 1992, only 68 square kilometres of land had been cleared, with 19 mine-removal staff killed and 62 others wounded.

By the end of 1992, there were 31 mine clearance teams of approximately 32 members each, and 16 four-man survey teams operating in Afghanistan.

It is estimated that it will take 15 years to clear just the priority zones.

At the current rate of removal, it is estimated that it would take 4,300 years to clear mines manually from just 20% of Afghan territory.

Over twenty types of mines have been identified in Afghanistan. Mines were produced in the former USSR and Eastern Bloc, Italy, UK, Pakistan and the US, among others.

Angola

Estimates regarding the number of landmines in Angola vary; but some of the mines date back to the independence struggle against Portugal. A British officer in charge of the British Military Mission in Angola estimates that there are 20 million mines covering one-third of Angolan territory.

The senior Angolan army officer for mine clearance states that 50,000 mines were cleared in a twelve-month period beginning May 1991. He cautioned that hundreds of thousands remained and estimated that it could take twenty years to clear them.

Of the 330,000 mines sown in Cuito Cuanvale municipality, the location of only 80,000 is known.

Thousands of acres of land are unusable due to landmines; for example, Mavinga valley, a fertile area in south-east Angola, is largely abandoned due to landmines.

Landmines restrict movement of returning refugees. For example, in Lunda Sul province, 30 major or strategic bridges are down, along with 58 secondary bridges. Most are surrounded by minefields which must be cleared before bridges can be repaired.

Despite the scope of the problem posed by landmines, there has been no systematic assessment of the problem nor serious attempts at systematic, coordinated mine eradication.

It is estimated that after 15 years of war, there are 20,000 amputees that are victims of mine blasts (one in every 470 people). By some accounts, 20,000 is a conservative estimate.

At least 5,000 prostheses will be required each year for the foreseeable future – a number far exceeding those currently manufactured.

The thirty-seven known types of landmines used in Angola were produced in Belgium, China, Italy, South Africa, US, former USSR, Czechoslovakia, and Germany. There may be others.

Bosnia/Croatia

Reports indicate civilian deaths in Croatia due to landmines now stand at 24/week (as of the end of 1992).

It is said that more than 2 children a day are being admitted to hospitals with major injuries from anti-personnel mines.

Croatians estimate 200,000 mines laid in their country alone; there are probably at least 800,000 mines in the former Yugoslavia. As with most other conflict situations, the minefields are not usually mapped.

Cambodia

The 4.5 million landmines in Cambodia are the legacy of more than 24 years of war. The use of landmines in Cambodia dates back to the Vietnam War.

Recent figures indicate a cost of more than $1,000 per mine for mine clearance. Cambodia, with some 4.5 million uncleared mines, has a population of 9 million and a gross domestic product (GDP) of $136 per person annually. Thus, mine clearance would cost for Cambodia, the equivalent of over 3 years its entire GDP.

According to US AID, 6,000 mines were found in a 1-kilometre stretch of road and 3,800 in another 2-kilometre stretch.

A mine survey was carried out in early 1992 at a site for resettlement of refugees. It estimated there were 1,200 Soviet anti-personnel mines, 1,800 stake mines and up to 5,000 unspecified types at the site. By April, mine clearers had removed 80 mines.

Despite the signed peace accord, mines continue to be laid in Cambodia. It is reported that more mines are now being laid than are being cleared (March 1993).

Landmines have injured, and possibly killed, more combatants and non-combatants alike than any other weapon in Cambodia's civil war.

There were 6,000 amputations due to landmine injuries in 1990 alone; there are some 30,000 amputees inside Cambodia, with an additional 5,000–6,000 in camps on the Thai border – this translates into one out of every 236 Cambodian amputees due to landmines compared to one in 22,000 US amputees due to traumatic injury in 1989.

Along with malaria and tuberculosis, landmines ranks as one of the three most serious public health hazards.

According to a United Nations official, landmines in Cambodia are one of the century's worst man-made environmental disasters.

Landmines deployed in Cambodia originated from the United States, China, France, Russia, Vietnam, Czechoslovakia, Singapore, Thailand and Italy.

Falklands-Malvinas

Mine clearance was begun immediately after the end of the war. Even though the war had lasted only two months and had been waged between conventional armies, mine field clearance was found to be so difficult that it was abandoned.

Iraqi Kurdistan

Millions of unrecorded landmines were sown in Iraqi Kurdistan which remain uncleared after the Iran–Iraq War and subsequently, the Gulf War.

For the four months prior to the start of the Gulf War precipitated by the invasion of Kuwait, Iraqi military forces used eight divisions to mine border areas with Syria and Turkey. One section commander contended that each division consisted of 15 mine-laying sections. His section would lay between 4,000–5,000 mines a day. This would result in a number of mines sown each day during that period of between 480,000–600,000. No maps were made of these minefields.

Sulaymanyah City Hospital treated over 1,650 mine injuries between March and September 1991; including 397 traumatic or surgical amputations, despite the fact that the hospital was looted of most of its equipment and supplies upon the withdrawal of the Iraqi army after the March 1991 Kurdish uprising.

A Red Crescent tent field hospital opened in May 1991 treated 200 mine victims in its first four months. Its chief surgeon estimated the total number of casualties to be nearer 500, including those who died without assistance and patients who went directly to Sulaymanyah Hospital. Approximately 20% of the victims were under fifteen.

Approximately 10% of the mine victims seen by the Red Crescent field hospital died due to shock and injuries to vital organs.

Between March and July 1991, Rawanduz Hospital in Erbil (which had previously served as a dispensary) received an average of four mine casualties a day. The majority of patients were men, but the hospital also treated 30 children under 15 and 15 women during that period.

Of 1,190 patients assessed by Handicap International (HI; France) by 1 June 1992, 55.2% were handicapped due to landmines. HI's workshop produced 423 prostheses between mid-November and the end of May 1992.

Prior to the Iraqi pullout from Kurdistan in 1991, three Iraqi army teams were working on mine clearance. Foreign military observers noted the serious lack of equipment and skills of these teams, with some officers unable to read maps.

In June 1992, Mines Advisory Group (UK) began training programs in Iraqi Kurdistan. The original objective of the program was to work with 160 Kurdish trainees. Turkish impoundment of the necessary mine-clearing equipment for the program resulted in adaptations and rescheduling of training with the result of 100 minefield operators having been trained by the end of 1992.

Mines were supplied to Iraq by the shipload. One Italian company was convicted of illegally shipping 9 million mines to Iraq from 1982–85. Other mines were produced in France, Germany, the former USSR, Jordan, Singapore, China and the US.

Kuwait

After the Gulf War, for purposes of demining, Kuwait divided itself into seven sectors and contracted mine clearance with Bangladesh, Egypt, France, Pakistan,

Turkey, the US and the UK. The US sector contained an estimated 500,000 mines over an area of 3,126 sq. km. The mine clearance contract awarded to one company was for $134 million.

At least 84 foreign and local experts in mine clearance were killed by mines in less than ten months after the war.

Mozambique

It is estimated that the number of mines in Mozambique could approach 2 million. At the beginning of 1993, when a French Médecins sans Frontières (MSF) doctor was wounded in a landmine explosion, MSF operations were frozen in the region. In a separate incident, a Renamo representative along with a member of Oxfam UK were killed by a landmine.

There are an estimated 8,000 amputees (one in every 1,862 people).

The United Nations has developed a comprehensive proposal for demining in Mozambique. Possible implementation of the plan has been stalled pending approval for the overall budget for the United Nations' peacekeeping operation for the country (as of March 1993).

Nicaragua

In February 1992, the Nicaraguan Foreign Minister said it would be easier to remove the estimated 600,000 mines from the Persian Gulf than to clear the mine fields in the mountainous terrain of Nicaragua.

The estimated 160,000 mines are distributed over some 800 sites, about 170 of which are unknown.

According to one official count, one mine exists for every 30 people in Nicaragua.

Somalia

There are hundreds of thousands of landmines in northern Somalia; most are anti-personnel devices left by Siad Barre forces.

Barre's forces deliberately mined wells and grazing lands to kill and terrorize nomadic herders viewed as supporters of opposition forces. Another mine warfare tactic used by the same forces was the deliberate mining of civilian homes.

Somaliland's Minister of Health estimated that before the end of the war in February 1991, two-thirds of mine-injured were military and one-third civilian. Since the end of the war, more than 90% have been civilian.

Landmine injuries in northern Somalia peaked between November 1990 and April 1991, with Hargeisa Hospital admitting two–three landmine casualties per day. The first five months of 1991 saw a surge in the number of casualties with the spontaneous return to Somalia of refugees from camps in Ethiopia.

Data from Hargeisa Hospital indicate 74.6% of mine-injured treated between February 1991–February 1992 were children between 5 and 15.

It is estimated that there are between 1,500 and 2,000 amputees in a total population of physically disabled of 9,000 in northern Somalia.

In early 1991, the Somali National Movement began demining in northern Somalia under the leadership of a soldier with demolition and demining training. The clearance force included 60 men; 40% were killed or injured by mines in the first six months of the year.

In August 1991, Rimfire, a UK-based organization, began operations to train and supervise demining personnel and defuse bombs and other ordinance. Rimfire has trained about 200 Somalis in basic mine detection and disposal techniques. In February 1992, Rimfire began training a second group of 220.

Landmines found in northern Somalia were produced in the former USSR, Italy, Pakistan, Egypt, China, and the US.

Vietnam

There are an estimated 60,000 amputees from the Vietnam War, in a population of 75 million (one in every 1,250).

One United Nations study states that many millions of mines and munitions remain hidden throughout Vietnam – and other countries affected during the Vietnam War.

According to a UNHCR official, in May 1992 nearly three decades after the start of the Vietnam War, land free from mines "is in very short supply."

* * * * *

While much of the emphasis on the problem posed by landmines is relatively recent, the problem itself is not a new one. Mines were developed and have been used since World War I. It is of note that where the commitment for mine clearance and the resources for it have been available, as in Western Europe after the Second World War, mine clearance began immediately with the end of the war and has continued as long as necessary.

In northern Africa, where millions of mines were laid during the war, the situation today is much more reflective of the contamination by mines seen in the other countries described above.

Second World War

Libya

According to a study by the United Nations, about 27% of the total arable land of the country is covered by minefields; a larger area of the arable land, 68%, is suspected of containing mines.

Much of its costal strip is severely hampered in its social and economic development because of Second World War mines.

By 1983, mines and munitions had killed about 4,000 and injured more than 8,000; in both categories most were children. Additionally, during the same period 460 clearance personnel had been killed and 650 injured. In a five-year period (1979–83), 30 to 40 were still being killed each year and 50 to 80 injured.

Since the Second World War, 125,000 domestic animals including camels, sheep, goats, and cattle have been killed.

France

By 1983, France still required some 90 mine clearance specialists working full time.

Poland

During the Second World War, approximately 80% of the country was mined. At the end of the war, the Soviet Union provided 17,000 minefield maps; maps of German minefields were captured and used in eradication.

Poland, with the concerted efforts of 10,000 military engineers, cleared 10 million mines in 1945 alone.

By 1977, some additional 15 million mines (and 74 million pieces of other ordinance) had been cleared from Poland. Nearly 4,000 civilians had been killed and 9,000 injured and 30 to 40 were still being killed each year.

Presentation and discussion

Mrs Williams said that she had recently had an interview with the representative of a mine manufacturing company in Washington whose only concern with regard to the moratorium on mine export imposed by the United States Congress was that it would cost his firm between 300 and 500 million dollars.

That man, an Army Reserve officer, had absolutely no qualms about the moral aspect of his company's activities, on the grounds that the so-called enemy was also designing weapons to kill and maim American soldiers.

Yet it was not the enemy that usually fell victim to landmines, but innocent civilians, mostly in areas where there were no ambulances for evacuating casualties, no hospitals capable of dealing with amputations and no prostheses; in Cambodia alone, over 30,000 people had been maimed by mines.

The countries concerned also lacked demining infrastructure, and their economies suffered from the fact that huge areas of land were rendered unusable for cultivation and animal grazing. The mine manufacturer in question had been impervious to her arguments, and she wished that she could conduct some producers on a tour of the war zones of Central America to show them some of the effects of mine-laying operations.

In response to questions and comments, **Mrs Williams** traced the background of the moratorium of one year which the United States Congress had imposed on mine exports and which would expire on 23 October 1993; efforts were being made to extend the moratorium.

Producers wanted self-destruct mines to be exempted from the ban, but some within the United States Defense Department were not inclined to grant any exemptions. The Landmines Campaign initiated about a year previously had now been joined by NGOs in Australia, New Zealand and many of the Western European countries, and an NGO conference was to be held to devise a strategy on how to ban the use and production of and trade in landmines.

It was pointed out that the European Parliament had adopted a resolution in December 1992 recommending all member States to impose a five-year moratorium on landmine trade.

ITEM (3)

THE PROLIFERATION OF ANTI-PERSONNEL LANDMINES IN DEVELOPING COUNTRIES: CONSIDERABLE DAMAGE IN HUMAN TERMS AND A DRAMATICALLY INSUFFICIENT MEDICO-SOCIAL RESPONSE

Report presented by: Dr Philippe Chabasse
Handicap International

Cambodia: 8 million inhabitants – 36,000 amputees.
Afghanistan: 10 to 12 million inhabitants – 20,000 to 30,000 amputees.
Angola: 12 to 15 million inhabitants – 20,000 amputees.

Such figures could also be lined up in the same manner for Laos, Sri Lanka, Iraqi Kurdistan, Nicaragua, Somalia, Mozambique and elsewhere.

Although these figures do not have the scientific value of our western statistics, the cultures of countries in the South bear the full weight of their meaning. They are rough estimates: terrifying rough estimates.

How and why has this happened? Are there more specific figures? What are and what should be the humanitarian, medical and social responses to aid the victims of this plague? These are the main questions that this paper is attempting to answer.

1. The size of the problem

The indiscriminate use of anti-personnel landmines is a relatively recent phenomenon. Following the dropping of American "bombies" on Laos up to 1973, it was the Soviet intervention in Afghanistan in the early 1980s that signalled the escalation of the problem.

Initially diverted in this way from their original purpose by a modern army, the low cost of anti-personnel landmines enabled them to be used massively from then on by regular armies and by guerrilla forces to hinder or channel the movements of civilian populations.

Such massive use does not however totally explain the enormous damage caused in human terms. Sadly, the specific nature of this weapon is also that it remains active well after the end of the conflict. Laid over wide tracts of agricultural land, the danger of anti-personnel landmines increases as demographic growth forces the population to cultivate larger areas.

Without claiming to produce precise statistics, it is possible to illustrate the size of the problem by using commonly accepted estimates.

In Afghanistan, even before intensified repatriation of the refugees brought with it a growth in the number of mine accidents, the UNOCA (the United Nations humanitarian aid coordination in Afghanistan) reported that 2 million people, that is one person per family, were handicapped.

In 1989 and 1990, in the district of Spin Boldak in Kandahar province, deaths caused by the explosion of mines represented 2% of the total population; those mutilated represented 3.5% (source: ICRC). A recent ICRC report, quoted by the HCR, states that in some areas of Afghanistan, 80% of mine victims die before arrival at hospital.

In Cambodia, at least 36,000 people have lost one or more limbs. Today, an average of 120 new amputees arrive every month in the existing reception centres (source: Handicap International).

In Angola, there are a recorded 15,000 amputees. However, the true figure is deemed to be higher than 20,000 (source: Human Rights Watch).

To complete this chapter, it should be noted that in September 1992, the Secretary-General of the United Nations set up a working group composed of representatives of the organization's main agencies to prepare a report on the subject. Up to now, the World Health Organization (WHO) has no usable information on the problem.

2. Some points of analysis

Because of their activities on behalf of mine victims, the NGOs and the ICRC would therefore seem to be the only organizations to have the basic information to make an initial analysis of the phenomenon.

The information that follows is not based on epidemiological research but compiled from 10 years of work in prosthetics and rehabilitation for physically handicapped populations by the volunteers of Handicap International.

Nine of the most significant countries or areas have been chosen. The common selection criterion is the percentage of amputees in relation to the number of physically handicapped patients who have visited the prosthetics and reeducation units set up by Handicap International. This information has been supplemented where possible. One can consider that, in the countries concerned, the number of amputees is a low estimate of the overall number of serious mine injuries. (Amputations for other reasons cannot compensate for the number of mine victims who do not manage to reach a medical unit.)

This percentage, which is confirmed in the field, enables us to estimate total figures in relation to the number of physically handicapped people with regard to the overall population.

The WHO considers that 10% of a typical population is handicapped. These are however figures concerning western situations where all types of handicap are registered. More often than not, light handicaps among poor populations in the developing world will go unnoticed and the people concerned will not complain. When the handicap is severe, it rapidly leads to death.

The experience of Handicap International in this context points to figures roughly half that size: between 1.5% and 2% suffer from physical invalidity, 3% to 4% from mental handicaps and between 1% and 2% from varying degrees of sensorial handicap.

This figure of 1.5% to 2% physically handicapped in proportion to the overall population allows an estimation of the total number of amputees in the following countries.

2.1 Cambodia–Thailand Border

Period in question: 1991
Areas concerned: 7 refugee camps
No percentage available, but in absolute terms:
– 1,300 new amputees admitted over the year.
– Total prosthetics production for 1980–1991: 10,929.

2.2 Cambodia

Period in question: 1992
Areas concerned: Kampot, Takeo, Kompong Thorn, Kompong Cham, Siem Reap, Pursat
– Total prosthetics production for 1992: 1,325.

2.3 Laos

Period in question: 1992
Areas concerned: Xieng Khouan, Kham Mouan, Savanakmet, Cham Passak
8% of amputees for 1,200 orthopaedic patients, that is an estimated 1.6 per 1,000 of overall population.

2.4 Southern Angola (UNITA statistics)

Period in question: 1988–1990
Area concerned: south-east
14% of amputees for 11,800 orthopaedic patients, that is an estimated 2.8 per 1,000 of overall population.

2.5 Afghanistan (Pakistan)

Period in question: 1985–1992
Areas concerned: Quetta workshop for Afghan refugees
9.4% of amputees for 18,000 orthopaedic patients, that is an estimated 1.9 per 1,000 of overall population.

2.6 Mozambique

Period in question: 1991
Areas concerned: Vilanculos, Inhambane, Chicuque, Nampula, Tete.
4.5% of amputees for 3,500 orthopaedic patients, that is an estimated 0.9 per 1,000 of overall population.

2.7 Djibouti

Period in question: 1991
Areas concerned: Pelletier Hospital (Djibouti) and Ethiopian refugee camp.

7.5% of amputees for 930 orthopaedic patients, that is an estimated 1.5 per 1,000 of overall population.

2.8 Iraqi Kurdistan

A Handicap International census took place in February 1992 in Suleimaniya.
Of a total of 1,130 amputees: 55% landmine victims
18% shell victims
9% bullet victims
of which: 10% women of whom 50% under 20 years old
90% men of whom 20% under 20 years old.

2.9 Ethiopia

Period in question: April–June 1991
Areas concerned: Hartisheck refugee camp.
9% of amputees for 400 orthopaedic patients, that is an estimated 1.8 per 1,000 of overall population.

If one omits southern Angola, for which the figures used by UNITA may not be representative, the other elements are quite coherent. Thus, in the countries that are particularly affected, the percentage of amputees is between 1.5 and 2 per 1,000 of overall population.

3. The stages of survival

Behind these figures lie everyday human and individual realities. The chain of events around the accident are dramatically similar.

The three first-hand accounts that follow were chosen at random among Cambodian victims.[8] They illustrate the main stages of and conditions for the survival and/or the quality of the future life of the patient.

At around 3 o'clock on 6 April 1991, ***Ken Kop****, 42 years of age and mother of 7 children, was walking as she always did along the path by the river bank to go to work in the paddy fields when she stepped on a mine. Her brother carried her in his arms to their house where other relatives helped him to make a stretcher out of a hammock and bamboo poles. They took her to the district hospital. Her right leg was so badly mutilated that the duty doctor immediately amputated above the knee. The operation took place without anaesthetic. Late that evening, she arrived at Battambang hospital where the wound was closed under general anaesthetic.*

Six-year old ***Chok Chuon*** *lost her left leg on a landmine while playing near the railway on the morning of 6 April 1991. Her mother heard the explosion and took her daughter to the house. According to Chok Chuon's mother, there was no warning sign. A relative applied a tourniquet to the little girl's left leg and carried her on a* *stretcher to the highway, 15 km away, where they found a taxi. She arrived at Battambang hospital at 2.30 in the afternoon. Two hours later, she went into surgery where her leg was amputated above the knee.*

Lach Pem, *55 years old, and his wife are "displaced Cambodians" who arrived at Site 2 border camp in March 1991. They are peasants from the district of Moung in the south of Battambang province. In 1987, Lach Pem had stepped on a mine while collecting firewood in the forest. Five friends carried him on a stretcher to Moung district hospital. On foot, it took them 20 hours. After amputation, he suffered a serious post-operative infection and remained in hospital for three months. Over this period, he spent 15,000 riels – almost $150 – for drugs. When he left the hospital, he bought crutches and went back to work in the paddy-fields.*

3.1 The place of accident and the human environment

When the victim is alone, which is often the case for peasants or herders, the chances of survival are minimal. In some parts of the world (Kurdistan for example) they are buried on the spot: such victims are never recorded in any statistics.

When a relative is immediately or rapidly on the spot, survival depends on that person's knowledge of first aid. In this field, there is need for improvement in education and training.

3.2 Transport to a medical unit

The quality of infrastructures and community solidarity are major factors here.

3.3 The medical unit

Distance to medical units, medico-surgical skills, the quality of instruments and the cost of health care are directly linked to the level of development of the health system in the country concerned.

3.4 Hospitalization and early functional reeducation

The presence of reeducation professionals and also the cost and therefore duration of such services are conditions for successful follow-up.

3.5 Orthopaedic care and functional rehabilitation

The same limits apply as for the previous point.

3.6 Psycho-social follow-up and professional rehabilitation are the two final stages.

4. Technical responses

It is therefore clear that an effective health-care policy for the victims of anti-pesonnel landmines should integrate these six stages which cover in effect a large part of the health system of a country.

A detailed study of the medico-social systems of the countries concerned is at present an impossible task. In the more specific field of care for the physically handicapped, various sources estimate that, all countries included, the existing means cover not more than 15% to 20% of the needs. (This estimate is not contradicted by the rehabilitation department of the WHO.)

We will here simply set out a list of the actions carried out in the field of prosthetics by NGOs that we know of.

4.1 Cambodia–Thailand Border

Six prosthetics workshops and 1 rehabilitation centre managed by Handicap International up to the end of 1992.

4.2 Laos–Thailand Border

Two prosthetics workshops managed by Handicap International.

4.3 Laos

4 prosthetics workshops set up by Handicap International, 1 public prosthetics and reeducation centre in Vientiane, 1 project in process by the NGO World Vision.

4.4 Cambodia

1 components production unit in Phnom Penh managed by the ICRC

1 reference centre for prosthetics and reeducation in Phnom Penh – Handicap International and the American Friends Service Committee.

6 provincial prosthetics workshops managed by Handicap International

1 workshop managed by the Vietnam Veterans of America Foundation (USA)

1 workshop managed by The Cambodia Trust (GB)

1 physiotherapy school managed by Handicap International

1 workshop managed by the American Red Cross.

4.5 Afghanistan

3 prosthetics workshops in Kabul, Mazar-i-Sharif and Herat managed by the ICRC

2 workshops in the Provinces of Badakhshan and Peshawar managed by the Sandy Gall Appeal (GB)

1 training centre for technicians in Peshawar managed by GTZ (Germany)

1 prosthetics, reeducation and training unit in Quetta and 4 prosthetics workshops in the South and South-East of Afghanistan managed or supervised by Handicap International.

1 public prosthetics centre in Mashad in Iran.

4.6 Iraqi Kurdistan

1 Handicap International prosthetics workshop in Suleimaniya

1 prosthetics workshop in Dyana (Erbil Province) and 1 project in El Habja managed by Voluntary Relief Doctors (Germany)

4.7 Sri Lanka

2 prosthetics workshops in Columbo and Jaffna managed by The Jaipur Foot Project.

4.8 Mozambique

3 ICRC prosthetics workshops in Maputo, Beira and Nampula

6 Handicap International workshops with associated training and professional rehabilitation projects in the main provinces.

4.9 Angola (information prior to the resumption of fighting)

1 public prosthetics unit in Luanda

ICRC prosthetics workshops in government zones

4 workshops managed by UNITA in the south-east after training by Handicap International up to 1987.

Although this list is certainly incomplete, it is significant and suggests two main conclusions:

- Artificial limbs for amputees who are victims of anti-personnel landmines are practically exclusively supplied by non-governmental structures. No country particularly affected by anti-personnel mines has, to our knowledge, developed a significant public programme in this field.
- The projects that exist are particularly centred around the production of orthopaedic aids. A wider vision incorporating prosthetics in the overall process mentioned above in point 3 is clearly needed.

5. Social responses

One very widespread idea is that of the acceptance of the handicap and of the handicapped person within the community in developing countries. It is true that this level of acceptance is much higher than in western countries, however, a number of factors reveal difficulties of integration and suggest that it is necessary to tone down this capacity for social acceptance.

A social study was conducted on the subject with a sample of the handicapped population of the town of Inhambane in Mozambique by Handicap International in May 1991.

Albeit confirming the ease of direct integration in a strongly community-based society, the study concludes that upsets are to be expected in the socio-cultural system caused by the acceleration of emigration towards urban centres.

A large proportion of handicapped people indeed take refuge within their family which remains the most effective system of social protection.

However, although it is still the strongest pillar of the structure, the family has less and less the financial and production capacity to support an entire group.

In an urban environment, faced with a consumer society, the handicapped person ceases to be a "member of the family" and becomes a "dependant on the family" just as old people and children. What the reaction of the group will be and how the society will deal with this rejection remains to be seen.

Education and training are thus the only responses to this evolution. By enabling handicapped people to gain and use other skills useful to the community, the vicious circle of the handicap can be broken.

Yet it is still necessary to encourage the integration of the handicapped into the traditional education and training system and at all costs avoid their being "pushed to the fringe" into a specialized system, the limits of which are evident today in the West.

6. Recommendations

6.1 Improvement of the initial care for patients

- Development of Mines Awareness Programmes together with training in first aid.
- Support for and equipment of medical first-aid centres.

6.2 Improvement of the medico-surgical environment

- Surgical training

- Anaesthetics techniques
- Instruments

6.3 Improvement of early assistance in functional reeducation

- Introduction of basic ideas of functional reeducation in nursing and medical courses.
- Training of reeducation professionals (physiotherapists) who should then be incorporated into surgical teams.

6.4 Increase in the production of orthopaedic aids

Three main concepts should guide action in this field:

- Cost-free health care and prosthetics is an illusion in the context. It is therefore necessary to reduce costs to render them accessible to direct payment by the patient by standardizing production and using materials available on site and by using local know-how.
- Prosthetics is a field that can only be analysed over the long term: needs in terms of maintenance, readjustment and replacement. Using standardized techniques and local materials is part of the response to this question. It would also be useful to decentralize the prosthetics structures in order to bring them closer to where the amputees live. This seems the only way to ensure proper follow-up.
- The need to design prosthetics within an overall context of support and assistance to the handicapped person should go hand in hand with adapted training structures for prosthetists. It is necessary to make these future technicians capable of ensuring the whole chain from the basic evaluation of the patient to getting him to walk again, that is to say not simply being limited to the construction of an artificial limb.

6.5 The question of social and professional rehabilitation

- Taking into account the psycho-social dimension of the handicap and its effect on the balance of the family cell as a whole.
- Taking into account the specific nature of the rehabilitation of veterans.
- Developing and following up micro-projects for rehabilitation where the amputees live.
- Developing the basic and professional training of handicapped people while avoiding the creation of reserved employment or activities structures.

7. Conclusion

There is an enormous difference between the needs, the complexity of the medico-social measures to be taken and the technical and financial responses of the countries concerned.

Determination is absolutely vital in this field to respond both to immediate needs and to integrate unavoidable budgetary constraints.

Two points need to be developed in order to do this:

- the development and incorporation of integrated medico-social projects (prosthetics, physical, social and professional rehabilitation) into public structures. This should be based on the use of locally available material and human resources, on enlargement of capacity and proximity to the beneficiaries.
- the training of local personnel based on a global approach to the handicapped person and able to incorporate the medical, technical and social aspects of rehabilitation.

Presentation and discussion

Dr Chabasse said that his organization's involvement in the campaign against the proliferation of anti-personnel landmines had originated in two periods of revolt.

In 1979, young doctors who had tried to help the amputee landmine victims who had been brought daily into Cambodian refugee camps had refused to accept the passive argument that nothing could be done in the absence of specialists, sophisticated equipment and a stable situation, and had decided to set up an organization specializing in helping handicapped people in poverty situations: Handicap International was now working in 26 countries, using simple and appropriate technology for rapid intervention and training of local technicians.

After ten years of hard work by over 120 volunteers, nearly 15,000 amputees were walking again, but there were now some 30,000 amputees in Cambodia, and the figures were growing at an increased rate in nearly half the countries where Handicap International was operating; that was the reason for the organization's second revolt, when it had decided to broaden its mandate and attack the source of the problem, in three main directions:

- First, in the field, participating by agreement with the Mines Advisory Group (MAG) in facilitating MAG activities, especially in Cambodia and Iraqi Kurdistan and soon in Mozambique, and also helping with the administration of programmes for training Khmer demining staff;

- Second, circulating information to decision-makers and the general public – arranging for the translation into French of the book entitled *The Coward's War*, by agreement with Physicians for Human Rights and Asia Watch, the publication of a book called *Hidden Death* about Iraqi Kurdistan and a campaign based on a petition, co-authored by Handicap International, Human Rights Watch and MAG, which had already obtained some 25,000 signatures;
- And third, concerted pressure was brought to bear by a number of NGOs on the United Nations and concerned governments to assume their responsibilities in the matter – for example, the French Government had been persuaded to propose the convening of an international conference to review the 1980 Convention on the Use of Certain Conventional Weapons.

In response to questions and comments, **Dr Chabasse** said that Handicap International had begun to collect data in the field and had been working for a year on the provision of more accurate statistics.

ITEM (7)

THE MINES PROTOCOL: NEGOTIATING HISTORY

Report presented by: Brigadier A.P.V. Rogers[9]
Brigadier Legal
British Army of the Rhine

Introduction

This paper tries to explain the negotiating history of the Mines Protocol to the United Nations Convention on Conventional Weapons of 1981. It is a review in narrative form of that history. It is not an attempt to comment on or analyse the current text of the Mines Protocol. This has been done by the author elsewhere.[10] At Annex A there is a detailed examination of the evolution of the text starting with the proposals originally tabled at the Lugano Conference of government experts in 1976.

Background

Landmines and booby-traps are defensive weapons employed to hinder enemy troop movements or to provide a screen of protection for a military position. There are also other devices of a more offensive nature, which are laid manually but which are set to explode at a certain time or are set off by remote control. The latter are often used in ambushes. Mines and booby-traps are dissimilar to other weapons because they are not aimed at a specific target. Mines are set off by persons stepping on them, vehicles running over them or by persons or vehicles being detected by a sensor. A booby-trap might be activated where, for example, a door is opened or a person falls into a trap. There was concern that because these weapons are not aimed at the target they are less discriminate than other weapons. They can be set off by civilians. Unless they have a self-neutralizing mechanism, they can still be dangerous long after they cease to serve any useful military purpose and even after an armed conflict has come to an end. Unless careful records are kept, it is very difficult later to locate and destroy these weapons.

The other devices referred to also pose special dangers for the civilian population.[11]

Early negotiations

Until 1981 there was no treaty dealing specifically with landmines, booby-traps and other devices.[12] For convenience these weapons will be collectively referred to as "delayed-action weapons." The Diplomatic Conference which negotiated Protocols Additional I and II of 1977 to the Geneva Conventions of 1949 set up an Ad Hoc Committee to study possible prohibitions or restrictions on conventional weapons, including delayed-action weapons. But no provisions on specific weapons were included in the final text of the Protocols.[13]

It was left to the United Nations to sponsor a formal conference on conventional weapons. This made use of discussions and proposals made at two conferences of government experts organized by the International Committee of the Red Cross (ICRC) at Lucerne in 1974 and Lugano in 1976. There were two preparatory sessions of the United Nations conference in 1978 and 1979 followed by two full sessions in 1979 and 1980. These culminated in a Convention on Conventional Weapons opened for signature in New York on 10 April 1981.[14] The convention has three protocols including one on "prohibitions or restrictions on the use of mines, booby-traps and other devices." For ease of reference the conference will be called the Weapons Conference and the convention the Weapons Convention.

Delegates at the Lucerne conference started off by considering the legal criteria that should apply to the use of conventional weapons. The three criteria most frequently discussed were contained in a United Kingdom paper and were as follows.[15]

a. Unnecessary suffering or superfluous injury: a balance between humanitarian considerations and military necessity. This required trying to ascertain the minimum force necessary to incapacitate an enemy combatant.

b. Discriminateness: the accuracy of a weapon in hitting the intended target and in reducing incidental loss of life and damage to civilians.

c. Treachery or perfidy: means of warfare that might be regarded as abhorrent because of its perfidious nature.

All of these considerations were relevant to the question of how delayed-action weapons and their use might be controlled.

The conference considered various different devices: time-fused weapons designed to go off after a set interval, very often in the form of cluster bomblets to hamper airfield runway repairs or to suppress anti-aircraft fire,[16] emplaced land-mines to create barriers to enemy troop movements; mines delivered remotely by rocket, artillery or aircraft which could be laid whenever or wherever the tactical situation demanded; booby-traps which performed a similar function to mines in slowing and hampering an enemy's movement. Mines were divided into two groups: anti-vehicle mines and anti-personnel mines. Discussion on control of these weapons concentrated mainly on the need to record the location of mines and booby-traps to enable clearance and on the use of self-neutralizing mechanisms to ensure that mines did not remain a danger once they had served their military purpose. There was a fair measure of agreement that there should be a prohibition on weapons that were perfidious by nature. That would include those that were made to look like inoffensive objects such as children's toys or which abused humanitarian rules, for example the booby-trapping of a dead soldier.[17]

Following the Lucerne Conference, the Ad Hoc Committee considered the subject of mines again and a working paper on the marking of remotely emplaced minefields was introduced.[18] There was also discussion about booby-traps and of making self-neutralizing devices mandatory and general agreement that the matter should be taken forward at the second conference of government experts at Lugano.[19]

At the Lugano Conference in 1976, France, the Netherlands and the United Kingdom put forward draft proposals[20] on the use of mines and booby-traps which formed the basis for all future discussions of the subject until the conclusion of the Weapons Conference. The main proposals were: that so far as feasible the location of minefields containing more than 20 mines should be recorded and that records of minefields in territory under the control of an adverse party should be published at the end of active hostilities; that mines delivered at a range of over 2,000 metres or from aircraft (remotely delivered mines) should only be used if either they were fitted with a neutralizing mechanism or were delivered into a marked area; that the use of mines and booby-traps in populated areas should be restricted; and that the use of certain booby-traps should be prohibited.[21] The text of these proposals will be found in Annex A.

Another proposal considered was a prohibition on the laying of anti-personnel mines by aircraft.[22] The Swiss experts thought that it was misguided to draw up a list of prohibited booby-traps and submitted a suggestion that the use for explosive booby-traps of objects in general use among the civilian population should be prohibited.[23] The French-Anglo-Dutch proposals formed the basis for discussion in a working sub-group of military experts.[24] Matters discussed included the feasibility of recording minefields, whether more than 20 mines was the right threshold for the recording obligation, and whether the proposal for publication of information after the end of hostilities was satisfactory. There was general consensus that, as proposed by New Zealand, the recording of *pre-planned* minefields should be mandatory, and this requirement, which obviated the need to specify any number of mines, was retained unchanged to the end of the Weapons Conference. The term "pre-planned" was never defined, though it was discussed at various stages by interested delegations. As for remotely delivered mines, some experts thought there should be time limits within which the self-neutralizing mechanism should operate, others considered that self-neutralizing mechanisms should be mandatory and one felt that the range of 2,000 metres in the definition of remotely delivered mines needed further examination. An idea by the experts of Mexico and Switzerland to prohibit time-fused weapons exploding more than 24 hours after delivery[25] was advanced but not exhaustively discussed. The experts of the Philippines suggested that references to non-explosive booby-traps should be deleted.[26] The definition of booby-trap was also considered.[27]

The report of the Lugano Conference was taken by the Ad Hoc Committee at their third session in 1976 and the Committee continued the work started at Lugano by examining the following proposals.

a. By Mexico, Switzerland and Yugoslavia[28] to restrict the use of booby-traps to military objectives. A similar proposal was made by Venezuela.[29]

b. By Mexico and Switzerland[30] to prohibit the laying of mines in areas where combat is not taking place or imminent unless laid close to military objectives and precautions are taken to protect civilians. This proposal also required the recording of "methodically laid" minefields and contained a prohibition on remotely delivered mines unless fitted with self-neutralizing mechanisms.

c. By the United Kingdom[31] and later co-sponsored by France, the Netherlands and Denmark. This was a modified version of the Lugano proposals. In addition, it suggested the prohibition of booby-traps specifically designed and constructed to look like inoffensive objects such as fountain pens and cameras.

The main discussions concerned the range at which delivery was regarded as remote (2,000 metres in the UK proposal); the need to fix a time limit for self-neutralizing mechanisms; the possibility of restricting the use of remotely delivered mines to areas in which combat was taking place or imminent, and the possibility of harmonizing the various proposals.[32]

The fourth session of the Ad Hoc Committee took place in 1977. The Mexican delegation put forward a proposal,[33] co-sponsored by Austria, Sweden, Switzerland, Uruguay and Yugoslavia, which harmonized many of the proposals previously advanced. An informal group of the sponsors of the two main sets of proposals then produced a consolidated text of those of the United Kingdom and co-sponsors and of Mexico and co-sponsors.[34] The consolidated text was used as a basis for further discussion.

Although there was a fair measure of agreement, some areas of disagreement remained and the texts in question were put in square brackets for later discussion. The main contentious issues were as follows. First, the definition of mine:

> "an [explosive or incendiary] munition placed under, on or near the ground or other surface area and designed to be detonated or exploded by the direct action, presence or proximity of a person or vehicle".

Should it include the words "explosive or incendiary munitions"? Some delegations thought it better to deal with incendiary mines in context of mines rather than in the context of incendiary weapons. One delegation thought the words "direct action"[35] in the definition of mines should be deleted since they would have the effect of including devices detonated by remote control which were regulated elsewhere. Secondly, because of the difficulty of rendering the term "booby-trap" in other languages, it was referred to as an "explosive and non-explosive device" and defined. Thirdly, the definition of remotely delivered mine:

> "any mine delivered by artillery, rocket, mortar or similar means at a range of over 1,000 metres or dropped from an aircraft".

The reason for mentioning a range was to exclude mines emplaced by short-range launchers[36] where the location of the mines so laid could be recorded as accurately as those laid manually.

There was general agreement that minefield records should be released for mine-clearing purposes after the cessation of active hostilities[37] though one delegation thought the obligation to make them public was too vague.

There was the most disagreement about the use of remotely delivered mines. One delegation thought that they should be banned altogether. Doubts were expressed about the reliability of self-neutralizing mechanisms and about the feasibility of marking remotely delivered minefields. The Mexican proposal only permitted the use of remotely delivered mines within the combat zone but references to the "combat zone" were dropped when it was pointed out that mines could be used against objectives in rear areas such as airfields. It was also understood that mining the path of a tank column was permitted.[38] It was left that remotely delivered mines could only be used in "an area containing military objectives".

One draft article in the consolidated text concerned the use of mines in populated areas. There was disagreement about whether this article should be applied to remotely delivered anti-personnel mines. The article also required "effective protections" to be taken to protect civilians if mines were laid at a time when combat was not taking place or imminent, but it was not clear to some delegates what these precautions should be.

While there was general agreement on the provisions relating to explosive and non-explosive devices (booby-traps), some delegations were unhappy about the specific inclusion of primitive devices which stab or impale the victim. Others wanted food and drink to be excluded from the list of items that were not permitted to be booby-trapped on the basis that it might be necessary to booby-trap military supplies of food and drink.[39]

The Preparatory Conference

By the time of the first preparatory conference in 1978 of the Weapons Conference, considerable work on mines had been done. There was available the ICRC Report of 1973, the reports of the Lucerne and Lugano conferences, the four reports of the Ad Hoc Committee and a reasonably well-developed text which formed the basis for discussion at the preparatory conference.[40]

By the end of the first preparatory conference various changes had been made. The term "booby-trap" reappeared to replace "explosive and non-explosive devices." The area within which remotely delivered mines were permitted to be used was extended to "an area which is itself a military objective or which contains military objectives." This enabled the mining by remote means of an area of land of tactical importance to deny its use to the enemy. Remotely delivered anti-personnel mines were deleted from the article dealing with restrictions on the use of mines and booby-traps in populated areas. The booby-trapping of food and drink was only to be permitted if it were done in military establishments, military locations and military supply depots.

Work continued unabated at the second preparatory conference in 1979. Further protection for the civilian population from the effects of remotely delivered mines was afforded by the introduction of a requirement to give effective advance warning of the delivery of such mines. The United Kingdom sought the qualification "unless circumstances do not permit", since it might not be possible to give warnings if remotely delivered mines were used in advance of the forward edge of the battle area, but this was objected to by the Syrian delegation who were reluctant to countenance the delivery of remotely delivered mines in "an area which is itself a military objective". The United Kingdom responded that it was sometimes necessary to divert an enemy away from an area containing military objectives.

Because of doubts about whether "effective" precautions could be taken to protect civilians from the effect of mines laid in populated areas owing to accidents or unforeseeable incidents, an alternative of "all feasible" precautions was proposed. On the insistence of Vietnam, India and other delegations specific references to primitive booby-traps which crush, strangle or impale were deleted. The passage was replaced by a more general prohibition on booby-traps designed to cause superfluous injury or unnecessary suffering. The list of items that were not allowed to be booby-trapped was expanded by the inclusion, suggested by Israel and supported by the Netherlands, of kitchen utensils and appliances, as suggested by Bulgaria, Morocco and Italy of cultural property and, as suggested by Mongolia, of animals and their carcasses. The Moroccan delegation expressed concern about improving the protection of children from the dangers of booby-traps.

There were two significant changes at the second preparatory conference. First, a new draft article, proposed by Sweden, applicable to mines, booby-traps and other devices, was designed to afford general protection to the civilian population by[41] prohibiting the use of these weapons against civilians and prohibiting their indiscriminate use. This draft was later expanded by the chairman of the working group to require all feasible precautions to be taken to protect civilians from the effects of delayed-action weapons. This article was intended to provide some protection for the civilian population from the use of anti-personnel remotely delivered mines in populated areas. Secondly, a much more comprehensive article was introduced on the making available of minefield records to the adverse party and the Secretary-General of the United Nations upon the cessation of active hostilities and on protecting United Nations forces and missions. The Libyan and Moroccan delegations were keen to see this article impose an obligation on the mine-laying party to provide technical and material assistance in removing mines and booby-traps. A sub-group was set up to deal with this issue.

The Weapons Conference

By the time the Weapons Conference got under way in September 1979, the delegates had the report of the chairman of the working group,[42] with annexed draft articles for a treaty on the regulation of the use of landmines and other devices.[43]

The main areas where agreement had not been reached were:

a. Whether mines delivered at short ranges or from aircraft should be subject to the controls applicable to remotely delivered mines.

b. What was meant by "feasible precautions" in the article requiring all feasible precautions to be taken to protect civilians from the effects of delayed-action weapons.

c. The precise obligation on the mine-laying party in assisting in the later clearance of mines.

d. Whether the delivery of remotely delivered mines into an area containing military objectives should be permitted.

e. Whether remotely delivered mines should be completely banned.

f. Whether a party should be allowed not to give effective advance warning of the delivery of remotely delivered mines which might affect the civilian population if circumstances did not permit.

g. Whether "effective" or "all feasible" precautions should be taken to protect civilians from the effects of delayed-action weapons emplaced away from military objectives at a time when combat is neither taking place nor imminent.

h. Whether kitchen utensils and appliances and animals and their carcasses should be exempt from booby-trapping.

First Session

Numerous proposals were submitted by the Moroccan delegation including definitions, special measures for the protection of children and technical details concerning the recording of minefields.[44] The Moroccan proposals on the protection of children were discussed with great sympathy in the mines working group on 25 September 1979 but it proved difficult to devise any special rules beyond those already in the draft protocol for the protection of the civilian population. It had been agreed as early as the Lugano conference that the booby-trapping of children's toys should be prohibited and it was also proposed that there should be a prohibition on the use of pre-manufactured booby-traps in the shape of attractive portable objects. However, it was agreed to extend the list of items exempt from booby-trapping to include objects used for the feeding, health, hygiene, clothing or education of children.

The United Kingdom delegation took the lead in trying to resolve the problem mentioned at point a above. The object was to try to accommodate nations who had systems which laid mines at short ranges or from helicopters where the location of the mines laid could be accurately recorded but whose mines were not fitted with self-neutralizing mechanisms.

Since this is one provision of the mines protocol that is not capable of sensible interpretation without recourse to the negotiating history, it is worth setting out the history more fully.

The simple solution to permit helicopter laying of such mines was to change "aircraft" to "aeroplane" in the definition of remotely delivered mines. However, this would have subjected helicopter-delivered mines which were fitted with self-neutralizing mechanisms to the additional controls applicable to the use of delayed-action weapons in populated areas. A solution was accordingly sought in the article dealing with remotely delivered mines.

During the first week of the conference the Italian and United Kingdom delegations worked out a formula which permitted, as an alternative to fitting self-neutralizing mechanisms, the delivery of remotely delivered mines from helicopters into a predetermined area which was recorded and marked.

It was ascertained from the Soviet delegation that they had also been considering tabling a proposal to enable mines without self-neutralizing mechanisms to be delivered from helicopters.

Early discussions with various delegations including those from India, Indonesia, Pakistan and Nigeria indicated that there were no real objections in principle to remotely delivered mines provided sufficient protection for the civilian population could be guaranteed.

Various different formulae were then discussed on a quadrilateral basis (Italy, USA, USSR, UK) before a paper was tabled in the Working Group. The actual text proposed[45] was as follows:

> "The use of remotely delivered mines is prohibited unless . . ., and unless:
>
> a) [self-neutralizing mechanisms] or,
>
> b) they are delivered from a helicopter for the purposes of laying a pre-planned minefield the location of which is recorded in accordance with Article 3(1)(a) above."

In the Working Group, the Mexican delegate expressed concern that the location of mines laid by helicopter in mountainous or wooded terrain could not be accurately recorded. The Italian delegate stressed that the wording did not permit the dropping of mines from helicopters in all circumstances and that one must act consistently with the recording requirement. The delegate of Argentina suggested that if the location of a mine could be accurately recorded, it should not be treated as remotely delivered. The Mexican delegate then conceded that it might be possible to reach an understanding. The Israeli and Moroccan delegates expressed some scepticism. The Pakistani delegate stressed that helicopter mine laying was accurate, was normally confined to roads, tracks and gaps in mountains and should not be seen as something sinister.

The Mexican delegate in an effort to reach a solution suggested the following wording.

> "The use of remotely delivered mines is prohibited unless . . ., and unless the location of which can be accurately recorded in accordance with Article 3(1)(a) above or if they are fitted with an effective neutralizing mechanism, . . ."

The delegation from the Soviet Union stressed the importance for humanitarian reasons of accurate recording. The United Kingdom delegate intervened to say that the Mexican proposal was very promising and suggested that "the location of which" should be changed to "their location". The Argentine delegate endorsed the Mexican proposal which embodied their idea and overcame the need to specify a range in the definition of remotely delivered mines. The Italian delegate also found the Mexican proposal acceptable and it was adopted with the slight alteration suggested above and some further drafting amendments.

The reference to Article 3(1)(a)[46] was a clumsy attempt by delegates (including the author!) suffering from the effects of a late night sitting towards the end of a busy conference to ensure that remotely delivered mines without self-neutralizing mechanisms should only be used if their location could be accurately recorded. It did not mean that remotely delivered mines without self-neutralizing mechanisms could only be used for laying pre-planned minefields. It also did not mean that remotely delivered mines without self-neutralizing mechanisms could be used without recording in all cases when such use was not pre-planned. Both of these interpretations might result from a normal construction of the text without reference to the negotiating history, but both would be erroneous as not reflecting the intentions of the conference.

The problem identified at point b above was resolved by adopting the "all feasible" precautions test and defining feasible precautions as "those which are practicable or practically possible taking into account humanitarian and military considerations".

The biggest area of contention was point c above, especially the requirement to disclose minefield records after the cessation of active hostilities, and particularly where part or the whole of a country was under adverse occupation.

The problem listed at point d above was resolved, despite some opposition from Vietnam and Egypt, by retention of the words objected to.

As for point e above, no real pressure emerged for a complete ban on remotely delivered mines.

The point at f above was resolved by allowing the words "unless circumstances do not permit" to remain in the text,[47] but there still remained concern that the civilian population needed greater protection from remotely delivered mines. The delegation of Yugoslavia thought there should be a ban on the indiscriminate use of remotely delivered mines and reserved its position on this article. Since indiscriminate use of mines was banned under another article, this proposal would have been tantamount to deleting the article on remotely delivered mines altogether.

The Soviet Union and the United Kingdom suggested the following form of words to overcome the problem mentioned at point g above:

> "measures are taken to protect civilians from their effects, for example, the posting of warning signs, the posting of sentries, the issue of warnings or the provision of fencing."

The United Kingdom suggested a revised version of the article on booby-traps which with minor modifications was finally adopted. Kitchen utensils and appliances, except in military installations, and animals and their carcasses were included in the list of objects exempt from booby-trapping, thus resolving point h above.

The Secretary-General of the United Nations proposed modifications to the article dealing with the release of minefield records. These were eventually included, with some amendments by a sub-group of nations contributing to United Nations forces,[48] in that article and in a new article dealing specifically with the protection of United Nations forces and missions.

The word "pre-planned" in connection with the mandatory recording requirement came in for some discussion. It was made clear by the United States delegate that the word implied more than a mere conscious act on the part of the mine layer and involved a coordinated plan for the laying of defensive minefields. Remotely delivered mines were unlikely to fall into this category since their use would depend on the tactical situation. Although this view was not challenged, the term was never defined and it was agreed informally between some delegations that it would have to be explained by way of examples in military manuals.

Second Session

At the second session of the conference in September 1980 a lot of work was done by delegations[49] and by the chairman of the working group to resolve the difficulties over the release of minefields records and to complete the work on the technical annex on recording.[50] The final text on the disclosure of minefield records only required complete compliance by a state once complete withdrawal from its territory of opposing forces had taken place. Some delegations were unhappy that this reduced the protection available to the civilian population and pointed out that the parties to the conflict must take whatever measures are open to them to protect civilians wherever they are, using available records for marking minefields or warning the civilian population. They encouraged the provision unilaterally, by mutual agreement or through the Secretary-General of the United Nations of information on the whereabouts[51] of mines and booby-traps.[52] Libya supported by several delegations expressed concern that the final text on cooperation over the clearance of mines at the end of a conflict did not go far enough.

After some attention by the drafting committee, the agreed text of the mines protocol was completed on 10 October 1980[53] and submitted to the United Nations General Assembly on 27 October 1980.[54]

Conclusions[55]

Years of discussion, patient negotiation and readiness to find a compromise solution to differences in an attempt to reconcile the almost irreconcilable concerns to protect humanity yet preserve an important military option resulted in only a small step forward. The law of armed conflict develops at best very slowly, but any advance that will help to save lives or prevent injury, however, modest, is to be welcomed.

Innovations achieved in the mines protocol are as follows:

a. The general rules of Geneva Protocol I of 1977 on the protection of the civilian population, indiscriminate attacks and precautions in attack were applied to delayed-action weapons. A definition of feasible precautions appears for the first time.

b. In populated areas mines (other than remotely delivered mines) and other delayed-action weapons are not to be used if combat is not taking place or imminent unless either they are laid in the close vicinity of a military objective under the control of an adverse party or measures are taken to protect civilians from their effects.

c. Remotely delivered mines are only to be used within an area which is itself a military objective or which contains military objectives and then only if they are fitted with self-neutralizing mechanisms or if their location can be accurately recorded. Warnings are to be given to the civilian population unless circumstances do not permit.

d. Pre-manufactured booby-traps in the form of apparently harmless portable objects are prohibited as are booby-traps attached to or associated with certain listed items.

e. Pre-planned minefields and all areas where there has been large-scale and pre-planned use of booby-traps are to be recorded. The parties to a conflict must endeavour to record the location of all other mines and booby-traps they have put into position. There are detailed rules on the compilation and disclosure of these records.

f. There are special rules on the protection of United Nations forces and missions and on international cooperation on the clearance of mines and booby-traps.

ANNEX A

TEXTS

ARTICLE 1. MATERIAL SCOPE OF APPLICATION

Previous Texts

1. This idea was introduced in the Ad Hoc Committee

   These proposals relate to the use in armed conflict on land of the mines and other devices defined therein. They do not apply to the use of anti-ship mines at sea or in inland waterways, but do apply to mines laid to interdict beaches, waterway crossings or river crossings.

2. 1st Preparatory Conference

   In essence the same: minor amendment consequent upon use of "This Treaty" instead of "These proposals".

3. 2nd Preparatory Conference

   Unchanged.

Final text

This Protocol relates to the use on land of the mines, booby-traps and other devices defined herein, including mines laid to interdict beaches, waterway crossings or river crossings, but does not apply to the use of anti-ship mines at sea or in inland waterways.

ARTICLE 2. DEFINITIONS

"Mine"

Previous Texts

1. Lugano

   "Mine" means an explosive or incendiary munition placed under, on or near the ground or other surface area and designed to be detonated by the presence or proximity of a person or vehicle but does not include an underwater mine;

   "Remotely delivered mine" means any mine delivered by artillery, rocket, mortar or similar means at a range of over 2,000 metres or dropped from an aircraft.

2. *Ad Hoc* Committee

   "Mine" means an [explosive or incendiary] munition placed under, on or near the ground or other surface area and designed to be detonated or exploded by the direct action, presence or proximity of a person or vehicle;

   "Remotely delivered mine" means any mine delivered by artillery, rocket, mortar or similar means at a range of over 1,000 metres or dropped from an aircraft.

3. 1st Preparatory Conference

   "Mine means any munition placed under, on" etc.

   Comment:

   ". . . explosive or incendiary" deleted.

4. 2nd Preparatory Conference

   "Mine" means any munition placed under, on or near the ground or other surface area and designed to be detonated or exploded by the presence, proximity or contact of a person or vehicle;

   Comment:

   ". . . the direct action" deleted, "contact" added. In the definition of RDM[56] the range was specified as [1,000] [2,000] metres" and all references to range were placed in square brackets.

Final Text

"Mine" means any munition placed under, on or near the ground or other surface area and designed to be detonated or exploded by the presence,

proximity or contact of a person or vehicle, and "remotely delivered mine" means any mine so defined delivered by artillery, rocket, mortar or similar means or dropped from an aircraft.

"Booby-trap"

Previous Texts

1. Lugano

   "Booby-trap" means an explosive or non-explosive device or other material deliberately placed to kill or injure when a person disturbs or approaches an apparently harmless object, or performs an apparently safe act.

   Comment:

   The Venezuelan proposal COLU/219 (Lugano report p. 191) "designed to kill after deceiving" was not adopted.

2. Ad Hoc Committee

   "Explosive and non-explosive devices" mean manually emplaced devices which are specifically designed and constructed to kill or injure when a person disturbs or approaches an apparently harmless object or performs an apparently safe act;

   Comment:

   A Swiss proposal (CDDH/IV/209) introduced at the 3rd session of the Ad Hoc Committee and a Venezuelan variant (CDDH/IV/212) were not pursued. The term "explosive and non-explosive devices" replaced "booby-traps" because of difficulties of definition.

3. 1st Preparatory Conference

   "Booby-trap" means a manually emplaced device which is specifically designed and constructed to kill or injure when a person disturbs or approaches an apparently harmless object or performs an apparently safe act;

   Comment:

   "Booby-trap" reinstated.

4. 2nd Preparatory Conference

   "Booby-trap" means any device or material which is designed, constructed or adapted to kill or injure and which functions unexpectedly when a person disturbs or approaches an apparently harmless object or performs an apparently safe act;

Comment:

"... manually emplaced" and "specifically" deleted; "adapted" and "and which functions unexpectedly" added.

Final Text

"Booby-trap" means any device or material which is designed, constructed or adapted to kill or injure and which functions unexpectedly when a person disturbs or approaches an apparently harmless object or performs an apparently safe act.

"Other devices".

Previous Texts

None.

Final Text

"Other devices" means manually emplaced munitions and devices designed to kill, injure or damage and which are actuated by remote control or automatically after a lapse of time.

"Military Objective"

Previous Texts

The text tabled at Lugano which remained unchanged through the Ad Hoc Committee and preparatory conferences was:

> "Military objective" means, so far as objects are concerned, any object which by its own nature, location, purpose or use makes an effective contribution to military action and whose total or partial destruction, capture or neutralization in the circumstances ruling at the time, offers a definite military advantage.

Final Text

"Military objective" means, so far as objects are concerned, any object which by its nature, location, purpose or use makes an effective contribution to military action and whose total or partial destruction, capture or neutralization, in the circumstances ruling at the time, offers a definite military advantage.

Comment:

The word "own" before "nature" was dropped.

"Civilian objects"

Previous Texts

None.

Final Text

"Civilian objects" are all objects which are not military objectives as defined in paragraph 4.

"Recording"

Previous Texts

This proposal was introduced at the first session of the main conference and is based on a Moroccan draft.

Final Text

"Recording" means a physical, administrative and technical operation designed to obtain, for the purpose of registration in the official records, all available information facilitating the location of minefields, mines and booby-traps.

ARTICLE 3. GENERAL RESTRICTIONS ON THE USE OF MINES, BOOBY-TRAPS AND OTHER DEVICES

Previous Texts

This was introduced at the second preparatory conference as follows:

1. This Article applies to a) mines, b) booby-traps and c) manually emplaced munitions and devices designed to kill, injure or damage and which are actuated by remote control or automatically after a lapse of time.
2. It is prohibited in any circumstances (including reprisals) to direct weapons to which this Article applies against the civilian population as such or against individual civilians.
3. [the same as the final text].
4. All feasible precautions shall be taken to protect civilians from the effects of weapons to which this Article applies. "Feasible precautions" are those which are practicable or practically possible [taking into account all circumstances ruling at the time, including those relevant to the success of military operations and the need to minimize incidental loss of civilian life, injury to civilians and damage to civilian objects] [taking into account military and humanitarian considerations].

Final Text

1. **This Article applies to:**

   a) mines;
   b) booby-traps; and
   c) other devices.

2. **It is prohibited in all circumstances to direct weapons to which this Article applies, either in offence, defence or by way of reprisals, against the civilian population as such or against individual civilians.**

3. **The indiscriminate use of weapons to which this Article applies is prohibited. Indiscriminate use is any placement of such weapons:**

   a) which is not on, or directed against, a military objective; or
   b) which employs a method or means of delivery which cannot be directed at a specific military objective; or
   c) which may be expected to cause incidental loss of civilian life, injury to civilians, damage to civilian objects, or a combination thereof, which would be excessive in relation to the concrete and direct military advantage anticipated.

4. **All feasible precautions shall be taken to protect civilians from the effects of weapons to which this Article applies. Feasible precautions are those which are practicable or practically possible taking into account all circumstances ruling at the time, including humanitarian and military considerations.**

ARTICLE 4. RESTRICTIONS ON THE USE OF MINES OTHER THAN REMOTELY DELIVERED MINES, BOOBY-TRAPS AND OTHER DEVICES IN POPULATED AREAS

Previous Texts

1. Lugano

1. This proposal applies to:

   a. mines (other than remotely delivered mines);

   b. booby-traps; and

   c. all other manually emplaced munitions designed to kill, injure or damage and for that purpose to detonate automatically after a lapse of time or to be remotely detonated.

2. In any city, town, village or other area containing a concentration of civilians in which combat between ground forces is not taking place or is not yet imminent. Devices to which this Proposal applies may not be used unless either:

   a. they are placed on or in the close vicinity of a military objective; or

   b. due precautions are taken to protect civilians from their effects.

1A Revised following Lugano

1. This Article applies to mines (other than remotely delivered mines) and all other manually emplaced devices (explosive and non-explosive) which are designed to kill, injure or damage and for that purpose to be actuated:

   a. By the pressure or proximity of a person or vehicles;

   b. When a person disturbs or approaches an apparently harmless object or performs an apparently safe act;

   c. By remote control; or

   d. Automatically after a lapse of time.

2. In any city or town, village or other area containing a concentration of civilians in which combat between ground forces is not taking place or does not appear to be imminent, devices to which this Article applies may not be used unless either:

   a. They are placed on or in the close vicinity of a military objective belonging to, or under the control of, an adversary; or

   b. Due precautions are taken to protect civilians from their effects.

Comment:

This is a re-draft of a version prepared by the sub-committee at Lugano (report p. 150). Paragraph 1 has been changed from manually emplaced mines to all mines other than RDMs.

Paragraph 2 now includes "belonging to, or under the control of, an adversary" after "military objective." A bare reference to military objectives would include one's own as well as those of an enemy. "Location" means any place and the laying of mines there would make it such objective. The amendment makes it clear that when mines are laid on territory under one's one control, due precautions must be taken to protect civilians. This does not inhibit military operations since it only applies in populated areas in which combat is not taking place or imminent and it would obviously be necessary to take precautions to ensure that mines are not detonated by friendly troops or civilians.

2. Ad Hoc Committee

1) This proposal applies to mines (other than remotely delivered [anti-tank] mines), explosive and non-explosive devices, and other manually emplaced munitions and devices designed to kill, injure or damage and which are actuated by remote control or automatically after a lapse of time.

2) It is prohibited to use any object to which this proposal applies in any city, town, village or other area containing a similar concentration of civilians in which combat between ground forces is not taking place or does not appear to be imminent, unless either:

   a) They are placed on or in the close vicinity of a military objective belonging to or under the control of an adverse Party; or

   b) Effective precautions are taken to protect civilians from their effects.

Comment:

Most of the discussion revolved round the word "anti-tank", since some delegates wanted all RDMs controlled by this Article, others were prepared to compromise by having anti-personnel RDMs controlled, yet others thought that there were sufficient controls in the RDM Article. Some wanted to retain the option to use anti-personnel RDMs mainly to prevent hand-lifting of anti-tank mines, although the provision had to be construed in the light of "in which combat is not taking place or does not appear to be imminent".

3. 1st Preparatory Conference

1) This Article applies to mines (other than remotely delivered mines), booby-traps, and other manually emplaced munitions and devices designed to kill, injure or damage and which are actuated by remote control or automatically after a lapse of time...

4. 2nd Preparatory Conference

1) as per 1st Preparatory Conference; 2) and 2)a) as per Ad Hoc Committee.

2)b) [effective] [all feasible] precautions are taken to protect civilians from their effects.

Comment:

Some favoured "all feasible" but this added nothing to the protection required by the Indiscriminate Use Article. Most delegates preferred "effective". Although the meaning of this term was understood it was difficult to define so as to exclude genuine accidents and unforeseeable incidents. At the second preparatory conference it was understood that "effective precautions" were those measures that can, at the time they are taken, objectively be expected to be effective. An accident, for example, resulting from a change of circumstances which could not be foreseen at the time the measures were taken and which resulted in the precautions becoming less effective could not of itself be taken to imply that effective precautions had not been taken. On the other hand "effective" implies more than "all feasible" since it might not, in some situations, be feasible to take any precautions at all.

Final Text

1. **This Article applies to:**

   a) **mines other than remotely delivered mines;**
   b) **booby-traps; and**
   c) **other devices.**

2. **It is prohibited to use weapons to which this Article applies in any city, town, village or other area containing a similar concentration of civilians in which combat between ground forces is not taking place or does not appear to be imminent, unless either:**

   a) **they are placed on or in the close vicinity of a military objective belonging to or under the control of an adverse party; or**

   b) **measures are taken to protect civilians from their effects, for example, the posting of warning signs, the posting of sentries, the issue of warnings or the provision of fences.**

ARTICLE 5. RESTRICTION ON THE USE OF REMOTELY DELIVERED MINES

Previous Texts

1. Lugano

   The use of remotely delivered mines is forbidden unless either each such mine is fitted with a neutralizing mechanism or the area in which they are delivered is marked in some distinctive manner.

   "Neutralizing mechanism" means a self-actuating or remotely controlled mechanism which *is* designed to render a mine harmless or cause it to destroy itself when it is anticipated that the mine will no longer serve the military purpose for which is was placed in position;

2. Ad Hoc Committee

   The use of remotely delivered mines is prohibited unless:

a) Each such mine is fitted with an effective neutralizing mechanism, that is to say a self-actuating or remotely controlled mechanism which is designed to render a mine harmless or cause it to destroy itself when it is anticipated that the mine will no longer serve the military purpose for which it was placed in position; or

b) The area in which they are delivered is marked in some definite manner in order to warn the civilian population,

and, in either case, they are only used within an area containing military objectives.

Comment:

Delegates had difficulties with time limits for neutralization and with the concept of remote marking. In any event use of RDMs is restricted to areas containing military objectives. This was chosen in preference to "Combat area" in CDDH/IV/222 so as to include airfields and other rear area targets. Some delegates felt that mining the path of a tank column was clearly covered (Ad Hoc Committee, 4th session, report, p. 27).

3. 1st Preparatory Conference

"... and, in either case, they are only used within an area which is itself a military objective or which contains military objectives."

4. 2nd Preparatory Conference

1) The use of remotely delivered mines is prohibited unless a) each such mine is fitted with an effective neutralizing mechanism, that is to say a self-actuating or remotely controlled mechanism which is designed to render a mine harmless or cause it to destroy itself when it is anticipated that the mine will no longer serve the military purpose for which it was placed in position, and b) such mines are only used within an area which is itself a military objective [or which contains military objectives].

2) Effective advance warning shall be given of any delivery or dropping of remotely delivered mines which may affect the civilian population [unless circumstances do not permit].

Comment:

Both sets of square brackets were insisted upon by Syria. The UK argued that both sets of square bracketed language should be retained since:

a. mines may be used to divert an enemy away from an area containing military objectives;

b. it may not be possible to give a warning if RDMs are used in advance of the forward edge of the battle area.

Final Text

1. **The use of remotely delivered mines is prohibited unless such mines are only used within an area which is itself a military objective or which contains military objectives, and unless:**

a) **their location can be accurately recorded in accordance with Article 7(1)(a); or**

b) **an effective neutralizing mechanism is used on each such mine, that is to say, a self-actuating mechanism which is designed to render a mine harmless or cause it to destroy itself when it is anticipated that the mine will no longer serve the military purpose for which it was placed in position, or a remotely controlled mechanism which is designed to render harmless or destroy a mine when the mine no longer serves the military purpose for which it was placed in position.**

2. **Effective advance warning shall be given of any delivery or dropping of remotely delivered mines which may affect the civilian population, unless circumstances do not permit.**

ARTICLE 6. PROHIBITION ON THE USE OF CERTAIN BOOBY-TRAPS

Previous Texts

1. Lugano

It is forbidden in any circumstances to use any booby-trap or similar device which:

a. is designed to kill or injure by a non-explosive means which stabs, impales, crushes, strangles, infects or poisons the victim; or

b. is in any way attached to or associated with:

(i) internationally recognized protective emblems, signs or signals;
(ii) sick, wounded, or dead persons;
(iii) burial sites or graves;
(iv) medical facilities, medical equipment, medical supplies and medical transport;
(v) children's toys.

1A. Revised following Lugano

1. It is forbidden in any circumstances to use any apparently harmless portable object (other than an item of military equipment or supplies) which is

specifically designed and constructed to contain explosive material and to detonate when it is disturbed or approached.

2. It is forbidden in any circumstances to use any explosive or non-explosive device or other material which is deliberately placed to kill or injure when a person disturbs or approaches an apparently harmless object or performs an apparently safe act and which is in any way attached to or associated with:

   a. Internationally recognized protective emblems, signs or signals;
   b. Sick, wounded or dead persons;
   c. Burial or cremation sites or graves;
   d. Medical facilities, medical equipment, medical supplies or medical transport; or
   e. Children's toys.

3. It is forbidden in any circumstances to use any non-explosive device or any material which is designed to kill or cause serious injury by stabbing, impaling, crushing, strangling, infecting or poisoning the victim and which functions when a person disturbs or approaches an apparently harmless object or performs an apparently safe act.

Comment:

This is a revision of the version agreed by the sub-group at Lugano (report p. 150). Paragraph 3 has been re-drafted to apply only to non-explosive devices designed to kill or cause serious injury. The previous draft covered any injury. The amendment covered objections that the previous draft related to non-explosive devices causing only minor injury whereas it was thought that it should only cover devices which are probably unlawful under Hague Regulations of 1907, Art. 23(e).

2. Ad Hoc Committee

1) It is prohibited in any circumstances to use:

   a) Any apparently harmless portable object which is specifically designed and constructed to contain explosive material and to detonate when it is disturbed or approached; [or

   b) Any non-explosive device or any material which is designed to kill or cause serious injury in circumstances involving superfluous injury or unnecessary suffering, for example by stabbing, impaling, crushing, strangling, infecting or poisoning the victim and which functions when a person disturbs or approaches an apparently harmless object or performs an apparently safe act].

2) It is prohibited in any circumstances to use explosive and non-explosive devices which are in any way attached to or associated with:

   a) Internationally recognized protective emblems, signs or signals;
   b) Sick, wounded or dead persons;
   c) Burial or cremation sites or graves;
   d) Medical facilities, medical equipment, medical supplies or medical transport;
   e) Children's toys;
   [f) Food and drink;] or
   g) Objects clearly of a religious nature.

Comment:

1)a) includes remotely delivered objects of that type.

2)f) some sponsors wanted to retain the right to booby-trap food and drink, at least if military supplies.

3. 1st Preparatory Conference

The square brackets were removed in paragraph 1b). "Booby-Traps" substituted for "explosive and non-explosive devices." Paragraph 2f) amended to read:

"f) Food and drink (except in military establishments, military locations and military supply depots);"

4. 2nd Preparatory Conference

1) It is prohibited in any circumstances to use:

   a) any booby-trap in the form of an apparently harmless portable object which is specifically designed and constructed to contain explosive material and to detonate when it is disturbed or approached, or

   b) any booby-trap that is designed to cause superfluous injury or unnecessary suffering.

2) It is prohibited in any circumstances to use booby-traps which are in any way attached to or associated with:

   a) Internationally recognized protective emblems, signs or signals;
   b) Sick, wounded or dead persons;
   c) Burial or cremation sites or graves;
   d) Medical facilities, medical equipment, medical supplies or medical transport;
   e) Children's toys;

f) Food and drink [kitchen utensils and appliances] (except in military establishments, military locations and military supply depots);
g) Objects clearly of a religious nature;
h) Historic monuments, works of art or places of worship which constitute the cultural or spiritual heritage of peoples;
[i) animals and their carcasses].

Final Text

1. Without prejudice to the rules of international law applicable in armed conflict relating to treachery and perfidy, it is prohibited in all circumstances to use:

a) any booby-trap in the form of an apparently harmless portable object which is specifically designed and constructed to contain explosive material and to detonate when it is disturbed or approached, or

b) booby-traps which are in any way attached to or associated with:

(i) internationally recognized protective emblems, signs or signals;
(ii) sick, wounded or dead persons;
(iii) burial or cremation sites or graves;
(iv) medical facilities, medical equipment, medical supplies or medical transportation;
(v) children's toys and other portable objects or products specially designed for the feeding, health, hygiene, clothing or education of children;
(vi) food or drink;
(vii) kitchen utensils or appliances except in military establishments, military locations or military supply depots;
(viii) objects clearly of a religious nature;
(ix) historic monuments, works of art or places of worship which constitute the cultural or spiritual heritage of peoples;
(x) animals or their carcasses.

2. It is prohibited in all circumstances to use any booby-trap which is designed to cause superfluous injury or unnecessary suffering.

ARTICLE 7. RECORDING AND PUBLICATION OF THE LOCATION OF MINEFIELDS, MINES AND BOOBY-TRAPS

Previous Texts

1. Lugano

So far as is feasible, the location of all minefields containing more than 20 mines shall be recorded. Such records shall be retained until after the cessation of active hostilities, at which time the location of all recorded minefields situated in territory controlled by an adversary party shall be made public.

Comment:

Some experts thought "so far as feasible" was too vague. Our counter-proposal worked out by the military sub-group was:

"The location of pre-planned defensive minefields shall always be recorded. So far as is feasible, the location of all other minefields containing more than 20 mines shall be recorded. Such records shall be retained . . ."

The main problem was the reference to 20 mines. This figure was found unacceptable to some delegations who wanted to reduce it (to e.g., 2). Consideration was given to removing the recording obligation where mines are laid in combat or in the immediate path of an advancing adversary. This rather vague concept would be clarified as part of the negotiating history.

2. Ad Hoc Committee

1) The parties to a conflict shall record the location of:

a) All pre-planned minefields laid by them; and

b) All areas in which they have made large-scale and pre-planned use of explosive or non-explosive devices.

2) The parties shall endeavour to ensure the recording of the location of all other minefields, mines and explosive and non-explosive devices which they have laid or placed in position.

3) All such records shall be retained by the parties and the location of all recorded minefields, mines and explosive or non-explosive devices remaining in territory controlled by an adverse party shall be made public after the cessation of active hostilities.

3. 1st Preparatory Conference

Unchanged except for the substitution of "booby-traps" for "explosive" and "non-explosive devices."

4. 2nd Preparatory Conference

1) and 2) same as final text.

3)a) All such records shall be retained by the parties, who shall:

(i) make available to each adverse party and to the Secretary-General of the United Nations as soon as possible after the cessation of active hostilities all information in their possession concerning the location of minefields, mines and booby-traps in the territory of such adverse party other than territory under the occupation or control of their own forces or allied forces;

(ii) whenever after the cessation of active hostilities their own forces or allied forces withdraw from the whole, or any part, of the territory of an adverse party which those forces had occupied or controlled, as soon as possible make available to such adverse party and to the Secretary-General of the United Nations all information in their possession concerning the location of minefields, mines and booby-traps in the area from which those forces had withdrawn; and

(iii) whenever it is possible to do so, having regard to their legitimate defence interests, make public after the cessation of active hostilities information concerning the location of minefields, mines and booby-mines in any parts of their own territory occupied or controlled by the forces of an adverse party.

5. First Session of Main Conference

3) All such records shall be retained by the parties, who shall:

a) as soon as possible after the cessation of active hostilities make available to each adverse party and to the Secretary-General of the United Nations all information in their possession concerning the location of minefields, mines and booby-traps in the territory of such adverse party [other than territory under the control of their own forces or allied forces]; and

b) as soon as possible, whenever after the cessation of active hostilities their own forces or allied forces withdraw from the whole, or any part, of the territory of any adverse party which those forces had controlled, make available, to such adverse party and to the Secretary-General of the United Nations, all information in their possession concerning the location of minefields, mines and booby-traps in the area from which those forces had withdrawn; and

c) when United Nations forces or missions perform functions in any area or areas, make available to the authority mentioned in Article 3 bis such information as is required by that Article.

Final Text

1. The parties to a conflict shall record the location of:

a) all pre-planned minefields laid by them; and

b) all areas in which they have made large-scale and pre-planned use of booby-traps.

2. The parties shall endeavour to ensure the recording of the location of all other minefields, mines and booby-traps which they have laid or placed in position.

3. All such records shall be retained by the parties who shall:

a) immediately after the cessation of active hostilities:

(i) take all necessary and appropriate measures, including the use of such records, to protect civilians from the effects of minefields, mines and booby-traps; and either

(ii) in cases where the forces of neither party are in the territory of the adverse party, make available to each other and to the Secretary-General of the United Nations all information in their possession concerning the location of minefields, mines and booby-traps in the territory of the adverse party; or

(iii) once complete withdrawal of the forces of the parties from the territory of the adverse party has taken place, make available to the adverse party and to the Secretary-General of the United Nations all information in their possession concerning the location of minefields, mines and booby-traps in the territory of the adverse party;

b) when a United Nations force or mission performs functions in any area, make available to the authority mentioned in Article 8 such information as is required by that Article;

c) whenever possible, by mutual agreement, provide for the release of information concerning the location of minefields, mines and booby-traps, particularly in agreements governing the cessation of hostilities.

ARTICLE 8. PROTECTION OF UNITED NATIONS MISSIONS FROM THE EFFECTS OF MINEFIELDS, MINES AND BOOBY-TRAPS

Previous Texts

The forerunner of the present text was the following extract from the Recording Article in the draft prepared at the 2nd preparatory conference:

(iv) when United Nations forces or missions are established to perform peace-keeping, observation, fact-finding or similar functions in any area, make available to the Secretary-General of the United Nations all information in their possession concerning the location of minefields, mines and booby-traps in that area or, in the case of a small United Nations fact-finding mission on a temporary visit to such an area, take such other measures as may be necessary to protect the mission from the effects of minefields, mines and booby-traps while carrying out its duties.

Final Text

1. When a United Nations force or mission performs functions of peacekeeping, observation or similar functions in any area, each party to the conflict shall, if requested by the head of the United Nations force of mission in that area, as far as it is able:

a) remove or render harmless all mines or booby-traps in the area;

b) take such measures as may be necessary to protect the force or mission from the effects of minefields, mines and booby-traps while carrying out its duties; and

c) make available to the head of the United Nations force or mission in that area, all information in the party's possession concerning the location of minefields, mines and booby-traps in that area.

2. When a United Nations fact-finding mission performs functions in any area, any party to the conflict concerned shall provide protection to that mission except where, because of the size of such mission, it cannot adequately provide such protection. In that case it shall make available to the head of the mission the information in its possession concerning the location of minefields, mines and booby-traps in that area.

ARTICLE 9. INTERNATIONAL CO-OPERATION IN THE REMOVAL OF MINEFIELDS, MINES AND BOOBY-TRAPS

Previous Texts

The following was contained in Article 3(3) of the text agreed by the 2nd preparatory conference:

b) After the cessation of active hostilities, the parties shall endeavour to reach agreement, both among themselves and, where appropriate, with other States and with international organizations upon the provision of information and technical and material assistance [including, in proper circumstances, joint operations] necessary to remove or otherwise render ineffective minefields, mines and booby-traps placed in position during the conflict.

Final Text

After the cessation of active hostilities, the parties shall endeavour to reach agreement, both among themselves and, where appropriate, with other States and with international organizations, on the provision of information and technical and material assistance – including, in appropriate circumstances, joint operations necessary to remove or otherwise render ineffective minefields, mines and booby-traps placed in position during the conflict.

TECHNICAL ANNEX: GUIDELINES ON RECORDING

Previous Texts

Proposal submitted by Morocco

The recording of the location of minefields, mines, booby-traps and other delayed-action devices shall be effected in the following manner:

1) With regard to pre-planned manually emplaced minefields and booby-traps:

   a) Maps and diagrams should be drawn in such a way as to indicate precisely the extent of the minefield and the location, nature, number and disposition of the mines and booby-traps laid;

   b) The mines and booby-traps laid should be described briefly with emphasis on the methods and means to be used for their neutralization or rapid destruction.

2) With regard to remotely delivered minefields, mines and booby-traps:

   a) The topographic or aerial data of the firing or dropping positions and the co-ordinates of the estimated points of fall should be specified;

   b) The nature of the mined ground and the atmospheric conditions prevailing at the time of delivery (wind speed and direction, etc.) should be indicated.

3) With regard to other minefields, mines, booby-traps and other devices laid or placed in position:

   All relevant information, however brief, should be provided so as to enable the nature and type of the mines, booby-traps or other devices used to be determined and the locations or areas in which they were laid to be identified.

Final Text

Whenever an obligation for the recording of the location of minefields, mines and booby-traps arises under the Protocol, the following guidelines shall be taken into account:

1. **With regard to pre-planned minefields and large-scale and pre-planned use of booby-traps:**

   a) **maps, diagrams or other records should be made in such a way as to indicate the extent of the minefield or booby-trapped area; and**

   b) **the location of the minefield or booby-trapped area should be specified by relation to the co-ordinates of a single reference point and by the estimated dimensions of the area containing mines and booby-traps in relation to that single reference point.**

2. **With regard to other minefields, mines and booby-traps laid or placed in position:**

   In so far as possible, the relevant information specified in paragraph 1 above should be recorded so as to enable the areas containing minefields, mines and booby-traps to be identified.

Presentation and discussion

Brigadier Rogers said that the legal approach in negotiating the Mines Protocol had been to seek common ground between the humanitarian and medical view that mines were a terrible weapon and the military view that mines were an effective weapon. Years of discussion had culminated in just a few pages of text. But there had been progress, even if it was almost imperceptible.

Some thought that a macabre calculation had been made of the greater strain placed on enemy resources by maiming rather than killing people outright, but humanitarian law, from the time of the Declaration of St Petersburg, had emphasized the principle of using the minimum force necessary to disable the enemy.

Some considered that mines caused unnecessary suffering and should therefore be banned. The issues raised by mines were the same as those discussed in relation to anti-personnel laser weapons, and it was generally concluded that those who had suffered trauma would probably rather be alive than dead. The outcome of the discussion on whether to seek a ban on mines or to attempt to control their use was, therefore, that the negotiations should focus on control.

Various matters were discussed, including the fact that mines were less discriminate than weapons pointed at a target, that unless equipped with self-destruct

devices mines posed a threat long after the end of a combat, and that it was difficult to find mines in the absence of records.

Suggested responses included the prohibition of weapons of a perfidious nature (such as booby-traps), the mapping and marking of minefields, the publication of records at the end of hostilities, and the use of self-destructing mines. A great many difficulties, however, remained.

Remotely delivered minefields were difficult to mark, and even though agreement had been reached that "pre-planned" minefields should be marked, no one really knew what they were. The question of the time-limits to be set for the self-destruction or self-neutralizing of mines remained open.

All agreed that effective precautions should be taken to protect civilians, but it seemed that little could be done when mines were used in populated areas. Stress was placed on the recording requirements and on ensuring, as for other weapons, that mines were not used indiscriminately or irresponsibly.

In the context of the requirement to give advance warning of attack (under Additional Protocol I), there was a discussion on giving advance warning about the delivery of mines but it was recognized that, under battlefield conditions, a commander could always claim that it had not been feasible to give such a warning.

The only useful responses to calls for the protection of United Nations forces and of children were, respectively, the suggestions to disclose minefield records and to ban booby-traps manufactured to look like attractive portable objects.

To put it briefly, after long and tedious negotiations, there had been some readiness to compromise between the humanitarian and military concerns, and a small step had been taken. The Law of War moved forward very slowly, but if that small step had protected some lives and prevented some injuries, then it had been worthwhile.

The question now was whether attitudes had changed and further progress could be made. He suspected that there had been no major change in attitudes and that, if negotiations were reopened, they would be as slow as the previous negotiations. Various steps could, however, usefully be taken. States could be encouraged to ratify the Mines Protocol.

The UK, for instance, had not yet done so. Efforts could be made to enhance the educational process, as much damage was done through ignorance. Consideration could be given to including educational material with the instructions for use, and approaches made to manufacturers.

The Law of War tended to be treated as a separate subject, whereas it would be much more accessible to soldiers if it were translated into tactical manuals. There should be more training in the Law of War in the less developed countries that used anti-personnel mines.

As regards the Law of War as such, perhaps progress could be made with respect to undetectable mines and, although many countries would say that self-neutralizing and self-destructing mines were too expensive, the subject could be discussed.

Replying to various comments, he said that the end of the Cold War had changed a perceived threat but that, for example in the UK, there was less certainty about how to deploy and arm troops. The Mines Protocol contained a provision prohibiting indiscriminate use; the question was how to ensure that the provision was implemented.

Perhaps it was true, as had been suggested, that the Mines Protocol as a whole was not applied because, in attempting to find common ground between military necessity and humanitarian ideals, it failed to satisfy either.

The Mines Protocol could not cover production and use, since those were outside its scope and were matters for the Disarmament Conference. It would, however, be possible to envisage a total ban under Humanitarian Law, as for chemical weapons, but such a discussion was outside the scope of the review conference.

The problem of how to ensure compliance, in particular by irregular armies, was perhaps insoluble. It was true that the Weapons Convention did not contain verification mechanisms; probably, many other aspects of it were unsatisfactory. But it should be recalled that the Weapons Convention was built on top of the general provisions of Humanitarian Law, which were to be applied.

He pointed out that the text of Article 3, paragraph 3(b), of the Mines Protocol differed from that of Article 51 of Additional Protocol I as the former had been deliberately adapted to cover weapons that were not directed at the target.

He agreed with the ICRC representative that awareness of the social and economic consequences of mine warfare, such as famine and the breakdown of infrastructure, might introduce a new element into the deliberations of the review conference.

ITEM (7)

BASIS OF DISCUSSION ON POSSIBLE FUTURE LEGAL DEVELOPMENTS

Report presented by: Louise Doswald-Beck
ICRC Legal adviser

Possible changes in legal regulation and compliance[57]

1. Changes needed for compliance with the present law:

   1.1. More widespread ratification of the 1980 Convention.

   1.2. Correct training.

   1.3. State interest in the law being correctly implemented.

2. Possible changes to the 1980 Convention:

   2.1. The inclusion of implementation provisions (France has suggested this in its request for a review conference to this Convention).

   2.2. The formal application of the rules in the 1980 Convention to non-international armed conflicts.

   2.3. Strengthening of some of the provisions of Protocol II of the 1980 Convention that are presently weak, e.g. more careful regulation of the duty to register minefields, stricter rules for the removal of minefields.

3. Problems that would remain even with these changes:

   3.1. Some minefields, even if duly registered, are very difficult to remove, e.g., those on the beaches in the Falkland Islands, or those which have moved following heavy rain.

   3.2. Increasing difficulty of detecting mines manufactured almost entirely of plastic.

   3.3. The very widespread use of mines renders their removal long, costly and dangerous.

   3.4. Lawless factions will continue to lay mines indiscriminately.

4. Alternative solutions that have been proposed by various persons

   4.1. Further prohibitions in the use of certain types of mines (i.e. stricter regulation in international humanitarian law), for example:

- all anti-personnel mines;
- scatterable mines;
- those without a neutralizing mechanism;
- undetectable mines;
- those specially manufactured so as to render their neutralization or destruction very difficult.

4.2. The introduction of prohibitions in the manufacture of certain types of mines (i.e., regulation in arms control/disarmament law).

The types of mines listed in 4.1 above could also be subject to prohibitions in manufacture. Any prohibition in the use of certain types of mines would have to be accompanied by a prohibition of their manufacture in order to be properly implemented.

4.3. A ban on the export of anti-personnel mines or other types of mines (e.g., the recent US Congress moratorium on the export of anti-personnel mines).

Presentation and discussion

Mrs Doswald-Beck said that two basic areas of law were applicable. First, international humanitarian law (IHL) applying to the use of weapons – whether they could be used and, if so, how. Secondly, to arms control/disarmament law which regulates the possession, manufacture, transfer and stockpiling of weapons. The international customary law of IHL which applies to all States irrespective of their treaty obligations, which differentiates between military and civilian objects and prohibits the use of weapons which are by their nature indiscriminate, as well as their use in an indiscriminate fashion.

Moreover, international customary law prohibits weapons causing unnecessary suffering. A number of these basic rules were restated in the 1980 Convention, more widespread ratification of which was certainly desirable. In addition, some possible changes in that Convention would be the inclusion of implementation procedures, a formal statement of the applicability of the rules of the Convention to non-international armed conflicts, and strengthening of some of the provisions of Protocol II to the Convention which were weak because of the compromises that had had to be made in 1980.

Other subjects for study were the problems that would remain even if the existing Convention was fully implemented and thus the possibility of further prohibitions of the use of certain types of mines. Such prohibitions would need to be accompanied by a ban on their manufacture. Further, the prohibition of the export of anti-personnel mines should be considered. **It should be understood that the ICRC was concerned only with matters governed by IHL, and that problems of ownership, manufacture and export of mines should be considered in other forums.**

In reply to questions, **Mrs Doswald-Beck** said that there was no obligation for a State which had laid mines to clear them, unless it was so agreed between the parties to the conflict.

Although the use of weapons having indiscriminate effects had been prohibited under IHL, there was a general unwillingness among States to admit that anti-personnel mines were by nature indiscriminate. Various United Nations bodies were beginning to examine the possibility of compensation for violations of human rights, and one could consider the extension of this to civilian victims of anti-personnel mines.

The date of the conference to review the 1980 Convention would be fixed by the Secretary-General of the United Nations; the French Government, which had requested the conference, wished some preparatory meetings to be held.

ESTABLISHMENT OF WORKING GROUPS

The ICRC representative suggested that six working groups should be formed to discuss particular approaches to the problem of anti-personnel mines. Each working group would select its own chairman and rapporteur, and would report back to the plenary meeting on its conclusions. The topics for discussion and the composition of the working groups, as proposed by the representative, were agreed.

Mines meeting – working groups

The purpose of the working groups was not to propose absolute solutions but rather to indicate the advantages and disadvantages of various possibilities and their viability from the point of view of applicability and the extent to which they would solve the practical problems presently created by mines.

1. Rehabilitation

- review of most serious present shortcomings compared with what is needed
- specific proposals for improvement.

2. Demining

- review of the most effective methods
- proposals on improvement on existing methods, if any
- review of existing organization of demining worldwide and specific proposals for improvement.

3. 1980 United Nations Convention

- proposals on how to improve its implementation
- what are the inevitable limits to its correct implementation
- to what extent are the rules of the Convention, even if correctly implemented, inadequate to prevent extensive civilian casualties.

4. Proposals for further humanitarian law rules

- possibility of the prohibition of the use of certain types of mines e.g.,
  - all anti-personnel mines
  - those without a neutralizing mechanism
  - those that are undetectable or otherwise particularly difficult to remove
  - any others
- other rules that are stricter than those in the present Convention, e.g., demining obligations.

This working group concentrated on the military possibilities rather than the purely technical aspects that were dealt with by Working Group 5.

5. Possible restrictions on the method of manufacture of mines

- advantages and disadvantages of different types of neutralizing or self-destruct mechanisms
- to what extent could an arms control treaty specify the type of neutralizing or self-destruct mechanism that should be compulsory
- are there mechanisms that are sufficiently tamperproof to prevent the probability of their being removed or otherwise rendered inoperable
- any other proposals.

6. Possible arms control measures relating to the commerce in mines and their stockage

- measures prohibiting or restricting exports
- destruction of existing stocks that are incompatible with possible new standards on manufacture
- prohibition of the manufacture of certain types of mines in an arms control treaty
- verification measures.

This working group looked at the viability of these arms control measures in order to assess how the subject could be approached in a Disarmament Conference context.

REPORTS OF THE WORKING GROUPS

REPORT OF GROUP 1

REPORT OF THE WORKING GROUP OF MEDICAL PROFESSIONALS

We, as medical professionals, consider the use of anti-personnel mines, which are inherently indiscriminate weapons, abhorrent. While we strongly endorse efforts to limit further the use of anti-personnel mines and, if possible, their ultimate prohibition, we offer the following recommendations to alleviate the suffering they cause.

The recommendations contained in this report fall into two areas: (1) the obvious lack of adequate medical care and rehabilitation for mine victims in most countries; and (2) the lack of objective and systematic information on the effects of mines on the individual and the community as a basis for future action.

RECOMMENDATIONS

1. The international community must recognize that only a small proportion of anti-personnel mine victims, whether combatants or not, receive adequate medical care and rehabilitation. Presently, resources for this care are woefully inadequate. We believe governments should bear the majority of costs for demining and increase their financial support to humanitarian agencies. Moreover, governments found to be responsible for abetting the indiscriminate use of anti-personnel mines, as defined under international law, should be required to make significant contributions to demining operations and the provision of medical care to victims of anti-personnel mines. We also believe governments should help humanitarian efforts by providing military medical personnel – which are otherwise under-utilized – to help care for victims of anti-personnel mines. Resources and personnel should be made available for first aid and evacuation, anaesthesia, surgery, nursing, hospital organization, rehabilitation, and psychological support.

2. There is an urgent need for epidemiological information on anti-personnel mine victims. Equally important is the need to gather objective and systematic information on the short- and long-term social and economic consequences of the use of anti-personnel mines. This should be conducted before the meeting to review the United Nations Convention on Conventional Weapons of 1981.

3. Because anti-personnel mine victims can easily overwhelm existing health care systems, there is a pressing need for reliable information on deficiencies and needs of these facilities. Such information should be gathered in a standardized manner and made available to all relevant institutions.

4. Resources should be provided for restoring and upgrading existing health facilities and training local health personnel in mined areas.

5. The planning of demining operations should include a provision for adequate medical care for personnel who may be injured.

6. Agencies that possess information on the location of anti-personnel mines should make it available to the local population, as well as responsible

demining agencies. In addition, first-aid courses should complement mine-awareness programmes.

Implementation of these recommendations will require adequate security measures in the affected areas.

REPORT OF GROUP 2

DEMINING

DIFFICULTIES TO BE BORNE IN MIND WHEN EMBARKING ON LANDMINE CLEARANCE

Distinction was drawn between "formal" wars and low-level insurgency operations. Mine clearance operations have in the past only tended to take place in the case of the first of these: e.g., Kuwait. Indeed, at times it has to be said it has not always been done even then: e.g., the Falkland Islands.

Most of the countries infested with anti-personnel landmines come into the category of low level (counter-)insurgency conflicts, and while the ordnance originates in many countries worldwide their governments do not feel under any onus to be involved in the clearing of these mines. In brief, where there has been no initial direct military involvement there is no subsequent mine clearance.

The scale of the problem is colossal. In this area accurate figures are as vital as they are difficult to come by. One estimate states that there are in excess of 200 million anti-personnel mines scattered worldwide. It is now suggested that the true figure is closer to 100 million. However, anything up to that number again could be held in stocks. This is not a wild figure, but a sensible guess-estimate. As stocks are constantly deployed it may be that it is not unreasonable to suggest that there are perhaps as many as 200 million anti-personnel landmines in existence, half of which are deployed in the field.

The technical distinction between anti-personnel mines and submunitions is an argument which is not appropriate here, but it is worth bearing in mind such elements as delayed-fuse bombs and munitions malfunctions in the context of this discussion. Note, too, that the greater the degree of electronic sophistication, the greater the likelihood of malfunction. Add to that the very short time it takes to plant a mine as against the time it takes to lift and disarm it, a ratio, given the variables, of 1:100.

Finally, there is the nexus of time, cost and security. A major consideration is that of insurance and along with this the necessity by any humanitarian criteria for a proper medical evacuation facility. It can clearly be seen that in addition to all the obvious hazards, landmine clearance is a slow and costly undertaking.

REVIEW OF THE MOST EFFECTIVE METHODS AND PROPOSALS ON IMPROVEMENTS

Mine Clearing

1. There are 4 main methods of clearing mines using exploding technics

(i) *In situ* charge

Pro	Effective once found. Safe. Easy to deploy. Relatively cheap.
Con.	Slow. Initial detection essential. Not environmentally friendly.

(ii) Explosive hose

Pro	Essentially a military technique for minefield breaching.
Con.	Very few available. Very expensive. Of limited effectiveness. Requires specialized deployment. Very environmentally unfriendly.

(iii) Foams and mixtures. (A long extendable bag up to approximately 40 m. filled with foam explosives. Can be scaled down for hand application.)

Pro	The explosive is made up of two substances which are only brought together for the purpose: relative ease and safety of transport. Self-destructs within a matter of hours.
Con.	Highly volatile once triggered. The bag is difficult to control in any wind situations. Of doubtful effectiveness.

(iv) Fuel/Air. (Major explosion which overpressures the mines underneath. An area technique.)

Pro	Helpful with windmines and booby-traps Terrain independent.

Con. Indiscriminate.
Needs a specialist team to store, handle and deploy.
Of doubtful effectiveness.
Very environmentally unfriendly.

(v) Note here the techniques of flame and red-fire systems as further possibilities.

2. Mechanical system

(i) Rollers

Pro Relatively safe for the operator.
Fairly reliable.
Best as a secondary system after the initial clearance has been done.

Con. Terrain dependent, though they can contour to a degree.
No use in, e.g., monsoon conditions.
No means of verification.
The land needs to be "proved" afterwards.
Very energy consuming.
Needs a large prime mover (tank, etc.).

(ii) Ploughs

Pro Strictly for mine breaching as it just lifts the mines to one side.

Con. If combined with (side-loaded) sieving system, it can be of some humanitarian use.

(iii) Flails

Pro Good general purpose device.
Good operator protection.
Catches tree mines and booby-traps.

Con. Limited by the effects of fragmentation and explosive impact.
Once started, the driver loses his vision and must rely on a gyroscope.

CONCLUSION

Desirable improvements to the flail system in particular are:

maintainability and robustness, mobility and effectiveness.

An element to bear in mind is the tension between seeking to clear a large area quickly but leaving a proportion of mines unexploded, and on the other hand being much more thorough but correspondingly slower.

3. **Manual methods.** (The prodder. Usually a spike pushed into the ground approximately every 2 cm. Ideally, should be done in the prone position. However, in some countries people prefer to squat. In any event, head, face and chest protection is essential. Furthermore the exercise should be carried out in "echelon" formation.)

Pro	Very precise. Highly efficient.
Con.	Slow. Manpower intensive. Maximum risk to the operator.

Mine detection

1. **Vapour detection (sniffing)**

(i) Dogs

Pro	Effective. Cheap. Fast.
Con.	Tire quickly. Weather dependent. Not failsafe. After an explosion they cannot sense for 48 hours.

(ii) Electronic devices (The Linköping system)

Pro	Not time-limited.
Con.	Explosive dependent. System is still at the prototype stage.

2. **Prodding (see above)**

3. **Electronic Metal Detectors**

Pro	Simple use. Effective against metal-cased mines.

Con. Will not reliably detect all mines (minimum metal mines). Has a high false alarm rate, particularly in certain ground conditions.
Generally agreed that this technology has now reached a ceiling.

CONCLUSION

There is clear and paramount need not only for an increase in detection capability, but also for a single system which can detect metal, minimum metal and non-metallic mines.

Although not exhaustive, the following are some of the technologies currently receiving attention in research establishments around the world:

- Ground penetrating radar
- Bio-detectors
- Thermal imaging (infra-red)
- Neutron excitation.

REVIEW OF EXISTING DEMINING ORGANIZATIONS AND SOME PROPOSALS

Demining organizations fall into five main categories:

- Military (to a limited extent only)
- United Nations
- NGOs
- Commercial companies
- National bodies (such as Afghan Technical Consultants).

The United Nations is not directly involved in practical mine-clearing activities.

Disadvantages of the present system are the lack of funding and a general lack of coordination.

The newly constituted Mine Clearance Center at the United Nations gives a helpful focus to all mine-clearing activities. The group was pleased to see the United Nations take this necessary and more active role towards mine clearance, specifically allocating funds for projects in a number of countries and play a greater coordinating role. The group also hoped to see, in due course, more specific proposals for further United Nations activities in mine clearance.

Demining Research

The group expressed concern at the reduction of funding within ministries of defence for research into demining which might reduce the benefit of new equipment to the humanitarian mine-clearing community. It noted that at the same time considerable research is conducted around the world often in isolation from similar undertakings in other countries. This is self-evidently wasteful of money and human effort. Furthermore, the group was agreed that there is a need for a central body which can monitor and coordinate research work where appropriate, although consensus was not reached on which particular body would be best suited to fulfil this role.

REPORT OF GROUP 3

POSSIBLE MEASURES TO IMPLEMENT THE 1980 CONVENTION

I. INCREASE THE LEVEL OF RATIFICATION

It is highly desirable to increase the overall level of ratification, in this respect it might be preferable at this stage to encourage ratification of the 1980 Convention exclusively, in order to achieve a higher ratification rate prior to the convening of the Review Conference of 1994. Ratification could be encouraged by pressure from:

a. Secretary-General of the United Nations.
b. International Committee of the Red Cross.
c. Non-governmental organizations.
d. The media. There should be instruction of the media on the effects of breaches of the laws of war with respect to anti-personnel mines.

II. INCORPORATION OF IMPLEMENTATION PROVISIONS IN THE 1980 CONVENTION

While fully recognizing the limited effectiveness in practice of implementation measures provided by international law, the group thought that some should nevertheless be proposed for incorporation into the 1980 Convention. Such provision should be included in the body of the 1980 Convention and not in its protocols.

1. **Incorporation of implementation provisions found in 1977 Protocol I**

   The group considered three possibilities:

   - A simple reference back to certain articles in the 1977 Protocol.
   - Reproducing wholesale the appropriate articles of Protocol I.
   - Reproducing the appropriate articles of Protocol I with suitable modification in the language so as to be obviously applicable to the use of weapons.

The group preferred the third alternative, and pointed out the following articles as being the most obvious examples:

- Article 82. Provision of legal advisers. We recommend that legal advisers should be incorporated at all levels down to Brigade or equivalent level, and be incorporated into the planning staffs.

- Article 83. Training in humanitarian law. The group thought that the following four measures need to be undertaken:

  a. Training in the use of weapons in accordance with humanitarian law in cadet academies, and all command and staff training.

  b. Incorporation of legal provisions in all manuals of weapon systems, in the languages of the user countries.

  c. Incorporation of warnings of legal limitations on the weapon packaging.

  d. Incorporation of training in the international laws of war in all military training of foreign nationals.

- Article 84. Translation of the 1980 Convention into local languages and the adoption of necessary national laws and regulations.

- Article 85–87. Criminal sanctions. The group recommended in particular the incorporation of Article 85, paragraphs 1, 2, 3a-d, 4d, 5, (suitably amended), recognizing that these provisions for the most part have not been enforced in the past.

As a comment on the enforceability of the last suggestion, the group proposed that grave breaches as defined in these Articles could in practice be identified or notified by the following measures:

a. By action by the appropriate authorities within the nation accused of the grave breach.

b. Notification to the Secretary-General of the United Nations.

c. By a finding of an ad-hoc fact-finding group, the International Fact-Finding Commission, or other fact-finding mechanisms.

As the preceding measures depend on the appropriate governmental will to carry them out, the group also proposed identification and exposure by NGOs and the media of grave breaches.

2. Provision for adjudication on violations

The group considered the possibility of providing for the compulsory jurisdiction of an adjudication body in the 1980 Convention, but had doubts as to the practicability of the suggestions:

a. The International Court of Justice (ICJ). The problem is that there exists no compulsory jurisdiction in other humanitarian law instruments, that the ICJ is only suitable for international conflicts, and that it does not cover individual accountability.

b. International arbitration. This has not been used in the past for humanitarian law disputes and might therefore be considered unacceptable. It also depends on a certain level of cooperation by the parties involved in order to create the arbitral tribunal and its regulations.

c. Other methods. The group could not identify any other compulsory adjudication mechanism that could be incorporated into the 1980 Convention. It considered it to be of relevance that the most that could be obtained in 1977 Protocol I was the fact-finding procedure which is based on both consent and confidentiality.

The group thought that in general States would not accept compulsory adjudication measures which would have jurisdiction on events occurring before the acceptance of these compulsory measures.

III. APPLICABILITY OF THE 1980 CONVENTION TO INTERNAL ARMED CONFLICTS

The group considered whether the 1980 Convention should be made applicable to internal armed conflicts. It believes this to be a good idea, but recognizes that this issue is sensitive for those States which resist any international involvement in internal armed conflicts. As an inferior alternative, an optional protocol may be considered. Nevertheless it is worth trying to persuade governments that it is in their interests to have this Convention applicable in internal armed conflicts. The major difficulty is that irregular forces (insurgents) may ignore this in any event.

IV. LIMITATIONS TO THE RULES IN PROTOCOL II EVEN IF IMPLEMENTED

<u>Article 1</u> – The Scope. The Scope of the Convention does not extend to internal armed conflicts.

<u>Article 2</u> – Definitions. The definitions included do not cover unexploded ordnance and submunitions.

<u>Article 3</u> – General Restrictions. This article depends on laying the mines so that they are directed against military objectives, but these mines become indiscriminate once the military objective concerned no longer exists. The precautionary measures that may be taken in accordance with this article are frequently removed or ignored by the local population.

Article 4 – Restrictions on mines (other than remotely delivered). The problems are the same as in Article 3.

Article 5 – Restrictions on remotely delivered mines. There are difficulties in recording accurately the locations of mines delivered by fixed-wing aircraft, artillery and rockets. Therefore the recording requirements in 5.1a in the absence of a neutralizing mechanism are not applicable when using these methods. Problems remain with the implementation of 5.1b, because self-destruct and self-neutralization devices are currently insufficiently reliable to guarantee the safety of a mined area; further there is no maximum time limit given for the active life of these mines.

Article 6 – Prohibition on booby-traps. This was not considered by the group.

Article 7 – Recording and publication of minefields, mines etc.

(1) There is no definition of a "pre-planned" minefield, which is the only type that requires recording.

(2) The reference to the recording of non pre-planned minefields is weak.

(3) Even if records are made, minefields may shift geographically if not removed shortly after being laid.

(4) Minefield records can be lost.

(5) Release of records is limited to periods after withdrawal from the territory.

Article 8 – Protection of United Nations forces. The provisions are weak.

Article 9 – International cooperation. This does not impose an international obligation to remove mines or address repatriation and land reclamation requirements.

V. GENERAL PROBLEMS OF IMPLEMENTATION

It is noted that existing agencies are powerless to adjudicate in the majority of the mine violation cases prevalent today.

The Security Council could, on an ad-hoc basis, set up a tribunal to deal with cases of international and non-international armed conflict. The Council could enforce suitable remedies for violations of humanitarian law, including compensation or other rectification measures. This, however, depends on the political will of the Council to carry this out.

A proposal to create further awareness of the need to abide by humanitarian law is the creation of a compensation fund to be paid to victims of mine incidents.

Possible contributors to the fund would include States providing or using mines, and manufacturers of the mine systems. The group recognized, however, that it is unlikely that this will be accepted.

In the last analysis, the group acknowledged that lawless States and other parties to the conflict will always ignore or evade these various measures.

REPORT OF GROUP 4

PROPOSALS ON FURTHER HUMANITARIAN LAW RULES

PART I

1. General discussion on approach

It was suggested that given the variety of opinions of the members of the working group, it would best serve our purposes to report on the points brought up in discussion (advantages and disadvantages to various proposals) than to attempt to achieve a consensus on any one proposal.

It must be emphasized that all suggestions and statements reported here are not necessarily the opinion of the whole group.

2. Total ban on anti-personnel (AP) mines

Some members of the group said that they would be dissatisfied with anything less than a complete prohibition of the use of anti-personnel mines. A discussion ensued as to whether this might result in anti-tank mines being adapted to include the same function as AP mines, and it was pointed out that these hybrid mines are already being manufactured. The military view is that if AP mines were banned, they would probably use adapted anti-tank mines. However, it seemed likely that the western military forces might accept such a ban, should the government be in favour of it, having weighed up the cost of replacing AP mines with another kind of weapon, both in financial and military terms. There would be a loss in military capability that would be taken into account in this decision. The counter-argument was put forward that the extra military cost must be set off against the long term humanitarian costs, and the costs of mine clearance.

Another military position put forward was that it would be unrealistic to ban mines altogether, but restrictions on the use and production of certain types of mines would probably be approved. It was, however, pointed out that the Chinese perceive mines as an expanding market, particularly low-cost, high-tech mines, and would therefore reject a ban, as would the Pakistani government.

The argument put forward for a complete ban on the use of AP mines was that other forms of restriction do not deal with the problem of the lack of civilian ability to use the land, which has far-reaching humanitarian consequences. Despite the counter-argument that such a ban might never be realized, and would not be ratified, some members felt that it was nevertheless necessary to see this as our ultimate goal.

3. Self-destruct mechanisms

Everyone saw the introduction into the law of a prohibition of the use of AP mines without self-destruct mechanisms as an improvement to the existing situation, although some members did not see it as a solution. The introduction of such a law might bring the problem down to a more manageable level. The major disadvantage lies in the failure rate of approximately 10%, and not knowing when the mine will self-destruct, some minefields being deliberately set to randomly self-destruct.

It was put forward that there could be a time limit set on the active state of the mines, though what that limit should be could not be clearly defined. Some manufacturers support the idea of inbuilt self-destruct mechanisms because it increases the market value of mines, but this would have to be a general demand before they would act upon it. It was suggested that it might be possible to persuade first and second world countries to comply with this idea, and since third world manufacturers are often subsidiaries of first or second world companies (excluding of course local "home-made" mines) this might be practicable.

The view was that the some western military forces would probably readily accept such a prohibition.

4. Self-neutralization mechanisms

It was felt that the inclusion of self-destruct mechanisms under the term "neutralization" was misleading, and that the two should be kept distinct from one another.

The argument against self-neutralizing mechanisms was that they still denied use of land by the fact that it was not obvious whether they were safe or not, therefore necessitating the same amount of time and cost in mine clearance.

Many members of the group felt that allowing the use of AP mines with self-neutralization mechanisms would not be acceptable, they must actually self-destruct in order to be considered safe. Added to this was the assertion that the explosive charge remains in the ground, and can over a period of time become more dangerous, or even be dug up and resold or re-used. Nevertheless, it was understood that self-neutralization mechanisms would be useful for anti-tank mines, because of the impracticability of using self-destruct mechanisms, due to the immense damage created by the explosion of these mines.

5. Anti-handling devices

This discussion was part of the discussion about the distinction between AP and anti-tank (AT) mines, but led on to the argument for banning the use of integral anti-handling devices, since they have no apparent military value, and mostly only hamper clearance operations. The only possible military advantage they present is in terms of the lowering of morale, and as a further deterrent to breaching a minefield.

6. Detectability

It was stated that making all mines detectable, without placing any other restrictions on the use of mines would not significantly help clearance operations, since metal detection is not the prime method used to find mines. However, detectability is still regarded as important in conjunction with the introduction of a prohibition of AP mines without self-destruct mechanisms because of the failure rate of those mechanisms. One opinion was that it was a move in the right direction, but not significantly helpful. This was contested, and a view was expressed that it would indeed be helpful to mine detection.

7. Conclusion on first part of discussion

Those members of the group that favour a complete ban of AP mines did not exclude the idea of supporting restrictions.

It was seen as necessary to approach the problem of landmines on all different fronts without excluding any possibilities.

PART II

1. General discussion on the mines protocol

Some members of the group saw the existing protocol as ineffective, since it was not substantially different *in effect* to customary law. However, it can be perceived as an advantage to have the protocol because it regulated the use of mines for the first time. There was the view expressed that the existing protocol provides loopholes that lend themselves to abuse. The discussion that ensued on particulars of changing or adding to the protocol hinged on the probability of countries refusing to ratify, or military pressure hindering these changes.

2. Definitions

It was felt that it may be necessary to include definitions of "anti-personnel mines" and "anti-tank mines", perhaps according to weight. This was because of the necessity to recognize that any insistence on attaching self-destruct mechanisms to anti-tank mines would be a bad idea, since the explosion would create too much damage.

3. Article 3

There was some general agreement that there needed to be an additional paragraph added to this article, that would cover the long-term effect of the use of mines. This long-term effect was perceived by us as the indiscriminate aspect of the use of mines. This might be the right place to introduce the prohibition of the use of AP mines without a self-destruct mechanism.

There was some discussion of changing point 4 to remove the words "all feasible" and the second sentence, because it created too large a legal loophole. There was some criticism based on the looseness of the definition of the word "feasible", what constitutes humanitarian and military consideration, and which should take priority. The counter-argument was that this puts the soldiers on the ground in an impossible position, and would most likely be rejected by the military.

4. Article 5

The wording of 1b might be changed from "neutralizing" to "self-destruct", in regard to anti-personnel mines. There is no definition of "effective advance warning" in paragraph 2. This could possibly be changed by adding a concluding sentence about notifying civilians of the laying of mines after the fact, should advance warning not have been possible.

5. Article 7

It was suggested that it would be useful to clearance procedures if parties should be required to record what types of mines have been laid.

A view was expressed that an international standardization of marking signs for minefields would be useful.

6. Article 8

It was generally agreed that this article should be changed to include other missions than the United Nations – such as the CSCE. This may be broadened to cover mine clearance organizations, and so forth.

7. **Article 9**

This was seen as being generally too weak in terms of the obligation to demine, but extremely problematic to formulate an acceptable provision.

A suggestion was made that a general principle has to be made that responsibility must be taken, and in specific cases the Security Council might be the body to determine who pays for the demining, and who carries it out.

This point was seen as a key problem in the whole issue of landmines, and it was further suggested that Working Group 6 might deal with manufacturer's responsibility for demining.

8. **Verification and implementation**

The suggestion was made that a fact-finding commission similar to that of the fact-finding commission of Additional Protocol I would be effective, if it was not required to wait for a request from a conflicting party to monitor possible violations of the protocol, but had an automatic right to do so.

Also suggested was the helpfulness of another organization conducting a technical survey during the conflict to ascertain what mines are laid and how, to enable mine clearance to be more effective afterwards.

Implementation might be enhanced if the measures on implementation from Additional Protocol I of the 1977 Convention were also added into the mines protocol.

9. **Internal conflict**

It was generally agreed that changing Article 1 to cover internal conflict is impracticable, because of the lack of definition of what constitutes a combatant or a civilian.

Nevertheless, importance was placed on linking this to human rights instruments. A separate treaty would probably be necessary to cover this issue.

REPORT OF GROUP 5

POSSIBLE RESTRICTIONS ON THE METHOD OF MANUFACTURE OF MINES

It was accepted that some countries do not respect the laws of war and that many fighting forces will continue to make and purchase types of mines which do not even conform to existing restrictions. However, a significant proportion of mine victims around the world are still being injured by mines deployed by disciplined forces, or by mines originating from or designed by companies operating in developed countries which are contracting parties to existing international agreements. Because these same governments frequently find themselves contributing large sums towards the process of mine clearance, or contributing resources to provide medical care for victims, the group feels that it is worthwhile to press for new restrictions on the design and manufacture of mines.

In an ideal world, a total ban on both the manufacture and use of anti-personnel mines would be desirable, but this group at least thought that the military would strongly resist such a move, since soldiers even in disciplined forces believe that anti-personnel mines have an important role in modern warfare. It follows that suggested restrictions on the method of manufacture, if they are to have any chance of being accepted, should not significantly reduce the military effectiveness of the mines.

Other criteria are that, ideally, any devices added to destroy or neutralize mines should be cheap, very reliable and tamperproof. In addition it would be a great advantage if at least some of the suggested designs could be fitted retrospectively to the vast numbers of mines now already in stock throughout the world. The group considered five types of mechanism:

ACID-BASED;
MECHANICAL;
ELECTRONIC;
BATTERY-BASED;
CHEMICAL.

In some areas hard knowledge existed within the group, in others people were only able to speculate on the possibilities. This was probably not because there are developments in progress which are secret, but more likely because to date there has been little pressure on or incentive for manufacturers to develop such systems.

There was some confidence that such systems could be devised and made reasonably cheap so long as the will to do so existed, and the group believed that the cost would certainly be less than the costs of demining. After considering the types of mechanism listed above, this report will comment on three other topics:

NEUTRALIZATION OR DESTRUCTION;
THE TIMING OF THE MECHANISMS;
DETECTABILITY.

AN ACID-BASED SYSTEM

This looks like a large safety pin which springs open when acid in a small container fitted around one arm erodes away the metal, this then forces out the trip-wire pin and sets off the mine. If the acid container is fitted directly to the pin, then the pin breaks and the trip wire fails, so neutralizing rather than detonating the mine.

ADVANTAGES: cheap; could be adapted to add on to existing mines; could be highly reliable; might be adaptable to pressure mines also; when used to neutralize shows an obvious visual indication of having worked; not likely to need years of trials before being introduced.

DISADVANTAGES: timing not likely to be precise; timing hard to adjust in the field; could easily be left out by soldiers arming the mines; an external add-on might not be liked by the military.

MECHANICAL

Mechanical clock mechanisms were not seen as practicable to give the long time delay required by the military before destruction or neutralization. However, the simple mechanism of a wire under tension slowly pulling itself through a small lead rod or plate is already used in some munitions to delay action by up to 800 hours. Some Czech anti-tank mines found in Kuwait spontaneously exploded when the glue and plastic assembly holding back their firing pins gave way, and such mechanisms might be designed deliberately to do so after a certain period. Considerable research would be needed to develop and try out such systems.

ELECTRONIC

Sophisticated electronic control systems are already incorporated into modern anti-tank mines. Such mines are relatively expensive and if the military can neutralize them they can be moved and re-used. Some systems can already be switched on and off remotely by the armies laying them, using radio or laser signals, and in principle this could be applied to anti-personnel mines.

ADVANTAGES: It would be easy to programme such circuits to cause the mine to self-destruct or self-neutralize. Timing could be precisely set when arming the mines, and changed to meet operational needs. The micro-chip based timing devices are extremely reliable. In theory electronic self-destruction systems could be added to existing mines.

DISADVANTAGES: The electro-mechanical devices needed to translate the timing signal into destruction or neutralization would probably not be as reliable as the chips themselves; the systems require batteries, not used in most mines in use today; initially the systems might be expensive as a proportion of the cost of anti-personnel mines (which are very cheap); the military might not accept them as an add-on if exterior to the mine itself, and anything built in would require extensive research and trials.

BATTERY-BASED

The idea would be to make the firing of mines dependent on continued battery power, and to install batteries with a guaranteed limited life.

ADVANTAGES: The batteries could be installed at the moment of arming, so the mines themselves would remain viable on the shelf for years; such systems could be very cheap; and fail-safe.

DISADVANTAGES: Could not be installed into existing mines and would require some time to develop and test. We have been told that some existing systems based on wet batteries designed to fail after approximately 30 days were found still not to have failed after one year! Timing would be approximate at best and hard to adjust. Soldiers could easily substitute the short-life batteries for long-life ones.

CHEMICAL

A tempting idea is to use explosives designed to decay over a specified period. Here a great deal of research would be needed, and some years would be required before any system could be approved. A basic problem is that the most commonly used explosive in mines for the main charge is TNT. This is extraordinarily stable and cannot easily be rendered inactive by water, acid or chemicals. It can be left for decades and still function perfectly. Research is under way to develop micro-organisms which could digest TNT.

This process could be initiated by dropping a capsule into the charge at the time of arming.

It was generally agreed that the military would not accept any system of planned obsolescence of explosive which might lead to the need to replace stocks of stored mines on a regular basis (even though manufacturers might see this as commercially desirable!).

The explosives used in detonators and booster charges are different. When they age, they get more sensitive and more likely to go off. Perhaps research could be done in this area because this would guarantee the destruction of the mine, while spoiling the main charge would only produce neutralization.

(Existing "green parrot" mines are offered for sale with a neutralizing capability, but nobody in the group or who advised the group knows if this operates on a chemical or mechanical basis.)

NEUTRALIZATION OR DESTRUCTION

Those in the group with experience of anti-personnel mines in the field expressed the strong preference for self-destruction rather than self-neutralization.

Even if mines may blow up unexpectedly, the chance of injury is much greater if people try to handle mines which they believe are neutralized. Mechanisms to indicate if a mine was neutralized or not might be complex and expensive and in any case could be tampered with to produce an excellent booby-trap.

TIMING

The general view amongst the people consulted with military knowledge was that for most purposes a maximum time limit for self-destruction or self-neutralization would be one year. After that time it would be unlikely that legitimate and disciplined fighting forces would want their mines to remain operational. The only exception suggested was for marked minefields laid alongside more or less permanent frontiers. Most members of the group agreed that some upper time limit should be written into any new rules or protocol.

QUANTITY OF EXPLOSIVE

The suggestion was made that some limit could be set on the amount of explosive in anti-personnel mines. But the military view was that this would not be acceptable as many types of mine, to fulfil their military task, must kill or seriously injure as many people as possible who happen to be nearby when they go off.

DETECTABILITY

It is very worrying that, probably because of cheapness in mass production, many of the latest designs of mine are all plastic and so cannot be detected.

There was one view expressed that the military might want their mines to be undetectable for good tactical reasons, but even in this case mines could be made which could be detected, but only in peacetime conditions and with sensitive detectors. All agreed that an international agreement ensuring that mines could be detected after the battle was over would be acceptable to the military in the developed world at least.

Although it was suggested that some developed world forces have already purchased all plastic mines which cannot be detected, all agreed that the international community should press for detectability to become mandatory.

It would be very easy to add a small piece of metal when the explosive charge was filled, or build a metal ring or other part into the mine itself, and in the case of existing mines, it would be easy to stick on strips of metal tape.

So we ended with at least one positive and immediately achievable proposal which appears to have no complicating factors!

REPORT OF GROUP 6

POSSIBLE ARMS CONTROL MEASURES RELATING TO THE COMMERCE IN MINES AND THEIR STOCKAGE

Procedurally the Working Group held a lively discussion on the four aspects outlined in the paper presented by the ICRC for the examination of possible arms control measures:

- measures prohibiting or restricting exports
- destruction of existing stocks that are incompatible with possible new standards on manufacture
- prohibition of the manufacture of certain types of mines in an arms control treaty
- verification measures.

The Working Group felt there was an urgent need to have an in-depth discussion with the military, government, and independent experts on the military need for continued use of anti-personnel mines.

One of the concrete proposals of the Working Group was that there be an expert meeting held on the above subject perhaps under the auspices of the ICRC.

In considering how anti-personnel mines fit into international security concerns, the Working Group discussed whether or not anti-personnel mines are destabilizing weapons.

This was raised in regard to the fact that widespread sowing of anti-personnel mines erodes the socio-economic infrastructure and leads to massive displacement of population, on a long-term basis.

In this context the military argument of cost-effectiveness of the weapon should be re-examined. While the weapon might be cost-effective in terms of immediate military use, the long-term costs to society in demining, rehabilitation and rebuilding might dwarf that initial cost-effectiveness.

A general thread that went through the discussion on all specific arms control measures was the critical importance of public opinion both from the point of view of the need to educate the general public on the human consequences of the use of anti-personnel mines and for the public to bring pressure to bear upon their national governments to seek arms control measures.

The general sense of the majority of the group, taking into account the open question of military need for anti-personnel mines, was that the ultimate objective should be a complete ban on the use, production and trade in anti-personnel mines. At the same time, in all of the measures discussed below, the question of whether to have an outright ban on all anti-personnel mines or exclusions for certain types of weapons such as those with self-destruct or self-neutralizing mechanisms or the newly emerging high tech "smart" mines was discussed.

The Working Group examined the advantages and drawbacks of measures that could be taken in the shorter term and those that might be better considered as long-term measures. Accordingly, the following possible short-term measures were discussed:

1. **Unilateral measures**

   a. The Working Group recognized the significance of the US one-year moratorium on anti-personnel mines. In order to build upon the US moratorium it was proposed that support be raised for its extension. In that context the group explored the process that led up to its passage and the possible difficulty that might be raised in its extension in that many manufacturers will seek exemptions for certain types of anti-personnel mines, that is self-neutralizing and self-destruct mines and high tech "smart" mines.

      Some members of the Working Group from the US were keenly aware of the need for public support for the moratorium's extension.

   b. It was felt that other States should be encouraged to adopt similar measures. This encouragement should come from both the national publics and from the US government.

      The group discussed the value of such a moratorium on exports as being one of setting a standard for state behaviour; focusing world attention on the use of anti-personnel mines and as a critical first step toward achieving more far-reaching limitations.

      Possible drawbacks raised included the actual effectiveness of such a moratorium in considerably alleviating suffering; the repercussions on domestic producers; the difficulties of verification; and that the moratorium does not eliminate the problem of clandestine exports.

   c. The possibility was raised of a multilateral voluntary regime along the lines of the Australia Group which limits exports on dual-use chemicals and the Missile Technology Control Regime.

The advantage of such a regime would be to regionalize and/or internationalize the control of exports of anti-personnel mines. This type of non-proliferation regime might have the drawback of creating North-South friction.

Another drawback is that it usually does not imply a control system and the possibility that the focus of exports will shift to non-participating States.

In respect to all three of the above-described unilateral measures, national control must be strengthened and the role of the customs authorities in this respect was particularly emphasized. The possibility of using independent organizations for such control was explored.

2. Multilateral confidence-building measures

The importance of openness and transparency measures was stressed frequently. Toward this end the following measure was discussed:

a) the need for an exchange of information in the form of a register/data base as has been agreed in recent arms control treaties, on production, stockpiles and export of anti-personnel mines to which public organizations would have input and thus access. Some participants felt that the need for financial transparency in anti-personnel mine exports was important.

b) as a follow-up control to this information exchange, confirmation visits could be envisaged.

In connection with the above, the group recognized the difficulties in reaching agreement on any international mechanism for follow-up control.

3. Longer-term arms control measures

Regarding these, once a thorough discussion as mentioned above on the military need for anti-personnel mines has been fully explored, the Working Group discussed the possibility of a multilateral agreement to ban the development, manufacture, transfer, use and the destruction of all existing stocks.

The convening of the review conference of the 1980 Convention was borne in mind throughout the Working Group discussion of an overall ban. In this context, the discussions of the other Working Groups are of particular interest. Some suggested that the review conference mechanism be used for consideration of an overall ban.

The question of overlap between international humanitarian law and disarmament treaties was recognized with regard to which international forum would take up the negotiation of a convention for a comprehensive ban on anti-personnel mines.

The fact that the French request for a review conference calls for a consideration of verification/compliance issues as well as possible sanctions would have relevance to any similar regime to be worked out for a complete ban.

The issue of non-governmental organization participation in the review conference was proposed. It was also stressed by some that it was essential to engage the mine producers in the consultative process.

With respect to an overall ban, several issues will have to be considered although the list below should not be considered exhaustive:

a) a technical definition of what exactly constitutes an anti-personnel mine, production facility, dual-use componentry, and mine delivery systems. In this connection the blurring of the line between anti-personnel and anti-tank mines needs special attention.

b) a routine verification regime, that would include declarations and inspections.

c) a special challenge inspection regime.

d) destruction of stockpiles in an extremely limited time period and on-site verification of that destruction.

e) sanctions provisions in case of non-compliance.

f) strict national legislation and enforcement to support the terms of the multilateral agreement.

g) allowance for certain permitted purposes, such as research for the improvement of demining equipment and for protection of troops.

The advantages of a comprehensive ban seemed obvious to members of the Working Group, although a representative of a government said he could not endorse such a ban until the question of the military need for anti-personnel mines had been resolved. The requirement for as wide an adherence as possible to an agreement was stressed. The problem of how to deal with continued use, production and trade by non-States parties was recognized as potentially serious. Clandestine commerce would still have to be controlled as well.

DISCUSSION OF THE WORKING GROUPS' REPORTS

Chairman: Mr Y. Sandoz (ICRC)

GROUP 1 – MEDICAL PROFESSIONALS

The **Rapporteur** of Group 1 introduced the report, pointing out that the first four recommendations set out desiderata without any clear indication of who should carry out the suggested measures, whereas recommendations 5 and 6 were easier to implement.

In connection with recommendation 1, there had been much discussion concerning the possible use of the enormous existing military medical staff to treat anti-personnel mine victims: some participants had favoured such use, while others considered that attachment to a military organization might hamper the humanitarian effort.

The suggestion that military medical personnel might welcome an opportunity to gain experience of war trauma conditions by treating mine victims was countered by the statement that although military medical resources were plentiful at a fairly low level, there were not so many military physicians and surgeons available.

The need for teaching special techniques for the treatment of severe war trauma was stressed, and it was suggested that the recommendation be reworded to take account of the responsibility of manufacturers, who were not necessarily associated with governments.

It was further emphasized that attempts to obtain compensation must be realistic, since funds could hardly be expected to be forthcoming from the Afghan Government, the warring Khmer factions or the Russian Federation as successor to the USSR.

Preliminary discussions with United States government officials had evinced a guarded willingness to accept responsibility in the future, but certainly not retroactively. It was suggested that an extragovernmental fund be established, independent of United Nations agencies, so that individuals could contribute to it unofficially; several participants considered that such a fund should be established under United Nations auspices.

The review conference would certainly wish to examine the information mentioned in recommendation 2.

GROUP 2 – DEMINING

After the **Rapporteur** had introduced the report, a discussion took place on the accuracy of the statement in the second paragraph that the ordnance might largely originate in the West, since countries such as China and Pakistan were also heavily involved in mine production, and it was agreed that the problem should be described as universal.

It was pointed out in connection with the penultimate paragraph of the report that the new Mine Clearance Centre was only part of the overall United Nations effort to deal with the problem, including its human rights, legal and financial aspects.

With regard to the last sentence of the report, it was pointed out that, since the need for demining would not diminish in the near future while military interest in it would decline, it would be particularly important for some body, perhaps the ICRC, to keep a watching brief on the situation.

Stress was laid on the need to provide insurance for all personnel actually involved in mine clearance – not only foreign specialists, but also local staff.

In reply to questions, **the Chairman of Group 2** said that no one knew the exact number of landmines scattered worldwide, but that the figure could be assumed to lie somewhere between 50 and 300 million; **it was suggested that an effort should be made to provide more accurate figures, in order to acquaint potential donors with the scale of the problem, possibly by calculating the area of land rendered unusable by the presence of mines, or assessing the approximate number of mines sold by all producing countries.**

Various sources of funding for demining research were mentioned, and attention was drawn to the enormous discrepancy between the cost of the arms supplied to the countries concerned and the amounts required to clear the mines left behind – in the case of Afghanistan, 2.8 billion USD for arms as against less than 30 million USD for clearance.

The need was stressed to insert in the 1980 Convention at least a moral obligation for the party which had laid mines to clear them. From the developmental point of view, it was suggested that information on the involvement of local communities in demining, mine awareness and rehabilitation would be of use to funders and United Nations agencies.

Finally, the amount of money available for mine clearance was declining, since the military accorded low priority to such operations; although it might seem advisable to assign those limited funds to actual mine clearance, the need for research to improve demining techniques must also be borne in mind.

GROUP 3 – POSSIBLE MEASURES TO IMPLEMENT THE 1980 CONVENTION

The **Rapporteur** presented the report, which contained five sets of proposals for implementation of the 1980 Convention. **In the first place, an all-out effort should be made to increase the number of ratifications, which currently stood at only 30, so that a large number of States could participate fully in the forthcoming review conference.**

Secondly, with regard to the incorporation of implementation provisions in the existing Convention, the Group had disagreed with the French Government's suggestion that the provisions be inserted in Protocol II and considered that they should appear in the Convention itself, so that they should be applicable to any further protocols that might be drawn up for new weapons.

Of three possibilities for incorporating the implementation provisions of Additional Protocol I of 1977, the Group had preferred that of reproducing the relevant articles of the Protocol adapted to make them obviously applicable to the use of weapons; the most clearly pertinent provisions were Articles 82, 83 and 84 and, where criminal sanctions were concerned, Articles 85 to 87; the grave breaches of IHL referred to in those articles could be identified or notified by action of the authorities of the responsible State, by notification to the Secretary-General of the United Nations, or by the finding of some fact-finding body.

The Group had noted, however, that provisions concerning grave breaches had practically never been implemented, so that NGOs and the media should be encouraged to publicize such breaches.

The Group had further considered **the possibility of inserting a provision in the 1980 Convention on the compulsory jurisdiction of an adjudication body, but had had doubts concerning the practicability of such a course**: using the International Court of Justice (ICJ) raised the problems that no compulsory jurisdiction provisions appeared in any other IHL instruments and that the ICJ dealt only with international conflicts and did not cover individual accountability; international arbitration had been suggested, but it had never before been used for disputes under humanitarian law, and the establishment of an arbitration tribunal and its regulations called for a considerable degree of cooperation between the parties involved.

The Group had therefore concluded that States would be unlikely to accept the insertion of compulsory adjudication measures in the Convention, particularly since the most that could be obtained in the 1977 Protocol had been a fact-finding procedure based on consent and confidentiality.

Thirdly, the Group had generally agreed on the advisability of extending the applicability of the Convention to non-international armed conflicts, but had recognized the difficulties involved.

Attempts could be made to persuade States that it would be in their interests to extend the applicability of the instrument, but even if governments accepted such a provision, it was most likely to be disregarded by irregular insurgent forces.

Fourthly, with regard to the limitations of the rules in Protocol II to the Convention, Article 3 on general restrictions was based on the generally

accepted distinction between military and civilian objectives, but came up against the difficulty of maintaining that distinction when a military target had moved from the mined area, leaving behind weapons which had become indiscriminate.

The provisions of Article 7 on recording and publication of the location of minefields, mines and booby-traps were also difficult to implement, since there was no definition of a pre-planned minefield, the only type that required recording; minefields might shift under certain conditions, so that records would no longer apply; records might be lost; and their release was limited to periods following withdrawal from the territory.

Finally, a major drawback was that Article 9 imposed no international obligation to remove mines and did not deal with repatriation and land reclamation requirements.

Fifthly and finally, the main problem with regard to implementation was the absence of a compulsory adjudication mechanism to deal with IHL violations with respect to mines. The Security Council might conceivably impose suitable remedies for such violations, but that would naturally depend on the political will of the members of the Council.

The Chairman invited participants to comment on the five sets of measures outlined.

On the subject of increasing the number of ratifications of the 1980 Convention, it was suggested that States which were already parties to that instrument, as well as countries where the problem of landmines was acute, should try to promote further ratifications; political declarations of support had been received from certain States which were contemplating ratification.

It was further suggested that, in view of the inadequacy of Protocol II, that instrument could be ratified with reservations, in order to hasten enforcement; it was pointed out, however, that States might delay ratifying the Convention pending its amendment.

In response to the question why States should be encouraged to ratify such a flawed instrument, attention was drawn to the advantage of enlisting the support of a large number of States, some of which might be well advanced with the ratification procedure and would look forward to the amendments that would be introduced at the review conference.

Two reasons for failure to ratify were suggested – first, that the Convention was too far-reaching and raised security problems for some States, and second, that the instrument was regarded as meaningless; more ratifications should

certainly be sought, but it was probably more important to introduce some really progressive amendments. One participant observed that his country would have had no difficulty in ratifying the Convention if it did not belong to a certain alliance.

In connection with the incorporation of implementation provisions in the Convention, the view was expressed that the United Nations General Assembly would be better suited than the Security Council to deal with international arbitration. It was further suggested that arbitration might be written into the dispute settlement mechanism of the Convention.

In response to the observation that, despite the enormous number of mines scattered worldwide, mine-laying was not in itself a grave breach of IHL and did not necessarily entail criminal sanctions, a participant drew attention to the complete prohibition of the use of booby-traps in Protocol II and to the studies under way in the United Nations International Law Commission on the establishment of a permanent criminal court.

Those studies were well advanced, and might be referred to at the review conference, so that amendments might be linked to the jurisdiction of the permanent court. Emphasis was laid on the value of intensifying the education of armed forces in IHL.

With regard to extending the applicability of the Convention to non-international armed conflicts, the view was expressed that this instrument, which had not been ratified by any party to a current or recent armed conflict, should not be overloaded.

It was pointed out that in such conflicts a large proportion of mines was laid by non-governmental insurgents with whom it was extremely difficult to enter into contact; accordingly, incorporation of the relevant provisions in legal instruments would have little effect, and a better course of action would be to concentrate on restricting manufacture of and trade in landmines.

It was further observed that such an extension of applicability might be objected to by States wishing to invoke national sovereignty. However, it would be more difficult now to argue that they wish to be in fact free to use mines indiscriminately in order to despoil parts of their own territory and to kill and maim their own nationals. Another problem could be that insurgent factions might use those provisions as a lever for inserting themselves into the recognized world community.

It was questioned whether there was in fact any such thing as an internal armed conflict, in view of the many and various sources of weapons, and it was pointed out that States which were oppressing their own nationals were contributing units to United Nations peace-keeping forces.

On the other hand, it was noted that treaty provisions applicable to non-international conflicts could slightly facilitate enforcement.

In the discussion on general problems of implementation, attention was paid to the possibility of compensating mine victims, either through a fund or through direct payment to the victims. A suggestion to use a tax on mine sales to finance compensation was criticized on the grounds that liability for damage caused rested with the perpetrator, by analogy with liability for environmental damage.

Exception was taken to the use of the term "lawless States" in the last paragraph of the report; it was explained that mine users could increasingly be divided into two categories, those operating within an official structure governed by law, and those using mines quite indiscriminately: **whereas legal provisions might have some effect on the first category, the only way of curbing the anarchical activities of the second was to restrict trade in and supply of mines.**

One participant considered that the term "lawless" could well be applied to a Western State which had allowed 9 million mines to be shipped to Iraq.

GROUP 4 – PROPOSALS ON FUTURE HUMANITARIAN LAW RULES

The **Rapporteur** said that the Group had begun by discussing the possibility of a total prohibition of the use of anti-personnel mines, and had concluded that it may be unrealistic to expect such a ban to be accepted and that the introduction of restrictions in Protocol II to the 1980 Convention would be more acceptable.

The advocates of complete prohibition did not exclude the possibility of such restrictions. Prohibition of anti-personnel mines without self-destruct mechanisms had received considerable support, and was said to be acceptable to some manufacturers.

The Group had suggested some amendments to Protocol II, especially to Article 9 on responsibility for mine clearance, in connection with which it had been suggested that the Security Council might be the most suitable body to determine who would pay for and carry out demining operations.

Finally, the Group had concluded that the issue of applicability to internal conflicts should be linked with human rights instruments, but not with the 1980 Convention or its Protocol II.

In the ensuing debate, several participants emphasized the urgent need for research into a method of detecting all anti-personnel mines, including non-metallic ones. It was pointed out that metal detection was not universally useful, since mines might well be located in terrains covered with scrap metal, and the view was expressed that it may be risky to invest too much in expensive detection equipment, if it is at the expense of actually carrying out the demining.

It was observed that self-destruct mechanisms were more effective than self-neutralization ones, which had to be treated as active mines, for the purpose of demining.

It was suggested that the ICRC might convene an expert group to hold a thorough discussion on the subject. Attention was further drawn to the advanced capability of the military to lay mines by artillery, helicopters, rockets and mortars, and also to the importance of mines in military strategy: for example, minefields had played a key part in the Gulf War.

In response to the suggestion that the military might be prepared to accept a prohibition on anti-personnel mines but not on anti-tank mines, it was observed that either ban would represent a loss of military capability. It was pointed out that the correct use of mines for military purposes only would already be a step forward, since the great majority of current mine victims were civilians.

In reply to questions some demining experts said that locally made mines tended to be detectable and that the incorporation of self-destruct devices greatly increased the cost of mines.

It was considered by some that Protocol II provided legal loopholes for abuse, as in the case of Article 3, which was made unduly loose by the use of the term "all feasible precautions"; it was suggested that this particular difficulty could be avoided by referring to the long-term effects of the use of mines, by analogy with the effects of environmental damage. The need was stressed for a careful study of the definitions of "anti-personnel" and "anti-tank" mines.

It was reiterated that the use of weapons was more difficult to control than their prohibition, and it was suggested that, since a total ban on anti-personnel mines was unlikely to be accepted, the meeting should confine itself to strengthening its message and filling in certain gaps, such as making self-destruct mines compulsory, banning the use of anti-personnel mines in populated areas and imposing on the mine-laying party the obligations to improve technology, to provide assistance in the removal of mines and booby-traps and to arrange for the compensation of civilian mine victims.

GROUP 5 – POSSIBLE RESTRICTIONS ON THE METHOD OF MANUFACTURE OF MINES

The **Rapporteur**, in presenting the report, outlined five possible mechanisms for the self-neutralization or self-destruction of anti-personnel mines and stressed that such mechanisms would have to be acceptable to the military.

The working group expressed a preference for self-destructing, rather than self-neutralizing mines but recognized that the development of a reliable mechanism to operate over the required timescale would be both lengthy and costly.

Given the availability of such a mechanism, however, there seemed to be no military objection to its use, although guerrilla forces might not be in favour of it. The working group agreed that the international community should press for detectability to become mandatory.

In the ensuing discussion, the comment was made that pressing for detectable mines should in no way jeopardize research into mine detection. Attention was also drawn to the legal aspects of biological research. There followed an exchange of views on the utility of self-destructing mines, the main points being made as follows.

It was observed that such mines would only be useful if they were 99.9% reliable. Otherwise, minefields would have to be cleared before civilians were allowed back into those areas. From a demining point of view, it took just as long to clear a minefield with only one mine in it – or even no mines in it – as it did to clear a minefield with, say, a hundred mines in it, because the ground had to be cleared inch by inch.

It was pointed out that a 10% failure rate had to be expected, but considered that a 90% success rate was better than nothing. In this context, the view was expressed that with a known minefield, self-destructing mines were no advantage since the minefield could be cleared. Where the location of minefields was not known, where mines were laid at random and the minefield was unfenced, it was surely an advantage if 90% of those mines were to self-destruct. Such a self-destruction rate would reduce the number of civilian casualties.

A representative of the ICRC challenged the working group's deference to the views of the military. The military were servants of the government; public opinion should exert pressure on government as to what was acceptable.

The **Rapporteur** said that, ideally, a total ban would be desirable. In practice, the military view had to be confronted. The suggestions made by the working group no more than scratched the surface of the tragedy.

GROUP 6 – POSSIBLE ARMS CONTROL MEASURES RELATING TO THE COMMERCE IN MINES AND THEIR STOCKPILING

The **Chairman** of this Working Group presented the report at the request of the **Rapporteur**. He outlined the discussions that had taken place within the working group, and presented the group's suggestions for measures that could be taken, as set out in the report.

Commenting on a suggestion that the issue should be brought before the Conference on Disarmament (Geneva), he pointed out that the Conference adopted decisions on the basis of consensus and that much diplomatic effort would be needed to achieve agreement on the unprecedented inscription of a conventional weapons system on the agenda of the Conference.

The importance of involving public opinion through non-governmental organizations and the press in order to bring pressure to bear to prevent the trade in anti-personnel mines was stressed. The destabilizing effects of such mines could easily be assessed through measures such as population movement, GNP and pollution control.

The evidence for illegal arms trade lay in the associated financial dealings; often it was discovered by customs officials. Mines should bear a certificate of origin, and the issuing of false end-user certificates should be a criminal offence.

It was said that greater attention should be paid to licensing agreements that might undermine the US moratorium. In this context, it was noted that the US moratorium covered licensing, but that there could be clandestine activities. It was equally pointed out that the moratorium did not cover components and explosives, which could then be assembled elsewhere.

The view was expressed that, for a ban on anti-personnel mines to be effective, the definition of such mines would have to be watertight. However, it would be difficult to draw up a definition that could not be circumvented. It was thought that even if there were a clear definition, it would not cover components and explosives, which could then be assembled into mines. Explosives could be used legally in other military ordnance, so it would be hard to impose restrictions.

The difficulty of drawing up a valid definition of anti-personnel mines was repeatedly stressed. Some anti-tank mines were composed of a block of explosive with an anti-personnel mine on the top.

Although the view was expressed that if it had been possible to ban chemical weapons, it should also be possible to close the loopholes and ban anti-personnel mines, it was considered that attention should in the first instance be directed to the supply of anti-personnel mines, since the number of producers of mines and even of explosives was not huge. A ban should, however, remain the ultimate objective.

RECOMMENDATIONS

Pursuant to the discussions on the reports of the six working groups, the Chairman stated that these reports and their conclusions were generally adopted and that some concrete suggestions for future action, which was one of the main objectives the Symposium had set itself, could in fact be identified. He recalled that the Symposium formed part of a growing movement, in which the ICRC has participated from the beginning (the symposia in Lucerne and Lugano, 1974 and 1976 respectively) to address the humanitarian concerns raised by the use of anti-personnel mines and that its conclusions could help maintain the momentum already reached, as they offered concrete suggestions, on the basis of a better understanding of the mechanisms involved, aimed at developing a coordinated strategy to tackle the medical, commercial, legal, technical, socio-economic and data-related aspects of the entire problem in order to alleviate the sufferings of mine victims. He then introduced the series of issues that the suggestions referred to, and asked participants to comment on them, in particular with a view to indicating the immediate follow-up action that needed to be taken as well as the appropriate bodies to do so. The various issues were not listed in any order of importance.

Possibility of using military medical facilities in the treatment of mine injuries

Further to the first working group's report, one expert alluded to the difficulty of finding complete medical units to provide treatment for mine-injured civilians, and noted that military medical facilities would provide a useful resource, with the additional advantage that military surgeons would appreciate the opportunity to gain field experience in dealing with trauma, even though working conditions would not be the same as in wartime. The ICRC recalled that it had published quite a few scientific papers and had renewed contact with various armies and medical corps. It was therefore suggested that the ICRC could act as a focal point and proceed to organize seminars and lectures, and develop materials for dissemination to involved surgeons and medical staff on the surgical treatment and rehabilitation of the mine-injured.

Collection of information on demining research and establishment of a compensation fund for mine-injury victims

Following the reports of working groups one and two, the need for creating a body to centralize this information in order to motivate funders was stressed. In this context, it was noted that a feasibility study on the setting up of such a body, including questions relating to its composition and sources of finance, needed to be undertaken. The participants agreed to investigate the possibility of the United Nations undertaking a centralized coordinating function and that a letter to the United Nations from the Symposium would be a useful basis for exploring possibilities. With regard to the establishment of an international compensation fund, it was generally felt that governments, manufacturers, buyers, sellers, licensors and violators of humanitarian law, could figure among the contributors. Besides compensation to mine-injury victims, the fund could finance rehabilitation, mine clearance activities, research and development, educational programmes and training.

Review of the 1980 United Nations Weapons Convention

The Chairman stressed the need for careful preparation of the review conference, and offered ICRC support in holding preliminary expert meetings, as well as in preparing documentation for the preparatory conference. He recalled that if such work has to be considered as official preparatory meetings for the review conference, then governments should specifically ask the ICRC to undertake it. Towards this end, an expert suggested that the United Nations Secretariat could ask the ICRC to organize a preparatory meeting, whose report could then be forwarded to States before the review conference. A representative of the ICRC pointed out that a series of expert meetings would in fact be required to thoroughly examine the different questions involved. The participants agreed that the review conference should be well prepared and welcomed the offer of the ICRC to hold expert meetings. It was felt that it would be interesting to have the views of the ICRC itself, in addition to those of the experts. The Chairman believed that such a discussion should perhaps not be limited to mines alone. He further thought that an effort should be made to have the Conference on disarmament consider the question of mines as well as the possibility of including them in the United Nations Register of Conventional Arms.

Estimation of the military necessity and the socio-economic cost of mines

The Chairman said that, although fragmentary evidence of the socio-economic cost of mines was readily available, it would be useful to have comprehensive information in order to be able to put into perspective military arguments based on the cost-effectiveness of mines. It might also be useful to get a wider military view of the effectiveness of anti-personnel mines and whether or not they were essential in military terms. Perhaps the ICRC should explore that avenue. The representative of the ICRC suggested that consideration could be given to the question of a total ban, but the necessity of resolving the issue of the military need for anti-personnel mines before governments could endorse such a ban was generally accepted.

The participants agreed to suggestions that appropriate non-governmental organizations take on the task of coordinating a study on the socio-economic costs of mines, of a multidisciplinary nature that would include costs relating to long-term follow-up of rehabilitation, land rendered unavailable for agriculture and cattle grazing, etc. It was equally mentioned that publication of articles in medical journals, which required no coordination, could generate enormous publicity.

Technical aspects of self-destruct and self-neutralizing mechanisms

These aspects would have to be looked into in case a total ban was not acceptable. A representative of a non-governmental organization offered to work in coordination with United Nations demining experts on this question. He underlined the fact that cooperation with industry would also be important because of the many different types of mines. Attention would equally have to be paid to existing stockpiles of mines and likely future developments in mine technology. An ICRC representative pointed out that the advice of the military as well as contributions from technical and research specialists would be needed. It was stated by a representative of the United Nations demining team that they were not permitted to contact the military. It was therefore suggested that the demining and humanitarian needs be looked at first, and that the ICRC then put those views to the military.

Collection of information on the trade in anti-personnel mines

In an effort to get the States to introduce the subject of mines in the Conference on disarmament, it was felt that public access to information contained in the United Nations Register of Conventional Arms would be helpful, but it was pointed out that this information, submitted by governments, was limited to governments only. However, there existed a possibility that governments might eventually agree to make this information available to the public. One participant thought that all information provided in the context of future treaties should be open to the public, but that it would be very difficult to have information already collected made public retroactively. Non-governmental organizations were a valuable source of information, but governments were unlikely to put information into a voluntary register. It was equally stressed that non-governmental organizations could not obtain information on a country-by-country basis as the task would be overwhelming, but that some of them could serve as a clearing house for information from all sources.

Sensitization of the public

One participant drew attention to a forthcoming conference in London of non-governmental organizations on how to educate the public and how to influence the military and governments to change the law. The idea of extending the involvement of non-governmental organizations was welcomed, and representatives of some of these organizations referred to their ongoing efforts, such as releasing of reports on landmines in Cambodia, Angola and Mozambique and compilation of a dossier on children and mines. The need for increased involvement of National Societies and the Federation in sensitizing the public was invoked, as also the role of UNICEF and UNHCR in doing the same even at school level. The need to keep the press informed about statistics on mine injuries was stressed. The ICRC was urged to present the results of the Symposium in various cities, for example in New York, as a means of sensitizing delegations. A suggestion was made that the President of the ICRC ask the Secretary-General of the United Nations to appeal to Member States to ratify the 1980 United Nations Weapons Convention.

CLOSURE OF THE SYMPOSIUM

The ICRC was applauded for holding the Symposium.

Dr Russbach (ICRC) said that the excellent reports prepared for the Symposium, together with the frank discussion during the meetings, had led to a better understanding of a very complex problem. Various solutions had been suggested, not least the importance of raising public awareness of the intolerable plight of innocent civilian mine victims. The Symposium was one step in a journey, and the ICRC would continue with the next steps. He hoped that participants would do so too, each in their own ways.

Montreux Symposium Report

Table of Contents
(of the full report)

"Overview of the problem of anti-personnel mines" by Kenneth Anderson

Presentation and discussion

ITEM (2)

The trade in mines
– manufacturers, exporters and importers
– the types of profits involved

"The trade in mines – manufacturers, exporters and importers" by Christopher Foss

"Anti-personnel mines production and trading" by Ferrucio F. Petracco

Presentation and discussion

"Global production and trade in landmines" by Steve Goose
Presentation and discussion

ITEM (3)

The humanitarian consequences:
– medical eVects
– extent of the casualties
– social consequences
– extent of rehabilitation needs compared with what is actually available

"Humanitarian consequences of mine usage" by Robin Gray

Presentation by Dr Coupland and discussion

"Social consequences of widespread use of landmines" by Jody Williams

Presentation and discussion

"ICRC rehabilitation programmes on behalf of war-disabled" by Alain Garachon

Presentation and discussion

"The proliferation of anti-personnel landmines in developing countries: considerable damage in human terms and a dramatically insufficient medico-social response" by Philippe Chabasse

Presentation and discussion

ITEM (4)

Technical characteristics of anti-personnel mines:
– different possibilities for self-destruct or neutralizing mechanisms
– methods for rendering mines detectable

"The mechanics of anti-personnel mines" by Terry J. Gander

Presentation and discussion

"Technical aspects of anti-personnel mines" by Paul Jefferson

Presentation and discussion

"Self-neutralizing and self-destruct mechanisms for anti-personnel mines" by Jean-Pierre Golay

Presentation and discussion

ITEM (5)

Demining (including mine detection):

- **technical means (which means of demining are available and what they can do; which means of demining are needed but not available; are there certain types of minefields that technically could never be demined?)**
- **organization of demining (financial, political, training and time factors)**

"Summary of United Nations demining" by Patrick M. Blagden

Presentation and discussion

"An overview of demining, including mine detection equipment" by Paul Jefferson

Presentation and discussion

"Demining – an operator's view" by Brian Halliwell and L. Malin

Presentation and discussion

"Mine detection and demining" by J. Alistair Craib

Presentation and discussion

"Practical demining in Afghanistan" by Kefayatullah Eblagh

Presentation and discussion

"Mine detection and mine clearance" by Terry J. Gander

"Radar, multispectral and biosensor techniques for mine detection" by Curt Larsson

Presentation and discussion

ITEM (6)

The professional military perception of the use of mines:

- **the use of landmines (purpose, method of laying)**

- **possibilities for recording, methods**
- **possibilities for self-destruction or neutralization of mines – the appropriate timing and methods**
- **what is the military use of undetectable mines?**

"Anti-personnel mine warfare – an outline" by Terry J. Gander

Presentation and discussion

"Mine detection – the military necessity to render anti-personnel mines non-detectable" by J. Alistair Craib

Presentation and discussion

"The military use of anti-personnel mines" by Lt. Col. N. Hamish Rollo

Presentation and discussion

ITEM (7)

The legal situation:

- **present international law applicable**
- **explanation for the limited participation in the 1980 Weapons Convention**
- **explanation of the limited legal provisions**

"The mines protocol: negotiating history" by Brigadier A.P.V. Rogers

Presentation and discussion

"Basis of discussion on possible future legal developments" by Louise Doswald-Beck

Presentation and discussion

"The deadly legacy: report on Western views on landmines and ways of restricting their indiscriminate use" by Media Natura

Presentation and discussion

ESTABLISHMENT OF WORKING GROUPS

REPORTS OF THE WORKING GROUPS

DISCUSSION OF THE WORKING GROUPS' REPORTS

RECOMMENDATIONS OF THE SYMPOSIUM

List of participants

NOTES

1 "ALAN MOOREHEAD" Biography. Tom Pocock. Bodley Head. London 1990. ISBN 0–7126–5031–8.
2 Dummy minefields are areas marked with mine warning signs which do not actually contain mines. The purpose being to achieve land denial without actually laying mines.
3 It should be noted that acceptance of this kind does not carry with it any guarantee of popular support. The Iraqi Army, for instance, was reviled by a large part of the population and had a high percentage of unwilling conscripts – nonetheless it was treated by the Coalition Force as a real army and strategy designed accordingly.
4 "Deep strike" and similar terminology are not necessarily accepted military terms, but are employed by the author for their clarity of meaning to non-military readers.
5 "Remote" as distinct from manually laid.
6 Karez – stone-lined underwater irrigation tunnels which tap the water table and provide a complex supply of irrigation and drinking water in many areas of Afghanistan.
7 Mines lifted by Mines Advisory Group teams and other organizations in Iraq, Cambodia, Afghanistan, Kuwait and other countries bear testimony to the fact that few military customers pay for the detection insert option.
8 These witness reports are taken from the *The Coward's War*, based on a survey by the British NGO Mines Advisory Group, available in French from HANDICAP INTERNATIONAL.
9 Brigadier A. P. V. Rogers, OBE, Army Legal Services, solicitor. Brigadier Rogers was a member of the United Kingdom delegation at the UN Conventional Weapons Conference 1979–1981 and spokesman for the delegation in the mines working group. He is a member of the Board of Directors of the International Society for Military Law and the Law of War and chairman of the committee for military instruction of the International Institute of Humanitarian Law. He has written various articles on the law of armed conflict and is currently working in his spare time on a thesis entitled Law on the Battlefield. This paper contains the personal views of the author and it does not necessarily reflect United Kingdom Government policy.
10 A Commentary on the Protocol on Prohibitions or Restrictions on the Use of Mines, Booby-traps, and Other Devices, Military Law and Law of War Review (1987) and Mines, Booby-traps and Other Devices, *International Review of the Red Cross* (1990).
11 The technical aspects and medical effects of delayed action weapons were discussed at a conference of experts called by the ICRC in 1973 and encapsulated in a report entitled "Weapons that May Cause Unnecessary Suffering or Have Indiscriminate Effects", ICRC, 1973.
12 Although mines at sea were regulated by Hague Convention VIII of 1907.
13 Article 35 of Protocol I merely lays down general principles.

14 Convention on Prohibitions or Restrictions on the Use of Certain Conventional Weapons Which May be Deemed to be Excessively Injurious or to Have Indiscriminate Effects (with annexed Protocols) published by HMSO London as Miscellaneous No. 23 (1981), Cmnd. 8370.
15 See Report on Conference of Government Experts on the Use of Certain Conventional Weapons (Lucerne), ICRC, 1975, pp. 7–13.
16 These were not included in the final version of the mines protocol.
17 Lucerne Report, above, pp. 63–71.
18 CDDH/IV/202.
19 CDDH/220/Rev.l dated 15 December 1975.
20 COLU/203, see Annex A2 to the Conference of Government Experts on the Use of Certain Conventional Weapons (Lugano), ICRC, 1976.
21 The architect of this proposal was Colonel (later Major General) Sir David Hughes-Morgan (now His Honour Judge Sir David Hughes-Morgan, Bt, CB, CBE).
22 CDDH/IV/201 submitted by Algeria, Austria, Egypt, Lebanon, Mali, Mauritania, Mexico, Norway, Sudan, Sweden, Switzerland, Venezuela and Yugoslavia, see Annex A21, Lugano Report.
23 COLU/206. The experts from Israel made a similar proposal in COLU/217.
24 Some proposals for amendment were put forward by the experts from Spain in COLU/215.
25 COLU/213.
26 COLU/214.
27 See the counter-proposal of the experts from Venezuela in COLU/219.
28 CDDH/IV/209.
29 CDDH/IV/212.
30 CDDH/IV/211.
31 CDDH/IV/213.
32 CDDH/237/Rev.l dated 31 December 1976, para. 46.
33 CDDH/IV/222.
34 The combined text was produced as CDDH/IV/GT/4 and Corr. 3.
35 These were not contained in the original UK proposal, but were included in a definition of "mines" introduced by Mexico and Spain – CDDH/IV/GT/1.
36 Such as the British Ranger system.
37 This is the time specified in Article 118 of the Geneva Prisoner of War Convention of 1949 for the release and repatriation of prisoners of war.
38 CDDH/408/Rev. l dated 1 December 1977, p. 27.
39 For a further report of the discussions at the fourth session of the Ad Hoc Committee, see CDDH/408/Rev.l dated 1 December 1977.
40 A/CONF.95/PREP.CONF/L.9 submitted by Australia, Austria, Denmark, Federal Republic of Germany, France, Mexico, Netherlands, Norway, Spain and the United Kingdom.
41 Taking language borrowed from Geneva Protocol I of 1977.
42 The late Robert J. Akkerman of The Netherlands.
43 A/CONF.95/PREP.CONF./CRP7 dated 6 April 1979.

44 The latter, A/CONF.95/CW/WG.1/L.5 dated 17 September 1979, eventually became the technical annex to the mines protocol and a definition of "recording" was included in the body of the protocol to support the annex.

45 A/CONF.95/CW/WG1/L.7 dated 24 September 1979.

46 Later this became Article 7(1)(a).

47 Indeed, this reflects the corresponding text of Article 57 of Geneva Protocol I of 1977.

48 E.g., Austria, Canada, Denmark, Poland.

49 Federal Republic of Germany, Sweden, Yugoslavia, United States, United Kingdom.

50 Australia, Federal Republic of Germany, United Kingdom, as well as a technical sub-group.

51 The word "whereabouts" was suggested by the UK delegation and met with some initial scepticism from the US delegation because in US English usage the word is apparently regarded as archaic.

52 A/CONF/95/CW/7 dated 2 October 1980, para. 17.

53 A/CONF.95/14/Add.

54 A/CONF/95/15 dated 27 October 1980 and A/CONF/95/15/Corr.2 dated 2 April 1981.

55 A detailed analysis of the various main texts studied before and during the Weapons Conference is at Annex A. Although the author has tried conscientiously to reflect the debate as accurately as his memory of events and the available documents permit, it is impossible to avoid subjectivity or, in the compass of a short work, to be completely comprehensive and the author hopes that readers will understand if he has not been able to do full justice to the many points of view, arguments and proposals advanced during the negotiations.

56 RDM = remotely delivered mine.

57 The various proposals in this outline are listed for indicative purposes and do not necessarily reflect the view of the ICRC.

3

Mines: A Perverse Use of Technology

May 1993

In many ways, the publication of *Mines: A Perverse Use of Technology* by the ICRC in May 1993 marked a turning point in the ICRC effort to tackle the humanitarian catastrophe wrought by mines. The language used in the condemnation of the indiscriminate use of landmines was uncompromising and the images – not presented here – were shocking in their graphic depiction of the suffering endured by virtually every mine victim.

The opinion of one former ICRC delegate included in the publication – and since reproduced widely – perhaps summed up the rationale for the entire ICRC effort against mines:

> Mines may be described as fighters that never miss, strike blindly, do not carry weapons openly, and go on killing long after hostilities are ended.
>
> In short, mines are the greatest violators of international humanitarian law. They are the most ruthless of terrorists.

The ICRC decided that it was essential to inform the public of the horrendous suffering inflicted by mines. With this course determined, a path was cleared to make the way for the major public advertising campaign in favour of a total prohibition of anti-personnel mines that the ICRC would launch – for the first time in its history – only two years later in the autumn of 1995.

MINES: A Perverse Use of Technology

Foreword by Cornelio Sommaruga

A tragedy too often forgotten among the many experienced by a population involved in armed conflict is the havoc wrought by mines, millions of which lie on or in the ground, spreading terror for years or even decades after the hostilities have ended.

Every year, in numerous war-torn countries, thousands of men, women and children are victims of mines. Some of the mines are simple devices, while others make use of advanced technology. But all of them kill or cause severe mutilation. That this form of warfare, often extremely pernicious, is used indiscriminately must be deplored. Fortunately, a number of organizations are now turning their attention to this tragic problem.

The role of the International Committee of the Red Cross (ICRC) is first to gain respect for international humanitarian law, which prohibits the indiscriminate use of mines, as well as the deliberate endangering of the lives or health of civilians, and all serious and lasting damage to the environment. In addition, in co-operation with other bodies, the ICRC endeavours to help the victims of mines by setting up surgical and orthopaedic centres in countries where there are conflicts, in order to treat and rehabilitate those who have been disabled and ensure that they are able to walk again. Furthermore, the institution considers that it has the duty to point out to the international community the immense effort needed to eliminate all mines in the areas affected.

This publication is issued as part of a campaign against the indiscriminate use of mines.

All States, humanitarian organizations and peoples of the world are urged to unite their energies to eradicate this scourge.

Cornelio Sommaruga
President of the International
Committee of the Red Cross

> **"Mines may be described as fighters that never miss, strike blindly, do not carry weapons openly, and go on killing long after hostilities are ended.**
>
> **In short, mines are the greatest violators of international humanitarian law. They are the most ruthless of terrorists."**
>
> *Opinion of a former ICRC delegate*

The use of all weapons is governed by certain fundamental rules of international humanitarian law that must be respected in armed conflicts:

- parties shall at all times distinguish between civilians and combatants in order to spare the civilian population;
- civilians shall not be the object of attack;

- indiscriminate attacks and the use of indiscriminate weapons are forbidden;
- it is prohibited to employ weapons or methods of warfare of a nature to cause unnecessary losses or excessive suffering;
- States have a responsibility to ensure that the use of a new weapon being developed or acquired is not prohibited by the rules of international humanitarian law.

These rules are a summary of the relevant provisions of the 1977 Protocols additional to the Geneva Conventions of 1949, of the Hague Conventions of 1907 and of international customary law.

Hazardous return

The return meant here is that of refugees to their own countries, from which they have had to flee and to which they have always hoped to return once the hostilities were over. But in the intervening period their land has become hostile, lethal, and destined to remain so for many years. Exile and the scars of conflict are compounded by further tragedy, arising from the massive and indiscriminate sowing of mines throughout the country.

One example: the return of Afghan refugees. According to the Office of the United Nations High Commissioner for Refugees (UNHCR), more than one million refugees returned to Afghanistan in the first nine months of 1992. Between April and June of this year, the number of mine victims in the ICRC's hospitals alone was two or three times higher than during the same period the previous year. This figure only partly mirrors the reality, since it does not take into account those who were killed and the injured who were unable to reach a surgical centre.

These millions of mines, in addition to killing, maiming and terrorizing, have the power to deny, by their very presence, the fundamental right of refugees to return to their native lands.

Mines are not only laid by hand; they can also be delivered from a distance, for example by being fired from a long-range gun or dropped from an aircraft, without exploding on hitting the ground.

Contempt for the rules

Originally, mines were used chiefly to protect military objectives. Regrettably, from being rudimentary deterrent devices, they have become unusually effective and deadly weapons, particularly because of their incredibly large numbers and the absence of any self-destruct mechanism in most of them. However, one of the most intolerable aspects of the use of mines is the way in which they are laid or scattered. For example, to drop them from aircraft or to lay them without marking their positions, that is, quite indiscriminately, amounts to a violation of the most elementary rules governing the conduct of hostilities.

Survival and the environment

In Angola, after 15 years of war, there are an estimated 20,000 amputees, all victims of mine blasts. The majority are women and children.

Yet that is only part of the story. In some regions, farmland cannot be cultivated because it is strewn with mines or thought to be so, a situation that during the war gave rise to chronic malnutrition.

Yet Angola is not the only case of this kind and probably not the worst. Indeed, any country that has large numbers of mines in the ground invariably suffers some degree of economic damage. For certain population groups, the inability to cultivate their land is a threat to their very survival.

The indiscriminate laying of mines, which makes land incultivatable, also causes damage to the natural environment. International humanitarian law affords protection for the environment, to safeguard the health and survival of the population.

Burden

For a country engaged in or recovering from war, the presence of thousands of people disabled by mines is an economic and social burden which soon becomes intolerable, not only for their families but also for the community. For the victims themselves, it is an unending tragedy.

A long, costly, dangerous – and perhaps impossible – task

After the war in the Falklands-Malvinas, mine-clearing operations were begun at once. But they were found to be so difficult that they were abandoned. Yet the war had lasted only two months and had been waged between two conventional armies. If mine clearance was abandoned in areas that had been clearly identified, what must the situation be in countries where fighting has continued for years and mines have been used indiscriminately and on a massive scale?

In Afghanistan, a country that is infested with mines, the UN Office for the Coordination of Humanitarian and Economic Aid to Afghanistan (UNOCHA) has launched a mine-clearance programme. It has 27 mine-clearing teams and it will require at least 15 years to clear the priority zones! At a rate of 30 sq km per year, it would take 4,300 years to clear mines manually from only 20 per cent of Afghan territory...

Mine clearance is not only costly and slow; it is also extremely dangerous. According to the *Arab Times* (Kuwait), as many as 84 foreign and local experts working in Kuwait in the aftermath of the hostilities were killed by mines in less than ten months. In Afghanistan, UNOCHA has had 16 of its staff killed and 20 injured in removing mines from only 68 sq km of territory.

In addition to those mentioned so far, a large number of other countries or regions world-wide are affected by the problem of mines. The following list is not exhaustive:

Mozambique, Somalia, Ethiopia, Sudan, Rwanda, Uganda, Nicaragua, Guatemala, El Salvador, Colombia, Peru, Honduras, Iran, Iraq, Kurdistan, the Golan Heights, Armenia, Azerbaijan, the former Yugoslavia, Viet Nam, Myanmar, Laos . . .

Appalling injuries

Mines produce damage by either blast or penetration of metallic fragments. Those triggered by foot pressure wound by blast. The tissues of the foot are exploded or with a larger charge the lower leg is simply blown off. Damage is rarely confined to one leg; lesser but still severe damage is frequently caused to the other leg, the genitals, arms, chest and face.

With the explosion dirt, mud and other debris is driven into the tissues. Bounding mines jump into the air to a variable height above the ground before exploding. Individuals within the lethal radius die from blast and fragmentation injury. Those more peripheral to the explosion are wounded by metallic fragments or spheres; stake and directional mines cause similar injury to those outside the lethal radius.

Surgeons recognize three patterns of injury produced by mines: traumatic amputation of part of a lower limb and less severe injuries elsewhere, multiple fragment wounds, and upper limb, chest and face damage caused by handling. Within these patterns the variable extent of damage and contamination by dirt and debris makes surgery for mine injury difficult. It is particularly problematical because injuries of this severity and degree of contamination are rarely seen in civilian practice and few surgeons have experience and skill in dealing with such wounds.

The surgical task is to cut away all the dead and severely damaged tissue and remove the debris and dirt. In principle it is easy, in practice not so. In the case of a traumatically amputated limb the blast pushes debris far up into the leg, which necessitates a higher amputation than is apparent to the inexperienced surgeon or untutored observer. Unwise attempts to preserve tissue and poor techniques of amputation result in much unnecessary suffering with multiple operations and poor stumps for the fitting of prostheses. Without physiotherapy and provision of a functional prosthesis a lower-limb amputee is sentenced to permanent incapacity and loss of the potential to be independent and self-supporting in society. Double lower-limb amputees are usually confined to wheelchairs, if available; otherwise many become beggars.

ICRC surgical hospitals

The war wounded can be helped by providing hospitals and surgical teams. Some 20,000 war victims were treated by ICRC surgeons in 1991. In the armed conflicts in which ICRC has a mandate to assist the victims directly, the enormity and complexity of providing for the needs is way beyond the capacities of the institution. Victims of mines often die without medical aid and those who survive may only receive token treatment. The ICRC was the first to introduce the concept of mine awareness into its first-aid courses in Pakistan. Medical activities are not the answer to the underlying problems.

ICRC rehabilitation units

The ICRC orthopaedic workshops in 14 countries made limbs for 7,876 amputees and produced 11,116 orthopaedic appliances in 1991. The majority of amputees were mine injured. Rehabilitation and provision of prostheses and appliances is far from easy. Materials and technology must be appropriate and a high level of expertise is required. In 1992, the ICRC had 51 expatriate specialists working worldwide. The objective, i.e. to provide limbs which allow full functional capacity and social integration, cannot be achieved without following up the initial emergency action by a long-term development programme.

The need for future action

Respect for existing legal provisions would go a long way towards reducing civilian mine casualties. Particular efforts need to be made to ensure the implementation of these provisions. This must be done by proper training and also by an active interest on the part of the international community in seeing the rules respected.

However, the dramatic situation that we are now faced with indicates that the following problems remain to be solved:

As mines are generally not equipped with neutralizing or self-destruct mechanisms, they continue to maim and kill when the military purpose for which they were laid no longer exists, and even long after the end of hostilities.

Mines are increasingly manufactured in such a fashion as to be virtually impossible to detect. The military value of such undetectable mines needs to be weighed against the human tragedy and environmental devastation that they cause.

The clearance of minefields is a long, costly and dangerous task, but it needs to be done as soon as possible by those who have the means.

The specific rules on the use of mines in Protocol II to the 1980 United Nations Convention apply to international armed conflicts, whereas the majority of today's conflicts are internal ones.

4

The call of the ICRC for a global ban on anti-personnel mines

Geneva, Switzerland
24 February 1994

On 24 February 1994, to the surprise of many, ICRC President Cornelio Sommaruga announced the ICRC's opinion that, from a humanitarian point of view, a worldwide ban on anti-personnel mines was the only truly effective solution. Apart from a growing body of non-governmental organizations, at that time no other major international organization had supported an outright ban on these weapons. President Sommaruga also called for a prohibition on blinding as a method of warfare and for universal adherence to the 1980 Convention on Certain Conventional Weapons.

Statement of Cornelio Sommaruga

President, International Committee of the Red Cross
Geneva, Switzerland
24 February 1994

ICRC: A Total Ban on Anti-personnel Mines and Blinding Weapons is the Best Option

The first preparatory meeting for the Review Conference on the 1980 United Nations Weapons Convention opens in Geneva on 28 February 1994.

It will be an important milestone in the long-term effort to control the use of the more destructive of modern weapons.

The Convention came into force in 1983 and has been ratified by 41 countries. This is the first opportunity – and the last for at least another 10 years – to review its workings.

It is clear, for all its good intentions, that the Convention has had little impact on restricting the use of the deadly weapons it regulated in 1980 – in particular mines. Mines are now proliferating so fast that there are perhaps as many as 100 million of them in 62 countries. Scattered like deadly seeds, they have turned whole swaths of many countries into deserted, no-go areas.

Moreover, the Convention should take a stand against the invention of new even more terrifying weapons such as lasers that can destroy the eyesight of a victim. Blindness is considered the most incapacitating and dreaded form of disability. Against lasers there is no defence and there is no cure for the damage they cause.

* We believe that the 1980 UN Weapons Convention should restrict the use of certain conventional weapons more effectively.

* From a humanitarian point of view, we believe that a worldwide ban on anti-personnel mines is the only, truly effective solution, and that blinding as a method of warfare has to be outlawed now.

* We believe that the Convention should be extended to internal armed conflicts, be given means of enforcement and have wide-reaching control mechanisms.

* Most importantly, we believe that the 1980 UN Weapons Convention has to be ratified by all the countries of the world.

5

Meetings of governmental experts to prepare the Review Conference 1994–1995

Four meetings of governmental experts were convened in preparation for the Review Conference of the 1980 Convention on Prohibitions or Restrictions on the Use of Certain Conventional Weapons Which May be Deemed to be Excessively Injurious or to Have Indiscriminate Effects (1980 CCW). The ICRC was asked to prepare a report on the issues it believed States needed to address with respect to weapons, particularly mines, within the context of the 1980 Convention. In its formal report submitted to the first meeting of the group of governmental experts, the ICRC set out the issues. Giving the example of the general respect for the prohibition of the use of chemical weapons, the ICRC pointed out that merely restricting the use of the weapon would be less effective than prohibiting it entirely.

Addressing the need for a total ban on anti-personnel mines, the ICRC warned States that a careful definition of anti-personnel mines was essential. Sadly, the issue of the definition of the weapon would become a point of contention in the subsequent negotiations. Short of a total ban, a series of possible options were outlined for consideration by States Parties, including making all mines detectable and self-destructing or self-neutralizing. Yet even these relatively modest measures would not be acceptable to all States Parties. Finally, the ICRC stressed the importance of incorporating implementation mechanisms in the 1980 Convention.

During the later meetings, the ICRC made several specific proposals. Most notable are its calls for a prohibition on the use of anti-personnel mines and a ban on the manufacture, stockpiling and transfer of anti-personnel mines and non-detectable anti-vehicle mines. An obligation to destroy stockpiles, and requirements upon parties to a conflict to take certain action for the safety of humanitarian workers and activities were also submitted.

Report of the ICRC for the Review Conference of the 1980 United Nations Convention on Prohibitions or Restrictions on the Use of Certain Conventional Weapons Which May be Deemed to be Excessively Injurious or to Have Indiscriminate Effects

March 1994

TABLE OF CONTENTS

PART I: THE SIGNIFICANCE OF THE REVIEW CONFERENCE IN THE CONTEXT OF PRESENT-DAY ARMED CONFLICTS AND WEAPON DEVELOPMENTS

I. The importance of the 1980 Convention and the role of the Review Conference

The International Committee of the Red Cross has first-hand experience of the real consequences of the many armed conflicts now taking place and of the actual use and effects of weapons. It therefore has a particular interest in ensuring that the law takes into account the realities of the use of weapons in order to effectively reduce the amount of suffering caused in armed conflicts.

International humanitarian law aims at reducing the suffering caused by the use of weapons by prohibiting indiscriminate attacks and by prohibiting the use of weapons that are by nature indiscriminate or of a nature to cause unnecessary suffering or superfluous injury. These international customary rules are universally applicable and are codified in major international humanitarian law treaties, in particular Protocol I of 1977 additional to the Geneva Conventions of 1949. The ultimate purpose is to mitigate the suffering and damage caused during armed conflicts as much as is practically possible. In order to achieve this effect, it is essential that international humanitarian law treaties, including the 1980 UN Convention on Certain Conventional Weapons, be ratified widely and implemented correctly.

The 1980 Convention has the purpose of codifying and developing specific rules on the use of weapons, either by totally prohibiting the use of certain weapons, or by regulating their use, so that the customary principles of international humanitarian law on the use of weapons are given concrete expression in treaty form. However, in many respects this Convention has not achieved its aim, not only because it has been insufficiently ratified or implemented, but also because in many ways it does not provide the means needed to prevent the excessive damage that is actually being caused in armed conflicts, the majority of which are non-international. In particular, the Convention relies too extensively on regulating behaviour in relation to the use of certain weapons, which is frequently difficult to enforce, rather than prohibiting the use of certain types of weapons altogether. Further, no parallel measures have been taken in the disarmament context, although they are proposed in the preamble to the Convention.

The Review Conference is a unique opportunity to make a careful assessment of the real problems caused by the use of certain weapons and of the reasons for these problems, so as to decide on the most effective measures to redress the situation. It is also an opportunity to decide on measures that may be necessary to prevent major problems arising from weapon developments in the near future.

Mines

The most urgent problem which the Review Conference must address is that of landmines. Despite the fact that the legal regulation of the use of mines was carefully discussed in the 1970s and that these deliberations culminated in Protocol II of the 1980 Convention, the situation that we are facing today as a result of landmine use is a disastrous one. It is estimated that there are about 100 million uncleared mines in the world, rendering huge expanses of land uninhabitable and uncultivable. It is estimated that every month landmines kill about 800 people and maim thousands, most of the victims being innocent civilians, especially children. The worst feature of mines is that they continue to cause damage for years or even decades after the end of hostilities. Mine clearance is a very slow and dangerous task and in some situations virtually impossible. It takes many years to clear very small areas and casualty rates among mine-clearing teams are appallingly high.

Part of the problem is that mine-laying has been undertaken in ways that are in violation of the law, and there would have been fewer casualties had the law been respected. However, Protocol II has serious shortcomings as it stands and it is clear that in order to try to find an effective way of improving the situation it is essential to consider much firmer measures, including complementary arms control measures. This issue will be examined in Part II of this report.

Weapon developments

The Review Conference is also a critical opportunity for a more forward-looking assessment of the likely problems that weapons production and use are going to create.

The situation caused by the use of modern landmines is a pertinent example in this respect. These weapons have always been considered normal conventional weapons and certainly not weapons of mass destruction that merited important international arms control measures. However, a certain amount of thought and foresight would have shown that the introduction of plastic mines which can be sown in large quantities, which are cheap and widely available, and which remain active for an indefinite period would lead to the grave situation that we now face.

The international community does not have to wait for catastrophes to happen, but can rather anticipate probable dangers. In this respect it needs to take into account the types of conflicts that actually occur and the way in which weapons proliferate. Once a weapon is fielded it is very difficult to stem its proliferation and widespread use. Therefore it makes sense to devote some time to taking preventive steps that would avert enormous problems at a later stage.

Part III, Section II of our report looks at some of these issues, including the present development of directed energy weapons which could well begin to be used in the near future.

Implementation and non-international armed conflicts

The Review Conference is also an opportunity to address the fact that most damage inflicted by weapons, frequently as a result of indiscriminate use, occurs during internal armed conflicts, that effective implementation mechanisms are necessary for achieving better respect of the law, and that the Convention is lacking in both these respects. These issues will be addressed in Part IV, Section I of this report.

II. The need for regular review of the 1980 Convention

The 1980 Convention was intentionally structured in the form of a basic Convention with annexed Protocols so as to provide for the addition of further Protocols to specifically regulate, or prohibit where appropriate, the use of new weapons. The use of weapons is, of course, subject to international customary law but it is clear that specific treaty regulation is preferable in that it favours clarity of legal obligations.

Article 36 of Protocol I of 1977 additional to the Geneva Conventions of 1949 requires States Parties to review new developments in weaponry:

"In the study, development, acquisition or adoption of a new weapon, means or method of warfare, a High Contracting Party is under an obligation to determine whether its employment would, in some or all circumstances, be prohibited by this Protocol or by any other rule of international law applicable to the High Contracting Party."

The 1980 Convention provides for a review procedure, which ought to be used regularly. Such reviews could evaluate the effectiveness of the provisions of the Convention and also take timely preventive measures in relation to new developments, whether entirely new weapons or new designs of existing weapons, that are likely to create problems.

III. The need for a reinforcement of the complementary roles of international humanitarian law and arms control law in the light of present circumstances

The proliferation of non-international and unconventional armed conflicts, in which the combatants have access to modern weaponry by various means, has resulted in conflicts that are far more murderous and damaging than in the past. As the case of mines has shown, the wide availability of small weapons has contributed to a situation which, if left unchecked, is likely to grow worse.

This fact requires a fundamental review of how to use humanitarian law and arms control law most effectively in order to limit the damage caused by these spreading conflicts.

Arms control and disarmament law

Arms control and disarmament law has for the most part concentrated on containing the threat caused by the existence of nuclear weapons and, for the last two decades, on biological and chemical weapons. More recently, however, international attention has been drawn to the dangers of the unsupervised trade in conventional weapons, although this is at present limited to an optional register of the transfer of certain conventional weapons.

This report will indicate that serious consideration should be given to extending disarmament and arms control measures to support new regulations on the use of mines and possibly other weapons.

International humanitarian law

International humanitarian law originally controlled the damage caused by weapons by altogether prohibiting the use of weapons that were perceived as excessively cruel or "barbaric". The centuries-old customary prohibition of the use of poison was based on the perception of its treacherous nature and the fact that poisoned weapons inevitably caused death. The prohibition of the use of explosive bullets by the St Petersburg Declaration was similarly based on the wish to outlaw weapons which inflicted excessively cruel injuries or which normally killed the victim. Subsequently humanitarian law prohibited the use of expanding bullets (dum-dum bullets) in Hague Declaration IV, 3 of 1899 and chemical and biological weapons in the Geneva Protocol of 1925.

Since 1925, however, international humanitarian law has not made any significant progress in prohibiting the use of specific weapons, but has instead concentrated on imposing limitations on their use in the hope of sparing the civilian population as far as possible.

However, this approach has grave shortcomings in that it assumes that all concerned will in fact abide by the rules regulating the use of weapons and that this will indeed spare civilians from the effects of the weapons in question. In reality neither of these assumptions is correct, for not only are weapons in practice used indiscriminately by a very large proportion of the persons that have them, but also, even if they are used correctly, civilians frequently suffer their "incidental" effects. The result is that unless the use of certain weapons is altogether prohibited, civilians will inevitably become victims of them. Further, the rule prohibiting the use of weapons of a nature to cause unnecessary suffering or superfluous injury to combatants is still a valid legal rule, but unless it is applied to new weapons it will fall into desuetude.

One should think very seriously, therefore, of returning to the system of altogether prohibiting the use of weapons whose effects are particularly cruel and whose use is not indispensable.

The complementary effects of humanitarian law and arms control law

Given the reality of the proliferation and transfer of weapons, it is evident that prohibiting the use of a certain weapon will not completely prevent its use if the weapon continues to be manufactured and stockpiled. Therefore a prohibition on use is more effective if it is accompanied by arms control and disarmament measures, which should include verification mechanisms.

Conversely, it is unrealistic to assume that certain restrictions on the transfer of weapons will in practice prevent these weapons from reaching prohibited destinations, and still less that these restrictions will be sufficient to prevent them from being used in the many types of conflicts around the world.

The example of the development of legal restraints on chemical weapons

The problem of the potential development of chemical weapons was first addressed at the First Hague Peace Conference of 1899, which adopted a Declaration (IV, 2) prohibiting the "use of projectiles the only object of which is the diffusion of asphyxiating or deleterious gases". This Declaration did not receive the complete support of the major nations at the time and, of course, was not accompanied by any verification mechanisms.

It was the use of chemical weapons in the First World War, and the fact that public opinion was horrified by the effects of chemical weapons on soldiers, that led to the move to firmly ban their use in the 1925 Geneva Protocol. It is interesting to note that most delegates to the diplomatic conference did not make a minute legal analysis of the effects of chemical weapons as compared with other weapons, or make a careful assessment of their military necessity as compared with the suffering they caused, but rather boldly stated that the use of these weapons was "barbaric" and "horrific" and therefore to be outlawed.

At the time it was suggested that legal measures taken should be limited to a ban on the export of chemical weapons. However, the majority of States thought it important to make a statement of principle that their use was prohibited. History has certainly proved that a mere export ban would not have prevented chemical weapons from being used like any other weapon, as it would not have had the effect of stigmatizing chemical weapons that outlawing them did.

It is certainly also true that if the treaty regulating chemical weapons had merely indicated that they should only be aimed at military objectives, with the usual provisions to limit incidental civilian injury, as may well have been the case had the issue arisen not in the 1920s but some decades later, the situation would be very different now.

Subsequent experience showed, however, the need for additional prohibitions on the manufacture and stockpiling of chemical weapons, together with effective verification mechanisms.

Therefore experience, which took almost a century to develop in the case of chemical weapons, has revealed the need to take probable new weapon developments seriously, to take preventive measures through the total prohibition in principle of weapons that are likely to be particularly damaging, and to back these up with effective disarmament and arms control measures.

IV. The role of the International Committee of the Red Cross

Pursuant to its mandate to work for the mitigation of suffering caused by armed conflicts, and in particular to work for the faithful application of international humanitarian law and to prepare its development, the ICRC has over the years taken a number of initiatives in relation to weapons.

One of these initiatives was the appeal made by the ICRC to governments and to the League of Nations to take action to prohibit the use of chemical weapons, which contributed to the adoption of the 1925 Protocol.

Work on the 1980 Weapons Convention was initiated at a Conference of Government Experts which was convened by the ICRC and which met for several weeks in 1974 in Lucerne and again in 1976 in Lugano. The ICRC had prepared a preliminary report for this Conference based on consultations with experts and subsequently published the report of the Conference, which was later used as a basis for the United Nations Conference that adopted this Convention.

Purpose and structure of this report

The present report, which the ICRC has prepared for the forthcoming Review Conference, is intended to serve as a working document for the Group of Governmental Experts that will prepare the Conference. It is divided into two main parts.

Part II concerns the problem of landmines and is an analysis of the advantages and difficulties of the various proposals that have been made to amend Protocol II in order to achieve better regulation of landmines. This analysis takes into account the recommendations of the participants in the Montreux Symposium and of the military Symposium which the ICRC hosted, as well as other proposals.

Annex I to this report contains a summary of the principal findings of the Montreux Symposium; the full report has been sent to all States and is available from the ICRC.

The results of the Symposium of military experts are reproduced in full in Annex II.

Part III briefly examines other issues of relevance to the Convention which could be examined once progress has been made on the issue of landmines. The ICRC is of the opinion that it would be appropriate for these subjects to be on the agenda of the Review Conference, even if they do not necessarily result in agreed amendments or further protocols, and that more complete documentation will facilitate careful analysis once the Review Conference turns its attention to these subjects.

PART II: MINES

Section I Humanitarian, legal and military aspects of landmines: need for thorough discussion

Every year, thousands of men, women and children are victims of anti-personnel mines. Landmines not only kill but mutilate horrendously, strike blindly at all human beings alike, and continue to spread terror for years or even decades after the hostilities have ended. The effects of mines are frequently inconsistent with certain fundamental rules of international humanitarian law which require parties to distinguish between civilians and combatants, prohibit the use of indiscriminate weapons, and also prohibit the use of weapons that are liable to cause excessive suffering.

The magnitude of the damage caused by anti-personnel mines, in terms of both human suffering and long-term socio-economic destabilization, as witnessed by ICRC delegates and medical teams, prompted the ICRC to hold the Montreux Symposium on anti-personnel landmines in April 1993. It was recognized that the problems created by the use of mines are complex and multifaceted and that there would certainly not be a single solution to the present situation. Therefore, in addition to the surgical and orthopaedic needs of mine victims, the Symposium discussed the difficulties of mine clearance, the actual military use of mines in different situations, the technical construction of mines and possible developments, the manufacture and trade in mines, and, of course, the present legal regulation of mines and its shortcomings. It was clear that further legal regulation would require consideration of all the factors concerned in order to arrive at the most effective solution.

During the Montreux Symposium, the need was felt to secure a wider and more detailed military viewpoint on the operational use of landmines. Towards this end, the ICRC organized a Symposium of military experts on the military utility of anti-personnel landmines, in January 1994.

Annex I contains the results of the Montreux Symposium. These have been grouped under five themes: humanitarian, medical and socio-economic costs of landmines; prohibition of the use of certain types of mines; proposals for modification of the 1980 Convention and of its Protocol II; possible arms control and disarmament measures; and information to the public.

Annex II reproduces the results and proposals of the Symposium of military experts.

Possible amendments to Protocol II of the 1980 Convention

Section II of this Part will now give an overview of the various amendments to Protocol II that were proposed during the two symposia organized by the ICRC and by other persons or bodies. All suggestions pertinent to the possible amendment of Protocol II have been included. They are not necessarily alternative suggestions but could be combined. This report analyses the advantages and possible disadvantages of each proposal and indicates the conditions that would be necessary to make the proposal most effective.

For ease of reference, the proposals are presented in two groups:

- proposals on the prohibition of certain types of mines;
- proposals for further regulations on how mines are used.

Section II Analysis of various proposed amendments to Protocol II

A number of proposals were made by experts during the two symposia the ICRC hosted as well as by other persons or bodies on possible approaches that could be adopted to improve the situation caused by landmines. Most of these proposals have their advantages and difficulties and need to be considered in the light of a number of relevant factors. With a view to helping the Review Conference find the most realistic and effective solution possible, this part of the report will briefly analyse each of the proposals bearing in mind the considerations that need to be taken into account by the Review Conference.

I Proposals suggesting the prohibition of the use of certain types of mines

The Review Conference could amend Protocol II of the Convention by introducing a prohibition on the use of certain types of mines or a prohibition on the use of mines which do not have certain features. As this Convention is a humanitarian law treaty, it can only address the regulation or prohibition of the use of certain types of mines.

However, given the low price and widespread availability of mines, it is clear that any new prohibition relating to use must be accompanied by appropriate arms control/disarmament measures in order to make the rule effective. Therefore discussions during this Review Conference should take into account the arms control measures that would be necessary in order to choose which would be in fact the most effective rule.

1. Prohibition of the use of all anti-personnel mines

This is the proposal that was supported by a number of participants at the Montreux Symposium and is being put forward by many non-governmental organizations, Senator Leahy of the USA and other influential individuals and organizations.

There is no doubt that from the humanitarian point of view this would be the best option, as a total ban would have the effect of stigmatizing the use of mines and a violation of the rule would be easily provable. Although it is recognized that lawless groups would be likely to make their own explosive devices if anti-personnel mines were not available, such improvised devices would not be available in the vast quantities that anti-personnel mines are and the problem would therefore be reduced. The advantage of this option over a ban on anti-personnel mines without self-destruct mechanisms is that the latter would still be around for the duration of their life and, in the case of internal armed conflicts that frequently last for years, if not decades, they are still likely to cause civilian casualties. It is recognized, however, that with self-destruct mines there would not be the cumulative effect that results from decades of laying live mines. Another advantage of a total ban is the fact that there would not be the danger of the technical failure of self-destruct or self-neutralizing devices and that refugees could immediately return home at the end of hostilities.

In order to be effective, however, this option would require the following:

(i) General agreement among States, which would need to weigh the advantages and disadvantages of the continued use of anti-personnel mines. The meeting of military experts came to the conclusion that anti-personnel mines are the most effective means of achieving the objectives for which they are used. Alternative systems would require greater resources and the armed forces would be likely to suffer greater losses during the conflict. However, these considerations are to be balanced against the loss of and damage to innocent civilian lives, the loss of agricultural land, and the enormous resources necessary for mine clearance and for care of the victims of mines. These negative effects of mines would not be altogether avoided by the use of mines that self-destruct or self-neutralize, for the reasons outlined above.

(ii) There should also be an arms control agreement to ban the manufacture and stockpiling of anti-personnel mines, together with appropriate verification procedures. This verification might not have to be quite as strict as would be necessary if it were agreed that mines need self-destruct and self-neutralizing mechanisms, as the stigmatization of mines would have its own effect (as with the banning of the use of chemical and biological weapons in 1925).

(iii) There would need to be a very careful definition of anti-personnel mines that are to be banned, especially as dual-use systems exist and any vagueness in the wording would allow the prohibition to be evaded.

(iv) It would be useful to have an additional requirement that anti-tank mines must be both detectable and fitted with neutralizing mechanisms, for although these mines are causing fewer casualties than anti-personnel mines they are still very dangerous and ought not be left around for an indefinite period of time.

2. Prohibition of the use of scatterable (remotely delivered) mines that are not fitted with self-destruct mechanisms

This was one of the proposals of the Symposium of military experts. The advantage of the incorporation of self-destruct mechanisms is that it would reduce the need for mine clearance and also cause fewer civilian casualties in the long term. Civilian casualties from mines would not be avoided as they would still occur during the active life of the mines and would continue to be caused in the longer term by mines whose self-destruct mechanism failed to function. Civilian casualties are particularly likely to persist in internal armed conflicts which typically last for years and sometimes even decades, and where widespread mining tends to occur. However, one would avoid the build-up effect of the laying and relaying over many years of minefields that remain active throughout the hostilities and for years afterwards.

In order to make this proposal both effective and realistic, the following conditions need to be met:

(i) There must be general agreement on the proposal, otherwise those who do not agree will continue to use scatterable systems without self-destruct mechanisms.

(ii) There must be an arms control agreement to the effect that only scatterable mines with self-destruct mechanisms can be manufactured, and appropriate verification measures instituted. This is essential as there would be a temptation to manufacture scatterable mines without self-destruct mechanisms, which would be cheaper and therefore assured of a market.

(iii) There must be a defined time-limit on the active life of the mine once it is laid. This is critical, as without such a specification the regulation is meaningless. From a humanitarian point of view, the time-limit should be as short as possible. If the mines in question continue to be active for years, there will be no reduction in the damage that mines inflict on the civilian population.

(iv) The mechanism chosen must not depend on the good faith of the user for its correct functioning. Practice has shown that one cannot rely on proper behaviour. The self-destruct mechanism must therefore be incorporated into the mine in such a way that it cannot be easily tampered with and that the mine will automatically self-destruct at the end of a given period once the mine is activated.

(v) The mechanism chosen must have an extremely low failure rate. As scatterable mines are laid thousands at a time, any failures will continue to pose a serious threat to the civilian population.

(vi) This proposal should not be taken in isolation, as the problems caused by mines emplaced by hand or vehicle are equally serious.

3. Prohibition of the use of anti-personnel mines that are not fitted with self-destruct mechanisms

This proposal, which was put forward by participants at the Montreux Symposium, would apply to all anti-personnel mines, whatever their method of delivery and intended use.

Self-destruction in the case of anti-personnel (AP) mines was considered preferable to self-neutralization for a number of reasons and the Symposium of military experts concurred with this, with the possible exception of jumping and directional fragmentation (Claymore) mines (see point 5 below).

There would need to be a definition of "anti-personnel mines" for the purpose of such a rule and, in order for the rule to be effective, the considerations outlined in point 2 above (scatterable systems) would also apply here.

As with the proposal to incorporate self-destruct systems in scatterable mines, this proposal would certainly reduce the numbers of civilian casualties, although it would not eliminate the problem.

4. Hand-emplaced anti-personnel mines used for tactical purposes and scatterable mines should have a self-destruct mechanism, but hand-emplaced mines used for long-term and barrier minefields need not have such a mechanism.

The participants in the Symposium of military experts made a distinction between hand-emplaced mines that are used for tactical purposes during a given conflict, on the one hand, and those used for long-term protective purposes, such as protecting a border, on the other. In their opinion, hand-emplaced mines that do not self-destruct should still be permitted for long-term and barrier minefields, but would have to be used under tightly controlled circumstances.

Although these are clearly two different military uses, there are serious objections to basing an international regulation on this distinction:

(i) If hand-emplaced mines continue to be manufactured, there is at present no means of controlling that they are sent only to countries that need them for such long-term barrier minefields and that they are indeed only used for that purpose.

(ii) Very few mines would need to be manufactured for these types of minefields because of their limited number. Further, if the mines do not self-destruct, they do not need regular replacement. How could one control that only this small amount is manufactured and by which companies? Again, how could one control to which countries they are sent, according to which criteria and on whose decision?

(iii) Given the above difficulties, and the very great danger from a humanitarian point of view of allowing such "dumb mines" to continue to be manufactured, it would be preferable to require that all AP mines have self-destruct mechanisms. This would mean that barrier minefields would need to be replaced regularly (probably every year). This is essentially a financial problem but it should be seen against the enormous cost of demining that would otherwise continue to be needed as well as the medical, social and infrastructural costs that result from the use of "dumb mines".

5. Directional fragmentation mines do not necessarily have to be fitted with a self-destruct mechanism, but jumping mines must be fitted with either a self-destruct or a self-neutralizing mechanism.

The military experts drew a distinction between these anti-personnel mines and point-detonating mines, as directional fragmentation mines and jumping mines have a much greater radius of lethality. They pointed out the disadvantage in equipping these types of anti-personnel mines with self-destruct mechanisms: there would be a greater danger to anyone who might be passing at the moment of self-destruction. The difficulty with self-neutralization is that one cannot be certain that the mechanism has functioned, and parts of the mine might still be available for reuse.

It was thought that directional mines would be reused and are therefore less likely to be left in place to pose a threat to civilians. It will have to be established whether one can assume that these mines will indeed not be left lying around, or whether it would be safer to equip them with a self-neutralizing mechanism that would take effect after a certain period if the mine is not switched off for reuse.

With regard to jumping mines, it will have to be assessed whether it would be safer to provide for the incorporation of a self-destruct mechanism or a self-neutralizing mechanism, taking into account the advantages and disadvantages of each system.

6. Prohibition of the use of mines that are not detectable

This proposal was put forward by the participants in both symposia. The military experts agreed that in future anti-personnel mines should be manufactured so as to be detectable. However, they were unable to agree on the feasibility of rendering existing stocks of mines detectable. The problem of future stocks will be considered separately below.

The proposal to require that mines be made detectable is a useful one, although it should be combined with one or more of the other proposals in this section in order to have a real impact on the worldwide problem of mines. The requirement must provide, however, that the detectable element in the mine cannot be easily removed. There must also be verification that mines not conforming to these specifications are not manufactured.

7. Prohibition of the use of anti-tank mines that are not fitted with self-neutralization mechanisms

Anti-tank mines were not discussed at any length during the ICRC-hosted symposia, as these concentrated on anti-personnel mines. However, it was indicated during the Symposium of military experts that anti-tank mines frequently have a self-neutralizing device. This is because these mines are expensive and therefore frequently need to be reused, and to equip them with self-destruct devices would be too dangerous for passers-by.

It would therefore be useful to require that all anti-tank mines be fitted with neutralizing mechanisms, and it would have to be verified that only this type is manufactured.

8. Prohibition of the use of mines with anti-handling devices

This was a proposal put forward at the Montreux Symposium. It was felt that the military purposes of these devices, that is, to lower the morale of the enemy and act as a deterrent to breaching a minefield, did not justify the difficulty they create for mine-clearance operations after hostilities.

It was pointed out during the meeting of military experts that anti-handling devices are fitted to anti-personnel mines used to protect anti-tank mines. This makes it far more difficult for the opposing forces to remove the anti-tank mines. However, the major problem is that these anti-handling devices render mine-clearance efforts extremely difficult and dangerous. The majority of the participants in this meeting agreed that if anti-personnel mines fitted with anti-handling devices were also equipped with self-destruct mechanisms there would be less of a problem, as instead of attempting to clear the minefield one would simply wait until the mines had self-destructed.

The difficulty with this approach is that:

(i) It may well be necessary to clear an area before the active life of the mines has expired, and the longer the active life of the mines, the more likely this will be.

(ii) If there is no agreement on and effective implementation of a new rule that all mines are to have self-destruct or self-neutralizing devices (self-destruct for point-detonating anti-personnel mines and self-neutralization for jumping and possibly Claymore anti-personnel mines and anti-tank mines) then the problem of anti-handling devices will remain as acute as ever.

This proposal would, of course, have to be combined with some of the other proposals as by itself it would not have any major effect on the problems caused by the use of mines.

9. Existing stocks should be modified to be in conformity with the new law or destroyed.

Although this is an arms control or disarmament measure and therefore is not actually a proposed amendment to Protocol II of the 1980 Convention, it is a subject that should be kept in mind during the Review Conference.

It is estimated that in existing stocks there may be up to one hundred million mines, most of which are undetectable and do not incorporate self-destruct or self-neutralization mechanisms. Should the Review Conference decide to render illegal the use of certain types of mines (e.g., anti-personnel mines without self-destruct mechanisms), the question will arise as to what should be done with these stocks. States which accept the new prohibition would not themselves be able to use them and, in the case of small anti-personnel mines, it would not be possible to add such mechanisms.

There is a great danger in allowing such stocks to continue to exist, for they are likely to be used in one way or another, making the massive problem that has been created worldwide by the use of such mines much worse. It would be simple to add a means of detecting the mine, by applying a metal strip for example, but this in itself would not prevent the enormous numbers of civilian casualties that will continue to occur before comprehensive mine-clearance is undertaken. There would also need to be verification that this modification to mines in stock has indeed been carried out, but given the limited result this measure would have in humanitarian terms, it is questionable whether such a proposal is worthwhile.

The only major difficulty relating to the destruction of stocks is financial. However, the cost of destruction and of possible restocking with the new mines has to be compared with the enormous cost of mine-clearance operations, which would have to be stepped up if existing stocks continue to be used. This is in addition, of course, to the cost of medical care, loss of agricultural land, etc.

II. Proposals on further regulations on how mines are used during an armed conflict and cleared after hostilities

The existing provisions of Protocol II of the Convention relate exclusively to the way mines are to be used during an international armed conflict and to clearance operations after active hostilities. The effectiveness of these provisions depends entirely on combatants behaving in conformity with the law.

Although this is the case with all humanitarian law rules, non-compliance with the law has particularly serious effects in the case of the use of mines because of the fact that they continue to be active for such long periods. Prohibition of the use of certain types of mines is therefore necessary, as one cannot ensure behaviour in accordance with the law even in areas where various implementation mechanisms exist.

However, the participants in both ICRC-hosted symposia felt that rules on the way mines are used are still necessary. The Montreux Symposium looked at some of the shortcomings of Protocol II as it now stands, and both this Symposium and that of the military experts made some suggestions on possible improvements.

1. Introduce implementation mechanisms

The experts noted that a major shortcoming of the 1980 Convention is its lack of implementation mechanisms. Although it may be possible to conceive of implementation mechanisms appropriate only for the use of mines and booby-traps, and therefore incorporated in Protocol II, it is proposed that the more logical place to introduce implementation mechanisms is in the body of the Convention itself. This proposal will therefore be examined in more detail in Part III of this report.

Possible arms control/disarmament measures relating to the manufacture, stockpiling and transfer of mines were examined by participants in the Montreux Symposium. Such measures would probably not be included in Protocol II to the

1980 Convention as it is a humanitarian law treaty, but they would be necessary in order to render new rules on the use of mines effective.

2. Extend the applicability of the law to non-international armed conflicts

As the majority of conflicts are non-international, and as the appalling situation caused by the widespread and indiscriminate use of mines has occurred primarily as a result of these conflicts, it would make sense to extend the applicability of the law on the use of mines to non-international armed conflicts.

However, such an extension of the applicability of the law would normally be effected by an amendment to the Convention itself which specifies its scope of application. This proposal will therefore be considered in Part III of this report.

It should be noted, however, that in regard to the use of mines a mere extension of the applicability of the Convention to non-international armed conflicts is even less likely to have the effect of ensuring respect for the rules than is the case in international armed conflicts. There will inevitably be more violations of the law in non-international conflicts, and the report of the military experts indicates why insurgents are likely to continue relying extensively on the use of mines, especially if they are easily available. It is of particular interest, therefore, to seriously consider a total ban on the use and manufacture of certain types of mines so that forces in these conflicts do not have access to them. It is recognized that this will not stop insurgent forces from making explosive devices by hand and that each of these devices is likely to be very dangerous. However, there is a limit to the number they could make and they therefore would not be able to strew vast numbers of them around as is presently the case with mines manufactured on a large scale.

3. Introduce stricter rules relating to the precautions that should be taken to protect civilians

The participants at the Montreux Symposium noted that the duty to take precautions to protect the civilian population that is contained in Article 3 of Protocol II is very weak in that it is limited to taking "all feasible precautions". Some participants suggested that the word "feasible" should be removed as it leaves too large a loophole, but others were of the opinion that this would place an impossible burden on the military in some situations.

The Symposium of military experts thought that all efforts should be made to mark minefields by fences or other means, even if they comprise mines that self-destruct or self-neutralize. They also suggested that there should be a duty to mark scattered minefields if at all possible, although they recognized that this is less feasible in most circumstances. Such provisions are in fact already included in Protocol II as it stands, although the reference to precautions in Article 3 could be rendered more specific by indicating that fences, etc., are to be used wherever possible for all types of minefields.

In any event, the report of the Montreux Symposium shows that one cannot put too much trust in the rule requiring such precautions, not only because they are frequently not taken, but also because experience has shown that markers such as fences and signposts tend to be removed by the local population for their own use as they are seen as valuable items.

4. Introduce stricter rules on the recording of minefields

The participants in the Montreux Symposium criticized the provisions of Protocol II relating to the recording of minefields. There is a duty to record "pre-planned" minefields (Article 7) but no definition of this term. With regard to all other minefields, parties are only required to "endeavour" to record. A more careful definition of the content of the legal duty to record should be considered; it was suggested that records should also indicate the types of mines used.

The Symposium of military experts suggested that the recording of minefields should be required even in the case of mines that self-destruct or self-neutralize. With regard to mines that are scattered by aircraft or artillery, their general area of use should be recorded. Self-destruct or self-neutralization times should also be recorded.

Although the situation would be improved if these proposals were carried out, one cannot place too much reliance on the duty to record for the following reasons:

(i) in the light of past experience, even if there were an absolute rule that all minefields (or at least their general area of use) had to be recorded without exception, it is unlikely that the rule would be generally respected, especially in non-international armed conflicts;

(ii) in the confusion of war, records frequently get lost;

(iii) the recording of minefields is less effective than is generally expected as mines (especially scattered ones) frequently move to quite different locations owing to rainfall (which can displace mines kilometres from their original position) and to the movement of the soil or sand.

5. Introduce the requirement that self-destruct times and other minefield information should be declared to all parties at the end of the hostilities

This would be a useful step if it were carried out, but there are reasons to doubt that such a provision could be relied on:

(i) States did not accept a rule to this effect during the negotiation of the 1980 Convention, and Article 7, para. 3(a), sub-paras. (ii) and (iii), indicate only that such information is to be given once territory is no longer occupied by the adverse party. This reticence may still exist.

(ii) Such information will be available only if records have indeed been made and kept, and will be useful only if the mines have not moved to any great degree.

6. Extend the duty to take measures for the protection of third-party forces or missions

The participants in the Montreux Symposium thought that the duty to take certain measures for the protection of United Nations forces or missions currently contained in Article 8 should be extended to include other missions, such as those undertaken by the Conference on Security and Cooperation in Europe (CSCE). Some of the provisions could also be of use for the protection of mine-clearance organizations or humanitarian agencies that are attempting to work in the area.

If complied with, these rules would help alleviate the problems that arise after hostilities. It should be noted that the duty created by Article 8, para. 1, is conditional on whether the party concerned "is able" to undertake the measures. It is uncertain whether this can be improved on, especially as a large proportion of mines are laid not by government troops but by insurgents.

7. Introduce stricter rules on the duty to clear minefields at the end of active hostilities

The participants in the Montreux Symposium stressed that at present no one has responsibility for clearing minefields, and that this is a major shortcoming of the law. They recognized, however, that finding an acceptable and effective solution in this respect may be difficult.

Some participants in the Symposium of military experts expressed the view during discussion that, in principle, those who laid the mines should be responsible for their removal. A suggestion was made during the Montreux Symposium that the Security Council could determine who should pay for mine clearance and who should carry it out.

8. Introduce rules to prevent the indiscriminate effects of unexploded sub-munitions

The symposia that the ICRC hosted did not deal directly with the problem of unexploded sub-munitions, and therefore did not make any particular proposals in that regard, as it does not fall into the category of mines. However, participants in both meetings pointed out the extensive danger caused by these remnants of war which in many respects constitute the same type of hazard to civilians and clearing difficulties after hostilities as mines do. However, some of the issues involved are different from the question of mines. This subject will be looked at again in Part III below.

PART III: SUBJECTS RELATED TO THE CONVENTION ITSELF AND TO POSSIBLE ADDITIONAL PROTOCOLS

Section I Possible amendments to the Convention

I. The introduction of implementation mechanisms

The total lack of implementation mechanisms in the 1980 Convention is a problem that should be addressed during the Review Conference.

The Symposium on mines that the ICRC hosted in Montreux, while recognizing the limits of the implementation mechanisms provided by international law, proposed that certain implementation provisions be incorporated into the main body of the 1980 Convention. With this in view, the participants looked first at implementation mechanisms in other humanitarian law treaties that could be used in the 1980 Convention and then at other international law mechanisms that could be useful.

(i) Proposals on implementation mechanisms stemming from those that exist in other humanitarian law treaties

Insofar as the provisions of the 1980 Convention reaffirm the rules of international humanitarian law found in other treaties, implementation measures provided for in those other treaties are naturally also relevant to the 1980 Convention. However, it may be desirable to specifically include such measures in the 1980 Convention.

Provision of legal advisers

This is presently required by virtue of Article 82 of Protocol I additional to the Geneva Conventions. A similar provision in the 1980 Convention could indicate that legal advisers should give guidance on matters relating to the use of weapons. The participants in the Montreux Symposium recommended that legal advisers be appointed at all levels down to brigade or equivalent level and be incorporated into planning staffs.

Specific requirements for training in humanitarian law

Several of the participants in the Montreux Symposium placed great stress on the importance of correct training. The requirement to instruct the armed forces in the law is provided in Hague Convention IV of 1907, the four Geneva Conventions of 1949 and their Additional Protocols of 1977. It was thought that such a requirement ought also to appear in the 1980 Convention. Certain specific suggestions were made in this regard:

– there should be training in the use of weapons in accordance with humanitarian law at cadet academies and in all command and staff training programmes;

– manuals on weapon systems should incorporate the law applicable to their correct use in the languages of the user countries;

– the packaging of weapons should include warnings of the legal limitations on their use;

– all military training of foreign nationals should include training in humanitarian law.

Incorporation into domestic law

The 1980 Convention should be translated into local languages, and appropriate national laws and regulations should be adopted. This is similar to the provision in Article 84 of Protocol I of 1977.

Liability and criminal sanctions

It is clear that the law of international responsibility applies in relation to violations of the law governing the use of mines. The difficulty lies in determining liability with respect to compensation for damage resulting from violations of the law, and establishing which body should be responsible for making such decisions. The possibility of compulsory adjudication will be considered below.

With regard to individual liability, the participants in the Montreux Symposium thought that as a matter of principle criminal sanctions ought to be obligatory for violations of the rules contained in the Protocols of the 1980 Convention. However, they recognized that similar provisions in the 1949 Geneva Conventions and Protocol I of 1977 have not usually been respected.

If Protocol II to the 1980 Convention were to be amended to prohibit the use of certain types of mines, violation of such a rule would be much easier to establish than violation of the present rules, which place certain constraints on behaviour only. The experts at the Montreux Symposium made some general suggestions as to how to improve implementation of the rule requiring the application of criminal sanctions.

International Fact-Finding Commission

It was suggested that the International Fact-Finding Commission provided for in Article 90 of Protocol I additional to the Geneva Conventions could also be used to investigate possible violations of the 1980 Convention. In the context of the 1977 Protocol, the competence of the Commission is based on consent that can either be given in advance, in the form of a declaration, or ad hoc. It would have to be decided whether the same formula would be appropriate for the 1980 Convention and whether it should also be based on confidentiality, as provided for in the 1977 Protocol. The participants in the Montreux Symposium pointed out that the Commission would be more effective as a law-enforcement mechanism if it had an automatic right to monitor possible violations of the 1980 Convention.

(ii) Other possible implementation mechanisms

Compulsory adjudication

The following are possibilities:

– International Court of Justice. The compulsory jurisdiction of the ICJ is provided for by several treaties and a similar provision could be incorporated into the 1980 Convention. The disadvantage is that the ICJ has jurisdiction only over international disputes and cannot cover individual accountability.

– International arbitration. This depends on a certain degree of cooperation by the parties involved in order to create the arbitral tribunal and its regulations.

– International criminal court. The United Nations International Law Commission is at present studying the possibility of setting up such a court. However, the suggestion of establishing such a court has existed for a long time and it is not likely to materialize in the near future.

– A court created especially for the 1980 Convention. A number of international treaties create courts for the implementation of their rules by deciding on allegations of violations and sometimes also by delivering advisory opinions. These courts are frequently very effective law-enforcement mechanisms as they often have jurisdiction not only over inter-State disputes but also over cases brought by individuals or organizations.

The participants in the Montreux Symposium thought that it was unlikely that States would accept compulsory adjudication, as this does not at present exist in any international humanitarian law treaty. However, there is no doubt that it could be a very effective method, especially if individuals or organizations were able to bring claims.

United Nations Security Council

It was suggested during the Montreux Symposium that, in the absence of a compulsory adjudication mechanism, the Security Council might be able to impose suitable remedies for violations of the 1980 Convention. However, this would depend on the political will of the members of the Council.

Creation of a supervisory body

A number of international treaties create specific supervisory bodies to help the implementation of their provisions. These bodies typically receive periodic reports submitted by States Parties on the measures they have taken to implement the treaty, receive complaints about alleged violations, undertake investigations and discuss the results of these activities with the States concerned. They also often undertake promotional activities in order to improve compliance with the law.

The Review Conference could consider whether it would be appropriate to create an analogous body for the 1980 Convention or whether the terms of reference of the International Fact-Finding Commission could be extended to cover these roles for the purposes of the 1980 Convention.

II. Extension of the scope of application of the Convention to non-international armed conflict

The 1980 Convention at present formally applies only to international armed conflicts, even though the majority of conflicts are internal. The laying of millions of mines during non-international armed conflicts has caused not only tremendous immediate suffering but also severe social and economic damage to the countries concerned.

The major rules of international humanitarian law already apply to non-international armed conflicts by virtue of Article 3 common to the Geneva Conventions of 1949, Additional Protocol II of 1977 and international customary law. However, specific rules applicable to the use of weapons in non-international armed conflicts would give useful precision to the law.

The participants in the Montreux Symposium thought that the 1980 Convention ought also to apply to non-international armed conflicts, although they recognized that this may be a sensitive issue. They felt that given the enormous damage that is frequently inflicted on the assets of a State by the widespread improper use of weapons in an internal armed conflict on its territory, there may be greater interest in making these international rules applicable in such conflicts.

The most obvious way to extend the application of the 1980 Convention to non-international armed conflicts is by an amendment to Article I of the Convention. Should this cause too much difficulty, another possibility is to create an optional protocol to this effect, or to introduce into the Convention an additional article which parties could accept by a declaration to that effect or, conversely, which would be applicable to parties unless they specifically opted out. All three methods are to be found in other international treaties when a specific provision is desired by some of the parties but not by others.

As indicated in Part II above, additional arms control measures are vital to limit the damage inflicted during internal armed conflicts.

Section II Specific weapons

I. Blinding weapons

The ICRC is of the opinion that blinding weapons should be on the agenda of the Review Conference with a view to the possible adoption of an additional protocol on this subject. Given the advanced stage of development of hand-held versions of this type of weapon, together with the real possibility of their appearance on the battlefield in the near future and their subsequent proliferation among all groups that use force, it is essential that the Review Conference use this last opportunity to take preventive action.

(i) Information gathered at expert meetings convened by the ICRC

Prompted by reports concerning the development of certain types of laser weapons which would result in permanent and incurable blindness, the ICRC convened four meetings of experts. The meetings were attended by leading specialists in laser technology, ophthalmology, military medicine and psychiatry, and international humanitarian law. The scientists described the nature and effects of these laser weapons and the physical, psychological and social effects of blindness as compared with other combat injuries. Subsequently, the legal and policy aspects of this issue were discussed, together with possibilities for future legal regulation.

Technical characteristics of new laser weapons and their effects in medical terms

The specialists gave information on laser weapons under development as reported in unclassified sources.

A number of weapons were said to be designed for anti-sensor or anti-personnel use. "Anti-sensor" use refers to the destruction of enemy optical viewing systems, whereas "anti-personnel" use refers to an intentional effect on people's eyesight. As the energy and wavelength of the laser necessary to destroy sensors is similar to those necessary to damage eyes, laser systems said to be designed for anti-sensor purposes could also be used for anti-personnel purposes.

With regard to current technical possibilities for the further development of anti-personnel laser systems, the experts stressed that lasers can be very small and pointed out that small, clip-on laser devices that can now be fitted to rifles for training purposes could easily be made non-eye-safe. At present the range of these training devices is relatively limited but more powerful ones are being designed. It was also indicated that lasers can be very cheap. The group further pointed out that range-finding systems (which are less powerful than the anti-sensor/anti-personnel lasers being developed) could be misused to blind intentionally and that some accidents have indeed already occurred with these.

With regard to the effect of these lasers on the eye, it was indicated that the extent of damage to the eye will depend on the energy and distance. The anti-personnel and anti-sensor weapons presently under development will permanently blind a person up to a distance of a kilometre or more. Beyond this distance a person may be flash blinded, or even further away may be dazzled if a visible wavelength is used. The exact distance at which there is no longer a permanent blinding effect is unpredictable because a laser beam is affected by atmospheric conditions and dust. The aiming of the beam does not appear to be particularly difficult as it can be diverged to an area of about 50 cm across at a range of one kilometre, and the very large number of shots in each battery pack means that it is possible to sweep the battlefield with the beam. The weapon is silent and the beam is invisible.

The specialists then studied the possibilities for medical treatment and means of protection and concluded that neither was adequate. Damage to the retina is permanent and irreparable; vision loss caused by haemorrhage might be successfully treated in only a small minority of cases and even in those cases the long-term outcome would be doubtful.

Protection by special goggles would also seem to be largely illusory, as they would only screen out a limited range of known wavelengths, whereas lasers can operate over a wide range of wavelengths.

Functional disabilities and psychological problems that would be caused by blinding weapons as compared with those caused by other weapons

In making this assessment, the specialists drew attention to a number of considerations specific to blindness:

There is no prosthesis to reduce the effect of the disability, and in functional terms blindness is an exceptionally severe handicap, even when compared with the worst of injuries.

Rehabilitation training for the blind is essential, but is not available everywhere, and it also gives rise to major difficulties:

a. the learning process is long and very complex;

b. a psychologically robust personality is needed to undertake this learning effort, but people who have been blinded usually suffer from severe depression and cannot do it well;

c. comparatively satisfactory results are seen only in persons with a good education and sound financial, family and social support;

d. successful rehabilitation allows recovery of only a fraction of the person's previous skills and he will always remain dependent to quite a large degree.

The experts stressed that blindness almost always causes very severe depression which in a large proportion of cases lasts for many years, if not permanently.

Another matter of importance in a war context is the prevalence of an extreme fear of blindness; for the majority of people it is the most dreaded injury and soldiers are no exception. If soldiers are aware of the existence of weapons that can silently and invisibly blind them, there will be an increased incidence of combat stress disorder during battle and such weapons will cause more mental illness in the long term.

The medical experts thought that public reaction to blindness caused by weapons especially used for that purpose is likely to be very negative, as the public in general tend to feel special pity for blind persons. They likened the fear of blindness and the probable reactions to blindness-inducing weapons to the fear and disgust aroused by chemical weapons.

Finally, the experts pointed out that large numbers of blind persons would put an exceptionally heavy burden on medical and social services and on society in general.

Foreseeable situation if there were to be widespread use of anti-sensory/anti-personnel laser weapons

There would evidently be an increase in the numbers of blind servicemen returning from war. The number of eye injuries has steadily increased from 0.5% in the last century to between 5 and 9 % in the Vietnam war. The increase is said to be due to the effects of fragmentation weapons. It has been estimated that if anti-sensor lasers were used, but not to target the human eye, eye injuries would nevertheless increase by 2–3%. If, however, lasers were to be used intentionally to inflict blindness, so that blinding as a method of warfare became common practice, serious damage to the eye might account for between 25% and 50% of all casualties.

The experts also pointed out that laser weapons could easily be used to cause terror outside armed conflict situations by repressive regimes, terrorists or criminals. Since such weapons are so light and easy to transport, proliferation would be inevitable.

Legal and policy considerations

The final expert meeting was attended by 37 government officials, participating in their personal capacity, from 22 countries. They considered the legal and policy implications of the information gathered by the scientists.

The present lawfulness of the use of blinding weapons was discussed mainly in the light of the rule prohibiting the use of weapons of a nature to cause unnecessary suffering and superfluous injury. One participant was of the opinion that any intentional blinding would violate this rule, including the use of lasers to blind the pilots of aircraft. The majority of participants, however, thought that the most controversial use of lasers would be against infantry, as the latter can easily be put out of action by means other than blinding. There was a division of opinion, however, as to whether such use is already illegal under existing law.

The majority of participants thought that whatever the assessment of the present lawfulness of such use, it should be subject to legal regulation because there are important policy reasons for prohibiting blinding as a method of warfare. Many thought that such a prohibition ought to be introduced simply because blinding weapons are horrific and therefore totally unacceptable. The various possibilities for legal regulation were discussed and are outlined below.

(ii) Possibilities for legal regulation

Humanitarian law

Several approaches have been used to prohibit or restrict the use of certain weapons in international humanitarian law; it is possible to consider which would be the most appropriate by analogy in the case of blinding laser weapons:

a. Prohibition of the use of a certain type of weapon

This was the method used for chemical weapons and dum-dum bullets, because it was recognized that the overall dangers represented by the use of such weapons outweighed their military utility. In the case of laser weapons, this could involve prohibiting the use of all or of some types of anti-sensor/anti-personnel weapons. The difficulty is that these weapons can be used for both anti-sensor and anti-personnel purposes, but it could be decided that those more obviously suited to anti-personnel purposes should be prohibited.

b. Prohibition of certain uses of a particular weapon

Examples of limitations of this type are seen in some military manuals which prohibit the use of incendiary weapons against unprotected soldiers, or state that explosive bullets may be used against objects but not persons. In the case of laser weapons, such a regulation could prohibit the use of lasers against persons, or against certain classes of persons, e.g., infantry.

c. Prohibition of the use of weapons which have a certain effect, without mentioning the weapon by name

An example of this type of provision is Protocol I to the 1980 Weapons Convention, which prohibits the use of any weapon the primary effect of which is to injure by fragments which cannot be detected by X-rays. In the case of laser weapons, a norm of this type could read as follows: "The use of weapons the primary effect of which is to damage eyesight is prohibited."

Such an approach would have the advantage of covering not only lasers whose primary effect is to blind but also any other future weapons which may have this effect. A disadvantage is that such a wording may give rise to arguments as to whether blinding is a primary effect, given that these lasers can also have other uses (anti-sensor in particular), and that at the end of their range they only have a dazzle effect. This wording would not cover intentional blinding by the misuse of other systems such as range-finders.

d. Prohibition of certain types of behaviour without any reference to the characteristics of a weapon

This alternative could concentrate on the prohibition of blinding or of the use of weapons with the primary intention or expected result of permanently damaging eyesight. A norm of this type could be worded as follows:

"blinding as a method of warfare is prohibited",

or

"blinding as a method of rendering a combatant hors de combat is prohibited".

Alternatively, the wording of the rule could be more specific, such as:

"weapons may not be used against persons with the primary intention or expected result of permanently damaging their eyesight".

Such an approach could also include rules that create a duty to take precautions to avoid accidental blinding by weapons that are particularly dangerous for eyesight.

Arms control regulation

States might wish to think about prohibitions or limitations on the production of certain types of lasers that could be too easily misused to blind because of their particular features, e.g., tunability, power, portability. Other possibilities would be regulations to prevent undesirable proliferation, or policies favouring eye-safe lasers for range-finding, etc., in order to prevent avoidable cases of blindness.

II. Unexploded sub-munitions

Unexploded sub-munitions are remnants of war that in many ways represent the same type of threat to the civilian population as anti-personnel landmines.

Sub-munitions are bomblets which are delivered by aircraft or by artillery, rockets or guided missiles. The bomblets are assembled in "clusters" of hundreds or even thousands and delivered from aircraft dispensers, artillery shells or rocket or missile warheads. The bomblets are small (typically under 800 grams and under 7 cm in diameter) and can contain various payloads for use against different targets, such as a high explosive inside a controlled-fragmentation casing, a shaped charge (with or without a fragmentation casing), or a high explosive combined with an incendiary material. They may be fitted with an impact fuse (with or without a delay mechanism) or a proximity fuse.

Unlike landmines, which cause casualties among the civilian population when they are functioning normally, these bomblets create a similar situation as a result of malfunction, namely, when they have not exploded on impact and are left lying on or near the surface of the ground in an unstable condition.

Bomblets have reportedly had very high failure rates – up to 40%, depending on the state of the ground (the rate is usually higher on soft ground) and on meteorological conditions (especially if the soil is covered with snow). Once on the ground and unexploded, some of these bomblets are extremely unstable. They are liable to explode at any time and can be triggered by even the slightest movement of the ground on which they are lying, such as vibrations caused by people walking or a moving vehicle.

Clearing this unexploded munition is very difficult. If the bomblets are in an unstable condition, is it not possible to touch them and they cannot be neutralized. They must therefore be destroyed, but this too can be difficult. Indeed, it is very risky even to approach them as this might disturb the ground and trigger their explosion.

The use of cluster bombs has increased tremendously over the last 30 years.

Like anti-personnel mines, they have been used as area denial weapons. This means that very large quantities of unexploded bomblets are now threatening the civilian population and, unless some solution is found, their numbers will increase.

Possible solution

This lack of reliability in exploding at the intended time has prompted many manufacturers of bomblets to include self-destruct devices in their new models now in production. It is suggested that as the incorporation of such self-destruct devices is clearly a technical possibility, and acceptable to manufacturers, the Review Conference should seriously consider making such a measure mandatory.

III. Small-calibre weapon systems

During the United Nations Conference that led to the adoption of the 1980 Convention, the governments of Mexico and Sweden submitted a draft protocol on the regulation of the use of small-calibre weapon systems.

The Conference felt that further research was necessary to establish more accurately the wounding effects of new types of bullets in order to prevent an unnecessary increase in their injurious effects. The Conference therefore adopted a resolution on small-calibre weapon systems, at its seventh plenary meeting on 23 September 1979, expressing the view that:

". . . such research, including testing of small-calibre weapon systems, should be continued with a view to developing standardized assessment methodology relative to ballistic parameters and medical effects of such systems."

The resolution also invited "Governments to carry out further research, jointly or individually . . . and to communicate, where possible, their findings", and to "exercise the utmost care in the development of small-calibre weapon systems, so as to avoid an unnecessary escalation of the injurious effects of such systems".

A considerable amount of research has taken place since the adoption of this resolution and has confirmed that energy transfer is the most important factor for wound severity. High energy transfer, resulting in more severe wounds, is often caused by early turning of the bullet once it hits the body and by the break-up of the bullet. These phenomena can be caused by poor stability and by the construction of the bullet itself, especially the materials used and the thickness and toughness of the jacket.

On the basis of this information, some States have taken steps to improve the design of their bullets, in particular to increase their resistance to fragmentation so as to conform to the letter and the spirit of the Hague Declaration of 1899 which prohibits the use of expanding bullets.

Standardization of the testing of bullets would be a very important step towards clarification of manufacturing specifications in order to ensure that bullets do not fragment easily. To this end, the Swiss government has offered (in a diplomatic note of November 1991) to put its anti-personnel weapon test facilities at the disposal of all interested States.

The Review Conference could consider the most appropriate way to take these developments into account.

IV. Naval mines

In November 1991, the government of Sweden submitted to the First Committee of the United Nations General Assembly a working paper and a draft Protocol on Prohibitions or Restrictions on the Use of Naval Mines. This draft was presented as an additional protocol to be attached to the 1980 Convention on Certain Conventional Weapons.

The only existing treaty regulating the use of naval mines is the 1907 Hague Convention Relative to the Laying of Automatic Submarine Contact Mines (Hague Convention VIII). Although this Convention is still in force and has the effect of preventing the indiscriminate use of naval mines, it is clear that it has become outdated in certain respects. In particular, it makes specific reference to automatic contact mines and does not take later technical developments into account.

It would therefore be appropriate to consider this draft during the Review Conference with a view to adopting a new protocol to the 1980 Convention.

The draft will need to be studied with care to make certain that it does not in any way provide less protection than the 1907 Hague Convention. In this respect, it should be noted that Article 5 of the Hague Convention provides very clear guidelines as to which party is responsible for clearing mines after the conflict. This question is a critical one from a humanitarian point of view and the new protocol should not be weaker in this respect. The other provisions can be studied in the light of suggested new rules relating to landmines, taking into account considerations peculiar to the naval context.

V. Future weapons

The Conference of Government Experts that met in Lucerne and Lugano in 1974 and 1976, and whose findings served as a basis for the United Nations Conference that adopted the 1980 Convention, discussed a number of futuristic weapons.

These included laser weapons, microwave, infrasound, and light-flash devices, environmental warfare and electronic warfare.

The experts recognized that at that time it was too early to consider specific restrictions on devices that were only at the research stage. However, the majority stressed the importance of keeping a close watch on developments in order to introduce specific prohibitions or limitations that might be necessary before the weapon in question became widely accepted. Several experts underlined the importance of national review measures, which are now required under Article 36 of Additional Protocol I of 1977, as well as of international review measures.

As regards the futuristic weapons discussed at the Lucerne/Lugano Conference, developments in laser technology have raised the possibility of one disturbing application, namely, the use of lasers as anti-personnel weapons to damage eyesight. This matter is referred to above under the heading "Blinding weapons".

There has also been further research into other new technologies, in particular directed energy weapons such as high-power microwave and infrasound devices. Although it may be too early to consider the need for specific regulation, it should be recognized that such future developments are subject to the standards of humanitarian law. In particular, it is important to ensure that new weapons do not have indiscriminate effects and that they do not contravene the rule prohibiting the use of weapons of a nature to cause unnecessary suffering or superfluous injury to combatants. With regard to the interpretation of this latter rule, reference can be made to the standard on which it was originally based, namely, the provision in the 1868 St Petersburg Declaration which states that weapons which "uselessly aggravate the sufferings of disabled men or render their death inevitable" are "contrary to the laws of humanity".

ANNEX I

RESULTS OF THE MONTREUX SYMPOSIUM ON ANTI-PERSONNEL MINES

The general objective of this Symposium, which was held in April 1993 in Montreux, was to collect the necessary facts and ideas to coordinate future action by bodies that are interested in improving the situation of mine victims and in taking preventive action. More specifically, the aims of the Symposium were to gain as accurate a picture as possible of the actual use of mines and the consequences thereof; to analyse the mechanisms and means currently available to limit this use and to alleviate the suffering of victims, as well as to identify any lacunae in this respect; to decide on the best remedial action; to establish a strategy for coordinating the work of different bodies involved in such action; and to write a report which could be used as a reference for future measures.

In order to ensure a multi-disciplinary approach, the participants invited to attend comprised established experts in different fields related to the whole issue of the use of anti-personnel mines and their effects, and included military strategists, mines specialists and manufacturers, experts in international humanitarian law and disarmament, surgeons and orthopaedists, representatives of demining organizations, concerned non-governmental organizations, and the media.

Twenty-five expert reports were distributed to the participants prior to the Symposium, and were discussed at its first plenary session. These reports have not been included in the present document but are reproduced in the full report of the Symposium, which was sent to all governments in August 1993. The participants were then divided into six working groups, each of which conducted an in-depth examination of various political, legal, military and technical aspects of the landmine problem, with a view to weighing the advantages and disadvantages of various remedies, including their feasibility, and to come up with proposals for action, both preventive and remedial. The conclusions reached by the working groups, as they stood after discussion in the final plenary session of the Symposium, are summarized below.

I. Humanitarian, medical and socio-economic cost of landmines

Several expert reports underlined the magnitude of this cost, and backed up their statements with figures. The figures highlight that most of the victims are non-combatants, especially women, children and agricultural workers, that 800 people worldwide die from mines each month, and that the scale of the problem is best illustrated by the case of Afghanistan. At the current rate of mine clearance achievable by over 25 United Nations teams working there, it is estimated that it would take 4,300 years to clear that single country of landmines. Other reports drew attention to the fact that there is a serious lack of medical expertise and equipment to cope effectively with the number of casualties, often resulting in much unnecessary loss of life and limb. Moreover, the surgical care of mine victims is particularly demanding and time-consuming when done properly, and places considerable strain on blood bank services where they exist. In most mine-affected countries today, inadequate or non-existent blood transfusion services add to the difficulty of providing medical care for the mine-injured. The infrastructure necessary for adequate rehabilitation needs attention. Organizing training programmes for local prosthetists so that continuity is assured after the withdrawal of foreign humanitarian agencies and medical teams is far from easy in most affected countries, owing in particular to lack of financial resources and the shortage of local personnel trained at least in the basics.

In many severely mine-affected countries, clearance costs represent the equivalent of many years' gross domestic product. Furthermore, massive and indiscriminate sowing of anti-personnel mines has rendered whole regions unfit for human habitation, cultivation or animal grazing, spelling economic disaster for the affected countries as most of the societies concerned are rural and agricultural. This in turn has led to substantial internal and cross-border population movements causing economic destabilization, compounded by social tensions, in neighbouring countries. There is, furthermore, little chance of the majority of refugees returning, and certainly not in the near future.

The Symposium proposed that the possibility of using military medical units and facilities for the treatment of mine injuries be explored, together with that of establishing an international compensation fund. Governments, manufacturers, buyers, sellers, and licensors of mines, and violators of humanitarian law, could figure among the contributors. Besides paying compensation to mine victims, the fund would finance rehabilitation and mine-clearance activities, research and development, educational programmes and training.

II. Prohibition of the use of certain types of mines

1. Prohibition of the use of all anti-personnel mines

It was generally felt that this would be the best solution, not only from the humanitarian point of view but also because restrictions on the use of weapons are more difficult to control than their total prohibition. However, this solution was considered by a number of participants as unrealistic, for two main reasons. First, as some of the expert reports pointed out, the global annual production of anti-personnel mines has averaged five to ten million over the past quarter century, meaning that there are possibly more than 200 million mines already existing in the world today. Moreover, the world trade in landmines, involving around 30 countries, is both flourishing and complex, is cloaked in secrecy and involves various institutions and agents who interact to circumvent the regulations on the trade in such weapons. Secondly, governments would agree to such a prohibition only if their military establishments found it acceptable. The military experts present pointed out that it would be difficult to conceive of military operations being conducted without anti-personnel landmines, as there would be a definite loss in military capability and, were such a prohibition to be imposed, adapted anti-tank mines would probably be used. However, restrictions on the use and production of certain types of mines, such as those not fitted with self-destruct or self-neutralizing mechanisms, or non-detectable mines, would probably be approved. In this context, there was a need for a wider military view on the overall effectiveness of anti-personnel mines and whether or not they were essential in military terms.

2. Prohibition of the use of anti-personnel mines not fitted with self-destruct or self-neutralizing mechanisms

The expert reports pointed out that the technological capability to produce mines with a comparatively reliable self-destruct or self-neutralizing mechanism is certainly available. Examples of self-destruct mines cited were the Russian POM-2S, the PFM-1S (butterfly mine), the American GATOR (anti-personnel and anti-tank variants available). As for self-neutralizing mines or mechanisms, the Chinese Type 72–B, the Russian MVE-72 break-wire fuse and the VP-13 seismic fuse or control unit were mentioned.

With regard to self-neutralizing mechanisms, the strongest argument against them was that they still denied the use of land because it was not obvious whether they had indeed been neutralized. They therefore necessitated the same amount of time and cost for mine clearance. Added to this was the assertion that the explosive charge remains in the ground, and can over a period of time become more dangerous, or even be dug up and resold or re-used. The one case clearly favouring self-neutralizing mechanisms was anti-tank mines, because of the immense damage created by the explosion of these mines.

Mines fitted with self-destruct mechanisms seemed the best solution, as after the mechanism has functioned the danger is completely eliminated. No explosive components remain. Moreover, the evidence of detonation (for example, craters, fragments and the explosions themselves) serve to alert inhabitants to the presence of mines, including those whose self-destruct mechanisms have failed. It was felt that prohibition of the use of anti-personnel mines not equipped with such a mechanism would present a definite improvement over the existing situation, principally by helping bring the problem down to a more manageable level.

However, certain questions remain to be resolved. First, existing mines thus equipped are sophisticated and at present expensive. It was nevertheless felt that with the introduction of such a prohibition, economies of scale would greatly reduce the price of such mines. Secondly, current failure rates average approximately 10%, a figure which would have to be considerably reduced; such an improvement is attainable given existing technological capability. Thirdly, such mechanisms could easily be fitted into anti-personnel landmines by industrialized countries, but there would remain the problem of simpler mines manufactured indigenously by nations or groups involved in low-intensity guerrilla warfare. It was, however, recognized that strict observance of the rule by the leading manufacturers would lead to a sharp decrease in the current widespread availability of anti-personnel landmines without self-destruct mechanisms. The full report of the Montreux Symposium contains expert reports on technical aspects of self-destruct and self-neutralizing mechanisms, including those that are activated mechanically, electronically, by battery, or by acids or other chemicals, as well as a discussion on the advantages and disadvantages of each type. One final question remained outstanding, that is, the delay time before self-destruction. This was a predominantly military issue, and it was recognized that the opinion of military experts was needed to establish a realistic and acceptable delay time.

3. Prohibition of the use of mines which are not detectable

It was found that among the main reasons for the production of all-plastic, non-magnetically detectable mines were ease and cheapness of manufacture as compared with those with metallic components, and increased longevity because of the absence of corrosion-sensitive components. However, it was pointed that the cost and inconvenience of fitting non-removable metallic detection rings or plates, ideally by casting them into the explosive fill to prevent removal, would be minimal, and that non-detectability of mines provided little if any military advantage.

Detectability would be of great help in mine-clearance operations, and was also regarded as important in conjunction with the introduction of a prohibition of anti-personnel mines without self-destruct mechanisms, because of the failure rate of those mechanisms.

4. Prohibition of the use of mines with anti-handling devices

There were strong arguments in favour of banning the use of integral anti-handling devices, since they have no apparent military value and their main effect is to hamper clearance operations. The only possible military advantage they present is in terms of lowering of morale, and as a further deterrent to breaching a minefield.

5. Existing stocks

Information furnished in one expert report indicated that at least 78 producers in 44 countries have manufactured at least 307 anti-personnel landmine products in recent decades. Production of landmines currently averages between five and ten million annually, implying that there exist, worldwide, well over 100 million and possibly more than 200 million mines. Today, landmines are deployed in more than 50 countries.

Given this situation, a complete ban on the use, production and transfer of all anti-personnel mines, and the destruction of all existing stocks, was seen as the ultimate objective, probably to be achieved through a multilateral agreement. Until such an agreement can be reached, however, existing stocks ought to be modified and fitted where possible with self-destruct or self-neutralizing mechanisms, or destroyed. The technical means to do so are certainly available, although many anti-personnel mines are too small to be fitted with self-destruct mechanisms. Two problems would still remain. The first would be that of effective verification of compliance by States. The second would relate to the financial implications of the destruction of existing stocks, but this could be weighed against the exorbitant cost of mine clearance, which in some areas approaches US$ 1,000 per mine.

III. Proposals for amendment of the 1980 Convention and of its Protocol II

A study of the negotiating history of the 1980 Convention, undertaken in one expert report, shows that an attempt had been made to seek common ground between the military view that anti-personnel mines were an effective and operationally almost indispensable weapon, and the humanitarian view that these weapons were terrible in their effects as they not only caused unnecessary suffering but were indiscriminate by their very nature, inflicting heavy casualties among the civilian population and continuing to do so well after the conflict in question had ended. However, the final result was a modest treaty which in many respects was the product of various compromises. This has greatly undermined the protection from the effects of anti-personnel landmines that the Convention was initially intended to afford. The Symposium discussed at length some of the main shortcomings of the treaty and came up with proposals to overcome them.

1. Introduction of implementation mechanisms

One of the major weaknesses of the 1980 Convention is its lack of implementation mechanisms. The participants, while fully recognizing the limited effectiveness in practice of implementation measures provided for in international law, were nevertheless convinced of the necessity and utility of incorporating some such measures into the Convention. Again, even though some measures could be considered appropriate only for mines and booby-traps, it was generally felt that it would be more logical to introduce such provisions in the body of the Convention itself and not in just one of its protocols.

The implementation provisions found in 1977 Protocol I additional to the Geneva Conventions were regarded as providing a model. Three possibilities were considered: a simple reference to certain articles of 1977 Additional Protocol I; wholesale reproduction of the appropriate articles of this Protocol; or reproduction of the appropriate articles but with suitable changes in the wording so as to make them obviously applicable to the use of weapons. The third alternative was preferred and the following articles were identified as being the most relevant:

Article 82. Provision of legal advisers. It was recommended that legal advisers be incorporated at all levels down to brigade or equivalent level, and be incorporated into planning staffs.

Article 83. Training in humanitarian law. It was felt that the following four measures should be taken:

a. Training in the use of weapons in accordance with humanitarian law in cadet academies, and in all command and staff training.

b. Incorporation of legal provisions in all weapon systems manuals, in the languages of the user countries.

c. Incorporation of warnings of legal limitations on weapons packaging.

d. Incorporation of training in the international law of war in all military training of foreign nationals.

Article 84. Translation of the 1980 Convention into local languages and the adoption of necessary national laws and regulations.

Articles 85–87. Criminal sanctions. It was recommended in particular that Article 85, paras. 1, 2, 3a–d, 4d, and 5 (suitably amended), be incorporated, recognizing that these provisions have not generally been enforced in the past.

In regard to enforceability, it was proposed that grave breaches as defined in these articles could in practice be identified or notified by the following measures:

a. Action by the appropriate authorities within the nation accused of the grave breach.

b. Notification to the Secretary-General of the United Nations.

c. Enquiries by an ad-hoc fact-finding group, the International Fact-Finding Commission, or other fact-finding mechanisms.

Finally, the possibility of inserting a provision on the compulsory jurisdiction of an adjudication body was considered, but serious doubts were expressed in this respect. It was pointed out that no compulsory jurisdiction provisions appear in other international humanitarian law instruments; the International Court of Justice has jurisdiction only over inter-State disputes and does not cover individual accountability; and the establishment of an arbitration tribunal and its regulations call for a considerable degree of cooperation between the parties involved. Participants felt that it was unlikely that States would accept the insertion of compulsory adjudication measures in the Convention.

2. Extending the applicability of the Convention to cover non-international armed conflicts

It was unanimously recognized by the participants that most of the human suffering caused by anti-personnel landmines occurred in the context of non-international armed conflicts. However, as it stands the 1980 Convention formally applies only to international armed conflicts. There was therefore general agreement on the advisability of extending the application of the Convention to non-international armed conflicts through an amendment to the provision of the Convention itself which specifies its scope of application. However, several difficulties inherent in such an extension of the applicability of the Convention were identified. One major difficulty would be objections by States invoking national sovereignty, especially States which generally resist any international involvement in internal armed conflicts. Attempts could nevertheless be made to persuade States that it would be in their interest to extend the applicability of the instrument by arguing that it could be their own countries being devastated and their own population the victim. Another less satisfactory alternative would be an optional protocol on applicability to non-international armed conflicts. It was felt, however, that this would be disregarded by irregular insurgent forces. The extreme difficulty to entering into contact with such forces drastically reduced the chances of persuading them to comply with the law.

3. Shortcomings of the rules in Protocol II even if implemented

Article 3 of 1980 Protocol II, imposing general restrictions, is based on the generally accepted distinction between military and civilian objectives, but such a distinction is difficult to maintain once a military target has moved away from the mined area, leaving behind the anti-personnel mines which then automatically became indiscriminate.

Moreover, the duty to protect civilians from the effects of these weapons is couched in very weak terms, as paragraph 4 of this article makes reference to all "feasible" precautions. The term "feasible" allows for great flexibility in interpretation, but on the other hand it was felt that removing this term altogether would place the military in a position they were highly unlikely to accept. Furthermore, the provision is weak because feasible measures would include the installation of fences or signposts, but experience has shown that these tend to be removed by members of the local population, either out of ignorance or for the profit they can derive from such items. It was felt that Article 3 might be the right place to introduce the prohibition on the use of anti-personnel mines without self-destruct mechanisms.

Article 4 on restrictions on mines other than those that are remotely delivered has the same shortcomings as Article 3.

Article 5 deals with restrictions on remotely delivered mines. There are difficulties in recording accurately the locations of mines delivered by fixed-wing aircraft, artillery and rockets. Therefore the recording requirements in Article 5 (1) (a) in the absence of a neutralizing mechanism are not applicable when using these methods. Problems remain with the implementation of Article 5 (1) (b), because self-destruct and self-neutralization devices are currently insufficiently reliable to guarantee the safety of a mined area; further, no maximum time limit is set for the active life of these mines. The wording of paragraph 1 (b) also creates confusion between self-destruct and self-neutralizing mechanisms. In paragraph 2 there is no definition of "effective advance warning".

Article 7 contains a major flaw in that there is no definition of a "pre-planned" minefield, which is the only type that requires recording. With regard to all other minefields, parties are only required to "endeavour" to record them, which is a rather weak provision. In practice, it was found that further difficulties arose. For instance, some regular armies have followed strict procedures with respect to mine-laying and there are clear rules for marking and recording minefields. However, history has shown that such records are properly drawn up and kept by very few armies. They are also quite frequently lost.

Even where such records are available, successful minefield clearance can rarely be guaranteed for a number of reasons. Mines tend to move, sometimes long distances, over a period of time owing to the effects of weather and soil erosion, and on occasion by the action of animals. This is especially true in the case of scatterable mines. Furthermore, even the most conscientiously maintained minefield record can be subject to human error by soldiers who may be tired or under stress. In non-international armed conflicts, to which the 1980 Convention does not apply, no records will have been kept, no maps made and no warning signs set up, whether through incompetence, lack of discipline or a wish to inflict as many indiscriminate fatalities and injuries as possible on the enemy population.

Article 8. It was felt that there was a need to extend the measures of protection included in this article to organizations other than the United Nations, such as CSCE missions and private demining agencies. In fact, the United Nations itself has acknowledged the necessity of coordinating demining activities, and for that purpose the Department of Peace-keeping Operations, which includes a Mine Clearance Centre, has established a data base to which it welcomed any contributions. Expert reports from other demining organizations pointed out that more often than not mine clearance is an extremely hazardous exercise principally because records are not properly kept, and there are often no maps or signposts. In

Afghanistan, for instance, mine-clearance activities have resulted in over 30 deaths and over 45 amputations, and no less than 29 operatives have been blinded.

All these factors render mine clearance expensive, taking into account the high cost of experts' fees, personnel insurance premiums, and such support expenses as medical and casualty evacuation costs. Article 8 should therefore be extended in order to afford protection to third-party missions and logically also to humanitarian organizations working in regions affected by mines.

Article 9, which deals with international cooperation, does not impose an obligation to remove mines, as the words used therein are "shall endeavour to reach agreement". Moreover, this agreement relates only to "the provision" of such information and assistance as "necessary to" remove or render ineffective mines and minefields, and thus in no way imposes a specific obligation to do so. This was recognized as a major shortcoming of the law. Again, this article does not deal with other issues of crucial importance to a mine-devastated country after the cessation of active hostilities, such as repatriation and land reclamation.

IV. Possible arms control measures relating to the trade in mines and their stockpiling

Participants examined various arms control measures such as prohibition of or restrictions on exports, destruction of existing stocks that are incompatible with possible new manufacturing standards, prohibition of the manufacture of certain types of mines in an arms control treaty, and verification measures.

The advantages and drawbacks of measures that could be taken in the shorter term and those that might better be considered as long-term measures were considered separately. Accordingly, the following possible short-term measures were discussed.

1. Unilateral measures

a. The significance of the export moratorium on anti-personnel mines that some States have instituted was recognized, although many manufacturers would seek exemptions for certain types, such as self-neutralizing and self-destruct mines and high-tech "smart" mines.

b. It was felt that other States should be encouraged to adopt similar measures, This encouragement should come both from the public and from governments.

The participants discussed the value of a moratorium on exports as setting a standard for State behaviour, focusing world attention on the use of anti-personnel mines and constituting a critical first step towards achieving more far-reaching limitations.

Possible drawbacks mentioned included the questionable impact of such a moratorium on alleviating suffering, the repercussions for domestic producers, the difficulties of verification, and the fact that a moratorium does not eliminate the problem of clandestine export.

c. The possibility was raised of a multilateral voluntary regime along the lines of the Australia Group, which controls exports of dual-use biological and chemical items, and the Missile Technology Control Regime. The advantage of such a regime would be to regionalize and/or internationalize controls on the export of anti-personnel mines. This type of non-proliferation regime might, however, create North–South friction. Another drawback is that it does not usually comprise a control system and the focus of exports might shift to non-participating States. All three of the above-described unilateral measures would call for the stepping-up of national control, and the role of customs authorities in this respect was particularly emphasized. The possibility of using independent organizations for such control was also explored.

2. Multilateral confidence-building measures

The importance of ensuring openness and transparency was stressed frequently. To this end the following measures were discussed:

a. The exchange of information on production, stocks and exports of anti-personnel mines should be facilitated by the compilation of a register or data base, as has been agreed in recent arms control treaties. Public organizations would have access and be able to contribute to this information. Some participants felt that the need for financial transparency in anti-personnel mine exports was important.

b. As a follow-up to this information exchange, confirmation visits could be envisaged.

In this regard, the difficulties in reaching agreement on any international mechanism for follow-up control were recognized.

3. Longer-term arms-control measures

The participants discussed the possibility in the long term of a multilateral agreement to ban the development, manufacture, transfer and use of anti-personnel mines and the destruction of all existing stocks. The arms control measures indicated below could also apply to an overall ban on certain types of mines and the destruction of stocks that do not comply with new requirements.

The Review Conference of the 1980 Convention was borne in mind throughout the discussion. Some suggested that the review conference mechanism be used for consideration of an overall ban. The overlap between international humanitarian law and disarmament treaties was recognized with regard to the question as to which international forum would negotiate a convention for a comprehensive ban on anti-personnel mines.

It was proposed that non-governmental organizations be allowed to participate in the Review Conference and it was also stressed by some that mine producers should be included in the consultative process.

With respect to an overall ban, it was felt that several issues would have to be examined. The list below should not be considered exhaustive:

a. Technical definition of what exactly constitutes an anti-personnel mine (or the types of mine to be prohibited), production facility, dual-use components, and mine delivery systems. In this connection the blurring of the distinction between anti-personnel and anti-tank mines needs special attention.

b. A routine verification regime that would include declarations and inspections.

c. A special challenge inspection regime.

d. Destruction of stockpiles within an extremely limited period, and on-site verification of that destruction.

e. Provisions for sanctions in case of non-compliance.

f. Strict national legislation and enforcement to support the terms of the multilateral agreement.

g. Allowance for certain permitted purposes, such as research for the improvement of demining equipment and for protection of troops.

It was stressed that as wide an adherence as possible to such an agreement was vital.

The problem of how to deal with continued use, production and trade by non-party States was recognized as potentially serious. Clandestine commerce would still have to be controlled as well. Evidence of illegal arms trading lies in the associated financial dealings, and is often discovered by customs officials. Mines should bear certificates of origin, and the issuing of false end-user certificates should be a criminal offence. The view was expressed that if it is possible to ban chemical weapons, it should be possible to close loopholes when banning certain types of mines. However, it was considered that attention should focus in the first instance on the supply of mines, since the number of producers of mines, and even of explosives, is fairly limited.

In this context, another issue that the Symposium addressed was that of collection of information on the trade in anti-personnel mines. With a view to having States introduce the subject of mines in the Conference on Disarmament, it was felt that public access to information contained in the United Nations Register of Conventional Arms would be helpful, but it was pointed out that this information, submitted by governments, was available to governments only. However, it was possible that governments might eventually agree to make this information available to the public. Non-governmental organizations were a valuable source of information, but governments were unlikely to supply data for a voluntary register. It was also stressed that non-governmental organizations could not obtain information on a country-by-country basis, as the task would be overwhelming, but that some of them could serve as a clearing-house for information from all sources.

V. Information to the public

The Symposium also recognized the crucial importance of alerting public opinion in order to increase awareness among the military and governments. This would be an invaluable contribution towards a much-needed change in the law. The need for increased involvement of National Red Cross and Red Crescent Societies and their Federation, as well as United Nations agencies such as UNHCR and UNICEF, was stressed. There was also a constant need to keep the press informed about statistics on injuries caused by mines.

ANNEX II

SYMPOSIUM OF MILITARY EXPERTS ON THE MILITARY UTILITY OF ANTI-PERSONNEL MINES

One of the recommendations of the April 1993 Symposium was the convening of a meeting of military experts in order to study the military use of anti-personnel mines and possible alternatives.

The ICRC hosted a Symposium for this purpose on 10–12 January 1994. The topics covered were as follows:

The military utility of anti-personnel mines:
- their use across the spectrum of conflict
- their military and cost effectiveness
- their means of delivery
- the military implications of marking/recording minefields.

Alternative systems:
- what alternative systems exist
- whether they meet military requirements

- their likely cost-effectiveness
- any other implications of their use.

Control measures:
- self-destruct versus self-neutralizing mechanisms
- suitable delay times prior to self-destruction or self-neutralization in different scenarios
- the likely cost penalty and whether it can be offset by increased effectiveness
- detectable versus non-detectable mines.

The majority of the participants in this meeting were professional military combat engineers familiar with current tactical doctrine and trends within their own armed forces. They drafted the following report during the Symposium.

Results of the symposium[1]

I. Military utility of anti-personnel mines

(i) Introduction

Mines are used in different types of armed conflict. Accordingly, these weapons are also used in different ways. The experts felt that there was a need to distinguish between conventional warfare, which is generally carried out in international armed conflicts where classical contemporary military doctrine prevails and where trained and disciplined soldiers are engaged, and civil war and counter-insurgency operations, where these conditions are seldom met.

Mines should be distinguished from unexploded ordnance, which was not considered by the experts as it is an entirely different entity. It is nevertheless a dangerous, uncontrollable and long-lived battlefield hazard.

(ii) Conventional situations

Landmines and anti-personnel (AP) mines in particular are to be seen as an integral part of a combined military plan. By combining the effects of artillery, direct fire weapons and electromagnetic warfare with shaping of the terrain through the use of mined obstacles, the maximum synergistic effect is achieved. A combination of AP mines, anti-tank mines and other appropriate weapons systems therefore increases cost-effectiveness over what is achievable if one weapons system is used alone. Mines thus enhance the effectiveness of other weapons systems across the spectrum of military operations and cannot be considered on their own.

Purpose of the use of mines

The military engineer supports his higher commander by altering or moulding the terrain in a manner that is synchronized with the commander's plan for conducting the operation. Mines are used in the defence to deny access to areas, to encourage the enemy to focus its movement into areas where it can be attacked effectively, and to restrict the enemy's mobility while being attacked. Mines are used in the offence to prevent the enemy from manoeuvring through an area to attack the flank of an advancing force and can be used to block an enemy retreat. Furthermore, mines can be used to hinder logistic traffic and create confusion among headquarters elements.

Military utility and cost-effectiveness of mines

Mines are very effective and efficient. Manpower employed in laying mines produces much more effect on the battlefield than manpower attempting to mould the terrain in other ways. Very little equipment is required in addition to that normally carried by soldiers. The main purpose of mines is to canalize and delay the enemy; but they have the added advantage of inflicting casualties. This inherent ability to inflict casualties also has a powerful demoralizing effect on the enemy soldier. Forces are likely to avoid areas where there is even a reasonable possibility of encountering mines.

There are two basic types of mines. Anti-personnel mines attack the enemy soldier on foot, while anti-tank mines attack mounted enemy and other vehicular platforms such as tanks. The enemy threat usually dictates what type of mine is needed. Anti-personnel mines are used to deny access to friendly positions, to help protect anti-tank minefields from being breached and to attack enemy foot soldiers accompanying mechanized forces through such anti-tank minefields.

Mines can also reinforce existing natural and man-made obstacles. This type of use compels the enemy to deploy multiple means to overcome the obstacle, thus delaying him in his planning and in his ability to overcome the obstacle.

Mines can be emplaced by hand. This is a very slow and deliberate process that offers the possibility of the most accurate recording. Soldiers laying mines by hand can emplace several per hour. The rate for vehicles laying mines by mechanical means is several hundred per hour per vehicle. Anti-tank mines can also be mechanically laid on the ground or even ploughed into the soil. Mines may also be scattered by artillery, aircraft or vehicle; these are known as scatterable or remotely delivered mines. Scatterable mines systems are technically capable of dispensing thousands of mines per hour but usually operate for very short periods of time. Scatterable mines are the most difficult to record and mark.

With the advent of scatterable mines the commander has more flexibility in employing minefields. A force can now employ mines over a greater distance in relatively short periods. It is, of course, still necessary to ensure that mine warfare is synchronized and complementary to other weapon systems on the battlefield. There is also the need to plan the logistics involved. Generally, once authorized by the overall commander, scatterable mines greatly enhance flexibility.

There are certain hazards associated with all minefields. The most immediate is the potential danger to friendly forces. Uninformed individual soldiers can wander into minefields, and routes needed for counterattack or resupply can be closed off by uncoordinated mining. This hazard can increase with the use of scatterable mines if they are not the type that self-destruct or self-neutralize.

Military doctrine requires that landmine warfare operations are highly controlled. Commanders of manoeuvre formations must give positive approval for subordinate units to lay mines. Only commanders have this authority, which is usually held at the formation or general-officer level. Small units may lay command-detonated mines for immediate perimeter security. These will be recovered when the unit moves.

During operations all feasible precautions should be taken to protect civilians from the effects of mines. In particular military units should:

- report intention, initiation and completion of minefields;
- record details of minefields;
- pass records of minefields on to other units that later become responsible for the terrain;
- mark and fence minefields, with the exception of protective minefields and air/artillery/rocket-delivered minefields.

The experts recognized that in practice precautions may be difficult to take in all circumstances, e.g., in surprise enemy contact or retrograde operations.

In summary, forces that properly integrate mine warfare with other systems to maximize weapons effects are capable of quick, decisive victory. Forces that report and record minefields properly enhance the safety of their own soldiers and of non-combatants.

(iii) Use of mines in internal and unconventional armed conflict

The experts thought that it would be useful to analyse the use of mines by insurgent forces in internal and internationalized guerrilla armed conflict. This analysis also includes the response by conventional forces engaged in counterinsurgent warfare in such situations.

Insurgents

Just as in conventional doctrine troops use mines to stop, delay and create psychological trauma among the opposition, insurgent forces use them to target the opposition not only to weaken its military capability, but also to weaken its economic and socio-political infrastructure.

Insurgents rely on mines to a great degree for the following reasons:

– their limited financial and material resources lead them to place greater reliance on mines, which are relatively cheap and are seen as force and materiel equalizers;

– mines are effective in spreading terror in order to influence the population for political gain.

Insurgent forces have no incentive to use detectable rather than non-detectable mines, as the latter may be perceived as being more effective in further slowing down the opposition military and creating more casualties.

As the availability of funds to insurgent forces is limited, the perceived utility of mines increases and therefore more are used as a substitute for other more expensive systems.

With regard to the means of delivery of mines used by insurgent forces, their typical lack of resources means that they rely on hand emplacement or at most scattering from vehicles. It is highly improbable that they would be able to use artillery or aircraft scattering systems.

With regard to marking minefields, insurgent forces almost never do so as they consider this as risking loss of effect against the opposition military.

On the other hand, they are more likely to keep records of minefields for their own use, but as there is usually a low level of training, expertise and discipline, this is frequently not done. Further, it is questionable whether such forces have an adequate infrastructure for the recording of minefields and the passing on of this information to their own troops or possibly to the civilian population as a warning.

Conventional forces engaged in counterinsurgency

Conventional forces are able to apply conventional doctrine in deploying landmines in counterinsurgency operations to stop, delay and create psychological trauma among the opposition forces. The casualties resulting from the use of landmines are particularly effective as opposition forces lack the medical infrastructure to care for their wounded.

Mines are also used by the conventional military to protect their own positions, as well as national assets and other installations and infrastructure such as power lines, water treatment plants, bridges and airports from interference by guerrilla fighters.

As the level of conflict grows, however, and the insurgents gain increased territorial control, the actual theatre of combat is extended and frequently encompasses the entire national territory. As this situation develops, the use of landmines by the counterinsurgent forces increases. This has the effect of placing economic and socio-political pressure on the insurgent forces, and the large numbers of mines also have dire effects on the local population. This situation is therefore worse for the civilian population than conventional warfare.

II. Alternative systems

(i) Introduction

In order to assess the viability of alternatives to anti-personnel mines, the experts first reviewed the utility of anti-personnel mines and identified alternative military systems. The alternative systems were then assessed against the military attributes of anti-personnel mines. The results of the assessment were incorporated into an evaluation of the general effectiveness and/or contribution of alternative systems in terms of the following:

a. fulfilment of military requirements
b. military cost-effectiveness
c. post-conflict implications.

(ii) Utility of mines

Purpose of laying mines

- Delay the enemy
- Canalize the enemy
- Disrupt the enemy
- Inflict casualties
- Divert enemy resources/effort
- Protect own positions
- Reinforce terrain/obstacles.

Desired military effects of mines

- Psychological effect of maimed casualties
- Surprise the enemy
- Enhance other weapons
- Flexibility of use and application.

(iii) Assessment of alternative systems

The alternatives to anti-personnel mines identified by the experts were assessed in terms of their ability to achieve the purposes of anti-personnel mines.

Anti-personnel mines

The use of anti-personnel mines calls for considerable resources to procure and store systems, to transport items to the battlefield and then to deploy and arm the mines. These demands are acceptable in view of the ease of deployment and the effects of mines on the enemy.

Anti-personnel mines are very effective in creating delay and in canalizing and disrupting the efforts of the enemy. Military experience is that the use of mines in defence reduces the number of casualties to own troops. Studies and war-game modelling support these findings.

A further important impact of anti-personnel mines is the psychological effect on soldiers of seeing their comrades injured and the logistic burden involved in treating casualties. Particular advantages of anti-personnel mines are that they are not affected by weather, they need no maintenance or logistic support once laid, they are constantly alert and are unaffected by morale. However, the laying of anti-personnel mines must be recorded accurately and the information disseminated widely to prevent own troops taking casualties on their own mines during operations. The major disadvantage of anti-personnel mines is their existence after hostilities and their possible effects on the returning civilian population.

Wire

Some of the effects of mines may be achieved using wire obstacles, but the disadvantages to the military are the costs involved in purchasing and transporting the materials, plus the intensive manpower effort required to build the obstacle. Wire has little impact after hostilities other than a nuisance effect.

Ditches

These can only partly replace mines. They are costly in terms of the time and machine effort required for construction.

Improvised devices

The experts were unanimous in their view that if anti-personnel mines are not available to combatants, the latter will improvise and make alternative exploding devices during hostilities with the armaments and equipment available. The cost of such devices will be greater in terms of the time and manpower required to make and lay them.

Improvised devices will be difficult to neutralize and lift as they will not be of standard design. Military casualties could be higher with "field-made" devices because of the excessive amounts of explosives used (e.g., improvised 155 mm shell mines). The need to construct such devices in the field could, on the other hand, reduce the overall number deployed.

Flooding/mud

Flooding can be highly cost-effective, but it is not reliable or flexible. Moreover, it is very difficult to control and may place the civilian population at risk. After hostilities there could be long-term effects on agriculture, especially when sea-water is involved.

Land force fire

While land force fire can achieve the desired effects of anti-personnel mines, it is costly in terms of maintaining alert troops and the amount of ammunition involved. It is also limited by weather and visibility. Artillery fire can increase the incidence of unexploded ordnance on the battlefield if it is used to compensate for the lack of mines.

Air power

Air-delivered weapons are extremely flexible and have considerable range. However, their accuracy is not sufficiently good for use near to own troops and they are costly in terms of skilled manpower and maintenance. If air-delivered weapons are used to replace anti-personnel mines, the occurrence of unexploded ordnance would increase.

Novel devices

The experts took note of research into new devices, such as glues, foam, infrasound, etc., but had little information on such devices and saw no indication that they would achieve the military objectives of mines. The possible collateral effects of novel devices, such as toxicity and the effect of weather, plus the side-effects of countermeasures, are unknown.

(iv) Conclusions

Having considered alternative systems and analysed their utility and shortcomings, the following assessment was made:

Do alternative systems meet military requirements?

No alternative meets military requirements in the way that anti-personnel (AP) mines do. The improvised explosive device (IED) comes closest to replacing the AP mine. If AP mines were not available, such devices would proliferate. Although there would be fewer of them than AP mines, the casualty effect of each IED could be very great, and post-war problems of clearance would remain.

Military cost-effectiveness

The AP mine is the most cost-effective system available to the military. The alternatives require more resources and are less effective. In particular, land force fire and air power would never be available in adequate quantities and would be extremely expensive. Their effectiveness is also subject to the weather and visibility.

Post-conflict implications

AP mines create the worst post-war effects unless they have self-destructed, self-neutralized or been removed. Massive demining operations are required to render areas safe for the civilian population and for agricultural use. IED would pose a similar problem, although there should be fewer of them. As for air power and land fire, the problem of unexploded ordnance (UXO) would significantly increase, as approximately 20 to 30 per cent of the munitions would not function on use. UXO would also have to be removed from the battlefield to make the area safe for civilian use. All other alternatives are safe in this respect as they have practically no post-war implications.

In summary, the military do not regard alternative systems as being viable.

III. Control measures

<u>(i) Introduction</u>

This subject comprises four specific issues that were treated separately by the experts:

- self-destruct versus self-neutralizing mechanisms;
- suitable delay times prior to self-destruction or self-neutralization in different scenarios;
- the likely cost penalty and whether it can be offset by increased effectiveness;
- detectable versus non-detectable mines.

The experts began by observing that mines used in an armed conflict situation cause considerable civilian casualties both during the conflict itself and for many years thereafter. It is regrettable that at times mines are intentionally employed against civilians, even though such use is illegal. The experts then defined the two categories of mines involved.

Anti-personnel mines are small, autonomous, victim-initiated explosive devices, usually designed to wound rather than kill. They may be point detonating, directional fragmentation (e.g., Claymore), or jumping fragmentation mines. They may be scattered from vehicles, artillery or aircraft, or hand-emplaced.

Anti-tank mines are larger, and normally require heavy pressure to set them off, although other means of initiation exist. Anti-tank mines can be fitted with anti-handling devices, designed to deter hand-clearance. They can also be modified and misused so that a man rather than a vehicle can initiate them. This practice was not considered further by the working group because it represented gross and uncontrollable misuse.

(ii) Self-destruct versus self-neutralizing mechanisms

Self-destruct mines (SD) contain an integral system that causes the mine to be spontaneously destroyed at the end of a predetermined period of time.

Self-neutralizing mines (SN) neutralize themselves so that they cannot be victim-initiated after a predetermined period of time.

With SD, there is nothing left after detonation. Therefore, should any mines fail to go off, they can be treated as live and dealt with accordingly. There is some danger when the mines detonate, but this was considered acceptable by the working group. The disadvantage with SN mines is that it is difficult to determine whether the system has functioned or not, resulting in no real danger but necessitating some form of clearance.

(iii) Economic aspects

The experts also discussed the possible reuse of scatterable SN mines for economic reasons, and concluded that this was too dangerous to be a viable option because failure of the self-neutralizing system might have catastrophic results if the mines were being carried in vehicles or kept in storage bunkers.

(iv) Failure rates

With modern electronic fusing, it should be possible to bring failure rates down to between one in a thousand and one in a million for both SD and SN, and to design the system in such a way that the mechanism becomes harmless after failure. Mechanical pressure fuse systems can use air or gas generators to initiate SD or SN with varying degrees of reliability. In the future, chemical or other forms of degradation may be possible.

(v) Suitable delay times prior to self-destruction or self-neutralization in different scenarios

a. The delay time is the time that a mine is on the ground and active and therefore capable of being victim-initiated. The term "delay time" does not relate to the time a mine can be kept in storage prior to use, nor the time between emplacement and the mine becoming active. After discussion, the experts agreed that the delay time was dependent on the tactical scenario.

b. Scatterable mines. In a tactical situation where scatterable mines are likely to be employed, the majority of military experts felt that at the most a 12-month delay time would be acceptable. This delay time would be programmable at time of launch and would be selected by the formation commander depending on his appreciation of the military situation. This would generally be considerably less than 12 months, but adoption of a compulsory shorter delay time-frame would be militarily unacceptable. This is because should mines be required for longer periods (i.e., up to the full year) they would have destroyed themselves and would have to be replaced at great economic and military cost. A minority of military experts expressed the view that the maximum delay time should not be specified but left to individual nations to decide.

c. Hand-emplaced mines. Hand-emplaced AP mines fall into two categories for the purpose of this paper: those deployed for relatively short periods of time, i.e., tactical; and those needed for an indefinite period of time, possibly many years, i.e., strategic, usually barrier minefields.

– AP mines used in a tactical scenario are not needed for long periods and should be SD with a limited, timed life. They would thus represent a hazard to innocent civilians for only a limited amount of time, even in the event of illegal misuse.

– Strategic AP mines guarding international borders or sensitive military sites will be required to remain active for many years. These cannot SD for military and economic reasons but must therefore always be subject to tight control, recording and marking. This may require international verification systems.

(vi) The likely cost penalty and whether it can be offset by increased effectiveness

It was recognized that any additional feature, such as SD capability, fitted to a mine will cost money without any direct increase in the military performance of the mine. It will, however, allow flexibility in military operations by aiding future mobility. Developments in technology and the inevitable reduction in cost of mass-produced electronics will bring down the cost of SD, but today that cost may be prohibitive to many nations. On the other hand, SD will reduce demining expenses should the mines ever be used in the future. However, as the acquisition costs of a mine system are critical to most nations, the possibility of saving money in the future may not be a compelling argument, particularly if the price is double or triple for a SD mine. Future developments may reduce this cost so as to make SD acceptable to all. Once the heavy initial cost of an electronic fuse is accepted, various features such as SD, SN, etc. are relatively cheap.

The experts did recognize that if humanitarian considerations and the enormous cost of mine removal are considered, a self-destruct capability becomes highly desirable and ultimately cost-effective.

(vii) Detectable versus non-detectable mines

Recent AP mines are made of plastic for cheapness, ease of manufacture and resistance to the effects of weather. Concealment is therefore not the main reason for the use of plastic in manufacture. In certain countries, mine doctrine requires that all anti-personnel mines be fitted with a metallic ring to allow detection by current electronic metal-seeking mine detectors. It is almost inevitable that electronic fuses will be detectable anyway. However, some participants felt that, from an operational point of view, non-detectability by electronic mine detectors played a crucial role in internal conflicts, as detectable mines could be recovered and used by other parties. They also considered that non-detectable mines offered an increased surprise factor. Other participants observed that "non-detectable" mines could be detected by other methods, such as prodding. Where minefields are covered by fire, the value of non-detectability of mines is drastically reduced.

From a humanitarian point of view, delectability by electronic metal-seeking devices is crucial to the whole process of mine clearance, and it was generally felt by the majority of participants that such detectability in mines would have few negative consequences.

(viii) Recommendations with regard to Protocol II of the 1980 Weapons Convention

a. All scatterable mines should self-destruct; however, it is highly recommended that even in the case of self-destruct mines the general area of their use be recorded in accordance with existing law and military doctrine, and that where possible they be used in fenced or marked areas.

b. At the end of hostilities, self-destruct times should be declared to all parties, together with all other minefield information.

c. For the foreseeable future there will be a military need for some forms of hand-emplaced mines that will not self-destruct, for use in long-term and barrier minefields, but they must be used in tightly controlled circumstances.

d. The experts acknowledge that directional fragmentation mines, such as Claymore mines, will not necessarily be fitted with a SD mechanism, because they are principally designed for reuse. It is suggested that jumping mines be either SD or SN after a timed life, as they are too difficult and dangerous to re-use.

e. Despite the increased acquisition costs, future AP mines should self-destruct, except in the circumstances already mentioned in these recommendations, thereby reducing the tragic post-conflict toll of human lives and social, economic, medical and mine clearance costs.

f. In the future, all AP mines should be detectable.

NOTE

1 These are the results of the Symposium as drafted by the military specialists during the meeting.

Proposals by the International Committee of the Red Cross on Prohibitions and Restrictions

Second Session
Meeting of Governmental Experts
May 1994

CCW/CONF.I/GE/CRP.24
27 May 1994
Original: ENGLISH

GROUP OF GOVERNMENTAL EXPERTS
TO PREPARE THE REVIEW CONFERENCE
OF THE STATES PARTIES TO THE
CONVENTION ON PROHIBITIONS
OR RESTRICTIONS ON THE USE OF
CERTAIN CONVENTIONAL WEAPONS
WHICH MAY BE DEEMED TO BE
EXCESSIVELY INJURIOUS OR TO
HAVE INDISCRIMINATE EFFECTS

Second session
Geneva, 16–27 May 1994

Proposals by the International Committee of the Red Cross
on prohibitions and restrictions

The following are two sets of alternative proposals on restrictions and prohibitions to be introduced into Articles 4 and 6.

Alternative A for Articles 4 and 6 are measures that the ICRC believe are necessary to most effectively deal with the problems caused by landmines.

Alternative B for Articles 4 and 6 are less far-reaching but should nevertheless have some beneficial effect if properly implemented.

Alternative A

Article 4
Restrictions on the use of mines

1. It is prohibited to use anti-personnel mines;

2. It is prohibited to use anti-vehicle mines that are not easily detectable (i.e. with less than ___ grams of metal) and which are not equipped with an effective integrated self-neutralizing mechanism together with an effective locating mechanism.

Article 6
Prohibition of the manufacture, stockpiling and transfer of certain mines

1. It is prohibited to manufacture, stockpile or transfer:

– anti-personnel mines;

– anti-vehicle mines that are not deflectable (i.e. with less than ___ grams of metal) and which are not equipped with an effective integrated self-neutralizing mechanism together with an effective locating mechanism;

2. The State Parties undertake to destroy weapons to which this Article applies and which are in their ownership or possession.

Alternative B

Article 4
Restrictions on the use of mines

1. It is prohibited to use mines that are not easily detectable. To this end all mines must incorporate at least ___ grams of metal that cannot be removed from the mine;

2. It is prohibited to use mines with anti-handling devices.

3. It is prohibited to use anti-personnel mines without an effective self-destruction mechanism. Anti-personnel mines must automatically destroy themselves when they no longer serve their military purpose but in any event no later than ___ days after their emplacement and activation.

4. It is prohibited to use anti-vehicle mines that are not equipped with an effective integrated self-neutralizing mechanism together with an effective locating mechanism.

Article 6
Prohibition of the manufacture, stockpiling or transfer of certain mines

1. It is prohibited to manufacture, stockpile or transfer:

– mines which are not detectable i.e. those with less than ___ grams of irremovable metal;

– mines with anti-handling devices;

– anti-vehicle mines that are not equipped with an effective integrated self-neutralizing mechanism together with an effective locating mechanism.

– anti-personnel mines that are not equipped with an effective destruction mechanism that will ensure the destruction of the mine after a maximum period of ___ days after their emplacement and activation.

2. The States parties undertake to destroy weapons to which this Article applies and which are in their ownership or possession.

Informal Working Paper

SECRETARIAT FUNCTIONS TO SUPPORT EFFECTIVE IMPLEMENTATION OF THE 1980 CONVENTION ON CERTAIN CONVENTIONAL WEAPONS (CCW) AND ITS PROPOSED AMENDMENTS

Third Session
Meeting of Government Experts
August 1994

States Parties may, at the January meeting of the Group of Experts, wish to request the UN Secretary-General to produce for the Review Conference a report on his ability to carry out the secretariat functions listed below which are considered necessary for effective implementation of the CCW. The report should contain estimates of the costs and personnel requirements for carrying out such functions.

produce information materials and conduct seminars which promote understanding, ratification and implementation of the CCW and its Protocols;

report annually to States Parties on progress in implementation based on analysis of [periodic] reports by States Parties and on information from other credible sources;

serve as a clearinghouse for technical assistance to States which request such assistance in implementing provisions of the Convention, particularly its additional provisions on anti-personnel mines;

establish and manage a fund for de-mining assistance for States Parties to Protocol II;

compile and maintain information on the possible use of prohibited mines and bring such information to the attention of the [Verification Commission] in accordance with provisions of the Convention;

compile and maintain a list of qualified experts available to serve on short notice as members of Fact-Finding Missions commissioned under provisions of the Convention;

develop and propose to the [Verification Commission] guidelines and procedures for Fact-Finding Missions;

provide secretariat, technical and, as necessary, logistical support to the [Verification Commission] and to Fact-Finding Missions;

support preparations for [regular] review of the CCW.

COMPLIANCE
(ICRC, August 1994)

1. International measures to promote and monitor compliance

a) promote ratification and understanding of the Convention and its Protocols;

b) monitor implementation:

- collect information from all useful sources e.g. mine clearance organizations, victims, organizations, peace-keeping forces etc.;
- file and analyse information;
- draft reports on the result of the analysis;

c) support the verification procedure:

- make recommendations on the need for fact-finding missions;
- develop guidelines and procedures for fact-finding missions;
- establish and keep up-to-date lists on experts;

d) help technical assistance for:

- better implementation of the treaty;
- demining programmes;

e) monitor technological developments in weapons;

f) prepare review conferences.

2. Review

- establish a regular review procedure for the Convention.

3. Measures to redress violations

a) international responsibility;
b) universal jurisdiction and criminal sanctions.

4. Domestic measures of implementation

a) incorporation of provisions into domestic law;
b) legislation providing criminal and civil responsibility;
c) national military legal advisers;
d) specific training of military in conformity with the Convention's provisions.

Working Paper Submitted by the ICRC

Fourth Session
Meeting of Governmental Experts
January 1995

CCW/CONF.I/GE/CRP.41
10 January 1995
Original: ENGLISH

GROUP OF GOVERNMENTAL EXPERTS TO PREPARE THE REVIEW CONFERENCE OF THE STATES PARTIES TO THE CONVENTION ON PROHIBITIONS OR RESTRICTIONS ON THE USE OF CERTAIN CONVENTIONAL WEAPONS WHICH MAY BE DEEMED TO BE EXCESSIVELY INJURIOUS OR TO HAVE INDISCRIMINATE EFFECTS

Fourth session
Geneva, 9–20 January 1995

PROTOCOL II, ARTICLE 8, PARA. 3

Working Paper submitted by the International Committee of the Red Cross

When the International Committee of the Red Cross [, the respective Red Cross or Red Crescent organizations,] [or any other impartial humanitarian organization] is working in an area where mines, booby traps or other devices are present, each party to the conflict shall, upon request:

(a) remove or render harmless those mines, booby traps or other devices that directly prevent the implementation of the humanitarian mission;

(b) take such further measures as may be necessary to protect the mission from the effects of minefields, mined areas, mines, booby traps and other devices while carrying out its duties; and

(c) make available to the head of the mission all information in the party's possession concerning the location of minefields, mined areas, mines, booby traps and other devices in the area where the mission is operating.

6

ICRC Position Paper No. 1 Landmines and Blinding Weapons: From Expert Group to the Review Conference

ICRC Briefing and Position Paper

February 1995

The International Committee of the Red Cross (ICRC) is concerned about the mixed results of the fourth and final meeting (in Geneva from 9 to 21 January 1995) of the Group of Governmental Experts that has proposed amendments for the Review Conference of the 1980 United Nations Convention on Certain Conventional Weapons (CCW), to be held in Vienna from 25 September to 13 October 1995.

On the one hand, the ICRC welcomes the Group's recommendation that the Convention's restrictions on the use of landmines be extended to apply to internal armed conflicts. The ICRC also welcomes the group of governmental experts' decision to send the draft text for a new Protocol, prohibiting the use of blinding laser weapons, to the Review Conference.

On the other hand, the ICRC is concerned that the proposed new restrictions on landmines which have achieved the most support to date are both too complex and too weak. When combined with the lack of agreement on verification and enforcement measures, there is a real danger that the steps taken by the Review Conference will fail to have a significant impact on civilian casualties. The ICRC remains convinced that a total ban on mines is the only effective means of containing the current global disaster.

EXPERT GROUP PROPOSALS TO THE REVIEW CONFERENCE

1. Landmines

The proposals which received the most support in the Group of Experts and which will be before the Review Conference are as follows:

1. All anti-personnel mines must be detectable;
2. Remotely delivered mines must contain a self-destructing mechanism; and

3. All hand or vehicle emplaced anti-personnel mines used outside of marked, guarded and fenced minefields should have a self-destructing mechanism.

Despite the considerable progress made in agreeing to these principles, there are several serious weaknesses:

☑ It will be difficult to enforce the restrictions on hand- or vehicle-emplaced mines, as the continued use of "dumb mines" (lacking self-destruct mechanisms), which are such a threat to civilians, will still be permitted. As the sale of "dumb mines" will continue it will be difficult to enforce the rule that they indeed only be used in marked, fenced and guarded minefields, particularly among non-State forces in internal conflicts;

☑ A provision was accepted by consensus among the expert group's recommendations, which suspends the guarding and fencing obligations for fields of dumb mines "in situations where direct enemy military action makes it impossible to comply". This will, in practice, have the effect of undermining the proposed rule that all mines used outside fenced and guarded areas should be self-destructing mines;

☑ Some States were reluctant to specify precisely a minimum metallic content sufficient to ensure that mines can be detected by currently available means;

☑ It will be difficult to ensure the reliability of self-destructing mechanisms and many States may not develop or buy them in the near future. Some have stated that they will need a "grace period" of 10–20 years to implement this requirement;

☑ A number of States object to a suggestion that anti-tank mines must also be detectable and not contain anti-handling devices which cause them to explode when detected. Such mines in practice threaten the lives of mine-clearing teams;

☑ There is disagreement as to whether the Convention should contain provisions limiting the export of landmines;

☑ A number of States do not wish to accept any verification provisions or enforcement measures.

The ICRC is concerned that if governments fail to overcome the above-mentioned obstacles to an effective and enforceable land-mine control regime, the proposed new rules may not strengthen substantially the protection of civilians from the suffering currently being inflicted on them by the indiscriminate use of mines.

2. Prohibition of laser blinding

The group of governmental experts agreed to send for consideration by the Review Conference the text of a new Protocol which would prohibit the use of laser weapons to blind persons as a method of warfare. The text, which built upon previous proposals by Sweden and the ICRC, reflected consultations among a wide range of States favouring a ban on laser blinding. Growing support for the prohibition of blinding laser weapons has now come from 26 major countries from Europe, Asia, Latin America and the Pacific.

3. Participation in the Group of Experts

The January meeting of the Group of Governmental Experts was attended by 31 of the 42 States party to the CCW and, as observers, 26 non-party States as well as the ICRC, the UN Department of Humanitarian Affairs, the UNHCR and UNICEF. The presence of 57 States represented an increase over the 51 that attended the Group's previous session in August.

THE ICRC'S POSITION

The ICRC is convinced that the only clear and effective means of ending the suffering inflicted on civilians by anti-personnel landmines is their total prohibition. While only a year ago this proposition was considered unachievable it has now attracted a growing list of advocates including Austria, Belgium, Cambodia, Colombia, Estonia, Ireland, Mexico, Sweden, the UN Secretary-General and many humanitarian organizations which see it as the only real solution to the global scourge of landmines.

The type of complex measures which have been considered by the Group of Governmental Experts might have a modest effect on the level of civilian mine casualties. However, if the exceptions and exclusions currently being suggested by some States are adopted, the new control regime risks having no substantial effect on the overall problem of landmines and could even lead to an increase in the use of and trade in anti-personnel mines.

If States are unable, in the short term, to agree to a total prohibition on the use of anti-personnel mines, the ICRC proposes, as a minimum, the banning of all anti-personnel landmines lacking effective self-destruct mechanisms.

In addition, **certain essential minimum steps must be taken to protect civilians and facilitate mine clearance:**

- prohibit all mines which are not easily detectable and prescribe specific technical characteristics which will achieve this;

- extend the 1980 Convention on Certain Conventional Weapons to cover all internal armed conflicts;
- incorporate into the Convention effective implementation mechanisms which make use of independent international supervision;
- find ways to encourage implementation of the Convention by States and compliance with it by all parties to armed conflicts;
- find ways, including positive incentives, to encourage universal adherence to the Convention which, at this date, has only 42 States Parties.

The ICRC appeals to governments and to the concerned public to give the highest priority to ensuring that the September–October Review Conference takes bold and effective steps to end the worldwide scourge of landmines. Excessive caution and unnecessary exceptions will be reflected in lost lives, limbs and livelihoods for many years to come.

The ICRC also believes that intentional blinding with laser weapons should be stigmatized in 1995 as an unacceptable form of warfare. Given the current state of laser technology and developments in laser weapons, the 1995 Review Conference of the CCW represents the last and only opportunity for the international community to address this issue. Laser weapons suitable for blinding could be produced on a large scale in the next several years and their proliferation among the world's armies as well as terrorist and criminal groups could be expected soon thereafter. Efforts to deal with the problem at that stage would be immensely more complicated and costly, and unlikely to succeed.

7

United Nations General Assembly, 1994

(Forty-Ninth Session)

In its Statement on the 1980 Convention before the First Committee of the UN General Assembly, the ICRC welcomed the growing support for a total ban on anti-personnel mines and called upon States to consider adopting vigorous measures to tackle the landmines crisis. As a minimum, the ICRC called for self-destructing mechanisms to be fitted to all anti-personnel mines.

Statement by the International Committee of the Red Cross to the First Committee of the United Nations General Assembly

Forty-Ninth Session
24 October 1994

1. The challenge facing the Review Conference of the 1980 Convention

Mr. Chairman,

Thanks to the initiative of the government of France, we are now in the process of discussing possible amendments to the 1980 Convention on Prohibitions or Restrictions on the Use of Certain Conventional Weapons Which May be Deemed to be Excessively Injurious or to Have Indiscriminate Effects (Certain Conventional Weapons Convention or CCW).

The ICRC would like to express its gratitude for being able to take an active role in these discussions and for having been asked to prepare two preparatory documents for the Group of Government Experts, one on the subject of landmines and the second on other issues relevant to the review of the 1980 Convention.

The challenge that is facing the Conference is that of agreeing on amendments that will transform the Convention into a dynamic instrument. This Conference has received widespread attention from the world's media and from groups that earnestly hope for meaningful measures to rid the world of the terrible suffering caused by mines and to prevent severe problems which could be caused by other weapon developments. It is therefore essential that the amendments agreed to have tangible results. This requires not only clear and effective rules on prohibitions or restrictions on the use of these weapons, but also the application of the Convention to all conflicts, successful implementation measures and complementary arms control measures.

2. Mines

The problems caused by mines around the world are growing worse at a dramatic rate. The figures contained in the recent report of the UN Secretary-General are impressive. He indicates that for every mine being cleared, another twenty are being laid. He estimates that it would cost about $33 billion to clear the 110 million currently buried mines around the world. However, only 100,000 mines were cleared last year, whereas approximately two million more were laid. There are enormous stocks of mines contained in various arsenals ready to join the others already littering the globe.

These figures do not describe the human suffering that our doctors regularly see, and who attest that mine wounds are the worst that they have to deal with in practice. Neither can these figures give any idea of the profound disruption that mines cause to family, society and the long-term development of the countries affected.

A disaster on this scale cannot be dealt with by half-hearted measures. First of all, vigorous measures need to be taken to clear as quickly as possible the mines that are presently in the ground. Secondly, a lasting solution needs to be adopted. The ICRC is firmly of the opinion that the only really effective measure is to ban the use and production of anti-personnel landmines. We are also of the opinion that there should be strict controls on the use and design of anti-vehicle mines which in practice have led to casualties among both local civilians and humanitarian workers, including ICRC delegates, who need to use the roads to reach victims.

The ICRC notes with satisfaction that several States have joined the call for a ban on anti-personnel mines and earnestly hopes that others will do so before the Review Conference itself. At present the Group of Government Experts is considering a number of proposals which are less far-reaching than a total ban. Of these we believe that as a minimum all anti-personnel mines should automatically and reliably render themselves harmless within a specified period of time. However, we believe that although such a regulation should reduce the amount of civilian victims, it will not prevent large numbers of civilian mine victims as these mines will continue to create victims during their active lives. This is particularly true as many conflicts last for years, if not decades, and therefore the continual relaying of so-called "short-lived" mines can be expected. Further, any exception to the rule that mines are to self-destruct will have the result that so-called "dumb" mines will continue to be lawfully manufactured. They will inevitably continue getting into the hands of irresponsible users, thus perpetuating the widespread indiscriminate use that we are now witnessing.

3. Blinding laser weapons

The ICRC is very pleased that a large number of States have either formally or informally indicated their support for a Protocol on the subject of blinding weapons. It is essential that this Review Conference seize this last opportunity to adopt this legal regulation, as a later review conference would certainly be too late. This preventive step will save the world from the horrifying prospect of large numbers of persons being suddenly blinded for life by certain laser weapons that could soon be both inexpensive and easily available.

For the purposes of the 1980 Convention, a Protocol could be agreed on without lengthy negotiation, as it could in simple language ban blinding as a method of warfare and outlaw the use of laser weapons for this purpose. This in itself would be a major and critical step. Arms control and disarmament measures involving a technical analysis of specific suspect systems could, if necessary, be undertaken in another context. It is to be hoped, however, that the new Protocol would in itself deter any further developments of such weapons, as the international community would have clearly indicated that it considers them unacceptable.

4. Other weapons

Proposals that deserve due consideration have been put forward on other weapons, namely, on naval mines and on bullets. As this is the first Review Conference that is taking place since the adoption of the Convention, it is a pity that there is a prevailing sense of lack of time during the meetings of the Group of Governmental Experts. The immense problem created by landmines has meant that States feel obliged, and rightly so, to devote most of their time to this problem. However, this ought not to be to the exclusion of discussion of other important issues. The 1980 Convention can only be a dynamic treaty if it deals with problems before they become overwhelming.

For this purpose a regular review process needs to be established which is able to address and deal with issues as they arise. For example, during an expert meeting that the ICRC held in spring this year (30 May–1 June), it was irrefutably shown that a major problem is being caused by the very large numbers of bomblets, dispensed from cluster bombs, that do not explode when intended and which therefore remain on the ground creating a danger at least as serious as that of anti-personnel mines. The experts were unanimously of the opinion that it would be sensible, inexpensive and simple to require that these bomblets be manufactured with self-destruct devices.

5. Non-international armed conflicts

Mr. Chairman,

It is all too well known that the majority of armed conflicts are non-international ones and that they are primarily responsible for the immense suffering caused by the indiscriminate use of weapons. It is therefore essential that the 1980 Convention, in order to be effective in practice, apply equally to non-international armed conflicts. In this regard we would like to stress the importance of extending the Convention to all non-international armed conflicts and not only to those which have reached a certain threshold. This is because it is difficult in practice to be certain that a given threshold has been reached, whereas it is important that there is no doubt as to the applicable law. Further, as the purpose of restrictions or prohibitions on the use of certain weapons is to prevent excessive suffering and destruction, it would be inappropriate to use particularly damaging weapons in less serious situations when they are prohibited in more intensive conflicts.

6. Implementation

It is generally agreed that a major weakness of the 1980 Convention is its lack of implementation mechanisms and that this problem must be rectified during this Review Conference. Given the importance of this issue, it is worthwhile to consider carefully which mechanisms would in practice be the most effective for this Convention.

The expert meeting that the ICRC held in spring this year analysed the experiences of different types of implementation mechanisms that have been created in various branches of international law, in order to assess which systems have proved the most effective. The experts were unanimous in their conclusion that a variety of measures that encourage implementation are the most successful and, conversely, that the least effective system is one which relies exclusively on an inter-State allegation of a breach. This is particularly so in relation to treaties that regulate individual behaviour and that do not involve critical issues of State security. With regard to this type of treaty they also stressed the fundamental importance of supervision by independent impartial bodies and the use of information from all credible sources.

The ICRC sincerely hopes that these experiences will be taken into account when choosing the means for implementing the 1980 Convention. In particular, it is important that any procedure involving investigation into allegations of breaches be undertaken by means that are seen to be free of any political influence. Finally, although the ICRC sympathizes with the desire to avoid a system that is costly and complex, we would like to underline the importance of measures that are genuinely cost-effective. When one considers the figures cited in the Secretary-General's report on the horrific price that the indiscriminate use of mines is in fact costing, it is worthwhile, even from a purely financial point of view, to ensure that the most effective implementation measures are adopted. We would hope, however, that the suffering caused by the violations of the law will also be an important motivating factor.

7. Arms control and disarmament measures

The enormous scale of the problems caused by mines has largely arisen because of the cheapness and easy availability of these weapons. The experience of the ICRC is that the majority of suffering in armed conflicts, especially non-international ones, is caused by the massive and indiscriminate use of small arms. Arms control and disarmament law has so far largely concentrated on containing the threat caused by the existence of nuclear weapons and, for the last two decades, on biological and chemical weapons. The fact that these have been little used or not used at all attests to the success of these efforts which clearly need to continue unabated. However, the global problems caused by the largely unregulated manufacture and trade in conventional weapons also need seriously addressing. We are pleased to see that a first step has been taken in the form of the optional register on the transfer of certain conventional weapons. The ICRC hopes that more attention will be given to the problem of the massive trade in small arms in order to introduce some workable limitations on their manufacture and trade. Until this is done, we will unfortunately continue to witness the carnage they are causing on a massive scale around the world.

Weapons which are indiscriminate in their effects or cause excessively cruel suffering should not only have prohibitions on their use but also on their manufacture. In this respect, we are of the opinion that there should be a much greater complementarity between international humanitarian law, on the one hand, and disarmament law on the other. The need for this has been particularly seen in the case of chemical weapons, the use of which was prohibited in 1925 on the basis of their excessively cruel effect, but which nevertheless needed to be further regulated in disarmament law. It is unfortunately the case that once weapons are manufactured, they will inevitably get into the hands of irresponsible users, and this is particularly the case with small arms. It is in the light of this consideration that the ICRC is taking so seriously the potential problem of blinding laser rifles.

The ICRC hopes that these factors will be taken into account both in the context of the 1980 Convention and in other fora.

8. Conclusion

Mr. Chairman,

The ICRC hopes that the Review Conference on the 1980 Convention will do much to render the Convention a dynamic and meaningful means of limiting the suffering and destruction caused by the use of certain conventional weapons in the conflicts that are all too prevalent in today's world.

It is critical that all States speedily ratify this Convention and actively take part in its review so that its impact is universal. We earnestly hope that more States will attend the final meeting of the Group of Government Experts that will take place in January next year, as well as the Review Conference itself. We also consider it of utmost importance that the Convention is subject to frequent and regular review in order to maintain its relevance and credibility in the face of developments.

8

Regional meetings in Africa, 1995

Organized by the ICRC and the Organization of African Unity (OAU)

As Africa is one of the world's most severely mine-affected regions, it was one of the first areas where the ICRC directed its efforts to encourage States to restrict or ban use of the weapons. In 1995, in cooperation with the OAU, regional seminars were held in Addis Ababa (23–24 February and 11–12 April), Harare (2–3 March) and Yaoundé (25–27 April). The purpose of these meetings was to increase support for a ban or greater restriction on the use of mines, encourage ratification of Protocol II to the 1980 Convention on Prohibitions or Restrictions on the Use of Certain Conventional Weapons Which May be Deemed to be Excessively Injurious or to Have Indiscriminate Effects (1980 CCW) and promote participation in the forthcoming 1980 CCW Review Conference. The efforts of this early work bore fruit, in particular at the Oslo Diplomatic Conference. It was the determination and resolve of the African States to conclude a strong and unambiguous comprehensive ban treaty that was instrumental in achieving that result. Reproduced here are the final declarations of those 1995 regional seminars.

Regional Seminar on Land-mines and the 1980 UN Convention on Conventional Weapons

Addis Ababa

23–24 February 1995

I. The International Committee of the Red Cross in cooperation with the OAU convened a regional seminar on Land-mines and the UN 1980 Convention on certain conventional weapons from 23rd to 24th February, 1995 in Addis Ababa.

II. The following countries participated in the two-day seminar: Egypt, Ethiopia, Gambia, Ghana, Kenya, Libya, Liberia, Nigeria, Sierra Leone, Sudan, Tanzania and Uganda.

III. The opening session heard statements and opening remarks from Dr. Chris Bakwesegha, OAU; Dr. Girma Amare, MFA, Ethiopia; Mrs. Louise Doswald-Beck, ICRC HQ Geneva. The full list of participants and the programme are annexed to this report.

IV. The views of the participants were made in their personal capacity and don't necessarily reflect the views of their governments.

V. At the end of the two day seminar, the participants, after having listened, debated and considered a broad spectrum of issues concerning the subject they reviewed came out with the following recommendations:

1. That all African States consider acceding to the 1980 UN Convention on Certain Conventional Weapons and its additional Protocols. This was already proposed in a resolution CM/Res. 1526 (LX) adopted by the OAU Council of Ministers in Tunis, Tunisia, June 1994. The group further requested that the issue be brought up again in a specific resolution on this Convention during the forthcoming Heads of State and Government Summit of June 1995.

2. OAU Member States were urged to promote peace inside their respective States and also good neighbourliness in order to avoid being tempted to buy or plant mines.

3. It was strongly recommended that all African States participate (whether they are a party to the treaty or not) in the review Conference of the 1980 Convention.

4. In addition to these general recommendations, a number of participants made some specific proposals (these references are not exhaustive):

   a) A proposal on the total ban, manufacture and use of land-mines, in particular, anti-personnel landmines. Other delegations had expressed doubts as to the realistic application of this proposal.

   b) A much greater control of the proliferation of land-mines and efforts to ensure a greater respect of existing law on land-mines.

   c) Introduce mechanism to ensure compliance with the existing law and possibly to introduce sanctions against any defaulters whether they be manufacturers or users.

d) Call on those who were responsible for planting mines during the World War II and other subsequent conflicts, to provide the responsible authorities with maps and documents which indicate the affected areas, and also to explore possibilities of providing technical and financial help to assist in de-mining exercises.

e) A number of practical steps were proposed:

– National Seminars should be held to sensitize nationals on the problems caused by land-mines. Documentation and visual material need to be made available for such seminars.

– much more intensive training for soldiers in mine-clearance techniques, bearing in mind the great difficulty in clearing mines which are not readily detectable;

– strengthening of national social welfare systems for the care of victims of land-mines.

f) Affected countries were called upon to make concerted efforts to sustain these programmes as international assistance will inevitably be of a short-term nature.

g) Views were expressed by some participants that countries will inevitably suffer from the effects of land-mines far more than any utility they would derive from using them.

h) It was also suggested that manufacturers ought to be informed of the terrible destruction caused by their products – land-mines.

Regional Seminar on Landmines and the 1980 Weapons Convention

Harare
2–3 March 1995

Report and Recommendations

I The International Committee of the Red Cross in cooperation with the Organisation of African Unity convened a regional seminar on landmines and the 1980 United Nations Convention on Certain Conventional Weapons in Harare on 2–3 March 1995.

II. The participants came from the following countries: Angola, Botswana, Cap Verde, Guinea Bissau, Lesotho, Malawi, Mozambique, Namibia, Sao Tome, South Africa, Swaziland, Zambia and Zimbabwe.

III. The opening session heard statements from Mr. William Godwin Nhara, OAU; The Hon. P.A. Chinamosa, Attorney General of Zimbabwe; His Lordship Archbishop Desmond Tutu, South Africa; Mr. Rene Kosirnik, ICRC Harare.

IV. The views of the participants were made in their personal capacity and do not necessarily reflect the views of their governments.

V. At the end of the seminar, the participants made the following recommendations:

1. All States that have not yet done so should ratify at the earliest opportunity the 1980 Convention on Certain Conventional Weapons.
2. All African States should try to attend the Review Conference of the 1980 Convention that is due to take place in Vienna on 25 September–13 October 1995. During this Conference, they should support amendments that place meaningful and strict restrictions on the use of mines, suggest proposals for the effective implementation of the Convention and propose amendments that would render the Convention's provisions effective in non-international armed conflicts. African States should also support the adoption of a Protocol prohibiting blinding laser weapons.
3. States should seriously consider working towards a total ban of anti-personnel landmines. Participants were of the opinion that the serious problems created by landmines, in particular, the immense suffering of the victims, the rendering of large areas of land unusable and the extremely negative effect on post-war reconstruction and development, outweigh any military benefit. Such a ban would be particularly beneficial to Africa which has suffered immensely from the widespread use of landmines.
4. Discussions amongst African States to encourage ratification of the 1980 Convention and to work towards a total ban of anti-personnel landmines could be usefully undertaken through sub-regional groupings such as SADC, PTA and others. Through such discussions, African States should try to agree as far as possible on a common African position for the Review Conference in Vienna this year.
5. African States should not start producing landmines themselves. The production of anti-personnel landmines should be criminalised and producers should pay compensation to victims.
6. Greater awareness of the problem of landmines should be created on the national level by, in particular, television programs and greater support

and information given to interested African non-governmental organisations.

7. The OAU should advocate and work towards Africa becoming a landmine-free zone.

Recommendations of the 2nd OAU/ICRC Seminar for African Ambassadors Accredited to Ethiopia

Addis Ababa
11–12 April 1995

Within the framework of the Cooperation Agreement concluded between them since May 1992, the ICRC and the OAU organized from 11 to 12 April 1995, in Addis Ababa, a Sensitization and Information Seminar for African Ambassadors accredited to Ethiopia. Many other institutions also participated in the Seminar as observers. The list is attached hereto.

Second of its kind, the Seminar made it possible to deal with several issues, some linked to the International Humanitarian Law and several concerns expressed regarding its implementation and others related to the specific problem of anti-personnel mines.

At the end of the two days of the Seminar and after indepth discussions, the following recommendations were made:

1 – <u>Teaching, Dissemination and Implementation of the International Humanitarian Law</u>

All the participants stressed the need for teaching the International Humanitarian Law and its wide dissemination. In that way conflict prevention could be facilitated. In that view, the following measures were recommended:

a – Regular consideration by the appropriate OAU organs of an item on the teaching, dissemination and implementation of the International Humanitarian Law, taking into account particularly the persistence of conflicts in the Continent.

b – Implementation of specific sensitization and information activities related to the International Humanitarian Law not only for decision makers (politicians, military officers . . .). but also for the people in general: in that regard, one could draw inspiration from the original experience of the ICRC in Burundi.

c – Ratification of or accession by the African States which had not yet done so, to the following legal instruments:

- The Geneva Conventions of 12 August 1949 on the Protection of War Victims and their Additional Protocols of 8 June 1977;
- The Hague Convention of 14 May 1954 on the Protection of Cultural Property in the Event of Armed Conflict;
- The U.N. Convention of 10 October 1980 on Certain Conventional Weapons.

d – Appeal to the States Parties to the Protocol I to the Geneva Conventions to recognise the competence of the International Fact Finding Commission provided for by Article 90 of the said Protocol.

e – Call for OAU member States to comply with the International Humanitarian Law and to have it respected.

f – Request to the African States to draw up national measures for implementation purpose, by resorting to, if need be, the advisory services of the ICRC.

2 – Humanitarian Action

After having expressed their support for the humanitarian efforts deployed by the various actors present on the field:

a – The participants particularly stressed the imperative need for the humanitarian action to remain neutral and impartial and consequently to maintain its specificity and dynamics. Within that framework, they reaffirmed their support for the activities of the ICRC and particularly its neutral and impartial intermediary role.

b – The participants also expressed their concern about the lack of coordination and sometimes the competition which mark the activities of the humanitarian actors. They underscored the need to put an end speedily to that situation.

c – The participants further stressed the urgent need for an increased African contribution to the humanitarian operations: that could take the form of greater involvement of the NGOs of the Continent and an encouragement to their activities.

d – The participants also called upon the African Governments to intensify their efforts in the search for durable solutions to the root causes of conflicts in Africa. In that connection, they recognised the important role that the OAU mechanism for Conflict Prevention, Management and Resolution should play and the OAU Commission of 20 on Refugees was already playing in that field.

e – The participants finally appealed to the OAU and the African legal experts and intellectuals so that they could initiate a genuine process to consider the problem of humanitarian assistance in Africa and further ensure effectively the defence of the interests of the Continent in all the fora where the international law was prepared.

3 – Antipersonnel Mines

The participants expressed their deep concern about the scourge of mines and their generalised and indiscriminate use and the attendant harmful consequences. They deplored the fact that the continent had the largest number of those death devices manufactured elsewhere. They finally made the following recommendations:

a – Ratification of or accession to the 1980 Convention on Conventional Weapons; at present only Benin, Niger, Togo and Tunisia were parties to that legal instrument.

b – Full and active participation in the Review Conference scheduled in Vienna from 25 September to 13 October 1995.

c – Establishment and adoption, within that perspective, of an African Common Position on the following issues:

- the total ban of the manufacture and use of mines;
- The extension of the scope of implementation of the 1980 Convention to Non-International armed conflicts;
- Inclusion, in the Convention, of mechanisms to guarantee an effective implementation of the Convention;
- Increased resource mobilization for demining operations, rehabilitation of affected areas and assistance to victims.

d – Submission of a specific report on the problem of mines to the forthcoming session of the OAU Council of Ministers. Considering the fact that the Central Organ would convene in Tunis in April 1995, the participants requested the OAU Secretary General to kindly raise the matter with the Heads of State and Government who would be present there, so as to enable them to give directives for the preparation of an African Common Position.

e – Request to the States which were responsible to contribute the necessary resources for the demining operations in the African Countries which were infested with that scourge of mines during the Second World War and/or during the conflicts that preceded their accession to independence.

f – Need for an increase in the human and financial resources of the African Rehabilitation Institute (ARI) so as to enable the latter to play fully its role in the assistance to victims: it was understood that the increase should result from the joint efforts of the ARI Member States, which should pay up their contributions, and of the International Community.

g – Appeal to the International Community to give increased support to the African national structures, including those which had a regional vocation, in charge of assistance to the victims of antipersonnel mines.

4 – Strengthening of OAU/ICRC cooperation

After expressing satisfaction at the convening of the present Seminar, whose highly edifying nature was stressed, the participants made the following recommendations:

a – Broadening and intensification of the OAU/ICRC cooperation. A greater involvement of the OAU Commission of 20 on Refugees, the organization of joint field missions, the systematisation of sensitization and information seminars on the International Humanitarian Law could be considered.

b – Active participation of the OAU and its Member States in the preparation and proceedings of the XXVIth Conference of the Red Cross and the Red Crescent.

c – Tribute to the ICRC for the courage and devotion it shows in the service of armed conflict victims and its contributions.

d – Gratitude to the other humanitarian organizations for their activities on the field.

5 – Follow-up and Implementation of the present recommendations

The OAU Commission of 20 on Refugees was requested to include the present recommendations in its Activity Report and to submit them, for an appropriate action, to the consideration of the forthcoming session of the OAU Council of Ministers.

Done in Addis Ababa, 12 April 1995

Recommandations du Seminaire OUA/CICR sur les Mines Terrestres et la Convention des Nations Unies de 1980 sur Certaines Armes Classiques

Yaoundé, Cameroun
25–27 avril 1995

1 Dans le cadre de l'accord de cooperation qui les lie depuis mai 1992, le CICR et l'OUA ont organisé, du 25 au 27 avril 1995 a Yaoundé, un Seminaire sur les

mines terrestres et la Convention des Nations Unies de 1980 sur certaines armes classiques à l'intention des représentants de 25 pays d'Afrique dont la liste est jointe en annexe. Ont également pris part à ce Séminaire, en qualité d'observateurs, le PNUD ainsi que des représentants de la presse tant locale qu'internationale.

2 Intervenant après les Séminaires d'Addis Abeba (23–24 février 1995) et de Harare (02–03 mars 1995), la présente rencontre a permis d'aborder les multiples problèmes posés par la prolifération des mines terrestres en Afrique: impact sur les pays affectés; état de la législation internationale; solutions à mettre en oeuvre; difficulté des opérations de déminage.

3 A l'issue des trois jours qu'a duré le présent Séminaire et après un débat approfondi, plusieurs recommandations ont été formulées par les participants. Il est entendu que les vues exprimées par ces derniers l'ont été à titre personnel et, partant, elles ne reflètent pas nécessairement celles de leurs Gouvernements respectifs.

4 Les participants ont d'abord tenu à exprimer leur profonde préoccupation devant le fléau que constitue la prolifération des mines terrestres en Afrique. Ils se sont en particulier inquiétés des effets on ne peut plus néfastes qui en résultent pour le Continent: augmentation, pour ainsi dire exponentielle, du nombre des handicapés; pression accrue sur les services de santé, lesquels sont déjà au bord de la rupture; paralysie de l'activité agricole; difficultés d'acheminement de l'aide humanitaire . . .

5 Les participants ont exprimé l'espoir de voir ceux des pays africains qui ne l'ont pas encore fait devenir parties à la Convention des Nations Unies sur l'interdiction à la limitation de l'emploi de certaines armes classiques ainsi qu'à son Protocole additionnel n° II sur l'interdiction ou la limitation de l'emploi des mines, pièges et autres dispositifs.

6 Les participants ont également invité les Etats africains, qu'ils soient ou non parties à la Convention des Nations Unies de 1980, à prendre une part active à la Conférence d'examen prévue à Vienne (Autriche) du 25 septembre au 13 octobre 1995.

7 Dans cette optique, ils ont exprimé l'espoir qu'une position commune africaine puisse être formulée. Celle-ci s'articulerait autour des points suivants:

a interdiction totale de la production, du commerce et de l'emploi des mines terrestres; à cet égard, les participants se sont étonnés de la différence de traitement qui existe entre les armes biologiques et chimiques dont la production et le stockage sont interdits et les mines antipersonnel soumises a une réglementation limitée seulement aux modalités de leur emploi;

b l'extension du champ d'application de la Convention des Nations Unies de 1980 aux conflits armés non internationaux;

c l'adjonction à la Convention de mécanismes d'application en vue d'en assurer la mise en oeuvre effective;

d la mobilisation de ressources accrues pour le déminage, la réhabilitation des zones affectées et l'assistance aux victimes.

8 Les participants se sont également prononcés en faveur d'une conférence d'examen automatique de la Convention de 1980 qui interviendrait tous les 5 ans.

9 S'agissant plus particulièrement du déminage, les participants ont exprimé l'espoir de voir les Nations Unies octroyer, par le biais du Fonds d'assistance mis en place en novembre 1994, un soutien accru aux Etats africains afin de leur permettre de s'impliquer plus activement dans les opérations de déminage. Ils ont également insisté sur la nécessité de recourir prioritairement à l'expertise africaine tant gouvernementale que non gouvernementale. Ils ont enfin mis l'accent sur l'urgence d'une coopération interafricaine en ce domaine sous la forme notamment d'échanges d'expériences nationales; eu égard au mandat qui est le sien, l'OUA devrait jouer ici un rôle de premier plan.

10 Les participants se sont par ailleurs félicité de la décision du Gouvernement du Royaume de Belgique de ne plus autoriser la production, le stockage, le commerce (exportation, importation et transit) et l'emploi des mines antipersonnel ainsi que de leurs composants spécifiques. Ils ont exprimé l'espoir que cet exemple soit suivi par l'ensemble des Etats producteurs.

11 Les participants ont aussi insisté sur la nécessité pour les états africains d'articuler, de concert avec l'OUA et le CICR, des programmes nationaux d'information et de sensibilisation sur les problèmes posés par la prolifération des mines antipersonnel en Afrique.

12 Les participants ont exprimé leur appui au projet de protocole interdissant les armes à laser aveuglantes.

13 Les participants ont enfin exprimé leur profonde reconnaissance à l'OUA et au CICR pour la tenue du Séminaire de Yaoundé dont ils ont souligné le caractère hautement instructif.

Fait a Yaoundé, le 27 avril 1995

LISTE DES PAYS PARTICIPANTS

SEMINAIRE FRANCOPHONE D'AFRIQUE SUR LES MINES TERRESTRES ET LA CONVENTION DES NATIONS UNIES SUR LES ARMES CLASSIQUES (1980)

CAMEROUN, YAOUNDÉ, 25–27 AVRIL 1995

ALGÉRIE
BENIN
BURKINA-FASO
BURUNDI
CAMEROUN
CENTRAFRIQUE (RÉPUBLIQUE)
COMORES
CONGO
CÔTE D'IVOIRE
DJIBOUTI (RÉPUBLIQUE)
GABON
GUINÉE (CONAKRY)
GUINÉE ÉQUATORIALE
MADAGASCAR
MALI
MAURICE
MAURITANIE
NIGER
RWANDA
SÉNÉGAL
SEYCHELLES
TCHAD
TOGO
TUNISIE
ZAÏRE

9

United Nations International Meeting on Mine Clearance

Geneva, Switzerland

6 July 1995

Seeking to raise additional funds for badly needed mine clearance operations, the United Nations Department of Humanitarian Affairs convened a meeting of some 100 States to stimulate both funding and an international exchange of technical expertise. In his statement to the meeting, ICRC President Cornelio Sommaruga told assembled delegates that during 1992 a quarter of those treated by the ICRC had been mine casualties, most of them non-combatants. He termed the 'mindless carnage' that mines had wrought on populations around the world an 'affront to humanitarian values' and he reiterated his call for a total ban on these weapons.

Statement of Mr. Cornelio Sommaruga

President of the International Committee of the Red Cross

Geneva

6 July 1995

Over the past ten years medical teams of the International Committee of the Red Cross (ICRC) have treated more than 140 thousand war-wounded. One in five was a mine victim. Each and every year more than twenty thousand men, women and children are injured or killed by anti-personnel mines.

The use of these pernicious weapons has resulted in an acute human tragedy. Apart from the appalling number of casualties they cause, anti-personnel mines inflict the most horrific wounds regularly treated by war surgeons, strike blindly at all human beings alike and continue to spread terror for decades after hostilities have ended.

The International Committee of the Red Cross has been charged by the international community with providing protection and assistance to the victims of armed conflict. Not only have landmines created victims on a massive scale but these weapons have also hindered the ICRC and other humanitarian agencies in providing vital aid to victims or even locked such efforts entirely. In many cases mines force food and medical supplies to be sent by air instead of overland, increasing our costs by as much as twenty-five times. In some cases the massive presence of mines has prevented us even from assessing the needs of civilian populations in war zones.

The ICRC medical teams hear the cries of pain of those who have lost limbs and of those who have lost loved ones to these instruments of blind terror. During 1992 a quarter of those treated by the ICRC were mine casualties – most of whom were non-combatants. Mine-injured patients require far more hospital time, blood, surgery and nursing care than do victims of other war injuries.

The ICRC's twenty-seven rehabilitation centres for war-disabled, spread across thirteen countries, have fitted some eighty thousand artificial limbs over the past ten years, the vast majority on mine victims. In these centres we see the courage of people struggling to relearn skills and rebuild lives. But we also know the desperation of those who cannot go on, of those who feel they have lost their dignity, who know they can no longer support their families and whose families and communities are too poor to support them.

We also know of the toll which landmines take on war-torn lands for decades after hostilities have ceased, of the roads which are impassable, wells which are unusable, fields which are unplanted. The results are poverty, hunger and death.

This mindless carnage is an affront to humanitarian values. In many regions it constitutes a threat to economic and social development, to stability and to peace. It must be ended.

The only effective long-term solution to the threat of anti-personnel landmines is a total ban on their production, stockpiling, transfer and use. The limited military advantage of such mines is far outweighed by their horrific consequences.

The ICRC welcomes the recent establishment of the UN Voluntary Trust Fund for Assistance in Mine Clearance and would like to encourage States to ensure that this and other mine-clearance programmes receive resources commensurate with the needs arising from the use of mines worldwide. We pay tribute to the commitment, courage and perseverance of United Nations personnel and those of other agencies who each day save lives through mine clearance.

But much more is needed. It is essential that the forthcoming Vienna Review Conference of the 1980 UN Convention on Certain Conventional Weapons reaches the goal, endorsed by the 49th UN General Assembly, of the elimination of anti-personnel mines.

The ICRC is concerned that the set of partial measures short of a total ban which are to be considered by the Review Conference may, if adopted, fall far short of the goal of protecting civilians from the indiscriminate effects of mines and are likely to result in continued civilian casualties on a large scale for many years to come.

An essential element of mine clearance is detectability. In addition to prohibiting anti-personnel mines the Review Conference should require that anti-vehicle mines be made detectable and should prohibit anti-detection devices, which cause a mine to explode when detected and constitute a death sentence to mine-clearance personnel. In practice detectability means specifying a minimum metallic content in a recognizable shape. This can be achieved for anti-vehicle mines at very low cost.

While legal efforts to reduce the suffering caused by mines could take years to become effective, the immediate needs of mine victims and other casualties must not be overlooked. For this reason the ICRC has asked that measures be adopted at the Review Conference which would ensure the clearance of a path through minefields or designation of a safe alternative route when its access to victims is blocked. This is the minimum necessary to ensure the access which States have undertaken, in the Geneva Conventions and their Additional Protocols, to provide.

This meeting is a hopeful sign that the international community has begun to take seriously the moral, political and financial responsibility for the damage which landmines have already incurred. But the depth of that commitment and the true value of the undertakings made here will not be known until Vienna. Bold action is needed.

At the current rate we are adding, each year, two or more decades to the eleven hundred years which the clearance of currently emplaced mines will require. If this continues unabated, the value of this important meeting will be greatly diminished. We would simply be running up a fast-moving down escalator. The lives of many tens of thousands of potential victims depend on the resolve of governments to confront the mine problem head on. The most effective and least costly way to do that is to ban anti-personnel mines forever.

Our legacy to our children must not be a mine-infested world.

10

First Session of the Review Conference of the States Parties to the 1980 Convention on Prohibitions or Restrictions on the Use of Certain Conventional Weapons Which May be Deemed to be Excessively Injurious or to Have Indiscriminate Effects

Vienna, Austria

25 September – 13 October 1995

The Review Conference of the 1980 Convention took place in Vienna over a three-week period in September and October 1995. In a powerful opening statement to the Conference, ICRC President Cornelio Sommaruga reminded States that the international community was not impotent in the face of the worldwide scourge of landmines, nor was it helpless against the advance of abhorrent technologies. He asked States to prohibit anti-personnel landmines, prevent the horror of blinding laser weapons and reinforce a Convention which seeks to maintain a modicum of humanity, even in warfare, declaring that 'in so doing the public will surely support you'.

Sadly, however, the package of already limited measures laid before the Conference was not acceptable to all States Parties and negotiations ended without agreement on a strengthened Protocol II, although a new Protocol IV on blinding laser weapons was adopted by consensus. The ICRC welcomed the adoption of Protocol IV but deeply regretted the failure to achieve substantial restrictions on landmines. At this point, sixteen States were already supporting the call for a total prohibition of anti-personnel mines.

The Issues – the ICRC's Position

Review Conference of the 1980 United Nations Convention on Prohibitions or Restrictions on the Use of Certain Conventional Weapons, Vienna, 25 September – 13 October 1995

1 July 1995

The Review Conference of the 1980 United Nations Convention on Prohibitions or Restrictions on the Use of Certain Conventional Weapons Which May be Deemed to be Excessively Injurious or to Have Indiscriminate Effects will be held in Vienna from 25 September to 13 October 1995.

This Conference offers a unique opportunity for a thorough analysis of the problems caused by the use of certain weapons, with landmines heading the list. It should also specify measures to be taken to prevent the manufacture and use of new weapons from creating serious problems in future.

Finally, the Conference will examine the means necessary to prevent the excessive damage resulting from present-day armed conflicts, most of which are internal.

1. The issues

To lay the groundwork for this Conference, a Group of Governmental Experts set up by the United Nations Secretary-General has met four times in Geneva between February 1994 and January 1995. The participants gave special attention to the Convention's Protocol II on Prohibitions or Restrictions on the Use of Mines, Booby Traps and Other Devices, and a large majority stressed the need to amend Protocol II by incorporating mechanisms for its implementation and especially by extending it to cover internal conflicts.

Other subjects considered by the Group of Experts were self-destructing and self-neutralizing mines, mine detection and the production and export of prohibited weapons.

At its last meeting the Group of Experts agreed to submit the following proposals to the Review Conference:

1. all anti-personnel mines must be detectable;
2. remotely delivered mines must contain a self-destruct mechanism; and
3. all hand or vehicle-emplaced anti-personnel mines used outside of marked, guarded and fenced minefields should have a self-destruct mechanism.

There is still considerable disagreement, however, over the export of landmines and measures for implementation of the Protocol.

The experts also examined the issue of new weapons. They agreed to submit for consideration by the Review Conference the text of a new Protocol which would prohibit the use of laser weapons to blind persons as a method of warfare. The text is based on previous proposals by Sweden and the ICRC and reflects the opinions expressed during consultations of a wide range of States favourable to a ban on laser blinding. Growing support for the prohibition of blinding laser weapons has now come from 26 major European, Asian, Latin American and Pacific countries.

The United Nations held an International Meeting on Mine Clearance in Geneva from 5 to 7 July 1995, during which the United Nations Secretary-General, addressing representatives of 97 countries, called for a ban on the production, stockpiling and use of landmines, and ICRC President Cornelio Sommaruga reiterated his appeal to the forthcoming Conference in Vienna, to outlaw anti-personnel mines.

2. The ICRC's role

The 1980 Convention on Prohibitions or Restrictions on the Use of Certain Conventional Weapons is indisputably part of international humanitarian law and the specific prohibitions and restrictions it contains are in fact an implementation of principles and rules laid down by 1977 Protocol I additional to the 1949 Geneva Conventions. When the 1980 Convention was introduced, the ICRC, for which the question of weapons of mass destruction had always been of considerable concern, "realized it could best render service to the international community in this domain by bringing together experts from all specialized fields to examine every feature of weapons whose use could be prohibited or restricted".

Indeed, the ICRC has been required to act as a catalyst in this area with growing frequency. In 1993, for instance, it hosted a Symposium on anti-personnel mines (Montreux, Switzerland, April 1993) whose objectives were to gain as accurate a picture as possible of the current use and consequences of mines, and to analyse the mechanisms and means available to limit this use and to alleviate the suffering of victims.

The Montreux Symposium was followed by a Symposium of Military Experts (Geneva, January 1994) which examined in depth the military use of, and possible alternatives to, anti-personnel mines.

On the subject of blinding weapons, the ICRC has also hosted four meetings of experts on laser weapons that cause permanent and irreversible blindness.

A report based on the results of these meetings was drafted by the ICRC for the Review Conference of the 1980 United Nations Convention on Prohibitions or Restrictions on the Use of Certain Conventional Weapons and was officially submitted in February 1994 to the first meeting of the Group of Governmental Experts.

The United Nations Secretary-General invited the ICRC to attend preparations by the Group of Experts for the Review Conference, and allowed it to speak, to submit proposals and to distribute documentation. In addition, the ICRC was asked to prepare two working papers, one presenting ways and means of amending Protocol II on mines and the humanitarian and military considerations involved in such amendments, and one containing observations on other proposals concerning the 1980 Convention itself and its present and future protocols.

3. The ICRC's position

The following is a summary of the position adopted by the ICRC on the issues to be discussed by the Review Conference:

The ICRC is convinced that the only clear and effective means of ending the suffering inflicted on civilians by anti-personnel landmines is their total prohibition. While only a year ago this proposition was considered unachievable it has now attracted a growing list of advocates, including Afghanistan, Belgium, Cambodia, Colombia, Estonia, Iceland, Ireland, Lao People's Democratic Republic, Malaysia, Mexico, Norway, Peru, Slovenia, Sweden, the UN Secretary-General, the European Parliament, the Organization of African Unity and many humanitarian organizations which see it as the only real solution to the global scourge of landmines.

The type of complex measures which have been considered by the Group of Governmental Experts might have a modest effect on the level of civilian mine casualties. However, if the exceptions and exclusions currently being suggested by some States are adopted, the new control regime risks having no substantial effect on the overall problem of landmines and could even lead to an increase in the use of and trade in anti-personnel mines.

In addition, certain essential minimum steps must be taken as follows:

– extend the 1980 Convention on Certain Conventional Weapons to cover all internal armed conflicts;
– incorporate into the Convention effective implementation mechanisms which make use of independent international supervision;
– find ways to encourage implementation of the Convention by States and compliance with it by all parties to armed conflicts;
– find ways, including positive incentives, to encourage universal adherence to the Convention which, at this date, has only 50 States Parties.

The ICRC appeals to governments and to the concerned public to give the highest priority to ensuring that the Septembe–October Review Conference takes bold and effective steps to end the worldwide scourge of landmines. Excessive caution and unnecessary exceptions will be reflected in lost lives, limbs and livelihoods for many years to come.

The ICRC also believes that intentional blinding with laser weapons should be stigmatized in 1995 as an unacceptable form of warfare. Given the current state of laser technology and developments in laser weapons, the 1995 Review Conference of the CCW represents the last and only opportunity for the international community to address this issue. Laser weapons suitable for blinding could be produced on a large scale in the next few years and their proliferation among the world's armies as well as terrorist and criminal groups could be expected soon thereafter. Efforts to deal with the problem at that stage would be immensely more complicated and costly, and unlikely to succeed.

Statement of Mr. Cornelio Sommaruga
President of the International Committee of the Red Cross

The ICRC and the Review Conference of the 1980 UN Weapons Convention
26 September 1995

"I was asleep in front of the house when I was awakened by the sound of an explosion and my son's voice calling for help. My grandson was lying in the road, his left leg shattered by the mine blast. My son ran off to seek help. I was there looking at the child who was writhing in pain and I took him into my arms. When I started to get up I lost my balance a little and my right foot hit something. My right leg was amputated at mid-thigh. My grandson's left leg was cut off a little higher up. A few years ago my elder son and my daughter-in-law were killed by mines. Now I can no longer feed my family and this makes me ashamed." [Testimony of You Eng, 65, Cambodia. Photo, with grandson, is projected as text is read.]

* * *

This conference has an historic moral, political and legal obligation: to put an end to the "mass destruction in slow motion" caused by anti-personnel mines. It can also prevent a similar horror through the prohibition of blinding laser weapons. Sixty-five year old You Eng, whose testimony you have just heard, sought help in an ICRC orthopaedic centre in Battambang, Cambodia. He has seen three generations of his family ravaged by landmines. These same horrific scenes are replayed many thousands of times over, month after month. Our surgeons and nurses have seen more than enough suffering from these perverse weapons. They appeal to you today to stop this slaughter of innocents.

There is little further need to establish the facts, which were fully acknowledged just ten weeks ago at the International Meeting on Mine Clearance in Geneva. The production, transfer and use of anti-personnel mines is out of control. When available, in modern conflicts, they are used indiscriminately with horrific results. These weapons have caused a global epidemic of staggering proportions and have torn apart the social and economic fabric of dozens of societies.

There is no need to attribute responsibility. Mines have been produced and sold by some fifty States from both north and south. They have been used indiscriminately in many others.

This conference is the moment for States to face their responsibilities and to take actions which they know will be effective in the shortest possible time. The landmine issue may not again be addressed in such a forum for many years. What this means is that governments, through you, are deciding the fate of some hundred thousand potential mine victims over the next half decade, at least. Inaction, or ineffective action, will have tragic consequences which can be avoided.

The ICRC is convinced that a dramatic reduction in the number of landmine victims will only occur through the adoption and implementation of a comprehensive set of measures, including:

– total prohibition of the production, stockpiling, transfer and use of anti-personnel mines,
– extension of the Convention on Certain Conventional Weapons (CCW) to cover non-international armed conflicts,
– a requirement that anti-vehicle mines be detectable and do not contain anti-handling devices, and
– an effective regime to monitor compliance and punish violations.

The ICRC has carefully studied proposals aimed at alleviating the landmine problem through the increased use of self-destructing mines and new requirements for detectability. Whilst this might seem to be an advance, we are deeply concerned that a regime based on these measures could lead to an overall increase in the use and transfer of mines, particularly if users believe such mines to be less threatening to civilians or compensate for their short life through use in larger numbers. We are also not confident that an acceptable maximum failure rate can be achieved, that the reliability of self-destruct mechanisms will be internationally verified or that States and insurgents in likely conflict zones are prepared to pay for their higher cost. The "grace period" for transition to such a regime is likely to be measured in decades of continued civilian suffering.

I encourage delegates honestly to ask themselves if the combination of partial measures short of a total ban, which are currently being considered by many States, will be rapidly and effectively implemented and lead to a significant reduction in civilian landmine casualties. If not, I appeal to your governments to join a growing number of States, the Organization of African Unity, the European Parliament, the UN Secretary-General, scores of humanitarian agencies and many hundreds of NGOs in calling for a total ban on anti-personnel mines. This solution is simpler and more effective. It can be more rapidly implemented and far more easily verified than complex alternatives. The limited military advantage of anti-personnel mines is far outweighed by their horrific consequences.

Because even measures as strong as a total ban will take time to have an effect on the ground the immediate needs of mine casualties and other war victims urgently need to be addressed in the CCW. For this reason the ICRC has called for amendments to Article 8 which would ensure the clearance of a path through minefields or designation of a safe alternative route when its access to victims is blocked. This is the minimum necessary to ensure the access which States have undertaken, in the Geneva Conventions and their Additional Protocols, to provide. Similar protections are needed for other humanitarian organizations.

This conference also has the unique opportunity to prohibit a new and abhorrent method of warfare – the use of blinding laser weapons. As has now been widely acknowledged, laser weapons suitable for blinding large numbers of soldiers or civilians at long distances are on the verge of large-scale production and export. Once produced in large numbers these small arms will cost little more than an ordinary rifle and will proliferate rapidly not only to traditional armies but to terrorists and criminals. Whatever the intention of producers may be, like landmines, once they proliferate laser weapons are likely to be used indiscriminately.

The ICRC appeals to this conference to agree on the adoption of a new Fourth Protocol to the Convention on Certain Conventional Weapons which would prohibit the use of laser beams to blind persons as a method of warfare. We also call on States to refrain from the production of weapons suitable for such use and to begin vigorous efforts to prevent their proliferation. It is essential to address this problem now. Later may simply be too late.

Ensuring the continued relevance and effectiveness of the 1980 Convention, and of the decisions of this conference, will require sustained efforts and concerted diplomatic action over many years. For this reason we believe that a decision to provide for regular review of the Convention could be one of the most lasting contributions this conference can make to the development of international humanitarian law.

The landmine issue is but a part of a phenomenon of increasing concern to the ICRC: the virtually unrestricted transfer of vast quantities of weapons, particularly small arms, throughout the world and their consistent use in flagrant violation of the basic norms of international humanitarian law. The ICRC intends to actively study, as requested by the Intergovernmental Group of Experts for the

Protection of War Victims, the relationship between arms availability and violations of humanitarian law and to initiate a process of dialogue within the Red Cross and Red Crescent Movement on these matters.

The international community is not impotent in the face of the worldwide scourge of landmines. It is not helpless against the advance of abhorrent technologies. Your predecessors in 1925 largely stopped the use of poison gas in warfare. Your colleagues in 1972 and 1993 forever banned biological and chemical weapons. Public horror at the effects of nuclear weapons and fear of their possible use has been one of the principal forces which has prevented their use and inhibited their proliferation.

You and your governments can, in the coming weeks, prohibit anti-personnel landmines, prevent the horror of blinding laser weapons and reinforce a Convention which seeks to maintain a modicum of humanity, even in warfare. In so doing the public will surely support you. In so doing you might begin to rekindle public faith in international law and institutions at a moment when the fiftieth anniversary of the founding of the United Nations is being celebrated.

Over the past ten years ICRC medical staff have treated more than twenty-eight thousand mine victims and fitted some eighty thousand artificial limbs on those who have survived. They have too often held in their arms children like You Eng's grandson, whose limbs and lives have been shattered by mines.

It is unacceptable that ten years from now ICRC doctors will have to look into the eyes of You Eng's great-grandchildren, also crippled by a mine blast, and know that in October 1995 something could have been done to prevent it but wasn't. We will all lose something of our humanity if in future years ICRC medical staff must look helplessly into the eyes of soldiers or civilians whose retinas have been burnt by lasers, knowing something could have been done to stop it.

The world awaits a sign from Vienna that there are still certain minimum norms of humanity which civilized countries are unwilling to abandon. You can, in the coming weeks, prevent the unnecessary suffering of a new generation. On behalf of all potential victims I express to you hope and gratitude for your efforts.

ICRC's Informal Comments on the Chairman's Rolling Text

INTERNATIONAL COMMITTEE OF THE RED CROSS NON-PAPER
REVIEW CONFERENCE OF THE 1980 CONVENTION ON CERTAIN CONVENTIONAL WEAPONS
VIENNA, 25 SEPTEMBER–13 OCTOBER 1995

Introduction

This is a brief analysis prepared by the Legal Division of some of the provisions in the Chairman's rolling text that will be considered during the forthcoming Review Conference. It is submitted without prejudice to the ICRC's formal position in favour of a total ban on the use, production, stockpiling and transfer of anti-personnel mines.

While firmly believing that a total ban is the only effective solution, it is recognized that agreement on this may not yet be possible at the forthcoming Review Conference. Therefore, we are submitting this document in order to give an indication of which alternative proposals within the present rolling text are considered preferable, together with a brief indication of the reasons for this choice. This document also contains some informal suggestions in relation to some difficult issues that arose in the group of governmental experts, and points out some technical legal points relating to the framework Convention that will require amendment during the Review Conference.

It is hoped that governmental experts will find this document useful for their preparation of the Review Conference.

Protocol II

Article 1

The recommendation of the meeting of governmental experts to extend the applicability of this Protocol to non-international armed conflicts is welcomed. Of the two alternatives, Alternative A is preferable as mines are frequently laid in peacetime, usually to protect borders, and the rules in the Protocol relating to the laying of such mines should be implemented. Typically mines also remain in place after the cessation of hostilities, frequently for very long periods, and it should be clear that the rules relating to the fencing and guarding of such minefields and of their removal continue to be applicable.

Whichever alternative is chosen, we are of the opinion that it should apply to the Convention as a whole (see our comments to Article 1 of the framework Convention).

Article 3

Paragraph 3: the square brackets should be removed around the word "mine" as this customary law rule applies to all weapons and therefore also applies to mines. The phrase "which is designed to" should be replaced by the phrase "of a nature to" because this is the customary law formulation that is correct. "Design" would appear to relate to intention at the time of development, whereas "of a nature to" is an objective term. Not only does Protocol I of 1977 additional to the Geneva Conventions use the phrase "of a nature to" (Article 35), but also the authentic French of the 1899 and 1907 Hague Conventions use the phrase "propres à" (Article 23e), which is the objective standard "of a nature to".

Paragraph 7: the square brackets should be removed from around this provision as it is an expression of international customary law. Although this formulation was primarily introduced into Protocol I of 1977 with a view to reaffirming the illegality of target area bombardment, it clearly applies to all attacks that are not aimed at individual military objectives. Attacks covering entire areas in which there are both military objectives and civilians are indiscriminate. This reasoning clearly also applies to the use of mines, the indiscriminate and widespread use of which has led to the shocking situation that we now have in the world.

Article 4

During the Fourth Expert Meeting, particular concern was expressed over the wording of paragraph 3 which, if interpreted widely, would result in dumb mines continuing to be laid in many tactical situations without fencing. As it is precisely this use that has resulted in the tragic situation that we now face, we would like to suggest a minor rewording of paragraphs 2 and 3 that takes into account the military exigencies that were raised in the fourth meeting of the group of governmental experts whilst clearly maintaining the duty in all circumstances to fence dumb mines:

Proposed rewording:

para. 2: *It is prohibited to use weapons, to which this Article applies which are not self-destructing, unless:*

- (a) *they are placed within a perimeter-marked area and protected by fencing or other means to ensure the effective exclusion of civilians from the area. The marking must be of a distinct and durable character and must at least be visible to a person who is about to enter the perimeter area;*
- (b) *the perimeter-marked area is monitored by military personnel; and*
- (c) *they are cleared before the area is abandoned, unless the area is turned over to the forces of another State that accept responsibility for the maintenance of the protection required by this Article and the subsequent clearance of those weapons.*

para. 3: *A party to the conflict is relieved from further compliance with the provisions of subparagraphs 2(b) and 2(c) above only if such compliance is not feasible due to forcible loss of control of the area as a result of enemy military action, including situations where direct enemy military action makes it impossible to comply. If* *the party to the conflict regains control of the area, it shall resume compliance with the provisions of subparagraphs 2(b) and 2(c).*

A military commander will always be able to comply with these requirements, if necessary, by putting the fence in place before laying the mines.

Article 5 bis

Requiring that mines be detectable is an step forward. Present Protocol II applies to all types of mines and therefore there should be a weighty reason for introducing into the Protocol rules that apply only to anti-personnel mines. There appears to be no important military reason for using non-detectable anti-vehicle mines, but they do cause serious clearance difficulties and civilian casualties. Humanitarian agencies, including the ICRC, have lost personnel as result of such undetectable mines on roads which had been already cleared. (This comment is also pertinent to paragraph 6 of Article 4. Comments relating to the specification of detectability are indicated below in relation to the Technical Annex).

Article 6 bis

Paragraph 1: the ICRC's position is in favour of the first alternative because it considers the catastrophic individual, social and economic consequences of anti-personnel landmines far outweigh their military value. Although there can be no doubt that anti-personnel mines have some military utility, the fact that modern armies can breach minefields quickly means that their effect is marginal. The military will, therefore, frequently be able to avoid injury by them, and they are left to kill or maim the civilian population.

The ICRC is also convinced that lesser measures will not be sufficiently effective. As long as dumb mines continue to be available, they will inevitably be used by armed groups that do not conform to the rules. Even if the law is respected, there is still the problem that will arise after loss of control of the area. The fact that mines frequently move from their original emplacement renders subsequent clearance efforts very difficult. With regard to self-destruct mines, there is the danger that their reliability will not be assured unless the reliability of the self-destruct mechanism is checked by a central body and they are manufactured so that they inevitably self deactivate in a short period of time. Further, there will be the temptation to consider self-destruct mines as relatively benign and therefore to use them on a wide scale without recording and fencing them.

The second alternative, which requires that all mines self-destruct, is preferable to the present Article 4 in the rolling text in that it would outlaw the use, development, stockpiling or transfer of dumb mines. However, in order for this second alternative to be reasonably effective, there will need to be a compulsory control of the reliability of all self-destruct mines, their active life will need to be short, the "grace period" for the change of mine stocks will need to be as short as possible, and all existing stocks of dumb mines will need to be destroyed (or transformed) as proposed in paragraphs 2 and 4.

Paragraph 3: see comments to Article 5 bis and Article 2 of the Technical Annex.

Article 6 ter

The ICRC does not hold a formal view as to whether rules on the transfer of mines should be inserted into this treaty or into another. If the latter course is adopted, however, it is suggested that the Review Conference adopt a clear recommendation that the Conference on Disarmament should put this subject on its agenda. The same is true for Article 6 bis if those provisions relating to disarmament are not adopted within the context of the 1980 Convention. Stringent interim controls on production, stockpiling and transfer of mines will be necessary if a total ban is not achievable at the Review Conference.

We would like to point out, however, that Article 6 ter is much weaker than most existing moratoria on transfers of anti-personnel mines and that a total ban on exports is the only way to ensure that such mines are not widely available. Such availability has in practice led to the disastrous situation that we now have and paragraph 4 will not ensure a respect for the law without a comprehensive implementation and verification mechanism to ensure compliance. Further, paragraph 1 will not in practice eliminate the acquisition of mines by non-State entities by various means.

Article 7

Paragraph 2 chapeau: it is important that the first alternative "without delay after the cessation of active hostilities" be adopted rather than the second alternative "the effective cessation of hostilities and the meaningful withdrawal of forces from the combat zone". The latter alternative is open to too many contradictory interpretations in each particular situation, whereas the first alternative is a formula that is well known. The concern which probably motivated the proposal of the second formulation has been since met by the amendment of paragraph (b) to the effect that information only has to be given in relation to areas no longer under that party's control. Not giving such information after the cessation of active hostilities will only have the effect of endangering civilians as the other party, which will have control of the area, will not be able to warn the civilian population of their presence, nor fence or remove them.

Article 8

Paragraph 2(b): we urge the adoption of paragraph 2(b) in its entirety, removing the square brackets.

The reference to the ICRC's mandate makes it clear that the ICRC is either undertaking an action which the parties are required to allow under the Geneva Conventions, or is doing so with the consent of the party or parties concerned and subject to the specifications indicated in the Geneva Conventions and/or Additional Protocols. Article 81 of Additional Protocol I of 1977 (to which 137 States are party) states that "the Parties to the conflict shall grant to the International Committee of the Red Cross all facilities in their power so as to enable it to carry out the humanitarian functions assigned to it by the Conventions and this Protocol in order to ensure protection and assistance to the victims of conflicts". The ICRC considers that Article 8(2)(b) should be a specification of how this duty is to be carried out when the ICRC is undertaking these humanitarian functions in mined areas.

We would also like to point out that National Red Cross and Red Crescent Societies also have specific rights and duties for the protection of victims of war under the Geneva Conventions of 1949 and their Additional Protocols of 1977 and these should be taken into account in Article 8.

Most humanitarian action in mined areas will, in practice, relate to the delivery of food, water or other supplies essential to the survival of the civilian population or the care of the wounded and sick. Parties are required to allow these actions, subject to certain specifications, under the Geneva Conventions and Additional Protocols, and the provision of such essential relief actions is widely considered to be a part of international customary law.

Therefore the wording adopted needs to ensure that the necessary actions will be undertaken by the party in control of the mined area in order to assure that the humanitarian mission can be accomplished, and this will not be the case if the personnel of the missions are blown up by mines or are unable, because of the presence of mines, to venture into the areas which they need to reach. Paragraph 2(b) as it now stands, including the text that is now in square brackets, reasonably accomplishes this aim.

However, if doubts remain in the minds of some delegations as to this wording, any rewording will need to ensure that the most important and realistic provisions are included in order to effectively enable the accomplishment of the mission. This will require in practice a warning of the presence of mines in the area where the mission is due to work, and then either the designation of a safe route for that mission or, if necessary, the clearance of a lane so that the mission can reach its appointed destination.

Article 9

Paragraph 1: the proposed new Article 9 may well be the most important achievement of the Review Conference in that it will at last introduce clear and firm rules as to who is responsible for clearing mines at the end of hostilities. However, this achievement will be severely jeopardized if the wording makes it unclear as to when the responsibility begins. We therefore strongly support the use of the phrase "without delay after the cessation of active hostilities" rather than the alternative "after the cessation of hostilities and the meaningful withdrawal of forces from the combat zone". Not only is the latter phrase likely to cause difficulties of interpretation in specific situations, each party arguing that the other side has not meaningfully withdrawn, but also there are many situations where hostilities cease for years, or even decades, without withdrawal and keeping mines in place will only have the effect of endangering civilians.

Both formulae were proposed in the first and second sessions of the group of governmental experts, whereas in the third (August 1994) session a new phrase was introduced into Article 9 paragraph 1, namely, "or maintained in accordance with Article 3 and paragraph 2 of Article 4 of this Protocol". This takes into account possible needs in the case where there is no withdrawal from the combat zone and one or both sides believe that a recurrence of hostilities will take place in that it gives a choice between removing the mines or maintaining them in fenced and guarded areas. The latter maintains military readiness whilst protecting the civilian population to some degree. Therefore in this light the first alternative "without delay after the cessation of active hostilities" can be maintained.

Technical Annex

Article 2

Paragraphs (a) and (b): all delegations were in agreement that at least anti-personnel mines must be detectable (for our contents on anti-vehicle mines, see Article 5 bis above). This is because of widespread experience by mine-clearance teams that they are frequently unable to find modern mines that are mostly made of plastic. These teams have unanimously indicated that these undetectable mines actually do have a tiny amount of metal in them and that therefore, in perfect conditions, a very sensitive metal detector could find them. However, in real conditions, i.e., in naturally metal-rich soil or in ground which has lots of metal objects in it, e.g., spent cartridges, fragments, nails etc., metal detectors cannot detect these mines. Teams therefore have to resort to the time-consuming, expensive and dangerous method of prodding the ground every few centimetres.

It is assumed that it is this situation that the Review Conference wishes to remedy and therefore it is clear that alternative (b) as it now stands is ineffective as it will not change the present situation. We strongly urge that the Review Conference accept testimony from mine-clearance teams with extensive worldwide experience as to the criteria that should be adopted to ensure that mines are detectable in real conditions. The purpose of a technical annex is to indicate precise technical specifications, including precise figures. Therefore it would be appropriate to indicate a minimum amount of metal (of certain characteristics) with a minimum surface area, as urged by mine-clearance specialists.

Discussions in the group of governmental experts frequently referred to new technologies that are emerging to detect mines. Such technologies could be taken into account in a future amendment to the Technical Annex once they have been tried and tested in all parts of the world and proved effective.

Paragraph (c): this proposal has been put forward with a view to protecting mine-clearance teams and we strongly support it. Much mine clearance is undertaken around the world by different types of mine-clearance teams, whether non-governmental, commercial, United Nations, or other international or national bodies. These groups have indicated that they frequently come across more than twenty different types of mines in any one area coming from many different nations, and this renders their work particularly difficult and dangerous. The existence of mines that detonate by the operation of standard mine-sensing devices would amount to an intentionally inflicted death sentence on these persons who are trying to undertake such important work.

The first areas which such mine teams have to clear are usually roads in order to enable access to various areas and to help the return of refugees and the reconstruction of the country. Typically, anti-vehicle mines will be placed on these roads and it is therefore absolutely essential that they are not constructed so as to detonate when teams are trying to detect them.

There is no important military reason why anti-vehicle mines should be constructed this way. Anti-vehicle mines would be used in order to delay the advance of a mechanized army and according to military doctrine, the minefield would be covered by fire. An advancing army in these conditions would not send out individuals with mine detectors to find and remove these mines one by one but would rather use their mine-breaching machines or choose another route.

The effect on mine-clearance teams working after the end of hostilities would, on the other hand, be certain death. We appeal to the humanitarian conscience of nations not to let short-term financial considerations prevent the prohibition of such anti-vehicle mines.

Article 3

This is a critically important provision, and its content will determine whether self-destruct mines provide some improvement to the present situation or not. As indicated above (comments to Article 6 bis), the major dangers of self-destruct mines are their possible lack of reliability and the temptation to use them without any safeguards because of the perception that they are not dangerous to the civilian population. With regard to the first problem, we support the idea of a back-up self-deactivation feature. However, we are also of the view that the reliability of the primary self-destruct mechanism is of paramount importance as the timing of the self-deactivation feature (e.g., a battery) is likely to be highly unreliable. Although it is appropriate to indicate a maximum failure rate, it should be indicated how this is to be tested in practice and there should be an impartial body to assess this. With regard to the second problem, i.e., the likely use of them without safeguards, it is important that the maximum active life is as short as possible, otherwise they will continue to cause widespread civilian casualties. It is also essential to specify, for the self-destruction and self-deactivation periods, the point at which the period begins.

Appendix I: Proposals relating to verification and compliance

All delegations have expressed the importance of including implementation measures in, at least, Protocol II, but at the moment there is disagreement as to the provisions to be adopted. In its second official report to the Group of Governmental Experts, the ICRC suggested that the Group benefit from a study of which implementation measures have in practice proved the most effective in the context of other treaties. Based on the findings of an expert meeting held by the ICRC in May–June 1994, the ICRC report concluded that a variety of international supervisory mechanisms are appropriate to encourage correct implementation (preventive measures), to investigate possible violations and to provide for redress following findings of violations. It recommended in particular the creation of a permanent independent body which would undertake a variety of tasks to help the ratification and implementation of the treaty, including the collection of data from all credible sources.[1]

As particularly severe and widespread violations tend to occur in non-international armed conflicts, it is incumbent on the Review Conference to provide for mechanisms to improve implementation in these conflicts and to provide redress and sanctions for violations. Thus far, this problem has not been thoroughly discussed although a number of delegations have alluded to the difficulty involved.

Alternative A

This proposal has the great merit of providing for automatic, regular supervision that concentrates on encouraging the correct implementation of the Protocol. The regularity of meetings ought to be specified, and ought not be less than once a year. We are convinced that a permanent body is essential as the Commission of

States Parties will only be able to undertake effective work if a non-political group of persons (a Secretariat) collects the appropriate information and reports to the Commission. The group should be able to receive information from all credible sources and collate this in publicly available reports. This system also has the merit of being appropriate for non-international armed conflicts.

Alternative B

Paragraph 1: This provides for national measures of implementation that need to be taken to implement the Protocol. There is merit in specifying these concrete steps and they could usefully be discussed in the context of the Commission suggested in Alternative A. (Although it is understood that Alternative B is not meant to be complementary to the other proposals, we believe that it usefully could be). It could also be mentioned here that it would be most appropriate to insert general national measures of implementation in the framework Convention itself.

Paragraph 2: It is useful to specify the information that needs to be included in State reports, and this could be added in the context of paragraph 3 of Alternative A. In order for such State reports to play an effective role, it is important that they are discussed by the supervisory body which could ask supplementary questions and make recommendations.[2]

Alternative C

This is the only proposal that clearly aims at investigating and rectifying or redressing violations of the Protocol and it is important that the Convention contains a means to suppress violations. The Verification Commission as it stands in Alternative C has the advantage of being compulsory in that it is required to meet if one State requests the Depository to convene it. However, it has the disadvantage of not representing a permanent body. It is a system that is most appropriate for international conflicts as its existence depends on a formal request by a State Party. Various studies[3] have shown such requests to be rare and that a far more effective system would be to allow complaints to come from the Secretary-General and/or from other sources. The ICRC expert meeting referred to above[4] also stressed that implementation mechanisms should not be based solely on adversarial proceedings and therefore Alternative C would be of greatest value if it is complementary to the other proposals.

Article 12, paragraph 4: This proposed provision in Alternative C is of great importance as it would enable the sanctioning of individuals who violate the Protocol. This would be of particular value in the context of non-international armed conflicts where other means of implementation may well be very difficult to ensure. It is therefore hoped that, whatever happens to the rest of Alternative C, this provision will be kept.

In general the ICRC is strongly of the view that the framework Convention itself should contain various national measures and international mechanisms to ensure the implementation of all the Protocols, even if some provisions only appropriate for landmines are additionally inserted in Protocol II. See comments below in the context of the framework Convention.

The framework Convention

General comment

In the Diplomatic Conference that led to its adoption, this Convention was structured intentionally in the form of a framework Convention that includes, *inter alia*, a provision on its scope of application, a description of treaty relations and a provision on implementation (at present only the duty to disseminate). The specific rules in the Protocols are all meant to be dependent on these basic provisions. If the Review Conference intends to change this, there should be a pressing and logical reason for doing so, and if it chooses to do so, it will need to take care that there are no legal inconsistencies as a result.

Article 1

As indicated in its report to the Group of Governmental Experts, the ICRC believes that the extension to non-international armed conflicts, now agreed to for Protocol II, should apply to all of the Convention's Protocols as it cannot be argued that the rules in Protocols I and III and in draft Protocol IV are inappropriate for non-international armed conflict. There is, in particular, the danger that if the extension only applies to Protocol II, it could be argued, by an *a contrario* reasoning, that it is acceptable to use incendiary weapons, for example, against civilians or indiscriminately in non-international armed conflicts. This is surely not a result that any delegation would wish and in any event such a result is probably contrary to international customary law.

Whatever is chosen, Article 1 of the framework Convention will need to be modified in order to be consistent with the Convention (i.e., including the Protocols) as a whole. Care will also need to be taken that Article 7 is appropriately worded, especially paragraphs 2 and 4.

Implementation provisions

As already indicated in the ICRC's formal report to the Group of Governmental Experts, it would be appropriate for the Convention to contain both national measures and international mechanisms to ensure its implementation. In particular, it would be inappropriate to indicate national measures of implementation in Protocol II only as it is incumbent on States to undertake such measures for all treaties to which they are party. Suggestions as to what could be included in the framework Convention are indicated in the aforementioned report.[5]

Appendix II

Articles 5 and 9

We consider that these proposals to strengthen the Convention are useful and should be supported.

Article 8

The question of the review procedure of the Convention is a critically important one and the Review Conference will need to amend Article 8 both for technical and substantive reasons.

Paragraph 3, sub-paragraphs (a) and (b) have already served their purpose and are now irrelevant are they applied only to the first review conference. Once a decision has been made under sub-paragraph (c) of paragraph 3, that too will cease to have any purpose, as it only applies to the decision to be taken at the first review conference. As will be indicated below, we are firmly of the view that the next review conference should take place in five years' time and not ten as suggested in sub-paragraph (c).

Serious thought should be given to amending and simplifying paragraphs 1 and 2 that will remain operative. **It is suggested that the new Article 8 read as follows:**

A Review Conference shall be convened by the Depository every five years. The Conference shall review the implementation of the Convention and its Protocols, and shall consider any amendments or additional Protocols proposed by a High Contracting Party. All High Contracting Parties shall be invited and States not parties to this Convention shall be invited as observers.

The rules of procedure could be decided on by the Conference itself rather than including confused rules in the treaty, as is the case in the present Article 8.

The most obvious reason for providing for automatic review conferences is that it will encourage universal adherence. The poor ratification level of this Convention was remarked on by all States Parties at the first meeting of the group of governmental experts. However, ten new ratifications have taken place since the Review Conference was announced and there can be no doubt that these have taken place because of the Review Conference.

The other reason for providing automatic Review Conferences is that it will be an indication to the international community that this is an important Convention that needs to be kept alive and up to date. In this respect it may be mentioned that this Convention can be seen, in some respects, as specifying rules applicable to particular weapons so that the general rules of international humanitarian law continue to be observed in practice.

Annex II: Draft Protocol on Blinding Laser Weapons

Scope of application

If Article 1 of the framework Convention is not amended so that all Protocols apply to both international and non-international armed conflicts, it will be important to add another Article to this Protocol. The danger of blinding laser weapons is particularly acute in non-international armed conflicts as all the experts that the ICRC consulted stressed that portable laser weapons will be particularly attractive to rebel groups and to terrorists. It is therefore proposed that the Protocol specify that it is to be applicable at all times or at least to both international and non-international armed conflicts.

Article 1

The ICRC believes that this statement of principle is important. Some lasers may be designed for use against optical systems, but will also be capable of anti-personnel use and it is important that the latter use is prohibited. The phrase "as a method of warfare" will exclude from censure the occasional incidental case of blindness caused but will act as a norm to the effect that anti-personnel use is not acceptable. The phrase "method of warfare" is a term that is to be found in other treaties and is well understood to mean the common systematic use of a certain weapon in a certain way.

With regard to whether "serious damage" should be added, it could be assumed that "blindness" means legal blindness and would therefore include cases where some peripheral sight may be left. However, if this is not evident, then we would either favour the inclusion of the additional phrase "serious damage" or alternatively the Protocol could make a reference to the World Health Organization criteria for blindness and low vision (WHO, International Statistical Classification of Diseases and Related Health Problems).

Article 2

We support the inclusion of the words in brackets, i.e., "produce and" as it is important that such weapons do not exist at all. This is because of the unacceptability of their existence given the principle in Article 1, and also because of the major danger of terrorist and criminal use. We would also prefer to exclude the word "permanently" as specialists have confirmed that it is impossible to design a laser that can only blind temporarily.

It would also be preferable to use the formula "the primary purpose of which is to blind" rather then "primarily designed to blind". This is because "design" normally refers to intention at the time of manufacture, which is difficult to prove, whereas purpose can be established not only at the time of manufacture but also by looking at the principal use made of the weapon.

During the fourth meeting of the Group of Governmental Experts, two delegations suggested using the word "specifically" rather than "primarily" in this Article. The ICRC strongly advises keeping the word "primarily" as the use of the word "specifically" would remove all sense from this Article. This is because all low-energy laser weapons can have a secondary use, namely, against optical instruments or for dazzling (all lasers using visible wavelengths blind for a large part of their range and dazzle towards the end of their range).

Article 3

This Article appears to be acceptable as it stands as it protects soldiers who might incidentally blind someone whilst correctly using a range-finder, target-designator, etc., against allegations of violating this Protocol. The use of the words "as a method of warfare" in Article 1 supports this.

NOTES

1 CCW/CONF.I/GE/9, at pp. 4–6. The value of a broad-based trigger mechanism is confirmed in the study by Andrew Latham "Towards an effective verification regime for the Convention on Certain Conventional Weapons", distributed by the Canadian delegation to the Group of Governmental Experts, at pp. 7–8.

2 For the effectiveness of a reporting system, see a study commissioned by the ICRC entitled "A comparison of self-evaluating State reporting systems" by Elizabeth Kornblum, printed in the *International Review of the Red Cross* 304 and 305 (1995). This study concludes that a reporting system is only effective if supported by a permanent secretariat and if reports are subject to analysis, questions and recommendations by a permanent independent body. This body also needs to be able to provide technical assistance to States in order to help them establish an effective national reporting system and to help them carry out its recommendations.

3 See footnote 1 above.

4 In its introductory comments to Appendix I.

5 CCW/CONF.1/GE/9 at pp. 2–6.

Chairman's Rolling Text

CCW/CONF.I/GE.23
CCW/CONF.I/1

ANNEX I

Chairman's Rolling Text

Article 1

[Material] Scope of Application

ALTERNATIVE A:

[1. This Protocol relates to the use on land of the mines, booby-traps and other devices defined herein including mines laid to interdict beaches, waterway crossings or river crossings, but does not apply to the use of anti-ship mines at sea or in inland waterways.

2. With the main purpose of protecting the civilian population, this Protocol shall apply in all circumstances including armed conflict and times of peace.

3. Nothing in this Protocol shall be invoked as affecting the purposes and principles contained in the United Nations Charter.

4. The application of the provisions of this Protocol to or by parties to a conflict which are not States parties shall not change their legal status or the legal status of a disputed territory, either explicitly or implicitly.]

ALTERNATIVE B:

[This Protocol relates to the use on land of the mines, booby-traps and other devices defined herein including mines laid to interdict beaches, waterway crossings or river crossings, but does not apply to the use of anti-ship mines at sea or in inland waterways.

2. This Protocol shall apply to situations referred to in Articles 2 and 3 and common to the Geneva Convention of 12 August 1949. This Protocol shall not apply to situations of internal disturbances and tensions, such as riots, isolated and sporadic acts of violence and other acts of a similar nature, as not being armed conflicts.

3. In case of conflicts referred to in paragraph 2 above that take place in the territory of a High Contracting Party that has accepted this Protocol, the dissident armed groups in its territory shall be automatically bound to apply the prohibitions and restrictions of this Protocol on the same basis.

4. Nothing in this Protocol shall be invoked for the purpose of affecting the sovereignty of a State or the responsibility of the government, by all legitimate means, to maintain or re-establish law and order in the State or to defend the national unity and territorial integrity of the State.

5. Nothing in this Protocol shall be invoked as a justification for intervening, directly or indirectly, for any reason whatever, in the armed conflict or in the internal or external affairs of the High Contracting Party in the territory of which that conflict occurs.

6. The application of the provisions of this Protocol to Parties to a conflict which are not High Contracting Parties that have accepted this Protocol shall not change their legal status or the legal status of a disputed territory, either explicitly or implicitly.]

Article 2

Definitions

For the purpose of this protocol:

1. "Mine" means a munition placed under, on or near the ground or other surface area and designed to be exploded by the presence, proximity or contact of a person or vehicle.

2. ["Remotely-delivered mine"] means a mine not directly emplaced but delivered by artillery, missile, rocket, mortar, or similar means, or dropped from an aircraft. [Mines delivered from a land-based system from less than 500 metres are not considered to be "remotely delivered".]

3. "Anti-personnel mine" means a mine [designed to be] exploded by the presence, proximity or contact of a person and that will incapacitate, injure or kill one or more persons.

4. "Booby-trap" means any device or material which is designed, constructed, or adapted to kill or injure, and which functions unexpectedly when a person disturbs or approaches an apparently harmless object or performs an apparently safe act.

5. "Other devices" means manually emplaced munitions and devices designed to kill, injure or damage and which are actuated [by remote control or] automatically after a lapse of time.

6. "Military objective" means, so far as objects are concerned, any object which by its nature, location, purpose or use makes an effective contribution to military action and whose total or partial destruction, capture or neutralization, in the circumstances ruling at the time, offers a definite military advantage.

7. “Civilian objects” are all objects which are not military objectives as defined in paragraph 6.

8. “Minefield” is a defined area in which mines have been emplaced and “Mined area” is an area which is dangerous due to the presence [or suspected presence] of mines.

9. “Recording” means a physical, administrative and technical operation designed to obtain, for the purpose of registration in the official records, all available information facilitating the location of minefields, mined areas, mines, booby-traps and other devices.

10. “Self destructing mechanism” means an incorporated automatically functioning mechanism which secures the destruction of a munition.

11. “Self neutralizing mechanism” means an incorporated automatically functioning mechanism which renders a munition inoperable.

[12. “Self deactivating” means automatically rendering a munition inoperable by means of the irreversible exhaustion of a component that is essential to the operation of the munition.]

[13. “Remote control” means a control by commands from a distance.]

[14. “Anti-handling device” means a device by which a mine will explode when an attempt is made to remove, neutralize or destroy the mine.]

or [“Anti-handling device” means a device to protect a munition against removal.]

Article 3

General restrictions on the use of mines, booby-traps and other devices

1. The Article applies to:

   (a) mines;

   (b) booby-traps; and

   (c) other devices.

2. Each State party or party to a conflict is, in accordance with the provisions of this Protocol, responsible for all mines, booby-traps, and other devices employed by it and undertakes to clear, remove or destroy them as specified in Article 9 of this Protocol.

3. It is prohibited in all circumstances to use any [mine,] booby-trap or other device which is designed to cause superfluous injury or unnecessary suffering.

4. [All weapons] to which this Article applies shall meet the relevant standards [for armed period, reliability, [detectability,] design and construction] as specified in the Technical Annex.

5. It is prohibited in all circumstances to direct weapons to which this Article applies, either in offence, defence or by way of reprisals, against the civilian population as such or against individual civilians.

6. The indiscriminate use of weapons to which this Article applies is prohibited. Indiscriminate use is any placement of such weapons:

(a) which is not on, or directed against, a military objective; or

(b) which employs a method or means of delivery which cannot be directed at a specific military objective; or

(c) which may be expected to cause incidental loss of civilian life, injury to civilians, damage to civilian objects, or a combination thereof, which would be excessive in relation to the concrete and direct military advantage anticipated.

[7. Several clearly separated and distinct military objectives located in a city, town, village or other area containing a similar concentration of civilians or civilian objects cannot be treated as a single military objective.]

8. All feasible precautions shall be taken to protect civilians from the effects of weapons to which this Article applies. Feasible precautions are those precautions which are practicable or practically possible taking into account all circumstances ruling at the time, including humanitarian and military considerations. These circumstances include, but are not limited to:

(a) the short and long term effect of landmines upon the local civilian population for the duration of the minefield;

(b) possible measures to protect civilians (e.g., fencing, signs, warning and monitoring);

(c) the availability and feasibility of using alternatives; and

(d) the short and long-term military requirements for a minefield.

9. Effective advance warning shall be given of any emplacement of mines, booby-traps and other devices which may affect the civilian population, unless circumstances do not permit.

[10. Restrictions and prohibitions in this Protocol shall facilitate the ultimate goal of a complete ban on the production, stockpiling, use and trade of anti-personnel landmines.]

Article 4

Restrictions on the use of anti-personnel mines other than [remotely delivered mines,] [booby-traps] and other devices

1. This Article applies to:

(a) Anti-personnel mines other than [remotely delivered mines];

(b) [booby-traps;] and

(c) other devices.

2. It is prohibited to use weapons to which this Article applies which are not self-destructing,[1] unless:

(a) they are placed within a perimeter-marked area that is monitored by military personnel and protected by fencing or other means, to ensure the effective exclusion of civilians from the area. The marking must be of a distinct and durable character and must at least be visible to a person who is about to enter the perimeter-marked area; and

(b) they are cleared before the area is abandoned, unless the area is turned over to the forces of another State that accept responsibility for the maintenance of the protection required by this Article and the subsequent clearance of those weapons.

3. A party to the conflict is relieved from further compliance with the provisions of subparagraphs 2 (a) and 2 (b) above only if such compliance is not feasible due to forcible loss of control of the area as a result of enemy military action, including situations where direct enemy military action makes it impossible to comply. If the party of the conflict regains control of the area, it shall resume compliance with the provisions of subparagraphs 2 (a) and 2 (b).

4. If the forces of a party to the conflict gain control of an area in which weapons to which this Article applies have been laid, such forces, shall, to the maximum extent feasible, maintain and, if necessary, establish the protections required by this Article until such weapons have been cleared.

5. States parties shall take all feasible measures to prevent the unauthorized removal, defacement, destruction or concealment, of any device, system or material used to establish the perimeter of a perimeter-marked area.

6. [To facilitate clearance, it is prohibited to use [anti-personnel] mines which are not in compliance with the provisions on detectability in the Technical Annex.]

Article 5

[Restrictions on the use of remotely delivered mines
It is prohibited to use remotely delivered mines which are not self-destructing.]

Article 5 bis

[Prohibitions on the use of [anti-personnel] mines
which are not detectable[2]

It is prohibited to use [anti-personnel] mines which are not in compliance with the provisions on detectability in the Technical Annex.]

Article 6

Prohibitions on the use of booby-traps and other devices

1. Without prejudice to the rules of international law applicable in armed conflict relating to treachery and perfidy, it is prohibited in all circumstances to use booby-traps and other devices which are in any way attached to or associated with:

(a) internationally recognized protective emblems, signs or signals;

(b) sick, wounded or dead persons;

(c) burial or cremation sites or graves;

(d) medical facilities, medical equipment, medical supplies or medical transportation;

(e) children's toys or other portable objects or products specially designed for feeding, health, hygiene, clothing or education of children;

(f) food or drink;

(g) kitchen utensils or appliances except in military establishments, military locations or military supply depots;

(h) objects clearly of a religious nature;

(i) historic monuments, works of art or places of worship which constitute the cultural or spiritual heritage of peoples;

(j) animals or their carcasses.

2. It is prohibited to use booby-traps [and other devices] in the form of an apparently harmless portable object which is specifically designed and constructed to contain explosive material.

[3. It is prohibited to use booby-traps in armed conflicts not of an international character.]

Article 6 bis[3]

[Prohibition of the use, development, manufacture, stockpiling and transfer of certain mines and booby-traps]

[1. It is prohibited to use, develop, manufacture, stockpile or transfer, directly or indirectly:

- Anti-personnel mines defined in Article 2, [paragraph 3] of this Protocol; and]
- [Anti-personnel mines without self-destruction or self-neutralizing mechanisms]
- Booby-traps defined in Article 2, [paragraph 4] of this Protocol.

2. The States parties undertake to destroy the weapons to which this article applies and which are in their ownership and/or possession.]

[3. It is prohibited to use [, manufacture, stockpile or transfer] [anti-personnel] mines which cannot be detected, that is, which cannot be identified using widely available equipment such as electro-magnetic mine detectors [as specified in the Technical Annex].

[4. The States parties shall notify the Depositary of all stockpiles of weapons to which this Article applies and undertake to destroy them within a period of . . . years. The States shall report annually on the progress made regarding implementation of paragraph 3 of this Article.]

Article 6 ter

[Transfers[4]]

[In order to prevent the use of mines contrary to the purposes of this Protocol, each High Contracting Party:

1. Undertakes not to provide any mines to non-State entities;

2. Undertakes not to transfer[5] any mines to States which are not bound by this Protocol;

3. Undertakes not to transfer to any other High Contracting Party any mines the use of which is prohibited in all circumstances;

4. Shall ensure that in transferring to other High Contracting Parties bound by this Protocol any mines the use of which is restricted under this Protocol, the

receiving High Contracting Party agrees to comply with the relevant provisions of international humanitarian law.]

Article 7

Recording and use of information on minefields, mined areas, mines, booby-traps and other devices

1. All information concerning minefields, mined areas, mines, booby-traps and other devices shall be recorded in accordance with the provisions of the Technical Annex.

2. All such records shall be retained by the parties, who shall without delay after [the cessation of active hostilities] [the effective cessation of hostilities and the meaningful withdrawal of forces from the combat zones]:

(a) Take all necessary and appropriate measures, including the use of such information, to protect civilians from the effects of the minefields, mined areas, mines, booby-traps and other devices and,

(b) Make available to the other party or parties to the conflict concerned and to the Secretary-General of the United Nations all such information in their possession concerning minefields, mined areas, mines, booby-traps and other devices laid by them in areas no longer under their control.

3. This Article is without prejudice to the provisions of Article 8 of this Protocol.

Article 8

[Protection from the effects of minefields; mined areas; mines; booby-traps and other devices

1. When an operation covered by the [Convention on the Safety of United Nations and Associated Personnel] is taking place in any area, each party to the conflict, if requested by the head of the operation, shall make available to the head of the operation all information in the party's possession concerning the location of minefields, mined areas, mines, booby traps and other devices in that area and in order to protect personnel covered by the above-mentioned Convention who are participating in such operations shall, as far as it is able;

(a) remove or render harmless all mines, booby traps or other devices in that area; and

(b) take such measures as may be necessary to protect such personnel from the effects of mines, booby traps and other devices.[6]

2. (a) When a mission of a [regional arrangement or agency acting under Chapter VIII of the Charter of the United Nations] performs functions in any area with the consent of the parties to a conflict, each party, if requested by the head of that mission, shall make available to the head of that mission all information in the party's possession concerning the location of minefields, mined areas, mines, booby traps and other devices in that area and shall, as far as it is able, provide to the mission and its personnel the protections described in subparagraphs 1 (a) and (b);

[2 (b) When a mission of the International Committee of the Red Cross performs functions assigned to it by the Geneva Conventions of 1949 and their Additional Protocols of 1977, or a humanitarian mission of the United Nations system not otherwise covered by this article performs functions with the consent of the parties to the conflict, each party, if requested by the head of that mission, shall, to the extent feasible, provide to that mission and its personnel the protections described in subparagraphs 1 (a) and (b) and shall, as far as it is able, identify to the head of that mission minefields, mined areas, mines, booby traps and other devices in the area where those functions are being performed [and provide safe access either through the clearance of a lane through minefields or by designating an alternative land route that will permit the accomplishment of these mandated missions].]

2. (c) When the mission of an [impartial humanitarian organization] not otherwise covered by this article, performs functions with the consent of the parties to a conflict, each party, if requested by the head of that mission shall, to the extent feasible, provide to that mission and its personnel the protections described in subparagraphs 1 (a) and (b) and shall, as far as it is able identify to the head of that mission all areas where minefields, mined areas, mines, booby traps and other devices which may impede the performance of those functions are known or believed to be located.

3. When a United Nations fact-finding mission or other fact-finding mission with the consent of the parties, not otherwise covered by this article performs functions in any area, each party to the conflict concerned shall provide protection to that mission except where, because of the size of such mission, it cannot adequately provide such protection. In that case it shall make available to the head of the mission the information in its possession concerning the location of minefields, mined areas, mines, booby-traps and other devices in that area.

[4. Nothing in this Convention shall affect the rights and obligations of United Nations and Associated Personnel as set out in the Convention referred to in paragraph 1 above.]

Article 9

Removal of minefields, mined areas, mines, booby-traps and other devices [and international cooperation]

1. [Without delay] after [the cessation of active hostilities] [the effective cessation of hostilities and the meaningful withdrawal of forces from the combat zone] all minefields, mined areas, mines, booby-traps and other devices shall be cleared, removed, destroyed or maintained in accordance with Article 3 and paragraph 2 of Article 4 of this Protocol.

(a) Each party bears such responsibility with respect to minefields, mined areas, booby-traps and other devices in areas under its control.

(b) With respect to minefields, mined areas, mines, booby-traps and other devices laid by a party in areas over which it no longer exercises control, such party shall provide to the responsible party pursuant to paragraph 1 (a) above, to the extent permitted by such party, technical and material assistance necessary to fulfil such responsibility.

2. At all times necessary, the parties shall endeavour to reach agreement, both among themselves and, where appropriate, with other States and with international organizations, [on the provision of technical and material assistance,][7] including, in appropriate circumstances, undertaking of joint operations, necessary to fulfil such responsibilities.

Article 9 bis

Technological Cooperation and Assistance in Mine Clearance and Implementation of Protocol II

1. Each State party shall undertake to facilitate [and shall have the right to participate in] the [fullest possible] exchange of equipment, material and scientific and technological information concerning the implementation of this Protocol and means of mine clearance. [The States parties shall undertake not to maintain or impose any restrictions on the transfer of equipment or technology for mine clearance.]

2. Each State party undertakes [to give careful consideration to providing] [to provide] such assistance through the United Nations, international bodies,[8] or on a bilateral basis.

Mine Clearance

3. The States parties shall undertake to provide information concerning various means and technologies of mine clearance to the data bank established within the United Nations system.

[4. The coordinated mine-clearance programme established within the United Nations as per in the UNGA Resolution 48/7 adopted without a vote, shall also, within the resources available to it, and at the request of a State party, provide expert advice and assist the State party in identifying how its programmes for the mine clearance could be implemented.

5. Each State party undertakes to provide assistance through the United Nations coordinated programme and other relevant United Nations bodies and to this end to elect to take one of the following two measures:

(a) to contribute to the voluntary fund for assistance, established by United Nations coordinated programme;

(b) to declare not later than 90 days after the amended protocol II enters into force for it, the kind of assistance it might provide in response to an appeal by the United Nations coordinated programme. If, however, a State party subsequently is unable to provide the assistance envisaged in its declaration it is still under the obligation to provide assistance in accordance with this paragraph.]

6. Requests by States parties for assistance, substantiated by relevant information, may be submitted to the United Nations, to other appropriate bodies or to other States. These requests [may be provided] to the Depositary, which shall transmit them to all States parties and relevant international organizations. [Subsequently after the receipt of the request an [investigation] [assessment by the United Nations coordinated programme] [shall] [may] be initiated in order to provide foundation for further action.] The Depositary shall [, as appropriate,] provide a report to States parties on the facts relevant to these requests, as well as the type and scope of assistance that may be needed.

Implementation of Protocol II

7. The States parties shall undertake to provide information [to the Depositary] [to the Commission] concerning the implementation of this Protocol, including meeting the requirements for self-destructing and other features, as specified in this Protocol.

[8. Upon receiving the request from the State party for any technical assistance, [the Depositary] [the Commission] will render this assistance free of cost.

It will employ all possible means at its disposal to ensure:

(a) Transfer of technology from advanced nations to the developing countries for acquisition on no cost basis;

(b) Allocate requisite funds for the assistance through United Nations coordinated programme.]

Technical Annex

1. Recording

(a) The recording of the location of mines other than [remotely delivered mines,] minefields, mined areas, [areas of] booby-traps and other devices shall be done in accordance with the following:

(i) The location of the minefields, [mined areas], [areas of] booby-traps and other devices shall be specified accurately by relation to the coordinates of at least two reference points and the estimated dimensions of the area containing these devices in relation to those reference points.

(ii) Maps, diagrams or other records shall be made in such a way as to indicate the location of minefields, mined areas, [booby-traps] and other devices in relation to reference points, these records shall also indicate their perimeters and extent.

(iii) For purposes of detection and clearance of mines, [booby-traps] and other devices, maps, diagrams or other records shall contain complete information on the type, number, emplacing method, type of fuse and life time, date [and time] of laying and other relevant information of all the munitions laid. Whenever feasible the minefield record shall show the exact location of every mine; except in row minefields where the row location is sufficient.

(b) The estimated location and area of remotely delivered mines shall be specified by coordinates of reference points (normally corner points) and shall be ascertained and when feasible marked on the ground at the earliest opportunity. The total number and type of mines laid, the date [and time] of laying and the self destruction time periods shall also be recorded.

(c) Copies of records are to be held at a level of command sufficient to guarantee their safety [as far as possible].

2. Detectability of [anti-personnel] [mines]

(a) [A sufficient quantity of not easily removable material or any appropriate device, incorporating detectability equivalent to 8 grams of iron in a single coherent mass, to enable detection by commonly available technical detection equipment shall be placed in or on every [anti-personnel] [mine] emplaced.]

(b) [All [anti-personnel][mines] shall have irremovable metallic elements in their construction to enable detection and [clearance by standard mine-sensing devices].]

[(c) No [anti-personnel] [mines], [booby-traps] and other devices may be designed such that they will detonate by the operation of standard mine-sensing devices.]

3. Specifications for self-destructing anti-personnel mines

Anti-personnel mines required by Article 4, paragraph 2 and Article 5 of this Protocol to be self-destructing shall be designed and constructed so that no more than [1 in every 1,000] activated will fail to self-destruct [after no more than 7–90 days];[9] [and they shall have a [back-up feature] [self-deactivation feature], designed and constructed so that the mine will no longer function as a mine [30–365 days, with a reliability of 1 in every 1,000 surviving mines] [as soon as feasible] if the self-destruction mechanism fails.]

4. International signs for minefields and mined areas

Signs similar to the example in Annex A shall be utilized in the marking of minefields and mined areas. Each sign [shall] [should] meet the following criteria to ensure its visibility and recognition by the civilian population:

(a) Size and shape: a triangle or square no smaller than 28 centimetre (11 inches) by 20 centimetres (7.9 inches) for a triangle, and 15 centimetres (6 inches) per side for a square;

(b) Colour: red or orange with a yellow reflecting border;

(c) Symbol: the symbol illustrated in Annex A, or an alternative readily recognizable in the area in which the sign is to be displayed as identifying a dangerous area;

(d) Language: the sign should contain the word "mines" in one of the six official languages of this Convention (Arabic, Chinese, English, French, Russian and Spanish) and the language(s) prevalent in that area;

(e) Spacing: signs should be placed around the minefield or a mined area at a distance sufficient to ensure their visibility at any point by a civilian approaching the area.

[Drawing omitted]

APPENDIX I

Proposals relating to verification and compliance[10]

ALTERNATIVE A:[11 12 13]

[Commission of States parties

1. For the purposes of this Protocol, a Commission shall be established by the States parties. The Commission of States parties shall meet in Geneva regularly. Any State party may appoint a representative to the Commission. The ICRC shall be invited to participate in the work of the Commission as an observer. The Commission shall consider annual reports provided by the States parties on the implementation of the Protocol. The Commission shall take its decisions by consensus if possible, but otherwise by a majority of members present and voting.

2. Each State party undertakes to provide annually the relevant information to the Commission, i.e.

(a) Progress on implementation of the Protocol II;

(b) Information on mine clearance;

(c) Information on civilian casualties occurring due to deployment of mines in its territory.

3. Each State party undertakes to provide/exchange information with other States parties to promote transparency and credibility for wider adherence to this Protocol's requirements/restrictions.

[4. Each State party to this Protocol undertakes to facilitate the fullest possible exchange of technological information in order to assist States parties to comply with restrictions/requirements of this Protocol.]

5. The Commission shall also carry out other functions as are necessary for the implementation and review of this Protocol.

6. The costs of the Commission's activities shall be covered by the States parties in accordance with the United Nations scale of assessments, adjusted to allow for differences between the number of States Members of the United Nations and the number of States parties.]

ALTERNATIVE B:[14]

[Article 10
Compliance Monitoring

1. Each State party undertakes to protect civilians from the effects of the use of landmines and for that purpose undertakes to take necessary measures to prohibit and prevent the indiscriminate use of landmines. The measures shall include:

(a) legislation, if necessary;

(b) education of military personnel concerned on the relevant provisions of this Protocol;

(c) dissemination to the civilian population of the information on possible effects of landmines and on signs used for minefields and mined areas;

(d) appropriate measures to meet the technical requirements set out in this Protocol;

(e) measures to facilitate the exchange of technical information with other States parties on mine clearance and on the activities it conducted for the purpose of paragraph (d) in this Article;

2. Each State party affirms the recognized objective of prohibiting and preventing the indiscriminate use of landmines and to this end undertakes to provide annual report to the Depositary. The report shall contain the following:

(a) the relevant legislation;

(b) any measures it has taken to educate the military personnel and to disseminate the relative information for the purpose of this Protocol;

(c) any measure it has taken to meet the technical requirements set out in this Protocol;

(d) information on recovery, destruction or clearance after military use of landmines;

(e) information on casualty to civilian population occurred due to use of such mines in its territory and measures it has taken to redress the situation;

(f) measures it has taken on international technical information exchange and on international cooperation on mine clearance;

3. The Depositary shall distribute the above-mentioned report, upon request, to any other State party.]

ALTERNATIVE C:[15] [16]

[Article 10

Verification Commission

1. Each State party shall be entitled to ask the Depositary to convene a Verification Commission, within a period of one week, to conduct an inquiry in order to clarify and resolve any questions relating to possible non-compliance with the provisions of this Protocol concerning the use of mines, booby-traps and other devices. The request for an inquiry shall be accompanied by relevant information and evidence confirming its validity.

2. (a) The Verification Commission, which shall meet in New York, shall be open to the participation of all States parties. Subject to the provisions of both

paragraph 3 of this article and paragraph 1 of Article 11, the Verification Commission shall take its decisions by consensus if possible, but otherwise by a majority of members present and voting.

(b) The costs of the Verification Commission's activity shall be covered by the States parties in accordance with the United Nations scale of assessments, adjusted to allow for differences between the number of States Members of the United Nations and the number of States parties.

3. (a) An inquiry shall be held unless the Verification Commission decides, not later than 48 hours after it has been convened, with a two-thirds majority of its members present and voting that the information and evidence produced does not justify an inquiry.

(b) For the purposes of the inquiry the Verification Commission shall seek useful assistance and relevant information from States parties and international organizations concerned and from any other appropriate sources.

Article 11

Fact-finding missions

1. The inquiry shall be supplemented by evidence collected on the spot or in other places under the jurisdiction or control of the party to the conflict concerned unless the Verification Commission decides with a two-thirds majority of its members present and voting that no such evidence is required. The Verification Commission shall notify the party to a conflict concerned of the decision to send a team of experts to conduct a fact-finding mission at least 24 hours before the team of experts is expected to arrive. It shall inform all States parties of the decision taken as soon as possible.

2. For the purposes of paragraph 1 of this article, the Depositary shall prepare a list of qualified experts provided by States parties, and constantly keep this list updated. The experts shall be designated in view of the particular fields of expertise that could be required in a fact-finding mission concerning the alleged use of mines, booby-traps and other devices. The initial list as well as any subsequent change to it shall be communicated, in writing, to each State party without delay. Any qualified expert included in this list shall be regarded as designated unless the State party, not later than 30 days after its receipt of the list declares its non-acceptance, in which event the Verification Commission shall decide whether the expert in question shall be designated.

3. Upon receiving a request from the Verification Commission, the Depositary shall appoint a team of experts from the list of qualified experts, acting in their

personal capacity, to conduct a fact-finding mission at the site of the alleged incident. Experts who are nationals of States parties involved in the armed conflict concerned or of States parties which requested the inquiry shall not be chosen. The Depositary shall dispatch the team of experts at the earliest opportunity taking into account the safety of the team.

4. The party to a conflict concerned shall make the necessary arrangements to receive, transport and accommodate the team of experts in any place under its jurisdiction or control.

5. When the team of experts has arrived on the spot, it may hear a statement of information by official representatives of the party to a conflict concerned and may question any person likely to be connected with the alleged violation. The team of experts shall have the right of access to all areas and installations where evidence of violation of this Protocol could be collected. The party to a conflict concerned may make any arrangements it considers necessary for the protection of sensitive equipment, information and areas unconnected with the subject of the fact-finding mission, or for any constitutional obligations it may have with regard to proprietary rights, searches and seizures, or other constitutional protection or for the protection of the conduct of military operations. In that event, it shall make every reasonable effort to satisfy the legitimate needs of the team of experts through other means.

6. After having completed its fact-finding mission, the team of experts shall submit a report to the Depositary not later than one week after leaving the territory of the State party in question. The report shall summarize the factual findings of the mission related to the alleged non-compliance with the Protocol. The Depositary shall promptly transmit the report of the team of experts to all States parties.

Article 12

Compliance

1. The States parties undertake to consult each other and to cooperate with each other in order to resolve any problems that may arise with regard to the interpretation and application of the provisions of this Protocol.

2. If the Verification Commission concludes, based on the inquiry, including any report of the team of experts referred to in Article 11, paragraph 6, that there has been a violation of the provisions of this Protocol on the use of mines, booby-traps and other devices, the Verification Commission shall, as appropriate, request that the party responsible for the violation take appropriate measures to remedy the situation.

3. If weapons covered by this Protocol have been used in violation of its provisions, the States parties shall consider measures designed to encourage compliance, including collective measures in conformity with international law, and may, in accordance with the United Nations Charter, refer the issue to the attention of the Security Council.

4. The provisions of the 1949 Geneva Conventions relating to measures for the repression of breaches and grave breaches shall apply to breaches and grave breaches of this Protocol during armed conflict. Each party to a conflict shall take all appropriate measures to prevent and suppress breaches of this Protocol. Any act or omission occurring during armed conflict in violation of this Protocol, if committed wilfully or wantonly and causing death or serious injury to the civilian population shall be treated as a grave breach. A party to the conflict which violates the provisions of this Protocol shall, if the case demands, be liable to pay compensation, and shall be responsible for all acts committed by persons forming part of its armed forces. States parties and parties to a conflict shall require that commanders ensure that members of the armed forces under their command are aware of, and comply with, their obligations under this Protocol.]

APPENDIX II

Other Proposals[17]

RUSSIAN FEDERATION

[Article 5 of the Convention
Entry into Force

1. This Convention shall enter into force three months after the date of deposit of the sixth instrument of ratification, acceptance, approval or accession.

Paragraphs 2, 3 and 4 of this Article to be modified in accordance with the amendments to paragraph 1.]

[Article 9

(a) New paragraphs. Denunciation

1. Any High Contracting Party may, by so notifying the Depositary, denounce this Convention or any of the annexed Protocols upon the expiry of 10 years since the date on which the Convention and any of its Protocols came into force. Such denunciation shall take effect one year after the date on which it is registered.

2. Any High Contracting Party which ratifies this Convention and any of its annexed Protocols and does not, within the year following the expiry of the

10-year period mentioned in the preceding paragraph, exercise the right of denunciation provided for in this article, shall be bound for a further 10-year period and may thereafter denounce this Convention or any of its annexed Protocols upon the expiry of each 10-year period under the terms of this article.

(b) The first sentence of the existing paragraph 2 to be deleted.]

NEW ZEALAND, IRELAND, AUSTRALIA AND SWEDEN

[Article 8 of the Convention[18]

Article 8 (3 (c) of the Convention signals a need to consider at the first Review Conference the question of periodicity of review meetings. This issue could be addressed either through a decision of the Conference or an amendment to the Convention.]

NOTES

1 The chapeau of paragraph 2 will require reconsideration in the light of discussion on, *inter alia*, the Technical Annex and Article 6 bis.
2 Acceptance of this proposal would entail:
  (a) deletion of the word "detectability" from Article 3, para. 4;
  (b) deletion of Article 4, para. 6;
  (c) deletion of the square brackets in the Technical Annex around the word "mines" in the chapeau of para. 2, around the word "mine" in para. 2 (a), and around the word "mines" in para. 2 (b).
3 The inclusion of the issue of development, manufacture, stockpiling and transfer of mines, booby traps and other devices is not accepted by all delegations.
4 This Article is without prejudice to the position of delegations on the issue of prohibitions or restrictions on the production and stockpiling of certain conventional weapons.
5 It is understood that "transfers" involve, in addition to the physical movement of mines into or from national territory, the transfer of title to and control over the mines.
6 The following alternative drafting of paragraph 1 has been suggested:
  1. When an operation covered by the Convention on Safety of United Nations and Associated Personnel is taking place in any area, each party to the conflict, if requested by the head of the operation in order to protect personnel covered by the above-mentioned Convention who are participating in such operations shall, as far as it is able:
  (a) make available to the head of the operation all information in the party's possession concerning the location of minefields, mined areas, mines, booby traps and other devices in that area;
  (b) remove or render harmless all mines, booby traps or other devices in that area; and

(c) take such measures as may be necessary to protect such personnel from the effects of mines, booby traps and other devices.

Consequential reshuffling of wording and change in numbering in certain subsequent paragraphs may be necessary.

7 Paragraph 2 will be finalized in light of the final text of Article 9 bis.

8 The issue of a possible decision-making or a consultative mechanism will be further considered.

9 The self-destructing time needs to be further discussed in relation to the time of laying / time of activation.

10 Several delegations expressed the view that, whilst not agreeing to every provision of each proposal, the three alternatives A, B and C were not exclusive but complementary to each other.

11 Some delegations consider that elements of this text may be more appropriately addressed through amendment of the Convention, rather than of Protocol II. Further, this text is without prejudice to proposals for more frequent meetings of the Review Conference than currently provided for in the Convention.

12 The concept of a "Commission" proposed has not been accepted by a group of States.

13 A group of delegations considers that the concept of a "Commission" relates to and complements alternatives B and C.

14 Alternative B has been presented as an alternative text to alternatives A and C and is, according to several delegations, the most appropriate. It is not complementary to any other proposal.

15 The concept of verification for this Protocol is not accepted by a group of countries.

16 One delegation submitted in document CCW/CONF.I/GE/CRP.47 proposals elaborating on this text, which could be developed further.

17 The proposals in Appendix II require further consideration.

18 The proposal on Article 8 is further elaborated in CCW/CONF.I/GE/CRP.55.

Press Release

Vienna Conference on UN weapons Convention

ICRC Regrets Outcome on Landmine Issue but Welcomes New Ban on Blinding Laser Weapons

13 October 1995

The International Committee of the Red Cross (ICRC) deeply regrets that the Review Conference of the 1980 United Nations Convention on Certain Conventional Weapons, held in Vienna over the past three weeks, was unable to agree on measures to prohibit or substantially restrict the use of landmines. This unfortunate outcome reflects both the overly complex technical nature of many of the proposals considered and an unwillingness on the part of many States to place significant limits on landmine use, in order to achieve the humanitarian goals of the Conference. The ICRC does, however, welcome the adoption today of a new, legally binding agreement which will prohibit the use of laser weapons to intentionally blind soldiers or civilians.

The new protocol prohibits both the use and the transfer of laser weapons specifically designed to cause permanent blindness. It also requires States to take all feasible precautions, including training of armed forces, to avoid permanent blinding through the legitimate use of other laser systems.

Despite the deadlock on the landmines issue at the Vienna Conference, the ICRC sees the Conference as a step towards the stigmatization and elimination of anti-personnel landmines. The Conference clearly demonstrated that the international public outcry about these weapons has broken the consensus that they are legitimate weapons of war. By the first week of the Conference sixteen States had joined the ICRC and the Secretary-General of the United Nations in calling for a total ban.

The ICRC encourages all the States that will participate in the resumed session of the Review Conference to rise above their narrow national interests in the general interest of humanity.

11

ICRC Position Paper No. 2
Landmine Negotiations: Impasse in Vienna Highlights Urgency of National and Regional Measures

ICRC Briefing and Position Paper
November 1995

The International Committee of the Red Cross (ICRC) deeply regrets that the recent Review Conference of the 1980 Convention on Certain Conventional Weapons (CCW) was unable to agree on new measures to prohibit or severely restrict the production, use and transfer of anti-personnel landmines. The Conference, which adjourned on 13 October after three weeks of negotiations in Vienna and nearly two years of preparation, was seen as an important element in international efforts to address the humanitarian crisis caused by landmines. This unfortunate outcome reflects both the overly technical nature of many of the proposals considered and an unwillingness on the part of many States to place significant limits on landmines to achieve the humanitarian goals of the conference.

The ICRC appeals to governments and the concerned public to ensure that humanitarian considerations are put in the centre of negotiations at the resumed sessions of the Review Conference, to be held in Geneva from 15–19 January and 22 April–3 May 1996. Furthermore, it is calling for increased efforts on the national and regional levels to ensure that humanitarian responsibilities are met even if agreement on far-reaching measures is not possible in the near future at the international level. On 22 November the ICRC, together with national Red Cross and Red Crescent Societies, launched for the first time in its history an international media campaign aimed at mobilizing the public conscience for the stigmatisation of anti-personnel mines.

PROGRESS AND NEW OBSTACLES IN VIENNA

Despite adjourning in deadlock on a number of important technical issues, the Vienna session of the Review Conference did achieve a large measure of provisional agreement on new measures which the ICRC considers important steps forward. These include:

- Extension of the scope of the CCW's landmine restrictions to cover internal as well as international armed conflicts;

- Assignment of responsibility for the clearance of landmines to those who lay them;

- Increased obligations on the part of combatants to protect humanitarian workers, including the ICRC, national Red Cross and Red Crescent Societies and other humanitarian workers, from landmines so that they can reach people in need;

- A requirement that all, rather than only certain types of minefields, be recorded; and

- A prohibition on the use of mechanisms which cause a mine to explode when an electromagnetic detector, such as those used by mine-clearance teams, comes near it.

However, there was no agreement on the key restrictions on the use of landmines which were prepared for the Review Conference over the last two years by a Group of Governmental Experts charged with its preparation. These included requirements that:

- All anti-personnel mines must be detectable;

- Remotely delivered mines must contain a self-destruction mechanism; and

- All hand- or machine-emplaced anti-personnel mines used outside of marked, guarded and fenced minefields should have a self-destructing mechanism.

The deadlock in Vienna was in part due to the fact that some governments argued for far less restrictive measures than they had appeared willing to accept in previous meetings of the Group of Governmental Experts. Although most countries accepted the above requirements in principle, or in many cases supported stronger restrictions including a total ban, the disputes which led to a deadlock involved technical measures for implementing these requirements. Disagreements centered on:

- Whether self-neutralizing mines, which remain in the ground indefinitely and must be treated by civilians and clearance teams as if they are live, could be substituted for self-destructing ones;

- Whether self-destructing mines should destroy themselves within thirty days or could rather remain live for as long as a year;

- Whether the maximum permissible failure rate for self-destructing mines would be as stringent as 1 in 1000 (0.1%) or as lenient as 100 in 1000 (10%);

- Whether a minimum metallic content, such as 8 grams of metal, should be specified so that mines are detectable under actual conditions in post-conflict terrain; and
- Whether the technical requirements above (a) should be met immediately for all new mines used, (b) should be subject to a grace period of up to 15 years or (c) should be implemented "as soon as feasible".

Most of the technical disputes described above reflect an inability or unwillingness on the part of particular countries to adapt the type of mines they produce or use to achieve the humanitarian objectives of the Review Conference. On the other hand it represents an unwillingness on the part of States promoting new mine technologies to consider simpler, but more far-reaching measures.

The ICRC regrets that proposals were blocked which would have required that anti-tank mines be detectable and which would prohibit their use with anti-handling mechanisms – which cause a mine to explode when clearance teams attempt to remove it. It also regrets that no verification provisions were agreed upon.

THE ICRC'S POSITION

The ICRC remains convinced that the only effective means of ending the scourge of anti-personnel landmines is to entirely prohibit their production, transfer and use. The difficulties encountered in the Vienna negotiations demonstrate, as the ICRC had feared, that complex and costly technical measures will not solve the landmine crisis. Because many States are either unable or unwilling to make the technical changes suggested, and because promotion of self-destructing mines could lead to an overall increase in number of mines used, simpler and more far-reaching measures should now be considered. In addition to being far more effective, such measures are likely to be more easily verified than the complex regime which was considered in Vienna.

In addition to continuing its efforts to increase support for a global ban on anti-personnel landmines, which has now been supported by sixteen States, the UN Secretary-General, the heads of numerous UN Agencies, the Council of Ministers of the Organization of African Unity, the European Parliament and Pope John Paul II, the ICRC will actively promote two new initiatives:

- A **ban on all transfers** of anti-personnel mines in the context of the 1980 Convention; and
- **National and regional measures** – Stopping landmines doesn't depend only on the success of international negotiations. States can take their own moral and

political responsibility to end this scourge either unilaterally on their own territory or cooperatively in various regions of the world. The prohibition of the production, import and use of' anti-personnel mines and a commitment to clear and destroy existing mines, in the field and in stockpiles, would be an important step in protecting one's own population and territory from the devastating effects of their use. In post-conflict areas such undertakings could strengthen a country's case for mine clearance assistance from the international community. Such national measures would also be an important step in promoting the elimination of anti-personnel mines worldwide.

At the national level the ICRC urges States to begin implementing immediately and unilaterally the types of measures for the protection of civilians which they advocated at the Review Conference. In addition, enhanced efforts at the national level will be needed to ensure:

- The maintenance and strengthening of existing moratoria on the international transfer of anti-personnel mines (i.e., replacement of partial or temporary moratoria with comprehensive and permanent measures);
- For States which have not yet done so, accession to the 1980 Convention including its four Protocols; and
- Active participation in the 1996 sessions of the Review Conference and promotion there of the most stringent measures, including a total ban on anti-personnel mines.

The deadlock with which the Vienna landmine negotiations ended suggests that not enough political leaders yet understand the scope of the landmine crisis or consider the humanitarian, social and economic costs of these weapons to outweigh their limited military utility. States which participate in the resumed sessions of the Review Conference should be urged to put humanitarian interests squarely in the centre of their negotiating positions and to bring humanitarian experts into their delegations. These sessions will only succeed if States are able to rise above their narrow national interests in the general interest of humanity.

VIENNA'S HISTORIC SUCCESS: BLINDING LASER WEAPONS

The adoption in Vienna of a new fourth Protocol prohibiting blinding with laser weapons represents a significant breakthrough in international humanitarian law. **The prohibition, in advance, of an abhorrent new weapon the production and proliferation of which appeared imminent is an historic step for humanity. It represents the first time since 1868, when the use of exploding bullets was banned, that a weapon of military interest has been banned before its use on the battlefield and before a stream of victims gave visible proof of its tragic effects.**

The new Protocol prohibits both the use and transfer of laser weapons specifically designed, as one of their combat functions, to blind permanently. It also requires States to take all feasible precautions, including training of their armed forces, to avoid permanent blinding through the legitimate use of other laser systems. This is the first time that both the use and transfer of a weapon has been entirely prohibited under international humanitarian law.

Efforts to achieve this Protocol were initiated by Sweden and Switzerland at the 1986 International Conference of the Red Cross and Red Crescent and pursued by the ICRC which, between 1989–91 convened four international meetings of experts on this issue. The results of these meetings were published in Blinding Weapons – the primary reference work on the subject. In recent years the issue has been addressed by a growing number of non-governmental organizations, including Human Rights Watch, and organizations representing the blind and war veterans.

Although the scope of application of the new Protocol currently extends only to international conflicts, it was generally agreed in Vienna that it should also apply to conflicts of a non-international character. It was understood that the wording of the Protocol's scope provisions on internal conflicts would in the future be the same as that adopted for the landmines protocol.

The ICRC stresses the importance of vigorous national efforts to ensure that the new Protocol is widely accepted by States and effectively implemented. Such efforts include:

- Ensuring that States declare themselves bound by the Protocol at the earliest possible date;
- The adoption of national measures to prevent the production, transfer, use and proliferation of blinding laser weapons.

12

United Nations General Assembly, 1995
(Fiftieth Session)

In its statement before the United Nations General Assembly, the ICRC reiterated its disappointment at the failure of the Review Conference to achieve agreement on a strengthened Protocol II and warned of the dangers of allowing long grace periods for the introduction of the modest measures under discussion by States. The ICRC asked States to consider whether the limited military utility of anti-personnel landmines was really worth the tragedy they were causing.

Statement by the International Committee of the Red Cross

General and Complete Disarmament: Convention on Prohibitions or Restrictions on the Use of Certain Conventional Weapons Which May be Deemed to be Excessively Injurious or to Have Indiscriminate Effects

First Committee of the General Assembly
26 October 1995

Mr. Chairman,

A few weeks ago, we had assumed that we would be speaking at this session of the First Committee on the results of the first Review Conference of the 1980 Convention on Prohibitions or Restrictions on the Use of Certain Conventional Weapons Which May be Deemed to be Excessively Injurious or to Have Indiscriminate Effects (CCW).

As we know the Conference has been adjourned as it was unable to reach agreement on amendments to Protocol II on landmines and we share the disappointment that was felt in Vienna when this decision had to be taken. However, we are of the opinion that several important gains were made during what may now be termed the first session of the Conference, in particular the adoption of the

Protocol on Blinding Laser Weapons and agreement on certain aspects of Protocol II.

The ICRC would like to express its gratitude for being able to take such an active role in the Review Conference. In so doing, we strive to fulfil our mandate to promote the development of international humanitarian law in a way which gives due weight to humanitarian concerns. Our comments and suggestions are based on our wide practical experience of armed conflicts and the problems they engender.

Protocol on Blinding Laser Weapons

The adoption of Protocol IV on Blinding Laser Weapons on 13 October 1995 is a major achievement. Although the ICRC, like many delegations, would have preferred clearer and stronger provisions, this Protocol is an important breakthrough. It is the first time since 1868 that a weapon has been prohibited before it has been used on the battlefield and thus humanity has been spared the horror that such blinding weapons would have created. Quite apart from the actual wording of the instrument, the effect of its adoption is a strong message that States will not tolerate the deliberate blinding of people in any circumstance. As such it is a triumph of civilization over barbarity. It is also a major achievement that this Protocol includes a prohibition on the transfer of blinding laser weapons thus incorporating, for the first time in a humanitarian law treaty, arms control measures which help ensure the respect of the ban on use.

The ICRC sincerely hopes that States will adhere to this Protocol as quickly as possible and will take all appropriate measures to ensure respect for its provisions.

Landmines

Mr. Chairman,

During the three-week session of the Review Conference in Vienna, 36 people were killed and 243 maimed by landmines in Cambodia alone and about 1600 people worldwide suffered the same fate. During the same period, medical workers also paid a heavy price, namely, seven killed and twenty-one severely injured as a result of the explosion of anti-vehicle mines in Zaire, Rwanda and Mozambique.

These appalling statistics illustrate the urgency of dealing effectively with the landmine crisis. All delegations in Vienna were certainly aware of the importance of reaching agreement on amendments to Protocol II in order to prevent the slaughter and mutilation that landmines are causing daily in many countries. They expressed their disappointment at the difficulty in reaching an agreement and we share in this disappointment.

The problem centred on the criteria that should be preponderant in making a decision. During the last few days of the session, many delegations began to speak more openly of the difficulties they had, and these illustrated in particular the shortcomings of a technical solution. Some delegations indicated that they would need grace periods of up to fifteen years in order to fit their mines with a minimum metal content and equip them with self-destructing, self-neutralizing and/or self-deactivating systems. If mines continue to be sown at the present rate, up to seventy-five million mines could be added in such a period to the existing 110 million. Even more disturbing, however, is the fact that there is uncertainty as to the reliability that may be expected from the so-called "smart mines" to be developed. When used in their millions, hundreds of thousands are likely to remain active for a long period, thus continuing to cause casualties and preventing access to the areas in which they have been used. There is the added danger that they will be perceived as relatively benign and therefore be used widely without fencing or recording.

An equally great concern is that the present chairman's text does not call for the elimination of so-called dumb mines, but allows for their continued use in certain circumstances, including without fencing when "direct enemy military action makes it impossible to comply". As long as these are available there remains the very real danger that they will continue to be used indiscriminately.

The ICRC appeals to States to evaluate whether measures short of a total ban on anti-personnel landmines will in fact put a stop to the present situation. At least seventeen States are now calling for a total ban and they are joined by the Secretary-General of the United Nations, the Council of Ministers of the Organization of African Unity and the European Parliament. Is the limited military utility of anti-personnel landmines really worth the tragedy they are causing? Should not also strict controls be placed on anti-vehicle mines which regularly kill or maim civilians, including humanitarian workers who are trying to help the victims of war? We earnestly hope that States will rise above short-term national interests in favour of the general interest of humanity as a whole.

Future developments

Mr. Chairman,

The Review Conference is due to reconvene in January and again in April 1996. We hope that during this period, many more States will ratify or accede to the Convention, and also that those States that were unable to participate at the Vienna session will be able to do so at the next sessions due to be held in Geneva.

We trust that the gains made at the Vienna session will remain, namely, the agreement to extend the application of Protocol II to non-international armed conflicts, the assignment of responsibility for the clearance of mines at the end of active hostilities and measures to enable humanitarian personnel accomplish their work in favour of victims of conflicts in mined areas. In this regard, we are particularly grateful to States for their willingness to give specific protection to personnel of the ICRC and of Red Cross and Red Crescent organizations.

The work of the Conference, however, will have an effect beyond the regulation of the use of landmines. The discussions that will take place on implementation mechanisms are of major importance for the real effect of the amended Protocol. However, the Convention as a whole must be seen to be a living and effective instrument. We earnestly hope that a frequent and regular review of the Convention will take place so as to enable the international community to evaluate the effectiveness of its existing provisions, to encourage further accessions and to allow for amendments or additional Protocols as the need arises. In this way the spirit of international humanitarian law could remain alive and we could be proud of a Convention that spares humanity from the tragedy of weapons that are indiscriminate or excessively cruel.

Arms transfers

Mr. Chairman,

Our concern about landmines and blinding weapons is rooted in our experience with a much larger phenomenon – the virtually unrestricted flow of vast quantities of weapons, particularly small arms, around the world and their consistent use in flagrant violation of the norms of international humanitarian law. Our first-hand experience in the dozens of conflicts which are raging in various regions is that enormous quantities of small arms are available to almost any organization which seeks them and that when these arms are used humanitarian law is either unknown or simply not respected.

The ICRC strongly encourages this Committee to make the issue of global arms transfers a matter of high priority and to consider both the inclusion of small arms transfers in the United Nations Register of Conventional Arms and possible restraints on such transfers. For its part the ICRC intends to actively study, as requested by the Intergovernmental Group of Experts on the Protection of War Victims, the relationship between arms availability and violations of international humanitarian law and to publish a report on this in late 1996. It also intends to initiate a process of dialogue on this issue within the Red Cross and Red Crescent Movement as a whole.

One important step this body has taken in regards to arms transfers has been its resolution (A/RES/49/75D) encouraging national moratoria on the export of anti-personnel mines. Given the disappointing outcome of the Vienna Review Conference this resolution deserves to be reaffirmed and strengthened in 1995. An estimated 100 million landmines remain stockpiled throughout the world and the protection of civilians from the continued spread of these indiscriminate weapons requires a massive increase in mine-clearance efforts and expanded national moratoria on their export. The low level of pledges at the July 1995 International Meeting on Mine Clearance convened by the UN Secretary-General demonstrates that international commitments are insufficient to ensure the rapid removal of mines already in place. Any relaxation of attempts to bar exports of anti-personnel mines will only exacerbate an already catastrophic situation.

Chemical, biological and nuclear weapons

The March gas attack on civilians on the Tokyo underground and several subsequent incidents, remind us of the urgency of controlling the threat of chemical and biological weapons. They also demonstrate the need to ensure that weapons, the use of which is prohibited, do not nonetheless proliferate and become available for use in violation of the law. When this happens, as was tragically demonstrated in Tokyo, civilians are the most common victims. We urge States which have not already done so to ratify the Chemical Weapons Convention and to ensure its early entry into force. We welcome efforts to introduce a verification regime into the Biological Weapons Convention and encourage non-party States to adhere at the earliest opportunity. For the same reasons we encourage early negotiations, in the disarmament context, to ensure that blinding laser weapons as well as any landmines which are entirely prohibited by the Vienna Review Conference are not produced and do not spread.

Finally, we would like to recall, on the occasion of the fiftieth anniversary of the nuclear age and the commencement of considerations by the International Court of Justice of the legality of the use and threat of use of nuclear weapons, the position of the ICRC on this matter. Any use of weapons which would violate the norms of existing international humanitarian law, including customary law, is already prohibited. In addition, we hope that any deliberations on nuclear weapons will take into account what would probably happen if the threshold were breached and nuclear weapons were actually used. The ICRC has already indicated its opinion that the only effective solution for particularly dangerous weapons is their total prohibition and this has been achieved for chemical and biological weapons and for blinding laser weapons. We hope that the end of the cold war will allow States to work towards achieving the same result for nuclear weapons.

Thank you Mr. Chairman.

13

Launching of the International Media Campaign against Anti-personnel Landmines by the ICRC and National Red Cross and Red Crescent Societies

Geneva, Switzerland
22 November 1995

The impasse at the Review Conference in Vienna had two important consequences for the ICRC. First, it convinced the organization that it was imperative to launch an international advertising campaign, the first in its history in favour of the prohibition of a weapon – the anti-personnel mine. Second, as the negotiations on a revised Protocol II had foundered largely on the question of the perceived overriding military utility of anti-personnel mines, the ICRC decided to commission a historical study of the military use and effectiveness of anti-personnel mines.

In launching the campaign, ICRC President Cornelio Sommaruga emphasized that the dictates of public conscience would be instrumental in finding a solution to the landmine crisis and called upon the media to play an important role in this regard.

Statement of Cornelio Sommaruga
President of the International Committee of the Red Cross

Press Conference
Geneva
22 November 1995

Dear friends and colleagues from the international media,

I have invited you here today not only to speak of the humanitarian tragedy of anti-personnel landmines, but also to announce an historic step for the International Committee of the Red Cross: the launching of an international media campaign to stigmatize these barbarous weapons. Despite your sustained work and ours, the scourge of anti-personnel landmines continues unabated. In the hour we meet here, and in every hour which passes, three people will be killed or crippled for life by mines. Together we must find new ways to end this horrible scourge.

In February 1994 the ICRC came to the conclusion that the production, transfer and use of anti-personnel mines were out of control and called for their total prohibition. Since then we have been joined in this appeal by sixteen States, the UN Secretary-General, the heads of numerous UN Agencies, the Council of Ministers of the Organization of African Unity, the European Parliament and Pope John Paul II. Anti-personnel mines are increasingly being stigmatized as an abhorrent and unacceptable means of warfare.

Many had hoped that the recent Vienna Review Conference of the 1980 Convention on Certain Conventional Weapons would take dramatic steps to end the landmine crisis by placing stringent conditions on their production, transfer and use, or even prohibiting them entirely. Unfortunately, the humanitarian concerns which were at the forefront in the Conference's opening days were brushed aside under the pressure of national military and commercial interests. The Conference adjourned without results on landmines, to reconvene in Geneva next January and April.

By the time the Review Conference completes its work in April another five thousand people will have been killed by landmines and some eight thousand maimed for life. This mindless carnage is an affront to humanitarian values. It is an affront to civilization. It can and must be ended. The ICRC appeals to you, in the media, to political leaders and to our humanitarian colleagues to ensure that these negotiations do not continue in an environment of "business as usual".

The impasse of the Vienna Conference shows that there is little political will for dramatic change and that most military powers, North and South, still resist the elimination of anti-personnel landmines from their armouries. In such circumstances, the ICRC believes that a solution to the landmine crisis will have to rely on the dictates of public conscience. The deadlock of Vienna will itself increase pressure on governments to achieve results by the final session in April 1996.

This is why the ICRC is launching today, for the first time in its history, an international campaign in print, television and radio media for distribution worldwide. Its message is "Landmines must be stopped!" Its aim is to mobilize public opinion and political will for the stigmatization of anti-personnel mines and for an increased commitment to the care and treatment of victims and to mine clearance. This campaign, which the ICRC begins today in the international media, will be taken up throughout 1996 by National Red Cross and Red Crescent Societies in their national media.

The ICRC has taken this unprecedented step in recognition of the increasingly crucial role of the media and of public opinion in changing the course of modern history. Appeals to the norms of civilization and humanity which at the beginning of this century were successfully made directly to statesmen must now find resonance across a broad spectrum of public opinion before governments find the will to rise above narrow national interests and act in the interest of humanity as a whole.

I take this opportunity to congratulate and extend my gratitude to Abbott Mead Vickers – BBDO, the agency that has developed the concept of the ICRC campaign. They were also successful in realizing with the support of Archbishop Desmond Tutu an appeal for free advertising time. To its representatives who have joined us today in Geneva, thank you.

Stopping landmines doesn't only depend on the success of international negotiations. States can take their own moral and political responsibility to end this scourge. The ICRC considers that the establishment of "landmine-free-zones" in various regions of the world may be a positive transitory solution. Such zones, which do not require lengthy global negotiations, would be areas where States, or groups of States, prohibit the production, import and use of anti-personnel mines and in which existing mines are cleared or destroyed. The establishment of such zones in post-conflict areas could strengthen a country's case for mine clearance assistance from the international community and be an important step in promoting the elimination of anti-personnel mines worldwide.

The international community is not impotent in the face of brutality and injustice and you in the media have played a decisive role in ensuring this. The media have helped form the public conscience in its victory over apartheid and against chemical weapons, in its response to famine in western Africa and Eritrea and in its insistence that torture be made illegal. The media were also part of the success of the Vienna Review Conference when, on 13 October, it outlawed blinding laser weapons–only the second time in history that a weapon has been prohibited in advance of its use on the battlefield.

And we are not impotent in the face of landmines. I invite you to join with us now in redoubling your efforts to inform the public about this scourge on the world's poor, its children and its hungry and to frame the ethical debate required for its final resolution. As with chemical weapons success may take years or, as with apartheid, it may require decades. But together we will succeed. And in struggling to do so we will not only be upholding fundamental norms of civilization but also affirming our own common humanity.

Thank you.

14

The 26th International Conference of the Red Cross and Red Crescent

1995

The total elimination of anti-personnel mines was endorsed as a goal by the 1995 quadrennial International Conference of the Red Cross and Red Crescent, made up of States Parties to the Geneva Conventions, National Red Cross and Red Crescent Societies and their International Federation, and the ICRC. The Conference specifically welcomed the unilateral steps which some States had taken towards eliminating all types of anti-personnel landmines and the moratoriums on the export of anti-personnel landmines instituted by many States, and urged States that had not yet done so to take similar unilateral measures at the earliest possible date.

Resolution 2 Protection of the civilian population in period of armed conflict (excerpt)

The 26th International Conference of the Red Cross and Red Crescent, *deeply alarmed*

– by the spread of violence and the massive and continuing violations of international humanitarian law throughout the world,

– by the immense suffering this causes among the civilian population in cases of armed conflict or foreign occupation of a territory, and in particular by the spread of acts of genocide, the practice of "ethnic cleansing", widespread murder, forced displacement of persons and the use of force to prevent their return home, hostage-taking, torture, rape and arbitrary detention, all of which violate international humanitarian law,

– by the serious violations of international humanitarian law constituted by acts aimed at the expulsion of the civilian population from certain areas or even the extermination of the civilian population, or by compelling civilians to collaborate in such practices,

– by the serious violations of international humanitarian law in internal as well as international armed conflicts constituted by acts or threats of violence the primary purpose of which is to spread terror among the civilian population and by acts of violence or of terror making civilians the object of attack,

– by the difficulties encountered by humanitarian institutions in performing their tasks in armed conflicts, in particular when State structures have disintegrated,

– by the growing disparity between the humanitarian pledges made by certain parties to armed conflicts and the profoundly inhumane practices of those same parties,

– by the rapid expansion of the arms trade and the uncontrolled proliferation of weapons, especially those which may have indiscriminate effects or cause unnecessary suffering,

stressing the importance of full compliance with and implementation of international humanitarian law, and

recalling that international humanitarian law and international instruments relating to human rights offer basic protection to the human person,

recalling the obligation of States to repress violations of international humanitarian law and *urging* them to increase international efforts

– to bring before courts and punish war criminals and those responsible for serious violations of international humanitarian law,

– to establish permanently an international criminal court,

reaffirming that any party to an armed conflict which violates international humanitarian law shall, if the case demands, be liable to pay compensation,

aware that the urgency of alleviating the suffering of the civilian population in times of armed conflict should not distract attention from the pressing obligation to fight the root causes of conflicts and the need to find solutions to conflicts,

alarmed by the deliberate and systematic destruction of movable and immovable property of importance to the cultural or spiritual heritage of peoples, such as places of worship and monuments of architecture, art or history, whether religious or secular,

particularly concerned by the plight of women, children, dispersed families, the disabled and elderly, and civilian populations stricken by famine, deprived of access to water and subjected to the scourge of anti-personnel landmines as well as other weapons used indiscriminately,

(...)

G. With regard to anti-personnel landmines:

(a) *expresses deep concern and indignation* that anti-personnel landmines kill or maim hundreds of people every week, mostly innocent and defenceless civilians, obstruct economic development and have other severe consequences for years after emplacement, which include inhibiting the return and rehabilitation of refugees and internally displaced persons and the free movement of all persons;

(b) *takes note* of the fact that the Movement and a growing number of States, international, regional and non-governmental organizations have undertaken to work urgently for the total elimination of anti-personnel landmines;

(c) *noting* also that the ultimate goal of States is to achieve the eventual elimination of anti-personnel landmines as viable alternatives are developed that significantly reduce the risk to the civilian population;

(d) *welcomes* the unilateral steps which some States have taken towards eliminating all types of anti-personnel landmines and the moratoria on the export of anti-personnel landmines instituted by many States, *urges* States that have not yet done so to take similar unilateral measures at the earliest possible date, and *encourages* all States to take further steps to limit transfers;

(e) *regrets* that the Review Conference of States party to the 1980 United Nations Convention on Prohibitions or Restrictions on the Use of Certain Conventional Weapons Which May be Deemed to be Excessively Injurious or to Have Indiscriminate Effects, held from 25 September to 13 October 1995, could not complete its work;

(f) *urges* States party to the 1980 Convention and the Movement to redouble efforts to ensure that the resumed sessions of the above-mentioned Review Conference in 1996 result in strong and effective measures;

(g) *urges* all States which have not yet done so to become party to this Convention and in particular to its Protocol II on landmines, with a view to achieving universal adherence thereto, and further *underlines* the importance of respect for its provisions by all parties to armed conflict;

(h) *urges* all States and competent organizations to take concrete action to increase their support for mine-clearance efforts in affected States which will need to continue for many decades, to strengthen international cooperation and assistance in this field and, in this regard, to provide the necessary maps and information and appropriate technical and material assistance to remove or otherwise render ineffective minefields, mines and booby traps, in accordance with international law;

(i) *invites* the ICRC to continue to follow these matters in consultation with the International Federation and National Societies, and to keep the International Conference of the Red Cross and Red Crescent informed;

H. With regard to blinding and other weapons:

(a) *recalling* Resolution VII of the 25th International Conference of the Red Cross concerning the work on international humanitarian law in armed conflicts at sea and on land;

(b) *reaffirms* that international humanitarian law must be respected in the development of weapons technology;

(c) *welcomes* the adoption by the above-mentioned Review Conference of a new fourth Protocol on blinding laser weapons as an important step in the development of international humanitarian law;

(d) *emphasizes* the prohibition on the use or transfer of laser weapons specifically designed to cause permanent blindness;

(e) *urges* States to declare themselves bound by the provisions of this Protocol at the earliest possible date and to ensure they have in place necessary national measures of implementation;

(f) *welcomes* the general agreement achieved at the Review Conference that the scope of application of this Protocol should apply not only to international armed conflicts;

(g) *requests* States to consider, for example at a subsequent Review Conference, further measures on the production and stockpiling of blinding laser weapons prohibited by this Protocol and *requests* that other issues, such as measures concerning compliance, should be further considered;

(h) *underlines* that proper attention should be given to other existing conventional weapons or future weapons which may cause unnecessary suffering or have indiscriminate effects;

(i) *concerned* about the threat to civilian shipping posed by free-floating naval mines, and *notes* that a proposal to deal with problems such as this has been under discussion;

(j) *invites* the ICRC, in consultation with the International Federation and National Societies, to follow developments in these fields, in particular the expansion of the scope of application of the new fourth Protocol, and to keep the International Conference of the Red Cross and Red Crescent informed.

15

Second Session of the Review Conference of the States Parties to the 1980 Convention on Prohibitions or Restrictions on the Use of Certain Conventional Weapons Which May be Deemed to be Excessively Injurious or to Have Indiscriminate Effects

January 1996

The second session of the Review Conference was intended to bring together military and technical experts to seek agreement on the key substantive provisions of a revised Protocol II. In its statement to the second session of the Conference, the ICRC drew particular attention to the potentially dangerous ambiguity caused by the introduction of the word 'primarily' into the definition of an 'anti-personnel landmine'. The ICRC warned that this change was detrimental to the legal regime governing the use of anti-personnel landmines and declared that if a munition was designed so that it could be used both as an anti-personnel mine and for some other purpose, then it should be considered to be an anti-personnel mine for otherwise it might well escape all the restrictions introduced by the amended Protocol.

Statement by the International Committee of the Red Cross

Review Conference of the States Parties to the 1980 Convention on Prohibitions or Restrictions on the Use of Certain Conventional Weapons Which May be Deemed to be Excessively Injurious or to Have Indiscriminate Effects

Second Session, January 1996

Mr. President,

We earnestly hope that this renewed session of the Review Conference will find a way to resolve the deadlock of the Vienna session and we are confident that you, Mr. President, will make an important contribution to achieve this result.

Although this session of the Review Conference has frequently been referred to as a meeting on "technical issues", the provisions under discussion during this meeting, namely Articles 2–6 of Protocol II and its Technical Annex, are in fact the very heart of the landmine regime. Any decision taken on these provisions will determine whether there is to be a meaningful regulation of the use of landmines that will genuinely solve the problems caused by their use.

Mr. President, both you and the delegations present are aware that, in our opinion, only a total ban on anti-personnel landmines can solve the problem and that the introduction of technical specifications on the manufacture of mines will not do so. The reasons for this opinion were outlined in our President's speech in Vienna. However, for the purposes of this meeting and for the sake of brevity we will now limit ourselves to a few specific comments.

Definition

As the major problems are caused by anti-personnel mines, the definition of an anti-personnel landmine needs to be clear and unambiguous. In particular we see no reason why the definition of an "anti-personnel landmine" should differ from that of a "mine" other than making it clear that the intended victim is a person. The introduction of the word "primarily" (Article 2, para. 3 of the President's text) makes the definition weaker and therefore not only weakens the Protocol rather then strengthening it but also introduces uncertainty which is detrimental to the legal regime governing the use of anti-personnel landmines. In particular, if a munition is designed so that it can be used both as an anti-personnel mine and for some other purpose, then it should be considered to be an anti-personnel mine for otherwise it may well escape all the restrictions introduced by the amended Protocol. This is true not only for directional fragmentation mines but also anti-tank mines that are designed to have anti-personnel characteristics.

Detectability

On the issue of detectability, it is clear that the technical annex should specify the technical characteristics that will render a mine detectable using easily available means. However, it should be kept in mind that mine clearance specialists who have extensive experience in different parts of the world stress that it is particularly difficult to find mines in soils rich in iron, or in former battle grounds that contain very large numbers of metal fragments. They have all stressed to us that the shape of the metallic element in the mine is of greater importance than the weight alone. It would therefore be worth considering whether the present formulation, which refers to weight irrespective of its shape, has sufficient empirical proof of its efficacity in all soil types and situations.

With regard to the grace period to be allowed for rendering all mines used detectable, we would simply like to draw attention to the fact that every year another

2–5 million mines are being laid and each mine costs up to US$ 1,000 to remove, the cost being greater and the procedure being slower and more dangerous when the mines are difficult to detect.

We would also at this point like to record our disappointment that consensus could not be reached on assuring the detectability of anti-tank mines which would have considerably helped mine clearance teams and thereby protected both civilians and humanitarian workers.

Self-destruct/self-deactivation mechanisms

Mr. President,

We are aware that a great deal of attention will be given during this session to finding agreement on the type of self-destruct and/or self-neutralizing mechanisms that should be introduced for anti-personnel mines. It is clear that the reliability of such mechanisms needs to be assured, not only for humanitarian and environmental reasons but also because mines can severely hamper a country's recovery from armed conflict. So far this conference has not discussed how the reliability of such systems is to be demonstrated. Will it be experience alone? A major danger is that such mines will be used in large quantities without mapping or fencing as they will be considered safe. However, any failure will mean that the areas in which they have been used are not safe. Experience in different parts of the world has shown that the mere fear of the continuing presence of live mines can effectively prevent large areas of land being used.

We would also urge that the grace period for the introduction of such mechanisms be as short as possible in order to be able to establish as quickly as possible whether the new regime is effective. Each year of delay will add to the present appalling annual figure of approximately 24 thousand new landmine victims.

Still on this topic, Mr. President, we would like to reiterate our great concern with the fact that the present formulation of Article 5 of the President's text continues to allow "dumb" mines to be laid in unfenced and unmarked areas during times of direct enemy military action. This is the very time that the Convention's new restrictions are most needed but the present formulation undermines the entire purpose of the proposed new regime. We fear that the proposed amendments as they stand, if adopted, will not change the present situation very much.

Transfers

Finally, Mr. President, we would like to make a statement now about a topic that will be further considered in the April session. After careful consideration, the ICRC has decided to formally support the proposal made in Vienna for a total ban on the transfer of anti-personnel landmines. It is of great importance that the gains that have been made in this regard, i.e. the fact that 23 States have introduced comprehensive moratoria on the export of such mines, not be undermined. A weaker provision in the amended Protocol could be used as a basis for the reintroduction of exports which would be a tragic result. Rather, other States should be encouraged to follow the example of these 23 countries. Such a total ban on transfers would also be in keeping with the General Assembly resolution adopted on 12 December 1995 on the Moratorium on the export of anti-personnel landmines (resolution O in UN Doc. A/50/590) and with the objective of the eventual elimination of landmines declared in the same resolution.

In this regard we are particularly pleased with the resolution adopted on 12 December 1995 by the Meeting of Foreign Ministers of the Organization of the Islamic Conference (OIC) which supports the "complete elimination" of anti-personnel landmines. Taken together with the support for a total ban by the Council of Ministers of the Organization of African Unity, the European Parliament and twenty individual States, the OIC resolution reflects a growing awareness that a ban on anti-personnel landmines is the only solution to the current landmine crisis.

Thank you, Mr. President.

16

Anti-personnel Landmines: Friend or Foe? A Study of the Military Use and Effectiveness of Anti-personnel Mines

Commissioned by the International Committee of the Red Cross, March 1996

The study, commissioned by the ICRC on the military use and effectiveness of anti-personnel mines, was to become one of the key tools in the ICRC's campaign in favour of the total prohibition of anti-personnel mines, counteracting the widely held perception that landmines were essential weapons of high military value. The study was largely written by retired Brigadier-General Patrick Blagden and includes conclusions by an international group of military commanders. These military commanders concluded that 'The limited military utility of AP mines is far outweighed by the appalling humanitarian consequences of their use in actual conflicts. On this basis their prohibition and elimination should be pursued as a matter of utmost urgency by governments and the entire international community.'

The study's conclusions were adopted by consensus by a dozen military officers and have since been endorsed in a personal capacity by more than fifty senior military officers from nineteen countries. The information in the study was drawn only from open sources, as the ICRC does not have access to classified material. The ICRC continues to welcome written comments and additional information, particularly case studies, to corroborate or contradict the material presented in the study, for use in future discussions. Presented below is the executive summary of the study. The full document can be obtained through the ICRC.

Anti-personnel Landmines – Friend or Foe?

Executive Summary

March 1996

It has been generally assumed that anti-personnel landmines are an indispensable weapon of war, and that their indiscriminate effects can be moderated through compliance with military doctrine and the rules of international humanitarian law. This study examines the military case for continued use of these weapons in light of their employment in actual conflicts since 1940, whether by professional armed forces, by insurgents or in counter-insurgency operations. It has been undertaken in the absence of other publicly available studies on the actual use and effectiveness of anti-personnel mines.

In the 26 conflicts considered, few instances can be cited where anti-personnel mine use has been consistent with international law or, where it exists, military doctrine. The historical evidence indicates that during hostilities such mines are rarely used "correctly", whether by "developed" armies, "third-world" armies or insurgents and that their effects cannot easily be limited as law and doctrine presume. Such evidence as is available is most often of "incorrect" use, whether by intention or inadvertence or because of the impracticability of observing specific rules in the heat of battle. The study suggests that it would be unwise to justify the continued use of anti-personnel mines on the premise that they will be deployed in a carefully controlled manner.

Whether employed correctly or not, one must also ask whether the use of anti-personnel mines has achieved a legitimate military purpose. Here again the evidence considered indicates that, even when used on a massive scale, they have usually had little or no effect on the outcome of hostilities. No case was found in which the use of anti-personnel mines played a major role in determining the outcome of a conflict. At best, these weapons had a marginal tactical value under certain specific but demanding conditions which are described in the conclusions.

An often overlooked aspect of landmine warfare is also addressed, namely, the cost and dangers for forces employing anti-personnel mines. The price of properly laying, marking, observing and maintaining minefields is high, in both human and financial terms; it involves significant investment, risk to one's own forces and the loss of tactical flexibility. Even when these costs are assumed, the effects of anti-personnel mines are very limited and may even be counter-productive.

Technological innovation, such as the introduction of remotely delivered mines, has already begun to change the nature of military doctrine and landmine use. The increased use of such mines could dramatically alter the character of future mine warfare and increase its scale. The implications of these and other developments, including the introduction of seismic fuses, fuel-air anti-personnel mines and hybrid mines for dual anti-personnel and anti-tank use, are examined from both military and humanitarian viewpoints.

Proposed technical solutions to the humanitarian problems caused by anti-personnel mines, in particular the increased use of self-destructing and self-deactivating models, are analysed. For a variety of reasons these solutions are considered unlikely to significantly reduce civilian casualties and the disruption of civilian life due to landmines.

In reviewing alternatives to anti-personnel mines, the study describes a number of options such as fences, physical obstacles and direct fire, as well as improved intelligence, mobility and observation. These means have already been employed and found effective by forces facing a variety of tactical situations. Technological developments have also opened the way to promising alternatives, considered in Section IX, which merit examination in preference to the pursuit of new mine technologies. Improved clearance techniques and reliance on more resistant mine-protected vehicles are suggested as measures which could further reduce the incentives for anti-personnel mine use.

The study's conclusions were drawn up by a meeting of active and retired senior military commanders from a variety of countries and were unanimously endorsed by all participants in their personal capacity.

XI – Conclusions

(Unanimously endorsed in their personal capacity by members of a Group of Military Experts, 12–13 February 1996 and by other military experts whose names appear in the annex to this report.)

1. The military value of landmines, as used in actual conflicts over the past 55 years, has received little attention in published military studies. The specific added value of AP mines, as compared to that of anti-tank mines, has barely received any attention. There is also little evidence that dedicated research on the value of AP mines, based on historical experience, has been carried out within professional military organizations.

2. The material which is available on the use of AP landmines does not substantiate claims that AP mines are indispensable weapons of high military value. On the other hand, their value for indiscriminate harassment when used by irregular forces can be high. Their use for population control has regrettably been all too effective.

3. The cases reviewed in this study, together with the personal experience of members of the Group of Experts, provide a basis for a number of initial conclusions regarding traditionally emplaced mines:

   Establishing, monitoring and maintaining an extensive border minefield is time-consuming, expensive and dangerous. In order to have any efficacy at all they need to be under continuous observation and direct fire, which is not always possible. Because of these practical difficulties some armed forces have entirely refrained from using such minefields. Moreover, these minefields have not proved successful in preventing infiltration.

Under battlefield conditions the use, marking, and mapping of mines in accordance with classical military doctrine and international humanitarian law is extremely difficult, even for professional armed forces. History indicates that effective marking and mapping of mines has rarely occurred.

The cost to forces using AP mines in terms of casualties, limitation of tactical flexibility and loss of sympathy of the indigenous population is higher than has been generally acknowledged.

Use in accordance with traditional military doctrine appears to have occurred infrequently and only when the following specific conditions were met:

- both parties to the conflict were disciplined professional armies with a high sense of responsibility and engaged in a short-lived international conflict;
- the tactical situations were fairly static;
- mines were not a major component of the conflict;
- forces possessed adequate time and resources to mark, monitor and maintain minefields in accordance with law and doctrine;
- mined areas were of sufficient economic or military value to ensure that mine clearance occurred;
- the parties had sufficient resources to ensure clearance and it was carried out without delay; and
- the political will existed to strictly limit the use of mines and to clear them as indicated above.

4. Although the military value of anti-tank mines is acknowledged, the value of AP mines is questionable. Their use to protect anti-tank mines is generally claimed to be an important purpose of AP mines, but there are few historical examples to substantiate the effectiveness of such use.

   Where minefields are cleared by roller, plough, flail, explosive-filled hose, fuel-air explosive or bombardment, the value of AP mines has not been demonstrated.

> The effect of AP mines against unprotected infantry is limited; a relatively small percentage of troops is rendered *hors de combat.* Infantry have in the past advanced through AP minefields, accepting the risk and casualties this entails.
>
> The use of AP mines for harassment, whether in international or internal conflicts, is of doubtful military value. Historically, this use has ultimately targeted civilians.

Remotely delivered AP mines are not solely defensive weapons. In practice they will probably be used in huge quantities to saturate target areas. Even so, the mobility of professional armies will not be significantly hindered.

5. Remotely delivered AP mines will almost certainly cause vastly increased civilian casualties, even if such mines are designed to be self-destructing and self-deactivating, for the following reasons:
   - they will be dangerous during their intended active life;
   - the marking and mapping of such mines will be virtually impossible;
   - in extended conflicts they may be re-laid many times;
   - self-destructing and deactivating devices may be unreliable;
   - inactive mines, like unexploded ordnance, can still be dangerous; and
   - the mere presence of mined areas will produce fear, keeping civilians out of areas important for their livelihood.
6. Some barrier systems and other tactical methods offer alternatives to AP mines. Additional alternatives should be pursued rather than further development of any new AP mine technologies. Developments which further increase the lethality of AP mines are to be deplored and are unnecessary.
7. Improved mine clearance technologies for military, humanitarian and civilian agencies should be vigorously developed with a goal of making AP mines progressively less useful.
8. The limited military utility of AP mines is far outweighed by the appalling humanitarian consequences of their use in actual conflicts. On this basis their prohibition and elimination should be pursued as a matter of utmost urgency by governments and the entire international community.

Endorsements
by military officers acting in their personal capacity
of the conclusions contained in

Anti-personnel Landmines: Friend or Foe?

commissioned by the
International Committee of the Red Cross
update of 12 May 1997

Austria

Major General Günther G. Greindl, Director General for International Policies, Austrian Ministry of Defence

Brigadier (ret.) Leo Jedlicka, former deputy head, Austrian Army proving ground; head of training for Austrian Army and UN peace-keeping forces.

Benin

Lieutenant Colonel Amoussa Chabi Mathieu Boni, General Staff of the Army

Colonel Feliqien Dos Santos, Head of General Staff of the Army

Lieutenant Colonel Florent Fagla, First Inter-Army Battalion

Colonel Paul Sagbo, Directorate, Army Health Service

Canada

Major (ret.) Ted Itani, Consultant on security and humanitarian policy, Ottawa, Canada; Technical Consultant for the War Crimes Tribunal at the Hague; former officer for humanitarian and mine clearance operations in Bosnia-Herzegovina, Pakistan/Afghanistan and Iraq; former artillery officer and instructor in combined arms operations in the Canadian Army

Major General (ret.) John A. MacInnis CMM, MSC, CD, Chief, Mine Clearance and Policy Unit, Department of Humanitarian Affairs, United Nations, New York

Major General (ret.) Lewis MacKenzie, MSC, CD, Canadian Army, first Commander of UNPROFOR forces, sector Sarajevo (1992)

General (ret.) Paul D. Manson; former Chief of the Defence Staff, Canadian Army

Cape Verde

Captain Arlindo Jose Rodrigues, Director, Direction of Operations, Joint Chiefs of Staff

Major Antonio Carlos Tavares, Director, Department of Operations, Joint Chiefs of Staff

Croatia

Major General (ret.) Ivo Prodan, former head of the Medical Corps, Croatian Army

General (ret.) Anton Tus, former Chief of Staff, Croatian Army

France

General (ret.) Jacques Saulnier, former head of the Joint Chiefs of Staff of French Armed Forces

Ghana

Colonel A.B. Donkor, Judge Advocate General, Armed Forces of Ghana

Germany

Brigadier General (ret.) Hermann Hagena, former Deputy Commander, Command and General Staff College, Hamburg

India

Major General Dipankar Banerjee, Deputy Director, Institute for Defence Studies and Analyses, New Delhi; former commander *inter alia* of a mountain division in counter-insurgency operations; author and researcher on national, regional and international security issues

Jordan

Field Marshal (ret.) Fathi Abu Taleb, former chairman of the Joint Chiefs of Staff, Jordanian Armed Forces

Brigadier General (ret.) Fawwaz B. al-Khriesha, former commander of engineering field battalions, Jordanian Armed Forces

Major General PSC. (ret.) Shafik Jumean, former director of the staff college

Major General PSC. (ret.) Yousef A. Kawash, former director of morale guidance, Jordanian Armed Forces

Netherlands

Brigadier General (ret.) Henny van der Graaf, Director, Center for Arms Control and Verification Technology, Eßindhoven, the Netherlands; member of Advisory Board on Disarmament to the UN Secretary-General, member of UN Mission to Mali on the control of small arms transfers

Norway

Vice Admiral (ret.) Roy Breivik, Norwegian Navy

Major General (ret.) Bjern Egge, Norwegian Army; President, World Veterans Federation

Peru

Vice Admiral (ret.) Jose Carcelen Basurto, former commander of naval zones
Admiral (ret.) Hugo Ramirez Canaval, former Commodore of the Navy
Major General (ret.) Cesar E. Rosas Cresto, former Minister of Housing
Major General (ret.) Eduardo Angeles Figueroa, Air Force of Peru
Major General (ret.) Julian Julia Freyre, former Minister of Defence and Commander in Chief of the Army
Colonel (ret.) Jose Bailetti Mac-Kee, former head of the National Planning Institute
Major General (ret.) Alfredo Rodriguez Martinez, former Commander in Chief of the Army
Lieutenant General (ret.) Pedro Sala Orosco, former Minister of Labour
Major General (ret.) Pedro Richter Prada, former Minister of Defence and Commander in Chief of the Army
* Major General (ret.) Alejandro Cuadra Rabines, former Minister of Defence and Commander in Chief of the Army
Major General (ret.) Otto Elespuru Revoredo, former Commander in Chief of the Army
Major General (ret.) Luis Alcantara Vallejo, former head of the National Defence Secretariat

Philippines

Colonel Alfonso Dagudag, Chief of Staff, Seventh Division, Armed Forces of the Philippines; member of Strategic Group on Modernization of AFP Weaponry
General Arturo T. Enrile, Chief of Staff, Armed Forces of the Philippines

Slovenia

Major General (ret.) Lado Ambrozic, Army
Colonel General (ret.) Ivan Dolnicar, Air Force
Colonel General (ret.) Rudolf Hribernik, Army
Major General (ret.) Lado Kocijan, Army; Professor of Defense Studies
Colonel General (ret.) Stane Potocar, Army
Major General (ret.) Jamez Slopar, Army
Colonel General (ret.) Avgust Vrtar, Army

South Africa

Colonel A.J. Roussouw, Senior Staff Officer, Combat Engineers, South African National Defence Force; former commander of field squadrons; mine warfare and clearance operations in Angola and Namibia

Switzerland

Colonel Marcel Fantoni, Federal Military Department, Bern, Switzerland; Chief of Staff, Light Infantry Division; Ecole de Recrue, Birmensdorf

United Kingdom

General Sir Hugh Beach (ret.), British Army, former Master General of the Ordnance, commandant of the Army Staff College, and involved in Royal Engineers mine clearance operations in north-western Europe (1944)

General (ret.) Sir Peter de la Billiere, Commander of British forces in Middle East, 1990–91, Commander of British forces in Falklands/Malvinas Islands, 1984–85.

Brigadier (ret.) Patrick Blagden, Senior Adviser on mine clearance, UN Department of Peace-keeping Operations (1992–95), former British Army officer responsible for weapons research, former defence industry executive

Brigadier (ret.) J.H. Hooper, OBE, DL, former Royal Engineer officer, British Army

General (ret.) Sir David Ramsbotham GCB, CBE; Chief Inspector of Prisons, Adjutant General 1990–93

Zimbabwe

Brigadier G.M. Chiweshe, Judge Advocate General, Ministry of Defence

Lieutenant Colonel (ret.) Martin Rupiah, Lecturer, University of Zimbabwe, writer on landmines in Zimbabwe, former director and unit commander, Army of Zimbabwe

Total endorsements, including those in published study:
55 active and retired officers from 19 countries.

NOTE

* deceased

Press Release

Anti-personnel Mines: Not an Indispensable Weapon of High Military Value

28 March 1996

There is no clear evidence that anti-personnel landmines are indispensable weapons of high military value. On the other hand, their use in accordance with military doctrine is time-consuming, expensive and dangerous and has seldom occurred under combat conditions. These are some of the main conclusions of the study "The Military Use and Effectiveness of Anti-personnel Mines" commissioned by the International Committee of the Red Cross (ICRC) which was released today. The conclusions, based on a survey of the actual use and effectiveness of these weapons in conflicts over the past 55 years, were drawn up by Brigadier Patrick Blagden, a former combat engineer and weapons researcher with the British Royal Army, and a group of high-ranking military experts from eight countries.

The military use of anti-personnel mines in actual conflict has so far received almost no attention in published military studies. Therefore the ICRC took the initiative to commission an expert study that presents a compelling set of conclusions on the actual use of anti-personnel mines since 1940. These conclusions were unanimously supported by senior commanders with broad experience in landmine warfare at an ICRC expert meeting in February 1996 and are being endorsed by a growing number of senior military officers from around the world.

The ICRC study concludes that properly establishing and maintaining an extensive border minefield is time-consuming, expensive and dangerous and has rarely occurred in actual conflicts. In order to have any efficacy at all they need to be under continuous observation and direct fire, which is not always possible and is often not done. Under battlefield conditions the use, marking, and mapping of mines in accordance with classical military doctrine and international humanitarian law is extremely difficult, even for professional armed forces.

The commanders who have endorsed this report found that the use of anti-personnel mines in accordance with the military doctrine which has justified their use has occurred infrequently and only when certain conditions were met: (a) both parties to the conflict were disciplined professional armies with high sense of responsibility and engaged in a short-lived international conflict; (b) the tactical situation was fairly static; (c) forces possessed adequate time and resources to mark, monitor and maintain minefields in accordance with law and doctrine, (d) mined areas were of sufficient economic or military value to ensure that mine clearance occurred and (e) sufficient political will existed to implement the above conditions.

The ICRC study points out that the emerging generation of remotely delivered anti-personnel landmines are not solely defensive weapons but will probably be used in huge quantities to saturate targets which are likely to include civilian areas. Even so, the mobility of professional armies will not be significantly hindered. Remotely delivered anti-personnel mines are likely to cause vastly increased civilian casualties, even if such mines are designed to be self-destructing. This is so for several reasons: e.g., they will be dangerous during their intended active life, the reliability of self-destructing devices is unlikely to be verified and is likely to be insufficient and it is virtually impossible to map and mark remotely delivered mined areas.

In addition to examining the use and effectiveness of anti-personnel mines in 26 conflicts the ICRC study also considers an often overlooked aspect of landmine warfare: the cost and dangers to forces using these mines. The study suggests that the cost to forces using anti-personnel mines, in terms of casualties and limitation of tactical flexibility, is higher than has been generally acknowledged. The implications of technological innovations in landmine design, which could have a dramatic effect on future mine warfare and on the level of civilian casualties, are examined from both military and humanitarian viewpoints. Possible alternatives to anti-personnel mines, including a number which are already in use among armies in both developed and developing countries, are also considered.

The results of this study confirm the ICRC's position that the military value of anti-personnel mines is far outweighed by their human and social costs and reinforces both its call for a ban and its worldwide campaign against this weapon.

17

Third Session of the Review Conference of the States Parties to the 1980 Convention on Prohibitions or Restrictions on the Use of Certain Conventional Weapons Which May be Deemed to be Excessively Injurious or to Have Indiscriminate Effects

April–May 1996

At the outset of the final session of the Review Conference, the ICRC once more called – unsuccessfully – for the word 'primarily' to be removed from the definition of an anti-personnel mine. Despite public opinion surveys in twenty-one countries showing enormous support for a total ban on anti-personnel mines, States Parties, after last-minute negotiations, adopted by consensus Protocol II as amended on 3 May 1996. This text imposed stricter rules on the use of anti-personnel mines but did not prohibit their use. The ICRC expressed strong disappointment with the modest restrictions on use and warned that they were likely to have little impact on reducing the level of civilian landmine casualties. In the words of the ICRC Vice-President, Eric Roethlisberger, 'The horrific numbers of landmine victims of recent years are set to continue unless governments squarely face their humanitarian responsibilities and do far more than required by the agreement adopted today.'

Statement of Cornelio Sommaruga

President of the International Committee of the Red Cross

Review Conference of the States Parties to the 1980 Convention on Prohibitions or Restrictions on the Use of Certain Conventional Weapons Which May be Deemed to be Excessively Injurious or to Have Indiscriminate Effects

Third Session, April/May 1996
Geneva, Switzerland
22 April 1996

I come here today to raise with distinguished delegates an issue which I believe is of fundamental importance to the success of this Review Conference and to global efforts to end the landmine scourge.

But first, I would like to underline the importance of this Conference and of the negotiations you are about to resume. The Review Conference and its preparatory process have already played an indispensable role in focusing governmental attention on the desperate need for action to stop the killing and maiming of innocents caused by landmines. This process has been a catalyst for the review by many governments of their policies on the production, use and transfer of landmines. Many dramatic steps have been taken by national governments, particularly since the adjournment of the first session in Vienna. We are told that eight States have suspended or permanently renounced the use of anti-personnel mines by their own armed forces. Since Vienna, the number of States supporting the total prohibition of these weapons has nearly doubled to include 29 countries. These actions reflect a clear trend, on the ground if not yet in the negotiating framework, towards the complete prohibition and elimination of anti-personnel mines.

As I stated in Vienna the actions you take here will decide the fate of a hundred thousand potential landmine victims over the next five years. Short of a total ban, the compromises reached here will be paid for in human flesh and human lives for a very long time. I urge you and your governments to do your utmost <u>both</u> to achieve dramatic results in this forum and to take additional national and regional steps to ensure that anti-personnel mines are no longer produced, used or transferred.

This Conference has rightly focused, above all else, on strengthening the restrictions on the use of anti-personnel mines. And yet the Conference appears set to adopt a definition of this weapon (in Article 2, paragraph 3 of the President's Text) which would introduce a dangerous ambiguity into the heart of the proposed regime. Unlike the definition of a "mine" in the same text, the definition of an anti-personnel mine speaks of a weapon "primarily designed to be exploded by the presence, proximity or contact of a person". If this definition is adopted, any other accomplishments of this Conference could over time be subverted by the confusion and possible abuse which this definition could produce. Efforts to put an end to the humanitarian crisis caused by anti-personnel mines may well be severely threatened or even totally undermined.

If a munition is designed so as to be capable of use as an anti-personnel mine <u>and</u> for some other purpose it should clearly be considered an anti-personnel mine and be regulated as such. The primary purpose of such a munition would be difficult or impossible to assess while its intended functions can usually be clearly demonstrated.

This problem is not merely legal or theoretical. A recent publication of Jane's Information Group, <u>Trends in Landmine Warfare,</u> notes that "the rise in scatterable mines has blurred the already thin line between anti-personnel and anti-tank weapons" and describes mines which have both anti-personnel and anti-tank characteristics. Such mines are expected to become smaller and cheaper in the future and can be remotely delivered in huge quantities. An additional problem arises with directional fragmentation mines which, in our view, should be considered as a mine unless they are only capable of being detonated by remote command. Other future technologies will present similar challenges and invite abuse of an ambiguous definition.

It is our view that anti-personnel mines should be defined as those "designed to be exploded by the presence, proximity or contact of a person" which is consistent with the general definition of a mine used in the Convention. Introduction of ambiguity in this crucial definition could over time weaken the very protections against anti-personnel mines which this Conference is mandated to strengthen. I strongly urge every delegation to carefully consider its position on this matter.

I would like briefly to address other issues which the ICRC considers to be important at the current stage of negotiations:

• Only the complete prohibition of anti-personnel mines, which is easily implemented and far more readily verified than other proposals being considered, will be effective. If this cannot yet be achieved by consensus in this forum States should consider taking unilateral action, which should not be seen as a security concession, but as a means of fulfilling their humanitarian obligation to protect their own population and territory in the event of armed conflict. The recent ICRC study on the military use and effectiveness of anti-personnel mines, the conclusions of which have now been supported by 43 senior commanders from 17 countries, clearly highlights the difficulties of using these weapons according to legal and doctrinal norms, their limited effectiveness and their negative effects on one's own forces.

• In keeping with existing moratoria in most mine-producing countries the transfer of anti-personnel mines should be prohibited within the framework of the 1980 Convention. Provisions on transfers which this Conference adopts should be as far reaching as possible so as not to be a retreat from present practice and should enter into force immediately upon adoption.

• Other amendments should enter into force in the shortest possible period. Transition periods of years or decades could compound the landmine crisis and add many decades to the hundreds of years which clearance of existing mines will require.

• To protect civilians and humanitarian workers anti-tank mines must be made detectable and anti-handling devices not permitted.

• The strongest possible protections should be provided, under draft Article 12, to missions of humanitarian organizations. In the case of the ICRC these provisions are an essential expression of the commitment, which States have undertaken by adhering to the Geneva Conventions and their Additional Protocols, to provide access to war victims.

• The scope of the Convention must be extended to non-international armed conflicts and effective measures for implementation added.

• To ensure further development and effective implementation of the Convention future Review Conferences should be held on a regular basis every five years. Following the end of this Conference an interim diplomatic process aimed at encouraging adherence, implementation and dialogue on further improvements would be highly desirable.

Recent actions by a wide range of States have demonstrated that neither the public conscience, nor Parliaments nor governments are powerless in the face of the affront to humanity which landmines present. Neither is this Review Conference. It has both the opportunity and a moral obligation to make an historic contribution to ending this scourge. It has already taken dramatic action against the threat of blinding laser weapons. We await similar action on anti-personnel mines.

ICRC News

International Poll on Anti-personnel Mines: Public Opinion Largely in Favour of Total Ban

24 April 1996

A poll conducted free of charge for the ICRC by institutes associated with the Gallup International Group and Isopublic Zürich has shown that public opinion in 21 countries selected from four continents is largely in favour of a ban on anti-personnel mines. Denmark ranks first with 92% of favourable replies (out of 1,009 people questioned), followed by Spain (89%), Austria, Italy and Switzerland (88%), Russia (83%) and India (82%). The "least favourable" results were obtained in the United States (60%) and Japan (58%). "Our campaign against anti-personnel mines has not yet been launched in those countries," explained Johanne Dorais-Slakmon at ICRC headquarters in Geneva, "and this relatively weak support may change appreciably when the States in question have expressed their views at the Review Conference of the 1980 Convention, which is now meeting in Geneva with a view to strengthening existing rules on the use of anti-personnel mines."

Protocol on Prohibitions or Restrictions on the Use of Mines, Booby-Traps and Other Devices as amended on 3 May 1996 (Protocol II as amended on 3 May 1996)

Article 1 – Scope of application

1. This Protocol relates to the use on land of the mines, booby-traps and other devices, defined herein, including mines laid to interdict beaches, waterway crossings or river crossings, but does not apply to the use of anti-ship mines at sea or in inland waterways.

2. This Protocol shall apply, in addition to situations referred to in Article I of this Convention, to situations referred to in Article 3 common to the Geneva Conventions of 12 August 1949. This Protocol shall not apply to situations of internal disturbances and tensions, such as riots, isolated and sporadic acts of violence and other acts of a similar nature, as not being armed conflicts.

3. In case of armed conflicts not of an international character occurring in the territory of one of the High Contracting Parties, each party to the conflict shall be bound to apply the prohibitions and restrictions of this Protocol.

4. Nothing in this Protocol shall be invoked for the purpose of affecting the sovereignty of a State or the responsibility of the Government, by all legitimate means, to maintain or re-establish law and order in the State or to defend the national unity and territorial integrity of the State.

5. Nothing in this Protocol shall be invoked as a justification for intervening, directly or indirectly, for any reason whatever, in the armed conflict or in the internal or external affairs of the High Contracting Party in the territory of which that conflict occurs.

6. The application of the provisions of this Protocol to parties to a conflict, which are not High Contracting Parties that have accepted this Protocol, shall not change their legal status or the legal status of a disputed territory, either explicitly or implicitly.

Article 2 – Definitions

For the purpose of this Protocol:

1. "Mine" means a munition placed under, on or near the ground or other surface area and designed to be exploded by the presence, proximity or contact of a person or vehicle.

2. "Remotely-delivered mine" means a mine not directly emplaced but delivered by artillery, missile, rocket, mortar, or similar means, or dropped from an aircraft.

Mines delivered from a land-based system from less than 500 metres are not considered to be "remotely delivered", provided that they are used in accordance with Article 5 and other relevant Articles of this Protocol.

3. "Anti-personnel mine" means a mine primarily designed to be exploded by the presence, proximity or contact of a person and that will incapacitate, injure or kill one or more persons.

4. "Booby-trap" means any device or material which is designed, constructed or adapted to kill or injure, and which functions unexpectedly when a person disturbs or approaches an apparently harmless object or performs an apparently safe act.

5. "Other devices" means manually-emplaced munitions and devices including improvised explosive devices designed to kill, injure or damage and which are actuated manually, by remote control or automatically after a lapse of time.

6. "Military objective" means, so far as objects are concerned, any object which by its nature, location, purpose or use makes an effective contribution to military action and whose total or partial destruction, capture or neutralization, in the circumstances ruling at the time, offers a definite military advantage.

7. "Civilian objects" are all objects which are not military objectives as defined in paragraph 6 of this Article.

8. "Minefield" is a defined area in which mines have been emplaced and "mined area" is an area which is dangerous due to the presence of mines. "Phoney minefield" means an area free of mines that simulates a minefield. The term "minefield" includes phoney minefields.

9. "Recording" means a physical, administrative and technical operation designed to obtain, for the purpose of registration in official records, all available information facilitating the location of minefields, mined areas, mines, booby-traps and other devices.

10. "Self-destruction mechanism" means an incorporated or externally attached automatically-functioning mechanism which secures the destruction of the munition into which it is incorporated or to which it is attached.

11. "Self-neutralization mechanism" means an incorporated automatically-functioning mechanism which renders inoperable the munition into which it is incorporated.

12. "Self-deactivating" means automatically rendering a munition inoperable by means of the irreversible exhaustion of a component, for example, a battery, that is essential to the operation of the munition.

13. “Remote control” means control by commands from a distance.

14. “Anti-handling device” means a device intended to protect a mine and which is part of, linked to, attached to or placed under the mine and which activates when an attempt is made to tamper with the mine.

15. “Transfer” involves, in addition to the physical movement of mines into or from national territory, the transfer of title to and control over the mines, but does not involve the transfer of territory containing emplaced mines.

Article 3 – General restrictions on the use of mines, booby-traps and other devices

1. This Article applies to:

(a) mines;

(b) booby-traps; and

(c) other devices.

2. Each High Contracting Party or party to a conflict is, in accordance with the provisions of this Protocol, responsible for all mines, booby-traps, and other devices employed by it and undertakes to clear, remove, destroy or maintain them as specified in Article 10 of this Protocol.

3. It is prohibited in all circumstances to use any mine, booby-trap or other device which is designed or of a nature to cause superfluous injury or unnecessary suffering.

4. Weapons to which this Article applies shall strictly comply with the standards and limitations specified in the Technical Annex with respect to each particular category.

5. It is prohibited to use mines, booby-traps or other devices which employ a mechanism or device specifically designed to detonate the munition by the presence of commonly available mine detectors as a result of their magnetic or other non-contact influence during normal use in detection operations.

6. It is prohibited to use a self-deactivating mine equipped with an anti-handling device that is designed in such a manner that the anti-handling device is capable of functioning after the mine has ceased to be capable of functioning.

7. It is prohibited in all circumstances to direct weapons to which this Article applies, either in offence, defence or by way of reprisals, against the civilian population as such or against individual civilians or civilian objects.

8. The indiscriminate use of weapons to which this Article applies is prohibited. Indiscriminate use is any placement of such weapons:

(a) which is not on, or directed against, a military objective. In case of doubt as to whether an object which is normally dedicated to civilian purposes, such as a place of worship, a house or other dwelling or a school, is being used to make an effective contribution to military action, it shall be presumed not to be so used; or

(b) which employs a method or means of delivery which cannot be directed at a specific military objective; or

(c) which may be expected to cause incidental loss of civilian life, injury to civilians, damage to civilian objects, or a combination thereof, which would be excessive in relation to the concrete and direct military advantage anticipated.

9. Several clearly separated and distinct military objectives located in a city, town, village or other area containing a similar concentration of civilians or civilian objects are not to be treated as a single military objective.

10. All feasible precautions shall be taken to protect civilians from the effects of weapons to which this Article applies. Feasible precautions are those precautions which are practicable or practically possible taking into account all circumstances ruling at the time, including humanitarian and military considerations. These circumstances include, but are not limited to:

(a) the short- and long-term effect of mines upon the local civilian population for the duration of the minefield;

(b) possible measures to protect civilians (for example, fencing, signs, warning and monitoring);

(c) the availability and feasibility of using alternatives; and

(d) the short- and long-term military requirements for a minefield.

11. Effective advance warning shall be given of any emplacement of mines, booby-traps and other devices which may affect the civilian population, unless circumstances do not permit.

Article 4 – Restrictions on the use of anti-personnel mines

It is prohibited to use anti-personnel mines which are not detectable, as specified in paragraph 2 of the Technical Annex.

Article 5 – Restrictions on the use of anti-personnel mines other than remotely-delivered mines

1. This Article applies to anti-personnel mines other than remotely-delivered mines.

2. It is prohibited to use weapons to which this Article applies which are not in compliance with the provisions on self-destruction and self-deactivation in the Technical Annex, unless:

(a) such weapons are placed within a perimeter-marked area which is monitored by military personnel and protected by fencing or other means, to ensure the effective exclusion of civilians from the area. The marking must be of a distinct and durable character and must at least be visible to a person who is about to enter the perimeter-marked area; and

(b) such weapons are cleared before the area is abandoned, unless the area is turned over to the forces of another State which accept responsibility for the maintenance of the protections required by this Article and the subsequent clearance of those weapons.

3. A party to a conflict is relieved from further compliance with the provisions of sub-paragraphs 2 (a) and 2 (b) of this Article only if such compliance is not feasible due to forcible loss of control of the area as a result of enemy military action, including situations where direct enemy military action makes it impossible to comply. If that party regains control of the area, it shall resume compliance with the provisions of sub-paragraphs 2 (a) and 2 (b) of this Article.

4. If the forces of a party to a conflict gain control of an area in which weapons to which this Article applies have been laid, such forces shall, to the maximum extent feasible, maintain and, if necessary, establish the protections required by this Article until such weapons have been cleared.

5. All feasible measures shall be taken to prevent the unauthorized removal, defacement, destruction or concealment of any device, system or material used to establish the perimeter of a perimeter-marked area.

6. Weapons to which this Article applies which propel fragments in a horizontal arc of less than 90 degrees and which are placed on or above the ground may be used without the measures provided for in sub-paragraph 2 (a) of this Article for a maximum period of 72 hours, if:

(a) they are located in immediate proximity to the military unit that emplaced them; and

(b) the area is monitored by military personnel to ensure the effective exclusion of civilians.

Article 6 – Restrictions on the use of remotely-delivered mines

1. It is prohibited to use remotely-delivered mines unless they are recorded in accordance with sub-paragraph I (b) of the Technical Annex.

2. It is prohibited to use remotely-delivered anti-personnel mines which are not in compliance with the provisions on self-destruction and self-deactivation in the Technical Annex.

3. It is prohibited to use remotely-delivered mines other than anti-personnel mines, unless, to the extent feasible, they are equipped with an effective self-destruction or self-neutralization mechanism and have a back-up self-deactivation feature, which is designed so that the mine will no longer function as a mine when the mine no longer serves the military purpose for which it was placed in position.

4. Effective advance warning shall be given of any delivery or dropping of remotely-delivered mines which may affect the civilian population, unless circumstances do not permit.

Article 7 – Prohibitions on the use of booby-traps and other devices

1. Without prejudice to the rules of international law applicable in armed conflict relating to treachery and perfidy, it is prohibited in all circumstances to use booby-traps and other devices which are in any way attached to or associated with:

(a) internationally recognized protective emblems, signs or signals;

(b) sick, wounded or dead persons;

(c) burial or cremation sites or graves;

(d) medical facilities, medical equipment, medical supplies or medical transportation;

(e) children's toys or other portable objects or products specially designed for the feeding, health, hygiene, clothing or education of children;

(f) food or drink;

(g) kitchen utensils or appliances except in military establishments, military locations or military supply depots;

(h) objects clearly of a religious nature;

(i) historic monuments, works of art or places of worship which constitute the cultural or spiritual heritage of peoples; or

(j) animals or their carcasses.

2. It is prohibited to use booby-traps or other devices in the form of apparently harmless portable objects which are specifically designed and constructed to contain explosive material.

3. Without prejudice to the provisions of Article 3, it is prohibited to use weapons to which this Article applies in any city, town, village or other area containing a similar concentration of civilians in which combat between ground forces is not taking place or does not appear to be imminent, unless either:

(a) they are placed on or in the close vicinity of a military objective; or

(b) measures are taken to protect civilians from their effects, for example, the posting of warning sentries, the issuing of warnings or the provision of fences.

Article 8 – Transfers

1. In order to promote the purposes of this Protocol, each High Contracting Party:

(a) undertakes not to transfer any mine the use of which is prohibited by this Protocol;

(b) undertakes not to transfer any mine to any recipient other than a State or a State agency authorized to receive such transfers;

(c) undertakes to exercise restraint in the transfer of any mine the use of which is restricted by this Protocol. In particular, each High Contracting Party undertakes not to transfer any anti-personnel mines to States which are not bound by this Protocol, unless the recipient State agrees to apply this Protocol; and

(d) undertakes to ensure that any transfer in accordance with this Article takes place in full compliance, by both the transferring and the recipient State, with the relevant provisions of this Protocol and the applicable norms of international humanitarian law.

2. In the event that a High Contracting Party declares that it will defer compliance with specific provisions on the use of certain mines, as provided for in the Technical Annex, sub-paragraph I (a) of this Article shall however apply to such mines.

3. All High Contracting Parties, pending the entry into force of this Protocol, will refrain from any actions which would be inconsistent with sub-paragraph I (a) of this Article.

Article 9 – Recording and use of information on minefields, mined areas, mines, booby-traps and other devices

1. All information concerning minefields, mined areas, mines, booby-traps and other devices shall be recorded in accordance with the provisions of the Technical Annex.

2. All such records shall be retained by the parties to a conflict, who shall, without delay after the cessation of active hostilities, take all necessary and appropriate measures, including the use of such information, to protect civilians from the effects of minefields, mined areas, mines, booby-traps and other devices in areas under their control.

At the same time, they shall also make available to the other party or parties to the conflict and to the Secretary-General of the United Nations all such information in their possession concerning minefields, mined areas, mines, booby-traps and other devices laid by them in areas no longer under their control; provided, however, subject to reciprocity, where the forces of a party to a conflict are in the territory of an adverse party, either party may withhold such information from the Secretary-General and the other party, to the extent that security interests require such withholding, until neither party is in the territory of the other. In the latter case, the information withheld shall be disclosed as soon as those security interests permit. Wherever possible, the parties to the conflict shall seek, by mutual agreement, to provide for the release of such information at the earliest possible time in a manner consistent with the security interests of each party.

3. This Article is without prejudice to the provisions of Articles 10 and 12 of this Protocol.

Article 10 – Removal of minefields, mined areas, mines, booby-traps and other devices and international cooperation

1. Without delay after the cessation of active hostilities, all minefields, mined areas, mines, booby-traps and other devices shall be cleared, removed, destroyed or maintained in accordance with Article 3 and paragraph 2 of Article 5 of this Protocol.

2. High Contracting Parties and parties to a conflict bear such responsibility with respect to minefields, mined areas, mines, booby-traps and other devices in areas under their control.

3. With respect to minefields, mined areas, mines, booby-traps and other devices laid by a party in areas over which it no longer exercises control, such party shall provide to the party in control of the area pursuant to paragraph 2 of this Article, to the extent permitted by such party, technical and material assistance necessary to fulfil such responsibility.

4. At all times necessary, the parties shall endeavour to reach agreement, both among themselves and, where appropriate, with other States and with international organizations, on the provision of technical and material assistance,

including, in appropriate circumstances, the undertaking of joint operations necessary to fulfil such responsibilities.

Article 11 – Technological cooperation and assistance

1. Each High Contracting Party undertakes to facilitate and shall have the right to participate in the fullest possible exchange of equipment, material and scientific and technological information concerning the implementation of this Protocol and means of mine clearance. In particular, High Contracting Parties shall not impose undue restrictions on the provision of mine clearance equipment and related technological information for humanitarian purposes.

2. Each High Contracting Party undertakes to provide information to the database on mine clearance established within the United Nations System, especially information concerning various means and technologies of mine clearance, and lists of experts, expert agencies or national points of contact on mine clearance.

3. Each High Contracting Party in a position to do so shall provide assistance for mine clearance through the United Nations System, other international bodies or on a bilateral basis, or contribute to the United Nations Voluntary Trust Fund for Assistance in Mine Clearance.

4. Requests by High Contracting Parties for assistance, substantiated by relevant information, may be submitted to the United Nations, to other appropriate bodies or to other States. These requests may be submitted to the Secretary-General of the United Nations, who shall transmit them to all High Contracting Parties and to relevant international organizations.

5. In the case of requests to the United Nations, the Secretary-General of the United Nations, within the resources available to the Secretary-General of the United Nations, may take appropriate steps to assess the situation and, in cooperation with the requesting High Contracting Party, determine the appropriate provision of assistance in mine clearance or implementation of the Protocol. The Secretary-General may also report to High Contracting Parties on any such assessment as well as on the type and scope of assistance required.

6. Without prejudice to their constitutional and other legal provisions, the High Contracting Parties undertake to cooperate and transfer technology to facilitate the implementation of the relevant prohibitions and restrictions set out in this Protocol.

7. Each High Contracting Party has the right to seek and receive technical assistance, where appropriate, from another High Contracting Party on specific relevant technology, other than weapons technology, as necessary and feasible, with

a view to reducing any period of deferral for which provision is made in the Technical Annex.

Article 12 – Protection from the effects of minefields, mined areas, mines, booby-traps and other devices

1. Application

(a) With the exception of the forces and missions referred to in sub-paragraph 2(a) (i) of this Article, this Article applies only to missions which are performing functions in an area with the consent of the High Contracting Party on whose territory the functions are performed.

(b) The application of the provisions of this Article to parties to a conflict which are not High Contracting Parties shall not change their legal status or the legal status of a disputed territory, either explicitly or implicitly.

(c) The provisions of this Article are without prejudice to existing international humanitarian law, or other international instruments as applicable, or decisions by the Security Council of the United Nations, which provide for a higher level of protection to personnel functioning in accordance with this Article.

2. Peace-keeping and certain other forces and missions

(a) This paragraph applies to:

(i) any United Nations force or mission performing peace-keeping, observation or similar functions in any area in accordance with the Charter of the United Nations;

(ii) any mission established pursuant to Chapter VIII of the Charter of the United Nations and performing its functions in the area of a conflict.

(b) Each High Contracting Party or party to a conflict, if so requested by the head of a force or mission to which this paragraph applies, shall:

(i) so far as it is able, take such measures as are necessary to protect the force or mission from the effects of mines, booby-traps and other devices in any area under its control;

(ii) if necessary in order effectively to protect such personnel, remove or render harmless, so far as it is able, all mines, booby-traps and other devices in that area; and

(iii) inform the head of the force or mission of the location of all known minefields, mined areas, mines, booby-traps and other devices in the area in which the force or mission is performing its functions and, so far as is feasible, make available to the head of the force or mission all information in its possession concerning such minefields, mined areas, mines, booby-traps and other devices.

3. Humanitarian and fact-finding missions of the United Nations System

(a) This paragraph applies to any humanitarian or fact-finding mission of the United Nations System.

(b) Each High Contracting Party or party to a conflict, if so requested by the head of a mission to which this paragraph applies, shall:

(i) provide the personnel of the mission with the protections set out in sub-paragraph 2(b) (i) of this Article; and

(ii) if access to or through any place under its control is necessary for the performance of the mission's functions and in order to provide the personnel of the mission with safe passage to or through that place:

(aa) unless on-going hostilities prevent, inform the head of the mission of a safe route to that place if such information is available; or

(bb) if information identifying a safe route is not provided in accordance with sub-paragraph (aa), so far as is necessary and feasible, clear a lane through minefields.

4. Missions of the International Committee of the Red Cross

(a) This paragraph applies to any mission of the International Committee of the Red Cross performing functions with the consent of the host State or States as provided for by the Geneva Conventions of 12 August 1949 and, where applicable, their Additional Protocols.

(b) Each High Contracting Party or party to a conflict, if so requested by the head of a mission to which this paragraph applies, shall:

(i) provide the personnel of the mission with the protections set out in sub-paragraph 2(b) (i) of this Article; and

(ii) take the measures set out in sub-paragraph 3(b) (ii) of this Article.

5. Other humanitarian missions and missions of enquiry

(a) Insofar as paragraphs 2, 3 and 4 above do not apply to them, this paragraph applies to the following missions when they are performing functions in the area of a conflict or to assist the victims of a conflict:

(i) any humanitarian mission of a national Red Cross or Red Crescent Society or of their International Federation;

(ii) any mission of an impartial humanitarian organization, including any impartial humanitarian demining mission; and

(iii) any mission of enquiry established pursuant to the provisions of the Geneva Conventions of 12 August 1949 and, where applicable, their Additional Protocols.

(b) Each High Contracting Party or party to a conflict, if so requested by the head of a mission to which this paragraph applies, shall, so far as is feasible:

(i) provide the personnel of the mission with the protections set out in sub-paragraph 2(b) (i) of this Article, and

(ii) take the measures set out in sub-paragraph 3(b) (ii) of this Article.

6. Confidentiality

All information provided in confidence pursuant to this Article shall be treated by the recipient in strict confidence and shall not be released outside the force or mission concerned without the express authorization of the provider of the information.

7. Respect for laws and regulations

Without prejudice to such privileges and immunities as they may enjoy or to the requirements of their duties, personnel participating in the forces and missions referred to in this Article shall:

(a) respect the laws and regulations of the host State; and

(b) refrain from any action or activity incompatible with the impartial and international nature of their duties.

Article 13 – Consultations of High Contracting Parties

1. The High Contracting Parties undertake to consult and cooperate with each other on all issues related to the operation of this Protocol. For this purpose, a conference of High Contracting Parties shall be held annually.

2. Participation in the annual conferences shall be determined by their agreed Rules of Procedure.

3. The work of the conference shall include:

(a) review of the operation and status of this Protocol;

(b) consideration of matters arising from reports by High Contracting Parties according to paragraph 4 of this Article;

(c) preparation for review conferences; and

(d) consideration of the development of technologies to protect civilians against indiscriminate effects of mines.

4. The High Contracting Parties shall provide annual reports to the Depositary, who shall circulate them to all High Contracting Parties in advance of the Conference, on any of the following matters:

(a) dissemination of information on this Protocol to their armed forces and to the civilian population;

(b) mine clearance and rehabilitation programmes;

(c) steps taken to meet technical requirements of this Protocol and any other relevant information pertaining thereto;

(d) legislation related to this Protocol;

(e) measures taken on international technical information exchange, on international cooperation on mine clearance, and on technical cooperation and assistance; and

(f) other relevant matters.

5. The cost of the Conference of High Contracting Parties shall be borne by the High Contracting Parties and States not parties participating in the work of the Conference, in accordance with the United Nations scale of assessment adjusted appropriately.

Article 14 – Compliance

1. Each High Contracting Party shall take all appropriate steps, including legislative and other measures, to prevent and suppress violations of this Protocol by persons or on territory under its jurisdiction or control.

2. The measures envisaged in paragraph 1 of this Article include appropriate measures to ensure the imposition of penal sanctions against persons who, in relation to an armed conflict and contrary to the provisions of this Protocol, wilfully kill or cause serious injury to civilians and to bring such persons to justice.

3. Each High Contracting Party shall also require that its armed forces issue relevant military instructions and operating procedures and that armed forces personnel receive training commensurate with their duties and responsibilities to comply with the provisions of this Protocol.

4. The High Contracting Parties undertake to consult each other and to cooperate with each other bilaterally, through the Secretary-General of the United

Nations or through other appropriate international procedures, to resolve any problems that may arise with regard to the interpretation and application of the provisions of this Protocol.

Technical Annex

1. Recording

(a) Recording of the location of mines other than remotely-delivered mines, minefields, mined areas, booby-traps and other devices shall be carried out in accordance with the following provisions:

(i) the location of the minefields, mined areas and areas of booby-traps and other devices shall be specified accurately by relation to the coordinates of at least two reference points and the estimated dimensions of the area containing these weapons in relation to those reference points;

(ii) maps, diagrams or other records shall be made in such a way as to indicate the location of minefields, mined areas, booby-traps and other devices in relation to reference points, and these records shall also indicate their perimeters and extent;

(iii) for purposes of detection and clearance of mines, booby-traps and other devices, maps, diagrams or other records shall contain complete information on the type, number, emplacing method, type of fuse and life time, date and time of laying, anti-handling devices (if any) and other relevant information on all these weapons laid. Whenever feasible the minefield record shall show the exact location of every mine, except in row minefields where the row location is sufficient. The precise location and operating mechanism of each booby-trap laid shall be individually recorded.

(b) The estimated location and area of remotely-delivered mines shall be specified by coordinates of reference points (normally corner points) and shall be ascertained and when feasible marked on the ground at the earliest opportunity. The total number and types of mines laid, the date and time of laying and the self-destruction time periods shall also be recorded.

(c) Copies of records shall be held at a level of command sufficient to guarantee their safety as far as possible.

(d) The use of mines produced after the entry into force of this Protocol is prohibited unless they are marked in English or in the respective national language or languages with the following information:

(i) name of the country of origin;

(ii) month and year of production; and

(iii) serial number or lot number.

The marking should be visible, legible, durable and resistant to environmental effects, as far as possible.

2. Specifications on detectability

(a) With respect to anti-personnel mines produced after 1 January 1997, such mines shall incorporate in their construction a material or device that enables the mine to be detected by commonly-available technical mine detection equipment and provides a response signal equivalent to a signal from 8 grammes or more of iron in a single coherent mass.

(b) With respect to anti-personnel mines produced before 1 January 1997, such mines shall either incorporate in their construction, or have attached prior to their emplacement, in a manner not easily removable, a material or device that enables the mine to be detected by commonly-available technical mine detection equipment and provides a response signal equivalent to a signal from 8 grammes or more of iron in a single coherent mass.

(c) In the event that a High Contracting Party determines that it cannot immediately comply with sub-paragraph (b), it may declare at the time of its notification of consent to be bound by this Protocol that it will defer compliance with sub-paragraph (b) for a period not to exceed 9 years from the entry into force of this Protocol. In the meantime it shall, to the extent feasible, minimize the use of anti-personnel mines that do not so comply.

3. Specifications on self-destruction and self-deactivation

(a) All remotely-delivered anti-personnel mines shall be designed and constructed so that no more than 10% of activated mines will fail to self-destruct within 30 days after emplacement, and each mine shall have a back-up self-deactivation feature designed and constructed so that, in combination with the self-destruction mechanism, no more than one in one thousand activated mines will function as a mine 120 days after emplacement.

(b) All non-remotely delivered anti-personnel mines, used outside marked areas, as defined in Article 5 of this Protocol, shall comply with the requirements for self-destruction and self-deactivation stated in sub-paragraph (a).

(c) In the event that a High Contracting Party determines that it cannot immediately comply with sub-paragraphs (a) and/or (b), it may declare at the time of its notification of consent to be bound by this Protocol, that it will, with respect to mines produced prior to the entry into force of this Protocol, defer compliance with sub-paragraphs (a) and/or (b) for a period not to exceed 9 years from the entry into force of this Protocol.

During this period of deferral, the High Contracting Party shall:

(i) undertake to minimize, to the extent feasible, the use of anti-personnel mines that do not so comply, and

(ii) with respect to remotely-delivered anti-personnel mines, comply with either the requirements for self-destruction or the requirements for self-deactivation and, with respect to other anti-personnel mines comply with at least the requirements for self-deactivation.

4. International signs for minefields and mined areas

Signs similar to the example attached and as specified below shall be utilized in the marking of minefields and mined areas to ensure their visibility and recognition by the civilian population:

(a) size and shape: a triangle or square no smaller than 28 centimetres (11 inches) by 20 centimetres (7.9 inches) for a triangle, and 15 centimetres (6 inches) per side for a square;

(b) colour: red or orange with a yellow reflecting border

Statement of Mr. Eric Roethlisberger

Vice-President of the International Committee of the Red Cross

Third Session of the Review Conference of the States Parties to the 1980 Convention on Prohibitions or Restrictions on the Use of Certain Conventional Weapons Which May be Deemed to be Excessively Injurious or to Have Indiscriminate Effects

Geneva, Switzerland
3 May 1996

This day marks the conclusion of an initial diplomatic response to the immense human suffering caused by landmines. The results are indeed modest given the scale of the horror they attempt to address. But, fortunately, these results do not tell the whole story. This Conference has focused the attention of governments and their military forces on the humanitarian responsibilities involved in landmine use and the need for dramatic changes in their approach to these weapons. The ICRC is grateful for the privilege of participating in and contributing to this process. The history of the development of humanitarian law is an ongoing dialogue between legitimate military needs and the humanitarian concerns of all civilized society. We have sought to keep the spotlight on the human implications of options under consideration and the humanitarian costs of the compromises sought. We look forward to active involvement in future annual conferences and the second Review Conference.

The results achieved do not tell the whole story for another reason. Public conscience and a growing number of States have already stigmatized, as with poison gas, anti-personnel mines. Though not yet reflected in a global consensus, movement towards the elimination of these weapons has proceeded rapidly as State after State has reviewed the balance between military utility and humanitarian concerns and announced support for a ban, unilateral renunciation of production, use and transfer and the destruction of stockpiles. The assumption of moral and humanitarian responsibilities need not be negotiated; it need not await a global consensus. We are greatly encouraged that many States and regional bodies are already considering further steps.

Of the measures adopted today the ICRC considers several to be of particular importance:

- extension of the scope of Protocol II to non-international armed conflicts;
- clear assignment of responsibility for mine clearance to those who lay them;
- improved recording requirements; and
- the introduction of protection for humanitarian workers.

The introduction of articles on transfers in Protocols II and IV are important steps in the development of humanitarian law. However the provision on mine transfers, if narrowly implemented, would represent for most States a step back from present practice. We expect therefore, that States will maintain their current comprehensive moratoria on transfers pursuant to recent resolutions of the UN General Assembly.

The limitations adopted on the use of landmines are, in our view, woefully inadequate. They will encourage the production, transfer and use of a new generation of mines while not prohibiting any existing types other than, eventually, non-detectable APMs. Taken together with the absence of verification measures for production, transfer or use, these measures are unlikely to significantly reduce the level of civilian landmine casualties. The victims of the landmine carnage of recent decades will find little solace in these results. The horrific numbers of landmine victims of recent years are set to continue unless governments squarely face their humanitarian responsibilities and do far more than required by the agreement adopted today.

The ICRC deeply regrets that, for the first time in a humanitarian law treaty, measures have been adopted which, instead of entirely prohibiting the use of an indiscriminate weapon, both permit its continued use and implicitly promote the use of new models which will have virtually the same effects, at least in the short term. Given its mandate and humanitarian responsibilities, the ICRC cannot promote the development and use of new weapons. We nonetheless support the purpose of civilian protection embodied in Protocol II and will encourage further adherence to the Convention with a view to achieving dramatic improvements including a complete ban on anti-personnel mines. For this purpose we continue to be of the opinion that anti-personnel mines must be understood to be any munition which is designed to be detonated or exploded by the presence, proximity or contact of a person, whatever other functions the munition may also have.

The adoption by this Conference of a new Protocol on blinding laser weapons was a landmark achievement for international humanitarian law. Its importance lies both in the prohibition of a particularly abhorrent weapon before its use on the battlefield and its inclusion of a complete prohibition on transfers. While it is regrettable that the scope of this Protocol could not be extended as was agreed in Vienna and reaffirmed in the 26th International Conference of the Red Cross and Red Crescent, we encourage all States to issue a "Statement of Understanding" at the time of adherence indicating that they consider the Protocol to apply at all times.

Because so much remains to be done, this day must also mark the beginning of intensified public and political efforts, at national and regional levels to end the landmine crisis. In this regard we warmly welcome the Danish conference on Mine Clearance Technology in July and the initiative by Canada to convene in Ottawa this September a conference of States supporting an immediate ban to consider which steps they should take to achieve this. The ICRC will also host a regional conference of Central American States on the landmine issue on 28–29 May in Managua and a meeting for south-east Asian States in Jakarta on 29–30 May at which the issue will be addressed. We suggest that future political efforts should integrate hitherto separate elements of the international response, namely: work towards a ban, renunciation of production, stockpiling and use, an end to all transfers and assistance in mine clearance.

The ICRC deeply appreciates the sincere efforts of delegates to this Conference and its preparatory meetings who have sought over two long and sometimes difficult years to achieve the maximum possible results from this intergovernmental process. We also pay tribute to the many non-governmental organizations without whom this Conference would not have occurred. They have had a profound effect in bringing the reality of the landmines into the negotiating framework and in shaping events far beyond it. The ICRC reaffirms its commitment to continuing our dialogue with the military and to working together on this issue over the coming years with governments and other humanitarian organizations. We look forward to meeting at the next Review Conference with a single unified purpose: the total prohibition of anti-personnel mines.

Press Release

ICRC Views Amended Landmines Protocol as "Woefully Inadequate"
3 May 1996

The International Committee of the Red Cross (ICRC) considers as woefully inadequate the initial diplomatic response to the devastating effects of mines, as represented by the results of the Review Conference of the 1980 UN Convention on Certain Conventional Weapons (CCW), to be confirmed at its closing session today in Geneva. The horror of the immense human suffering caused by landmines is set to continue, and the amended Protocol II will do little to change this situation.

Of the measures adopted today, the ICRC considers several to be of particular importance: extension to non-international conflicts of the scope of Protocol II on landmines; clear assignment of responsibility for mine clearance to those who lay them; improved recording requirements; and improved protection for humanitarian workers.

However, the limitations adopted on the *use* of landmines are, from the ICRC's point of view, very modest. They will encourage the production, transfer and use of a new generation of mines, while not prohibiting any existing types other than, eventually, non-detectable anti-personnel mines. The ICRC deeply regrets that, for the first time in a humanitarian law treaty, measures have been adopted which, instead of entirely prohibiting the use of an indiscriminate weapon, both permit its continued use and implicitly promote the use of new models which will have virtually the same effects. Given its mandate and humanitarian responsibilities, the ICRC cannot endorse this development.

In contrast, at its initial session in Vienna this Conference adopted a new Protocol IV banning blinding laser weapons. This represents a landmark for international humanitarian law: a particularly abhorrent weapon has been prohibited before being used on the battlefield and its transfer across borders has been completely outlawed.

Much remains to be done. The ICRC believes that this Conference has focused the attention of governments on their humanitarian responsibilities regarding landmine use, and public conscience worldwide has stigmatized – as with poison gas – anti-personnel mines. In addition, a growing number of States have reviewed the balance between military utility and humanitarian concerns, and declared support for a total ban as well as announcing their unilateral renunciation of production, transfer and use of anti-personnel mines and plans to destroy stockpiles. A number of countries and regional bodies are known to be considering such steps.

The victims of the landmine carnage of recent decades will find little solace in the results of the Review Conference. The horrific level of landmine casualties in recent years is set to continue unless governments do far more than required by the agreement adopted today. Intensified public and political efforts, at national and regional levels, must be undertaken to end the landmine crisis. The ICRC suggests that future political efforts should integrate hitherto separate elements of the international efforts: ongoing work towards a total ban; the renunciation by States of production, stockpiling and use; an end to all transfers; and assistance in mine clearance.

The history of the development of humanitarian law is one of ongoing dialogue between legitimate military needs and the humanitarian concerns of all civilized society. The ICRC has sought to keep the spotlight on the human implications of the options under consideration. The ICRC looks forward to working on this issue together with governments, military establishments and other humanitarian organizations and to uniting with them at the next Review Conference with a single purpose: the total prohibition of anti-personnel mines.

To date 35 countries have supported an immediate global ban on anti-personnel mines; 16 have renounced their use by their own armed forces; four have suspended use; and at least five are destroying their stockpiles.

Landmine Negotiations Conclude with Modest Results

Peter Herby, Legal Division, International Committee of the Red Cross, Geneva
Published in the *International Review of the Red Cross*, No. 312 – May–June 1996

After two years of tortuous negotiations and despite support for a total ban on anti-personnel mines by nearly half of the 51 States participating in the final session of the Review Conference of the 1980 United Nations Convention on Certain Conventional Weapons (CCW),[1] held in Geneva from 22 April to 3 May 1996, only minimal restrictions on the use of anti-personnel landmines were finally adopted.[2] Nine years after entry into force of amended Protocol II, anti-personnel mines will have to be detectable and those scattered outside of marked minefields, by air, artillery or other means, will have to self-destruct after 30 days. However, long-lived mines will remain available for production, export and use – including indiscriminate use. Regrettably, this modest legal response to a major international humanitarian crisis, though adopted by consensus, is unlikely to significantly reduce the horrendous level of mine casualties.

Amended Protocol II of the CCW, on mines, booby-traps and other devices, is the end of a negotiating process which began in February 1994 in the Group of Governmental Experts charged with preparing the revision of this Protocol and with considering the need for possible new Protocols on specific weapons. The results of the Group's efforts were submitted to the first Review Conference session, held in Vienna from 25 September to 13 October 1995, which ended in deadlock on the landmine issue. Negotiations resumed in Geneva in January and continued in the final session, covered by this report, of April/May 1996. In many respects, the results achieved by the Group of Governmental Experts and by successive Review Conference sessions became progressively weaker until the lowest common denominator required for consensus was finally reached with the adoption of Protocol II as amended.

The International Committee of the Red Cross (ICRC) was invited as an expert/observer to participate in all sessions of the Group of Governmental Experts as well as the Review Conference and contributed to the negotiating process by providing substantial background documentation on the humanitarian and legal aspects of the landmine issue.[3] Resolutions adopted in December 1995 by the Council of Delegates[4] and by the 26th International Conference of the Red Cross and Red Crescent[5] reflect the concern of the entire Red Cross and Red Crescent Movement with regard to the landmine crisis. In many cases the active commitment of National Red Cross and Red Crescent Societies helped to bring about important decisions by governments to change their policies on landmines.

Results of the negotiations

On 3 May 1996 the Review Conference adopted an amended Protocol II on mines, booby-traps and other devices,[6] which will enter into force six months after 20 States Parties declare their consent to be bound by it. This is expected to take two to three years. States Parties will continue to be bound by the original Protocol II[7] until entry into force, for them, of the amended version.

General provisions

The most notable steps forward are contained in amended Protocol II's general provisions. They include the codification of a number of new principles and the introduction of new provisions. Key elements are as follows:

• A specific *definition of anti-personnel mines*, which are now subject to stricter control than anti-tank or vehicle mines, was introduced. Anti-personnel mines are defined[8] as those "primarily designed to be exploded by the presence, proximity or contact of a person . . .". The inclusion of "primarily" in the definition could be interpreted to exclude any dual-use anti-personnel mines which can be claimed to serve another "primary" purpose. The ICRC objected vigorously to this wording, and many government delegations considered it unnecessarily ambiguous. Twenty mainly Western States, led by Germany, introduced an official interpretation of the word "primarily" indicating that it means only that anti-tank mines with anti-handling devices are not anti-personnel mines.

• Extension of the Protocol's field of application to *non-international armed conflicts.*

• Assignment of *clearance responsibility* to those who deploy mines. This obligation will nonetheless be difficult to enforce when parties do not have the resources or expertise for mine clearance – as is often the case in internal armed conflicts.

• The location of all mines must be *mapped* and *recorded* in all circumstances, rather than only when used in "pre-planned" minefields, as was stipulated in the original Protocol. This too will be difficult to implement in the case of remotely delivered mines, the accurate recording of which is virtually impossible.

• *Protection for Red Cross and Red Crescent personnel (ICRC, National Societies and International Federation) and other humanitarian missions*, including provisions to provide heads of missions with information on minefields and safe routes around them and, in certain cases, to clear a route through mined areas when it is necessary for access to victims.

• A new provision on *transfers of mines* prohibits the international transfer of non-detectable anti-personnel mines, and of any mine to entities other than States. However, transfers to non-party States are permitted if they "agree to apply" the provisions of the Protocol. A stronger prohibition on transfers to such States would have provided an incentive for adherence to the CCW.

• The provision on compliance requires States Parties to enact *penal legislation to suppress serious violations* of the Protocol.

• *Annual consultations* among States Parties shall be held to review the operation of the Protocol and prepare future review conferences.

The amended Protocol does not, however, contain any provisions for the verification of either the reliability of its technical requirements (described below) or of possible violations of its provisions on the use of landmines.

New restrictions on the use of anti-personnel landmines

The new restrictions adopted by the Review Conference on the use of anti-personnel landmines reflect only modest progress over existing law and may be more difficult to implement for States without adequate resources. The major innovations are as follows:

• *Long-lived anti-personnel mines ("dumb" mines)* may be produced, transferred and used as before, provided that:

– they are detectable (compliance with this provision becomes mandatory no later than nine years after entry into force of the amended Protocol), and

– they are placed in areas that are fenced, marked and guarded in order to keep civilians out (except when a party to conflict is prevented by direct enemy military action from taking these precautions).

• *Short-lived anti-personnel mines ("smart" mines)* may be produced, transferred and used as before, provided that (compliance likewise becomes mandatory no later than nine years after entry into force of the amended Protocol):

– they self-destruct within 30 days (with 90% reliability), if used outside marked, fenced and guarded areas;

– those mines which fail to self-destruct will self-deactivate within 120 days (with 99.9% reliability), and

– they are detectable.

No specific restrictions on the placement of "smart" mines were adopted, although the general rules of humanitarian law, including prohibitions on the targeting of civilian populations and civilian objects, still apply. A large proportion of such mines are likely to be remotely delivered models, the locations of which will be difficult or impossible to record. It may be argued that the production, transfer and use of "smart" mines are implicitly encouraged by the amended Protocol because fewer restrictions apply to them than to "dumb" mines.

• The rules on the use of *anti-tank/vehicle mines, including remotely delivered models,* have not been changed:

– no detectability requirement;

– no specific restrictions on placement;

– no prohibition of anti-handling devices;

– no maximum lifetime.

As a result, only the general rules of humanitarian law, such as the protection of civilians, and amended Protocol II's requirements for recording and removal apply to these types of mine.

• The use of devices which cause a mine to *explode when detected by an electronic sensor* is prohibited for all types of mine.

The ICRC's response

In its statement to the closing plenary meeting of the Review Conference the ICRC welcomed the strengthened general provisions of amended Protocol II, but pointedly described the restrictions on the use of landmines as "woefully inadequate". It indicated that such provisions alone were "unlikely to significantly reduce the level of civilian landmine casualties". The principal points made by the ICRC in its response to the Review Conference to date have been as follows:

• The Review Conference was an important process which has led to dramatic developments in the policies of many States on the production, transfer and use of anti-personnel mines.

• The general provisions of Protocol II as amended include a number of welcome improvements:

- extension of the scope of application to non-international armed conflicts;
- clear assignment of mine-clearance responsibility;
- a provision on the transfer of mines (a new and important element for international humanitarian law);
- new protection for humanitarian workers;
- an obligation to repress serious violations of the new rules;
- annual consultations among Parties to the Protocol.

• The definition of an anti-personnel mine is, however, unnecessarily ambiguous. The ICRC is of the opinion that such a mine must continue to be understood as any mine which is "designed to be exploded or detonated by the presence, proximity or contact of a person",[9] whatever other functions the munition may also have.

• The restrictions on use are inadequate and, on their own, are unlikely to have a significant impact on the level of civilian casualties. If the humanitarian crisis caused by landmines is to be effectively addressed, States must do far more individually than could be agreed by consensus within the framework of an amended Protocol II. This includes maintaining existing *comprehensive* moratoria on the transfer of anti-personnel mines and completely ending their production and use.

• The ICRC deeply regrets that Protocol II as amended not only does *not* prohibit the use of a weapon with indiscriminate effects but even indirectly promotes the development and use of *new* weapons which will have precisely the same effects, at least in the short term. This is the first time that a humanitarian law instrument may have the effect of promoting the use of a new weapon.

• However, the text adopted by the Review Conference does not tell the whole story:

– The Review Conference is only one aspect of what is happening politically in the world at large, where anti-personnel mines are being stigmatized in the public conscience, military forces are questioning the utility of such weapons and State practice is changing rapidly.

– Whereas the Review Conference yielded disappointing results, efforts to achieve a *total ban on anti-personnel landmines* are succeeding at a pace which was inconceivable only two years ago. Forty States now support a ban; seventeen of them have renounced and six have suspended the use of anti-personnel mines. Nine States are destroying existing stocks. This trend is likely to gather momentum, partly thanks to the results, modest though they may be, achieved in Geneva.

– It can be hoped that a global ban on anti-personnel landmines will be achieved once there is a critical mass of States reconsidering their own use of mines and supporting a ban as the only effective and verifiable solution to the problems caused by these weapons in humanitarian terms.

• It is essential to focus on national and regional initiatives to end production, use and transfer of anti-personnel landmines and to build further support for a global ban. The Canadian initiative to bring pro-ban countries together in Ottawa in September 1996 to consider further steps which they can take to this effect is the beginning of an important new process. A number of States and regions are considering unilateral steps towards a ban. In this regard, an "Anti-personnel Mines Free Zone" in the Americas will be under consideration by the members of the Organization of American States.

Protocol IV on blinding laser weapons[10]

New Protocol IV on blinding laser weapons, adopted by the Vienna session of the Review Conference, is a landmark achievement. It is particularly encouraging that the Protocol contains an absolute prohibition on both the *use* and the *transfer* of blinding laser weapons – a first in the history of international humanitarian law. Contrary to the proposal by the 26th International Conference of the Red Cross and Red Crescent and agreements reached at the first session of the Review Conference in Vienna, the scope of this Protocol was not extended beyond international armed conflicts. States should, however, be encouraged to declare, when adhering to Protocol IV, that they consider the Protocol to apply under all circumstances.

Concluding remarks

Looking beyond the Review Conference, Yves Sandoz, Director of the ICRC Department for Principles, Law and Relations with the Movement, reflected in an article the organization's hopes and determination with regard to the landmine issue: [11]

Taken together, the awakening of the public conscience, the beginnings of dramatic changes in State practice and authoritative questioning of anti-personnel mine use from within military circles could lead to an end to the use of these arms in large parts of the world in coming years. On that basis the next Review Conference of the 1980 Convention in 2001, or possibly another forum, could be expected to produce agreement on outlawing this indiscriminate weapon.

Though attention has recently focused on globally negotiated solutions, the landmine crisis will be ended through the insistence of the public, through decisions of States which seek to protect their population and territories from the terrible scourge of these weapons and by the decisions of individual commanders who judge their human costs unacceptable.

A global legal ban will be the result, not the cause, of such actions. It will be a victory of human compassion and solidarity. It is the only fitting response to the carnage which continues to cost the lives and livelihoods of two thousand victims each and every month. The ICRC, together with the entire Red Cross and Red Crescent Movement, will tirelessly continue its efforts with both military and humanitarian organizations to ensure that anti-personnel mines are banned sooner rather than later.

NOTES

1 Convention on Prohibitions or Restrictions on the Use of Certain Conventional Weapons Which May be Deemed to be Excessively Injurious or to Have Indiscriminate Effects, adopted on 10 October 1980; 59 States party (as at 30 April 1996).

2 Protocol on Prohibitions or Restrictions on the Use of Mines, Booby-Traps and Other Devices as amended on 3 May 1996 (Protocol II as amended on 3 May 1996). See text below.

3 See the ICRC's reports in the *International Review of the Red Cross (IRRC)*, No. 307, July–August 1995, pp. 363–367 (on its position on the issues discussed), and No. 309, November–December 1995, pp. 672–677 (on the first session of the Review Conference).

4 Resolution 10 – Anti-personnel landmines – adopted by the Council of Delegates, December 1995, in: *IRRC*, No. 310, January–February 1996, p. 151.

5 Resolution 2, part G – on anti-personnel landmines – of the 26th International Conference of the Red Cross and Red Crescent, Geneva 3–7 December 1995, in: *IRRC*, No. 310, January–February 1996, pp. 66–67.

6 See note 2.

7 Adopted on 10 October 1980.

8 Protocol II, Article 2, para. 3.

9 I.e. the definition as contained in new Article 2(3) but without the word "primarily".

10 Protocol on Blinding Laser Weapons (Protocol IV), adopted on 13 October 1995. See the comprehensive study by L. Doswald-Beck, "New Protocol on Blinding Weapons", in *IRRC*, No. 312, May–June 1995, pp. 272–298.

11 "Anti-personnel mines will be banned", May 1996 (to be published shortly in several journals).

18

ICRC Position Paper No. 3

ICRC Briefing and Position Paper
July 1996

Stopping the Landmines Epidemic: From Negotiation to Action

Since international negotiations for new restrictions on the use of landmines began in March 1994, some 56,000 persons, mainly civilians, are estimated to have been killed or maimed by these indiscriminate weapons. By the time the existing international treaty concerning the use of landmines is again reviewed in the year 2001, mines may have shattered another 120,000 lives unless far more dramatic steps than those required by the law are taken by States and the international community. Before the recently adopted legal rules take full effect around 2007, the toll could exceed 200,000. Such levels of preventable death and injury are morally unconscionable and can be stopped.

Fortunately, we may already be witnessing the beginning of the end of the use of anti-personnel landmines worldwide. In face of growing public abhorrence anti-personnel mines are well on the way to being stigmatized in the public conscience, as was poison gas after World War I.

A recent opinion poll conducted in 21 countries reflected large majorities aware of the landmine crisis and in favour of a ban on these weapons. Twenty-five States have renounced or suspended the use of anti-personnel mines by their own armed forces; more than 20 have prohibited production; 11 are destroying their stockpiles and more than 50 have halted exports. These unilateral measures reflect a growing awareness of the urgent need for dramatic steps to end the landmine crisis and the growing momentum of efforts to achieve a global ban. Over time these steps will save lives.

The International Committee of the Red Cross (ICRC), along with the entire Red Cross and Red Crescent Movement, is fully committed to continuing its worldwide efforts to end the unspeakable suffering caused by landmines. These will include the promotion of national, regional and global bans on anti-personnel mines, public information on the problem, advocacy on behalf of the victims, increased care for landmine victims through surgical and rehabilitation programmes around the world, mine awareness for populations at risk and support for enhanced mine-clearance efforts in mine-infested countries.

This paper presents the ICRC's views on what has been achieved to date in the international response to the landmine scourge and suggestions on where future efforts may be fruitful.

I. THE 1980 UN CONVENTION ON CERTAIN CONVENTIONAL WEAPONS (CCW)

The Review Conference of the CCW Convention has been the focus of considerable attention to date: on 3 May 1996, the Conference adopted a revised version of the Convention's Protocol II regulating the use of "mines, booby-traps and other devices".

The revision contains a number of welcome improvements including:

a) an extension of the Protocol to apply in both international and non-international armed conflicts;
b) a clear assignment of responsibility for mine clearance to those who lay the mines;
c) a requirement that the location of all mines be mapped and recorded;
d) new protection for humanitarian workers;
e) a prohibition on the transfer of non-detectable anti-personnel mines;
f) a requirement that States enact penal legislation to punish serious violations of the Protocol and;
g) annual consultations among Parties to the Protocol to review its operation.

Unfortunately, the new limitations on the use of anti-personnel mines are weak and do not go much further than the existing general prohibition of directing mines against civilians. *(However, the nature of mines makes them impossible to direct once they have been laid.)*

A centerpiece of the new rules is that all anti-personnel mines used must be detectable so as to facilitate mine clearance.

Long-lived mines, which can strike decades after they are laid, may continue to be produced, transferred and used – if placed in fenced, marked and guarded minefields. *(Such protections are difficult and costly to ensure, however, and often does not occur in practice).*

Self-destructing mines *(equipped with mechanisms which cause them to self-destruct within 30 days and, if self-destruction fails, to deactivate within 120 days)* may be used without any specific restrictions on their placement. Since such mines can be remotely delivered over long distances and in huge quantities by artillery or aircraft, their use could lead to an increase in civilian mine casualties. When mines are remotely delivered, their accurate marking and mapping becomes virtually impossible.

States are not required to implement these provisions **until nine years after entry into force** of the revised Protocol (i.e., probably the year 2007 or 2008).

The revised Protocol places no new restrictions on the use of **anti-tank or vehicle mines,** which, although used on a smaller scale than anti-personnel mines, are just as indiscriminate. These mines may continue to be used even if they are not detectable; neither their placement nor maximum lifetime is subject to specific control.

Although the ICRC welcomes the improvements in the Protocol's general provisions, it is deeply disappointed in the weak restrictions on the use of anti-personnel mines, the lack of specific restrictions on anti-tank mines, the excessively long "grace period" for implementation of key provisions on use, and the absence of a mechanism to verify the fulfillment of technical requirements for self-destructing mines and to investigate possible violations of the restrictions on use. Given these weaknesses, largely due to the need to adopt the revised Protocol by consensus, the new Protocol, in and of itself, is unlikely to lead to a significant reduction in the level of civilian landmine casualties.

The ICRC is concerned that the complex technical provisions for the use of various types of mine will not be implemented in the types of conflict where most recent mine use has occurred. These are often internal conflicts involving poorly trained and equipped forces which may be unable or unwilling to abide by a complex set of rules or to pay a higher price for self-destructing mines.

The disappointing results of the Review Conference have reinforced the ICRC's view that only the stigmatization, prohibition and elimination of anti-personnel mines will put an end to the humanitarian scourge they have caused.

The ICRC will encourage States to adhere to the revised Protocol II of the CCW in order to strengthen the minimum international legal norms which apply when mines are used. However, these norms do not oblige States to use mines or to invest in new types of mine.

We will therefore also encourage States to go far beyond their minimum legal obligations and to stop the production, transfer and use of anti-personnel mines on the basis of:

- their broader moral and humanitarian responsibilities,
- the indiscriminate nature of the weapon,
- the conclusions of a wide range of acting and former military commanders that the use of anti-personnel mines in accordance with law and doctrine is difficult, if not impossible, even for modern professional armies,[1] and
- a judgment that the limited military value of anti-personnel mines is far outweighed by their human, economic and social costs. Renunciation of these weapons is therefore a means of protecting one's own population and territory and an act of human solidarity with past, present and future victims.

In addition, the ICRC will actively promote State adherence to the new Protocol IV of the 1980 Convention prohibiting the use and transfer of **blinding laser weapons,** which it considers an historic achievement of the Review Conference. States should be encouraged when adhering to this new international instrument to declare their understanding that it applies "at all times" (as originally agreed by virtually all States but blocked in the final stage of negotiations) and to take national measures to ensure that such weapons are neither developed nor produced.

II. TOWARDS ELIMINATION

➢ Stigmatization

The ICRC and the International Red Cross and Red Crescent Movement as a whole, will continue efforts to inform the public, military circles and decision-makers about the human tragedy caused by landmines and to achieve the stigmatization of anti-personnel mines in the public conscience. The goal will be to ensure that, even before a legal ban is in place, combatants will choose not to use anti-personnel mines because of the abhorrence of their effects within their own societies, and that producing countries will choose neither to produce nor to transfer these pernicious weapons.

➢ National and regional initiatives

The end of the landmine crisis need not await a global, negotiated consensus. Governments and regional or sub-regional organizations can decide to end the crisis in their own territories and thus contribute to a global solution. **The *25* States which have renounced the use of anti-personnel mines by their own forces and the 11 which are destroying their stockpiles have begun the process of**

freeing the world from the menace of these arms. When a critical mass of States have taken such steps a *de facto* ban will have been achieved; a legal ban may follow as State practice regarding mines changes.

The most effective and permanent national measures are laws, such as that adopted by Belgium and under consideration in other countries, prohibiting the production, stockpiling, transfer and use of anti-personnel mines. In other States unilateral measures have taken the form of moratoria, national policy declarations and administrative orders under existing laws.

At the regional level, political measures can be adopted calling for an end to the production, transfer and use of anti-personnel mines by States and encouraging national laws and decisions to this effect:

The *Organization of American States* in June 1996 called for the establishment of an "Anti-personnel Mine Free Zone" in the Americas. An initiative of the *Central American Parliament,* in which national renunciation of anti-personnel mines is combined with enhanced mine-clearance efforts and assistance to victims could make Central America the first mine-infested region to free itself from this scourge. In February 1996 the Council of Ministers of the *Organization of African Unity* urged sub-regional organizations on the continent to launch initiatives for the prohibition of anti-personnel mines in support of the OAU's previous commitment to a total ban. Although this type of initiative has not yet reached governmental level in Europe, the *European Parliament* on 13 May appealed to all Member States to unilaterally ban the production and use of anti-personnel mines and to destroy existing stocks.

➢ **International initiatives**

Canada will host an international conference in Ottawa in autumn 1996, bringing together the more than 40 States which support a global ban on anti-personnel mines, to consider short- and medium-term steps to this end. As the first international meeting of pro-ban States, the ICRC and other Movement representatives and selected non-governmental organizations, the meeting could significantly increase momentum towards a global ban and promote the types of national and regional measures mentioned above. It should also integrate, at the political level, efforts towards a ban with measures to stop transfers and assistance in mine clearance. It is expected that this meeting will issue a strong political declaration committing participating States to an action plan and to a regular programme of meetings in the coming years.

In its autumn 1996 session the *UN General Assembly* will be considering, for the first time, a strongly worded resolution calling for the total elimination of anti-personnel mines, in addition to its now routine resolutions on export moratoria and the CCW. The resolution should declare the intention of its supporters to work for the elimination of these arms through national and regional initiatives and put forward a target date for the achievement of a total ban.

Proposals have been made to begin at an early date negotiations for a ban on anti-personnel mines either in the 60-nation Conference on Disarmament in Geneva or in a specially convened diplomatic conference of pro-ban States. **Given that recent negotiations by consensus on new legal restraints have produced only minimal results, the ICRC is concerned that new negotiations, particularly if conducted on the basis of consensus, may be premature. They might be more successful once political efforts, such as that being launched with the Ottawa Conference, have matured.**

➢ **Increased assistance to mine victims**

Currently only a small proportion (estimated at 15%) of mine victims have access to rehabilitation programmes. Greatly expanded resources to provide both emergency medical treatment and lifetime care to victims will be needed, even if mine use were stopped immediately. National and international agencies can support these efforts through ICRC and National Society surgical and prosthetic/orthotic projects, through direct bilateral assistance and through the efforts of other humanitarian agencies.

➢ **Enhanced mine-clearance efforts**

In 1995 the United Nations requested $75 million for mine-clearance funding, but received only $20 million; this compared to an estimated $33 billion required to clear all currently emplaced mines. A massive and long-term international effort is needed if future generations are to be spared paying the price for today's landmine legacy.

➢ **Prevention through mine-awareness programmes**

Mine-awareness programmes for civilian populations are now being run by the ICRC in five countries and by a variety of national agencies and humanitarian organizations in many more. Such efforts need to be rapidly expanded so that populations at risk from mines can better protect themselves.

III. WHAT CAN BE DONE NEXT?

Following the public outcry over the use of poison gas in World War I, it took seven years to achieve adoption of the 1925 Geneva Protocol outlawing the use of chemical weapons and nearly another 70 to ban their possession. In comparison, efforts to stigmatize and end the use of anti-personnel mines have achieved an impressive degree of success since the early 1990s.

The renunciation by State after State of the use of these weapons has begun creating a new international norm against their use. Success on a global level will require a sustained long-term commitment by individuals, public organizations, the media, military circles, parliaments and governments.

Whether you work at the local, national or international level, you can ensure that the next generation of children in war-torn lands are not terrorized by the threat of mines by taking crucial steps such as these:

- increasing awareness of the humanitarian crisis and helping to stigmatize the use of anti-personnel mines;
- promoting the renunciation and prohibition of these weapons at the national, regional and international levels;
- raising funds for the treatment and rehabilitation of mine victims, through the ICRC, National Red Cross and Red Crescent Societies or other humanitarian organizations;
- supporting generous government funding for mine clearance through nation-to-nation programmes and the UN's Voluntary Fund for Mine Clearance.

NOTE

1 Conclusions of *Anti-personnel Landmines: Friend or Foe?* (ICRC, March 1996) endorsed by 49 commanders from 19 countries: similar conclusions were published by 15 high-ranking US commanders.

19

Making Central America a Mine-free Zone
ICRC Seminar in Managua, Nicaragua

28–29 May 1996

In association with the Nicaraguan Red Cross, and under the auspices of the Minister of Foreign Affairs of Nicaragua, the ICRC organized a seminar on landmines in Managua on 28 and 29 May 1996. The meeting was attended by a total of seventy-five officials from the ministries of defence and foreign affairs of six Central American States and Mexico. The participants adopted a declaration supporting a complete ban on anti-personnel mines and containing proposals intended to make Central America the first mine-infested region to become an area freed from the devices. The declaration encouraged States in the region to promote immediate national bans on the production, possession, transfer and use of anti-personnel mines, and to encourage national legislation outlawing such acts and establishing rules guaranteeing employment opportunities for mine victims.

The Managua Declaration was instrumental in encouraging the Central American Foreign Ministers to call for the region to be made a mine-free zone and to commit their countries to prohibiting anti-personnel mines at the international level.

Recommendations adopted by the Managua Seminar

Managua, Nicaragua

29 May 1996

The States taking part in the Regional Seminar on anti-personnel landmines, mine-clearance and rehabilitation in Central America:

Considering the use of anti-personnel landmines to be a violation of international humanitarian law,

Expressing the desire of the Central American States and Mexico to establish a zone free from anti-personnel landmines on their territories,

Urge all States that have not yet done so to ratify or accede to the United Nations Convention on Prohibitions or Restrictions on the Use of Certain Conventional Weapons which May be Deemed to be Excessively Injurious or to Have Indiscriminate Effects,

Reaffirm their governments' determination to promote national policies in favour of a total and immediate ban on the production, possession, transfer and use of anti-personnel landmines,

Invite national parliaments to enact legislation prohibiting and punishing the production, possession, transfer and use of anti-personnel landmines, and to establish rules guaranteeing employment opportunities for the victims of these weapons,

Appeal to the international community to continue and to step up its support to the Central American States for their humanitarian mine-clearance programmes within the framework of the Organization of American States, in conformity with United Nations General Assembly Resolutions 50/70, 50/74 and 50/82 and pursuant to the Central American Parliament's Resolution AP/2–LX-96 of 30 April 1996,

Support the draft resolution on the establishment of a zone free of anti-personnel landmines in the Western hemisphere, a resolution which is to be submitted to the 26th round of sessions of the General Assembly of the Organization of American States taking place in Panama City from 3 to 7 June 1996, as a step towards achieving a total ban on the production, possession, transfer and use of these weapons,

Encourage governments and the media to disseminate the necessary information among populations at risk so as to protect them from falling victim to anti-personnel landmines,

Welcome the action taken to alert international public opinion to the problem of anti-personnel landmines,

Urge the international community, the International Committee of the Red Cross and the non-governmental humanitarian organizations to help support programmes in Central American States aimed at reducing the numbers of victims of these weapons and rehabilitating them so that they can resume normal and productive lives,

Welcome the initiatives taken to assist Central American States in pursuing their programmes for mine-clearance, preventive education of the civilian population and the physical and psychological rehabilitation of the victims,

Express their gratitude towards the intergovernmental and non-governmental organizations and towards the friendly States that contributed to the success of the regional seminar,

Thank the government and citizens of the Republic of Nicaragua, the International Committee of the Red Cross and the Nicaraguan Red Cross for their invaluable help in ensuring the success of the seminar.

Managua, 29 May 1996

ICRC News

Americas: States Determined to Achieve National Bans on Anti-personnel Mines

12 June 1996

Representatives of six Central American States and Mexico unanimously came out in favour of a total ban on the production, possession, transfer and use of anti-personnel landmines and proposed, at a seminar held in Managua on 28 and 29 May, the establishment of a "zone free from anti-personnel landmines on their territories". The meeting, devoted to mines, mine clearance and victim rehabilitation, was organized by the ICRC and the Nicaraguan Red Cross under the auspices of the Minister of Foreign Affairs of Nicaragua. It was attended by 75 officials, including military officers and foreign affairs officials from Costa Rica, El Salvador, Guatemala, Honduras, Mexico, Nicaragua and Panama, as well as representatives of United Nations agencies, the Organization of American States and NGOs involved in demining.

The seminar's recommendations, adopted by all the participating States, lay out a series of steps which could lead to Central America becoming the first mine-infested region to free itself entirely from this scourge. In those recommendations the States reaffirm their determination to promote immediate national bans on the production, possession, transfer and use of anti-personnel mines, to encourage national legislation outlawing such acts, and to assist and provide employment opportunities for disabled mine victims. The meeting further appealed to the international community to increase its support for humanitarian mine clearance and victim rehabilitation efforts in the region. It encouraged the local media to help in disseminating mine-awareness information among populations at risk so as to protect them from falling victim to these weapons.

The participating States also encouraged the adoption by the Organization of American States, which met in Panama City last week, of a resolution calling for a zone free of anti-personnel landmines in the Western hemisphere.

ICRC News

Americas: Organization of American States for a Continent Free of Anti-personnel Mines

12 June 1996

The Organization of American States (OAS), which met in Panama City from 3 to 7 June for its 26th General Assembly, discussed the question of anti-personnel mines as one of the problems of concern to the Western hemisphere, and adopted a resolution providing for the establishment of a continent-wide zone free of all landmines. Secretary-General Cesar Gaviria and numerous representatives of Member States referred to the issue in their statements. The ICRC, for its part, called for a total ban on the use of this indiscriminate weapon.

The resolution urges States to declare a moratorium on the production, use and transfer of all anti-personnel mines and to ratify the 1980 United Nations Convention on Certain Conventional Weapons and its amended Protocol II. It further provides for the opening of a register at the organization's General Secretariat, to record information on existing stocks, the current mine clearance situation and follow-up activities after each session of the General Assembly.

This text is in line with a resolution recently adopted by the Central American Parliament and with the recommendations on anti-personnel mines, mine clearance and victim rehabilitation endorsed by the Managua seminar at the end of May. The OAS General Assembly also adopted its third resolution on respect for international humanitarian law.

Resolution
The Council of Foreign Ministers of Central America

12 September 1996

Concerned by the fact that large areas of Central America are strewn with anti-personnel landmines which have affected and continue to affect the civilians living in or moving through those areas;

Recognizing that antipersonnel landmines are illegal under international humanitarian law, of which they violate the basic principles;

Considering that the task of mine clearance should be part of a policy decision directly involving the governments, civilian societies and communities affected in the region;

Recalling Central American Parliament Resolution No. AP/2–LX-96 on the clearance and deactivation of other explosive devices;

Taking into account the resolutions entitled “Support for mine clearing in Central America” and “The Western Hemisphere as an antipersonnel-landmine-free zone”, adopted by the General Assembly of the Organization of American States during its twenty-sixth regular session, held in Panama;

Bearing in mind the recommendations approved by the “Regional Seminar on Antipersonnel Landmines, Mine Clearance and Rehabilitation of Victims in Central America” organized by the International Committee of the Red Cross under the auspices of the Nicaraguan Ministry of Foreign Affairs; and

Considering the important work carried out by the Security Commission in this field;

HAS DECIDED

1. To declare the region an antipersonnel-landmine-free zone in which the manufacture, possession, acquisition, transfer and use of landmines is prohibited and subject to punishment;

2. To encourage all the region’s countries to start the constitutional proceedings required for their rapid ratification of or accession to the 1980 United Nations Convention on Prohibitions or Restrictions on the Use of Certain Conventional Weapons and its Protocols;

3. To urge other governments which have not yet done so to take the same steps, with a view to averting further casualties of these extremely harmful, cruel and indiscriminate weapons;

4. To renew its appeal to the international community for the latter’s continued resolute and courageous support in the work of clearing Central America of mines;

5. To give its full support to the campaign of the International Committee of the Red Cross, “Landmines Must Be Stopped”.

Signed in Guatemala City on 12 September 1996.

(signatures of the Foreign Ministers of Nicaragua, Honduras, El Salvador, Guatemala, Costa Rica and Panama)

This is an unofficial ICRC translation of the original Spanish.

PART 3

The Ottawa Process

From Regional Initiatives to an International Prohibition of Anti-personnel Mines

1

Introduction

At the closing of the First Review Conference of the 1980 Convention on Certain Conventional Weapons, the Canadian Ambassador announced his country's intention to convene a strategy meeting of States supporting a total prohibition of anti-personnel mines later in the year. The Canadian-sponsored strategy conference, 'Towards a Global Ban on Anti-Personnel Mines', took place in Ottawa in October 1996 with the active support of fifty governments, the ICRC, the International Campaign to Ban Landmines (ICBL) and the United Nations.

On 5 October 1996, the conference adopted the Ottawa Declaration, which committed the participants to working to ensure that a ban treaty was concluded at the earliest possible date and to carrying out a plan of action intended to increase resources for mine clearance and victim assistance. At the end of the conference, the Canadian government once again seized the initiative by inviting all governments to come to Ottawa in December 1997 to sign a treaty prohibiting the production, stockpiling, transfer and use of anti-personnel mines. The 'Ottawa process' had been officially launched.

International support for a ban on landmines continued to build. In December 1996, the United Nations General Assembly passed Resolution 51/45S, which called upon all countries to conclude a new international agreement totally prohibiting anti-personnel mines 'as soon as possible'. A total of 157 countries voted in favour of this resolution, none opposed it, and only 10 abstained from voting. To support the Ottawa process, the Austrian government prepared a draft text of the ban treaty and circulated it to interested governments and organizations. This draft, which was subsequently revised a number of times, was the basis of the ban treaty concluded in Oslo in September 1997.

International discussion on the draft text began in Vienna in February 1997 at a meeting hosted by the Austrian government. In addressing the meeting, the ICRC called for a comprehensive ban treaty based on an unambiguous definition of an anti-personnel mine. In April 1997, the German government hosted a special meeting to discuss possible verification measures to be included in a total ban treaty. Views were divided between those who stressed the central importance of establishing a humanitarian norm prohibiting anti-personnel mines and others who considered that a focus on effective verification mechanisms was essential to the success of the treaty.

The formal follow-up to the 1996 Ottawa Conference took place in Brussels from 24 to 27 June 1997. The Brussels International Conference for a Global Ban on Anti-Personnel Landmines was attended by representatives of 154 countries. It was the largest gathering of governments ever assembled for a conference devoted specifically to the issue of landmines. On the closing day, 97 governments signed the Brussels Declaration, launching formal negotiations on a comprehensive landmine ban treaty, greater international cooperation and assistance for mine clearance, and the destruction of all stockpiled and cleared anti-personnel mines. The Declaration called for the convening of a diplomatic conference in Oslo to negotiate such a treaty on the basis of the draft prepared by the Austrian government.

In accordance with the Brussels Declaration, which by then had been signed by 107 countries, formal treaty negotiations took place from 1 to 18 September 1997 at the Oslo Diplomatic Conference on an International Total Ban on Anti-Personnel Land Mines, hosted by the Norwegian government. A total of 91 countries took part in the negotiations as full participants while an additional 38 countries were present as observers, as were the ICRC, the ICBL and the United Nations.

The Oslo Diplomatic Conference proved to be a tremendous success. Propelled by its South African Chairman, Ambassador Jacob Selebi, on 18 September the Conference solemnly adopted the 'Convention on the Prohibition of the Use, Stockpiling, Production and Transfer of Anti-Personnel Mines and on their Destruction' – the 'Ottawa treaty'. The treaty was formally signed by 121 States in Ottawa on 3 and 4 December 1997. The treaty entered into force on 1 March 1999, six months after the fortieth State deposited an instrument of ratification with the UN Secretary-General.

2

The International Strategy Conference Towards a Global Ban on Anti-personnel Mines

Ottawa, Canada

3–5 October 1996

The Canadian-sponsored conference of 50 pro-ban States was the first time that the key actors within the international community – States, international organizations and non-governmental organizations – had come together formally to elaborate a strategy towards achieving a total global ban on anti-personnel mines. In his address to the conference, ICRC President Cornelio Sommaruga recalled the success of the Managua Regional Conference held in May 1996, which had led to a formal call within the Organization of American States for the Americas to become a zone free of anti-personnel mines. He therefore stressed the importance of promoting regional initiatives rather than initiating new consensus-seeking negotiations at the international level. President Sommaruga also reiterated that anti-personnel mines were inherently indiscriminate and questioned whether the injuries the weapons inflicted – by design – were not superfluous and excessive to the military need.

The Conference adopted a Declaration and Plan of Action and many practical commitments were made by participants. Canadian Foreign Minister Axworthy's dramatic invitation to States to return to Ottawa before the end of 1997 to sign a total ban treaty, however, inevitably attracted the greatest media attention. In the words of President Sommaruga, welcoming the Canadian Foreign Minister's initiative, 'The results of this conference signal the beginning of the end of the global epidemic of anti-personnel landmines.'

Statement of Cornelio Sommaruga

President, International Committee of the Red Cross, Geneva

4 October 1996

I would like to pay tribute at the outset to the Canadian government and in particular to Foreign Minister Lloyd Axworthy for undertaking this important initiative to bring the international community together, for the first time, in pursuit of the total prohibition and elimination of anti-personnel landmines. This conference is invested with the aspirations of many tens of thousands of potential civilian victims who simply wish to live their lives without fear that the land which feeds them will kill or maim them, that the rains and streams upon which they depend will carry the seeds of unspeakable suffering, that a step too far will be the last. This is not too much for a human being to hope for; this is why this conference can and must succeed.

Every single day our doctors and nurses have to look into the eyes of children writhing in pain from a limb turned into a bloody tangle of blood, dirt, plastic bits, bone fragments and flesh. Eyes which ask us, "Why, why, why?"; to which we have no coherent answer. Neither, so far, has the international community.

Given its mandate to care for and protect the victims of war the International Committee of the Red Cross (ICRC) would have been negligent if it did not act. In our field work we have made intensive efforts to develop effective surgical techniques for mine victims and to expand prosthetic and rehabilitative care. In 1995 alone the ICRC's 33 prosthetics programs fitted nearly 8 thousand amputees and manufactured some 11 thousand prostheses. Over the past decade we have treated over 30 thousand mine victims and cooperated with local and national medical personnel to assist many times that number. We are currently running mine-awareness programmes for civilian populations in six countries on four continents.

In addition to its specific operational mandate as an impartial humanitarian organization in situations of armed conflict, the ICRC is charged with the promotion and development of international humanitarian law. Based on our field experience we began consultations in 1992 with military commanders, diplomats, and legal and medical experts to develop a view of what could be done on the legal level. By early 1994 we were convinced that anti-personnel mines were too cheap, too small and too difficult to use according to the complex rules of the 1980 UN Convention. At that point we publicly stated our view that these mines are an indiscriminate weapon and that the only effective solution would be an absolute prohibition on their production, transfer and use. My first high-level political contact on this issue was here in Ottawa with the Honourable Prime Minister of Canada, Jean Chrétien, in May 1994.

Anti-personnel mines must not only be outlawed, but their use must also be stigmatized, so that whatever their understanding of the law combatants will choose not to use them because they are considered abhorrent to the societies in which they operate. Towards this end the ICRC, along with the entire Red Cross and Red Crescent Movement launched in 1995, for the first time in its history, an international media campaign seeking to stigmatize anti-personnel mines and call for their elimination. Global efforts to reach the public on this issue have been effective. A recent survey by the Gallup organization of public opinion in 21 countries, from both north and south, shows support for a total ban by 60 to 92 per cent of these populations, including – I am glad to say – 73 per cent of Canadians.

Since 1994 the ICRC has had the privilege, in keeping with its mandate as guardian of the Geneva Conventions, to participate and contribute background documentation to the preparatory process and meetings of the Review Conference of the 1980 UN Convention on Certain Conventional Weapons. The ICRC warmly welcomed a number of improvements in the landmines Protocol including its extension to apply in both international and non-international armed conflicts, clear assignment of responsibility for mine clearance, requirements that the location of all mines be recorded, new protections for ICRC and other humanitarian workers and a requirement that States enact penal sanctions to punish serious violations of its provisions.

Unfortunately, the new limitations on the use of anti-personnel mines, covering detectability and self-destruction of certain mines, are weak and overly complex. There is a danger that these provisions will not be implemented in the type of conflicts in which most recent use has occurred. Poorly trained or equipped forces may be unwilling or unable to abide by a complex set of rules or pay an increased price for self-destructing mines. It is indeed appalling that parties are not required to implement even these minimal restrictions on use until 9 years after entry-into-force of the revised Protocol, which means around 2007. By this time we expect that mines will have claimed well over 200,000 new victims – unless States do far more than is required by the law.

We are therefore greatly encouraged that more than forty States have come to Ottawa ready to do more; determined to go beyond what could be achieved by the lowest common denominator in a process of consensus and explore what must be done in the name of humanity, compassion and enlightened self-interest.

It is our belief that the Ottawa Plan of Action towards the elimination of anti-personnel mines can build upon four conclusions which many States have accepted, explicitly or implicitly, in supporting a total ban:

1. that States have a moral and humanitarian responsibility to protect their own populations and territories from the proven effects of anti-personnel mines;

2. that these weapons are inherently indiscriminate;

3. that, as agreed by a wide range of acting and former military commanders, the use of anti-personnel mines in accordance with law and doctrine is difficult, if not impossible, even for modern professional armies, and

4. that the limited military utility of anti-personnel mines is far outweighed by their human, economic and social costs.

In accepting these conclusions one is compelled to move beyond negotiation to independent action. The end of the landmines crisis cannot await a globally negotiated consensus. Indeed, few, if any, emergencies of this scale have been resolved by consensus. The ICRC is convinced that leadership by like-minded governments, non-governmental bodies and the concerned public is now indispensable in ending the landmines crisis. Let me indicate the kind of steps we have in mind:

1. National and Regional Initiatives – National governments and regional or sub-regional organizations can decide to eliminate anti-personnel mines from their own territories and thus contribute to a global solution. The twenty-five States which have renounced the use of anti-personnel mines and the eleven which are destroying their stockpiles have begun the process of changing State practice. When a critical mass of States have taken such steps a *de facto* ban will have been achieved; a legal ban may follow as State practice changes.

We welcome the resolution adopted in June by the Organization of American States which called for the establishment of an "Anti-personnel Mine Free Zone" in the Americas. A similar initiative of the Central American Parliament, in which national renunciation of AP mines is combined with increased assistance for mine clearance and victim assistance could make Central America the first mine-infested region to free itself from this scourge.

In February 1996 the Council of Ministers of the Organization of African Unity called on sub-regional organizations to launch initiatives for the prohibition of AP mines in support of the OAU's previous commitment to a total ban.

Although such action has not yet reached governmental level in Europe, the European Parliament, on 13 May, called on all Member States to unilaterally ban the production and use of AP mines and to destroy existing stocks.

2. At the global level the Ottawa Conference must clearly signal the beginning of the end of anti-personnel mines. It can only do so by committing States present

to a specific plan of concrete actions which they will take independently and encourage others to take. Renunciation of the use of anti-personnel mines by a specific early date and a permanent end to their production and transfer should be the hallmarks of the Ottawa Group and an example for others to follow. Indeed many of the States present have already undertaken such commitments. In taking on such political commitments States of the Ottawa Group will be in a stronger position to promote consideration of similar steps in resolutions of the UN General Assembly and regional fora.

The Ottawa Group can also commit themselves to specific forms of political cooperation and material assistance among themselves, for instance in the destruction of existing mines and mine clearance activities. We hope this Group will launch a process of regular meetings which will review progress in implementing the Ottawa Declaration and consider new means to promote a global ban.

3. Although it is essential to continue building support for a future global legal ban on anti-personnel mines it is our view that it would be premature to begin new global negotiations for a ban before regional and political efforts, such as that being launched here, have a chance to mature. Given that recent negotiations by consensus on legal restraints produced only modest results, the ICRC is concerned that new negotiations, particularly if conducted on the basis of consensus, would lead to further disillusionment with the negotiating process and could divert attention from national and regional decisions on how to achieve progress in particular geographic areas. In addition, there is a real danger that negotiations conducted exclusively in a disarmament context, as is now being considered, would quickly lose sight of the humanitarian purpose and humanitarian law basis of this exercise.

4. Progress in international humanitarian law is the result of an ongoing dialogue between military imperatives and humanitarian concerns. The ICRC sought to launch an in-depth dialogue on the military utility of anti-personnel mines through the publication, in March 1996, of a study by military commanders on the actual use and effectiveness of these weapons in 26 conflicts. The ICRC will seek in 1997 to broaden and deepen our dialogue on this issue with military officers and research institutes and would encourage efforts of others in the same direction.

5. Currently only a small proportion of mine victims have access to rehabilitation programs. Greatly expanded resources for emergency medical treatment and lifetime prosthetics care to victims are needed. National and international agencies must be encouraged to increase support for these essential efforts both through bilateral arrangements and through humanitarian agencies.

6. In 1995 pledges announced for mine clearance amounted to around $100 million of the estimated $33 billion required to clear all currently emplaced mines. A massive and long-term international effort is needed if future generations are to be spared paying the price for today's landmine legacy. Clearance efforts also need to be integrated into comprehensive national and regional efforts to ensure that new mines are not laid and that the needs of affected populations are addressed.

In closing I would like to return to the human face of the landmines crisis.

Much of the emphasis in efforts to ban anti-personnel mines has been on the fact, also stressed by the ICRC, that they injure combatants and civilians alike without discrimination. This focus on non-combatants is of great importance. However, it has stolen attention from another group of potential victims of war who are provided protection by international humanitarian law, namely, soldiers.

Article 35 of Protocol I additional to the Geneva Conventions of 1949 re-states a long-standing customary rule of humanitarian law: "It is prohibited to employ weapons . . . of a nature to cause superfluous injury or unnecessary suffering". This rule is intended to protect combatants. It is understood to prohibit the infliction, by design, of more injury than is needed to take a soldier out of combat.

If a person steps on a buried anti-personnel mine, his or her foot or leg is blown off. The force of the blast drives earth, grass, the vaporized mine case and portions of the victim's shoe and foot upwards into the tissues of the other leg, buttocks, genitals, arms and sometimes the eyes. With those mines which have a larger volume of explosive, including some fragmentation mines, death may be inevitable. If the wounded person gets to a hospital with the necessary facilities and expertise (both of which are rare in mine-affected countries) he or she will require several operations, will stay in hospital four weeks at least and will require a safe blood transfusion. Awaiting the survivor is permanent and severe disability with all the social, psychological and economic implications of being an amputee. Mines are designed to produce these effects.

Would not most people, including soldiers, describe the effects of mines just mentioned as superfluous and excessive to the military need?

The international response to the landmines crisis, the recent prohibition of blinding laser weapons, the well-established bans on chemical and biological arms, and indeed the whole history of humanitarian law are proof that humanity is not impotent in the face of its worst tendencies or the destructive uses of modern technology. Collectively the governments and organizations gathered here have the ability to ensure that anti-personnel landmines disappear from large parts of the world; that children in war-torn lands no longer have to fear the ground they tread upon. In the name of the victims we insist that Ottawa must mark a watershed in eradicating forever the plague of anti-personnel mines.

Press Release

4 October 1996

THE OTTAWA CONFERENCE: NEW HOPE FOR MINES VICTIMS

In a panel presentation to the International Strategy Conference entitled "Towards a Global Ban on Anti-personnel Mines" currently being held in Ottawa, Cornelio Sommaruga, President of the International Committee of the Red Cross (ICRC), today declared that only strong leadership on the part of like-minded nations willing to ban production and use of anti-personnel mines would help spare hundreds of thousands of potential mines victims. Mr Sommaruga said that it would be premature at this stage to begin new international negotiations for a legally binding treaty before national and political efforts for a ban have had a chance to mature.

President Sommaruga submitted to the Conference a proposal for a practical strategy to achieve a total ban on anti-personnel mines. It included the following steps:

1. States participating in the Ottawa Conference should make a total ban part of their national legislation; they should further encourage other States to follow their example.

2. National and regional initiatives such as that taken by the Organization of American States, calling for the establishment of landmine-free zones in the Americas, are a step in the right direction. Other regional organizations have called on their members to consider this possibility, which would set in motion a strong movement towards a total ban.

3. A massive and long-term international effort is needed in mine clearance if future generations are to be spared paying the price of today's landmine legacy. Clearance efforts should therefore be integrated into comprehensive national and regional measures to ensure that new mines are not laid and that the needs of affected populations are addressed.

4. A military study published by the ICRC in 1996 indicated that the actual use and effectiveness of anti-personnel mines is limited, particularly given the damage they do to individual victims, their community, environment and economy as a whole in the countries concerned. Dialogue on this issue should be

broadened and deepened with military officers and research institutes. Regional efforts in the same direction should be encouraged.

The ICRC is more than ever convinced that anti-personnel mines must be banned and completely eradicated, not only because States have a moral and humanitarian responsibility to protect their own populations and territories against the effects of these weapons, but also because anti-personnel mines are inherently indiscriminate and that their military utility is far outweighed by their human, economic and social costs.

And finally, ICRC President, Cornelio Sommaruga, stated that "the Ottawa Conference must succeed – both in adopting a politically binding declaration and a plan of action for the coming years. I expect the Ottawa Conference to be the beginning of the end. The beginning of an ongoing process, in which like-minded States will meet regularly at the global or regional levels until an international prohibition of these horrible weapons is achieved".

Declaration of the Ottawa Conference

5 October 1996

TOWARDS A GLOBAL BAN ON ANTI-PERSONNEL MINES

Following consultations with relevant international agencies, international organizations and non-governmental organizations, the States represented at the Ottawa Conference, the "Ottawa Group", have agreed to enhance cooperation and coordination of efforts on the basis of the following concerns and goals with respect to anti-personnel mines:

1. a recognition that the extreme humanitarian and socio-economic costs associated with the use of anti-personnel mines requires urgent action on the part of the international community to ban and eliminate this type of weapon.

2. a conviction that until such a ban is achieved, States must work to encourage universal adherence to the prohibitions or restrictions on anti-personnel mines as contained in the amended Protocol II of the Convention on Certain Conventional Weapons.

3. an affirmation of the need to convince mine-affected States to halt all new deployments of anti-personnel mines to ensure the effectiveness and efficiency of mine-clearance operations.

4. a recognition that the international community must provide significantly greater resources to mine-awareness programs, mine-clearance operations and victim assistance.

5. a commitment to work together to ensure:

– the earliest possible conclusion of a legally binding international agreement to ban anti-personnel mines;

– progressive reductions in new deployments of anti-personnel mines with the urgent objective of halting all new deployments of anti-personnel mines;

– support for an UN General Assembly resolution at its fifty-first session calling upon Member States, *inter alia*, to implement national moratoria, bans or other restrictions, particularly on the operational use and transfer of anti-personnel mines at the earliest possible date;

– regional and sub-regional activities in support of a global ban on anti-personnel mines; and,

– a follow-on conference hosted by Belgium in 1997 to review the progress of the international community in achieving a global ban on anti-personnel mines.

Global Plan of Action

7 October 1996

TOWARDS A GLOBAL BAN ON ANTI-PERSONNEL MINES
CHAIRMAN'S AGENDA FOR ACTION ON ANTI-PERSONNEL (AP) MINES

Participants in the Ottawa Conference have re-affirmed their commitment to seek the earliest possible conclusion of a legally-binding agreement to ban the production, stockpiling, transfer and use of anti-personnel (AP) mines. This agreement will be achieved most rapidly through increased cooperation within the international community.

The purpose of the Ottawa Conference was to catalyse practical efforts to move toward a ban and create partnerships between States, international organizations and agencies and non-governmental organizations essential to building the necessary political will to achieve a global ban on AP mines.

The following Agenda for Action captures the dynamism of the discussions in Ottawa, the recognition that movement toward a global ban has already begun and details concrete activities to be undertaken by the international community – on an immediate and urgent basis – to build upon the Ottawa Declaration and to move this process ahead in preparation for the follow-up meeting which will be hosted by Belgium in 1997.

This Agenda for Action reflects the interrelationship of the global ban, mine clearance and victim assistance agendas. It highlights the need to reach out beyond the already committed to engage the broader international community in the global ban effort. It also recognizes that action must be taken at the global, regional, sub-regional and national levels to achieve a rapid global ban on AP mines.

A. Global Action

Building the necessary political will for a new legally-binding international agreement banning AP mines will require more nations to adopt national bans or moratoria, on the production, stockpiling, use and transfer of AP mines. Nations which are not AP mine producers should also consider adopting bans on the imports of AP mines.

These actions will also have the effect of reducing the total number of new deployments of AP mines – deployments which would create new victims and increase the costs of mine clearance operations.

Global actions suggested by participants in this conference include:

1. The passage of a UN General Assembly resolution at its fifty-first session promoting an international agreement to ban AP mines.

Recognizing that a key vehicle for building international support for a global ban will be the development of overwhelming support for the resolution being proposed by the United States at the current session of the General Assembly, the following activities were identified as key opportunities to develop political support for the resolution:

- 'Potential co-sponsors' meeting – 10 October 1996, New York
- Inter-Parliamentary Union Meeting at the UN – 22 October
- Parliamentarians for Global Action – Annual General Meeting, October 1996, New York
- Landmine Panel, NGO, Committee on Disarmament, 24 October 1996, New York
- Work in regional or sub-regional groupings, as well as bilaterally, to build support for the resolution

2. Building public awareness and political will for a global AP mine ban.

Building increased public awareness of the social, economic and human costs of AP mines is essential to develop and sustain the necessary political will for a global AP mine ban. Opportunities for building political will and public awareness include:

• Launch of the Machel Study in response to Resolution A/RES/48/157 of the 48th session of the UN General Assembly on the Impact of Armed Conflict (and Land Mines) on Children, New York at the UN and by Archbishop Tutu in South Africa – 11 November 1996

• Adoption of the Machel Report by the UN General Assembly and implementation of its recommendations

• Reports on progress in the development of national AP mines policies in national reporting on the implementation of the Convention on the Rights of the Child to the Geneva-based Committee on the Rights of the Child

• Engaging military experts in the study of the military utility/humanitarian costs of AP mine use

• Adding the AP mine issue to the agenda of appropriate United Nations fora

3. Encourage rapid entry into force and universal adherence to the prohibitions and restrictions on AP mines as contained in the amended Protocol II of the Convention on Certain Conventional Weapons.

4. Increased exchanges of information and data on AP mines and national AP mine policies to build the confidence and transparency necessary for rapid progress towards a global AP mine ban, including:

• The development and publication of a global data-base on national AP mine policies (to be circulated by Canada in the fall of 1996)

• Studies by experts on the international production and legal and illicit trade of AP mines

5. To lay the necessary groundwork for a legally-binding international agreement to ban AP mines, Austria will produce a first draft and Canada will produce a possible framework for the verification of such an agreement.

6. Suggested follow-up conferences to the Ottawa Conference include:

• Belgium, June 1997

• Norway, Germany, Switzerland

B. Regional Action

Actions at the sub-regional and regional levels will be instrumental in catalysing the development of political will for a global ban on AP mines. To build upon the recent decision by the Central American Council of Ministers for Foreign Affairs to ban the production, use and trade in AP mines – thus creating the world's first regional AP mine-free zone – participants in the Conference suggested the following actions:

Increased funding for mine clearance and victim assistance for those regions and sub-regions which have taken concrete steps to create "AP mine-free zones".

Within Africa:

• Efforts to enhance the de-mining capacities of African countries with priority given to heavily mine-affected countries. This will include a Conference of African Experts in Demining and Assistance to Victims of Landmines (1997)

• Meetings to engage military/national security experts on AP mines issues at the sub-regional level – including an ICRC seminar in Southern Africa (1997)

• 4th ICBL Conference on Landmines: Toward a Mine-Free [Southern] Africa, 25–28 February 1997 Maputo, Mozambique

• Work towards the implementation of the three-part program of the Union Inter-africain des droits de l'homme

Within Asia:

• Meetings to engage military/national security experts on AP mines issues at the sub-regional level – including a planned ICRC/Philippines seminar (proposed for the first half of 1997)

• ICBL Conference, 1998

• Work toward consideration of AP mine issues within the ARF framework, including an ARF intersessional meeting on Demining for UN Peacekeepers, to be held in New Zealand in March/April 1997

Within the Americas:

• Defence Ministerial of the Americas, Bariloche, Argentina, 6–9 October 1996 – seek support for follow-up to the OAS resolution on "The Western Hemisphere as an Anti-personnel Land Mine-Free Zone"

• Special meeting at the end of October or early November 1996 of the Organization of American States' Committee on Hemispheric Security to promote implementation of OAS General Assembly Resolution "The Western Hemisphere as an Antipersonnel Land Mine-Free Zone" including:

– information exchanges on national AP mine policies

– provision of information to establish a hemispheric AP mine registry

• Regional ICBL Conference – Fall 1997

• Possible discussion in the Rio Group on AP mines under the topic of conventional arms control

• Meetings to engage military authorities on AP mines issues at the regional and sub-regional level

• Include anti-personnel landmines trade in discussions on illicit traffic in arms

• Encourage development of CBM regimes to replace AP mines in border areas.

Within Europe:

• Implementation by the European Union (EU) of the joint action on AP mines adopted by the EU on 1 October 1996, in which the EU clearly asserts its determination to pursue the total elimination of AP mines. To this end:

– the EU will pursue efforts to ensure full implementation of the results of the Review Conference of the 1980 Convention on the one hand, and support for international efforts to ban AP mines on the other hand;

– the EU is committed to the goal of the total elimination of AP mines and shall work actively towards the achievement at the earliest possible date of an effective international agreement to ban these weapons world-wide;

– the EU shall seek to raise without delay the issue of a total ban in the most appropriate international forum;

– the Member States of the EU shall implement a common moratorium on the export of all AP mines to all destinations and shall refrain from issuing new licences for the transfer of technology to enable the manufacture of AP mines in third countries;

– EU Member States shall endeavour to implement national restrictions or bans additional to those contained in Protocol II of the CCW Convention;

– the EU will reinforce its contribution to international mine clearance. A budget of 7 million ECU is to be provided for initiatives to be launched in the period up to the end of 1997, in the form of contributions to the UN Voluntary Trust Fund for assistance in mine clearance and/or specific EU actions providing assistance for mine clearance in response to the request of a regional organization or a third country's authorities. In addition, the Commission of the European Communities intends to continue the Community's support for activity in the field of mine clearance in the context of humanitarian aid, reconstruction and development co-operation.

• The EU will invite the Associate countries of Central and Eastern Europe, the Associate countries Cyprus and Malta and the EFTA countries members of the

European Economic Area to align themselves with initiatives taken in pursuit of the aims of its joint action

• Support will be sought within the OSCE for participating States to work towards a ban on all AP mines as soon as possible

• In addition, other European countries:

– have taken concrete steps in terms of destroying their stocks of AP mines or have made decisions to do so within a specific time-frame;

– are introducing national legal regulations prohibiting exports and imports of AP mines and their components;

– are strengthening their capacity to carry out demining activities;

– are making contributions to strengthen the ability of the UN to initiate and coordinate demining activities in other regions, and;

– in the field of developing demining technology, Norway has started a pilot mine clearance programme in the former Yugoslavia utilizing a new mechanical mine clearance machine.

C. Land Mine Clearance, Mine Awareness and Victim Assistance

Delegates highlighted the need to take special action to deal with the humanitarian crisis caused by AP mines, while recognizing that without a ban, mine clearance and victim assistance programs will always be insufficient to deal with the crisis.

In this regard, in addition to the announcement of many States of increased financial commitments to clearance, awareness and assistance efforts, the following specific initiatives and ideas were discussed to foster international technical co-operation and to make further progress to improve and share mine clearance technology, equipment and expertise; to improve mine awareness efforts and to enhance victim assistance programmes. These initiatives include:

• Meeting of Technical Experts on De-mining Technology, in preparation for the Tokyo meeting – Germany, early 1997

• Development of Canadian capacities in humanitarian demining and assistance to victims – Winnipeg, Canada – early 1997

• Demining and victim assistance – Tokyo, March 1997

• Cooperation on victim assistance (Canada–Mexico and Cuban, South African offer of their expertise)

• Increased international co-operation in AP mine stockpile destruction

• Efforts to develop standard procedures for mine-awareness education

• Include consideration of humanitarian mine clearance within peace accords

• Strengthening the efforts by Central America to achieve a land-mine free zone by the year 2000

• Establishment of a centre at James Madison University to act as a database to assist in co-ordinating international demining efforts

• Submission by the Presidency of the European Union of a UN General Assembly resolution at its fifty-first session on assistance with mine clearance

In addition to the above, a number of countries indicated that other events are being planned and that appropriate details will soon be forthcoming.

Press Release

6 October 1996

OTTAWA MINE CONFERENCE: THE BEGINNING OF THE END

"The results of this conference signal the beginning of the end of the global epidemic of anti-personnel landmines." That was how the President of the International Committee of the Red Cross, Cornelio Sommaruga described the outcome of the three-day international strategy conference "Towards a Global Ban on Anti-personnel Mines" which involved States from all regions of the globe and concluded on Saturday in Ottawa, Canada.

Mr Sommaruga said the ICRC and the entire International Movement of the Red Cross and Red Crescent warmly welcome the historic political commitments undertaken by fifty countries in the Ottawa Declaration and the impressive list of concrete engagements outlines in the plan of action.

The President of the ICRC said he was particularly happy to hear the invitation by Canada's Foreign Minister Lloyd Axworthy to Foreign Ministers to come to Canada in December of 1997 to sign a treaty banning anti-personnel mines. That treaty conference he said, "will situate the issue in the framework of international humanitarian law and place all of the efforts of the Ottawa action plan in their proper context. That context is the rapid achievement of a legally binding total ban on the production, use, export and stockpiling of anti-personnel mines."

Mr. Sommaruga said the invitation by Canada to other Foreign Ministers is a very strong signal that landmines will be banned. The specific date set by Canada, he noted, "highlights the importance and urgency of each of the elements of the action plan resulting from the three days of discussion in Ottawa this week."

That action plan includes promotion of a resolution being prepared for the United Nations General Assembly calling on all States to end the use of anti-personnel mines and to begin negotiations to outlaw them, regional and national initiatives to prohibit and eliminate landmines and ten follow up meetings to deal with various aspects of the problem. It also emphasizes the role of regional seminars being organized during the coming year by the ICRC. These seminars will involve political and national security experts from various countries in Asia and southern Africa to continue a dialogue on the military effectiveness of anti-personnel mines.

The ICRC was also encouraged that the Ottawa action plan integrates efforts for a ban with a commitment to increase resources for assistance to mine victims, mine awareness and mine-clearance operations.

"We leave this Ottawa Conference," said Mr. Sommaruga, "with the confidence that the unspeakable suffering of mine victims has finally touched the conscience of leaders of governments. We now have a firm message of hope for the victims of anti-personnel mines and for their children and grandchildren that the suffering from this plague will be eased."

3

United Nations General Assembly, 1996

(Fifty-First Session)

The United Nations General Assembly had previously called for the 'eventual elimination of anti-personnel mines' but had tied such elimination to the development of more humane alternatives. At its fifty-first session, however, the General Assembly endorsed a resolution urging States 'to pursue vigorously an effective, legally binding international agreement to ban the use, stockpiling, production and transfer of anti-personnel landmines with a view to completing the negotiation as soon as possible'. A total of 157 governments voted in favour of resolution 51/45S, none voted against and only 10 abstained. The United Nations Secretary-General later linked this landmark resolution to the growing momentum of the Ottawa process.

Statement of the International Committee of the Red Cross to the First Committee of the United Nations General Assembly

Fifty-First Session
18 October 1996

A great deal has occurred this year in relation to the regulation of both conventional weapons and weapons of mass destruction. Actually, there is no such dual categorization of arms in international humanitarian law, which regulates all weapons in accordance with certain generally applicable rules in order to prevent excessive suffering and destruction. All of the work and comments of the International Committee of the Red Cross with regard to weapons, whatever their nature from a strategic standpoint, are aimed at assuring the faithful and impartial application of these rules of international humanitarian law.

Landmines

On 3 May of this year, the Review Conference of the 1980 UN Convention on Certain Conventional Weapons (CCW) amended Protocol II, which regulates the use of landmines. The ICRC has had the privilege, in keeping with its statutory mandate, to participate actively in this process.

The ICRC warmly welcomes a number of improvements in the landmines Protocol, in particular its extension to apply in both international and non-international armed conflicts, clear assignment of responsibility for mine clearance, requirements that the location of all mines be recorded, new protection for ICRC and other humanitarian workers, annual meetings of States Parties and a requirement that States punish serious violations of its provisions.

Unfortunately, the new limitations on the use of anti-personnel mines are both weak and complex, and there is hence a danger that these provisions will not be implemented in the types of conflict in which most recent use has occurred. In particular, poorly trained or equipped forces may be unwilling or unable to abide by a complex set of rules or pay an increased price for self-destructing mines. Further, the implementation of new provisions on detectability and self-destruction can be delayed for up to nine years after entry into force of the revised Protocol, which means around 2007. By that time we expect that mines will have claimed well over 200,000 new victims – unless States do far more than is required by the Protocol.

The ICRC will promote adherence to the amended Protocol II of the CCW. This Protocol is intended to restrict the use of mines, but it is not meant to encourage States to use mines or to invest in new types of mines. We urge States to go far beyond the provisions of the Protocol and to renounce the production, transfer and use of anti-personnel mines.

In March of this year, the ICRC published a study commissioned by it on the use and effectiveness of anti-personnel landmines in past conflicts. This study was undertaken by high-level military officers and its conclusions are now endorsed by 52 senior commanders from 19 countries. The study found that the use of anti-personnel mines in accordance with law and doctrine is difficult, if not impossible, even for modern professional armies. This shows that the indiscriminate effects of landmines cannot be contained in most cases. Further it was found that the military utility of such mines is most often negligible or even counter-productive for the layer. The study therefore concludes that the limited military value of anti-personnel mines is far outweighed by their human, economic and social costs.

Many States have already demonstrated that the end of the landmines crisis need not await a globally negotiated consensus. The "Agenda for Action" prepared by the Ottawa Conference in early October 1996 demonstrates how much can be done towards ending the landmines crisis through moral and political leadership. We welcome the establishment of the "Ottawa Group", made up of some 50 States which committed themselves, in their final declaration, to promoting, and implementing initially on the national and regional levels, the global prohibition and elimination of anti-personnel mines. In this spirit, 25 States have already renounced or suspended the use of these weapons by their own armed forces and 11 are destroying their stockpiles. The Ottawa Group's dynamic agenda for the coming year stresses the urgency of both regional and global efforts, while highlighting the need to combine moves to achieve a ban with increased assistance for mine clearance and the care and rehabilitation of victims.

The ICRC strongly supports Canada's initiative in inviting Foreign Ministers to Ottawa in December 1997 to sign a new treaty totally prohibiting anti-personnel mines. We consider this step to be a major breakthrough and encourage States to respond favourably to the Canadian invitation. This initiative properly places the Ottawa Agenda for Action and other international initiatives in the context of urgent efforts to achieve a treaty banning these pernicious weapons. Even if such a treaty does not at first attract universal adherence, as has been the case with most new instruments, it will help create an important new norm.

Recent regional initiatives reflect the growing momentum towards a ban. A resolution of the Organization of American States, adopted last June, calls for the establishment of an "Anti-personnel Mine Free Zone" in the Americas. States of Central America, implementing an initiative by the Central American Parliament, have gone even further in pledging to prohibit the production, use and transfer of these arms. If such efforts are combined with generous assistance from the international community, Central America could become the first mine-infested region to free itself from this scourge. In February 1996 the Council of Ministers of the Organization of African Unity called on sub-regional organizations on the continent to move to prohibit anti-personnel mines, in keeping with the OAU's previous commitment to a total ban. In December 1995, Ministers of the Organization of the Islamic Conference also called for the "complete elimination" of this weapon.

In the context of this General Assembly the ICRC would advocate the adoption of the strongest possible resolutions, which:

1. unequivocally support a global ban on, and the elimination of, anti-personnel mines;

2. call on States to end the production, use and transfer of such arms by a certain date in the very near future;

3. encourage the establishment of regional zones free of these weapons, pending the adoption of a global ban; and

4. call for a significant increase in assistance for mine clearance and the care and rehabilitation of victims.

(. . .)

General Assembly resolution 51/45 S*

AN INTERNATIONAL AGREEMENT TO BAN ANTI-PERSONNEL LANDMINES

10 December 1996

The General Assembly,

Recalling with satisfaction its resolutions 48/75 K of 16 December 1993, 49/75 D of 15 December 1994 and 50/70 O of 12 December 1995, in which it, inter alia, urged States to implement moratoriums on the export of anti-personnel landmines,

Also recalling with satisfaction its resolutions 49/75 D and 50/70 O, in which it, inter alia, established as a goal of the international community the eventual elimination of anti-personnel landmines,

Noting that, according to the 1995 report of the Secretary-General entitled "Assistance in mine clearance" (A/50/408), it is estimated that there are one hundred and ten million landmines in the ground in more than sixty countries throughout the world,

Noting also that, according to the same report, the global landmine crisis continues to worsen as an estimated two million new landmines are laid each year, while only an estimated one hundred and fifty thousand were cleared in 1995,

Expressing deep concern that anti-personnel landmines kill or maim hundreds of people every week, mostly innocent and defenceless civilians and especially children, obstruct economic development and reconstruction, inhibit the repatriation of refugees and the return of internally displaced persons, and have other severe consequences for years after emplacement,

Gravely concerned about the suffering and casualties caused to non-combatants as a result of the proliferation, as well as the indiscriminate and irresponsible use, of anti-personnel landmines,

Recalling with satisfaction its resolutions 48/7 of 19 October 1993, 49/215 A of 23 December 1994 and 50/82 of 14 December 1995 calling for assistance in mine clearance,

Welcoming the recent decisions taken at the Review Conference of the States Parties to the Convention on Prohibitions or Restrictions on the Use of Certain Conventional Weapons Which May Be Deemed to Be Excessively Injurious or to Have Indiscriminate Effects, particularly with respect to the amended Protocol II 13/ to the Convention, and believing that the amended Protocol is an essential part of the global effort to address problems caused by the proliferation, as well as the indiscriminate and irresponsible use, of anti-personnel landmines,

Welcoming the adoption of the declaration entitled "Towards a Global Ban on Anti-Personnel Mines" by participants at the Ottawa International Strategy Conference on 5 October 1996 (A/C.1/51/10, annex), including its call for the earliest possible conclusion of a legally binding international agreement to ban anti-personnel landmines, and further welcoming the follow-on conference at Brussels in June 1997,

Welcoming also the recent decisions taken by States to adopt various bans, moratoriums or other restrictions on the use, stockpiling, production and transfer of anti-personnel landmines, and other measures taken unilaterally as well as multilaterally,

Recognizing the need to conclude an international agreement to ban all anti-personnel landmines as soon as possible,

1. Urges States to pursue vigorously an effective, legally binding international agreement to ban the use, stockpiling, production and transfer of anti-personnel landmines with a view to completing the negotiation as soon as possible;

2. Urges States that have not yet done so to accede to the Convention on Prohibitions or Restrictions on the Use of Certain Conventional Weapons Which May Be Deemed to Be Excessively Injurious or to Have Indiscriminate Effects (see The United Nations Disarmament Yearbook, vol. 5: 1980, United Nations publication, Sales No. E.81.IX.4, appendix VII) and Protocol II as amended on 3 May 1996, and urges all States immediately to comply to the fullest extent possible with the applicable rules of Protocol II as amended;

3. Welcomes the various bans, moratoriums or other restrictions already declared by States on anti-personnel landmines;

4. Calls upon States that have not yet done so to declare and implement such bans, moratoriums or other restrictions – particularly on operational use and transfer – at the earliest possible date;

5. Requests the Secretary-General to prepare a report on steps taken to complete an international agreement banning the use, stockpiling, production and transfer of anti-personnel landmines, and on other steps taken by Member States to

implement such bans, moratoriums or other restrictions and to submit it to the General Assembly at its fifty-second session under the item entitled "General and complete disarmament";

6. Requests Member States to provide the requested information for the report of the Secretary-General on steps taken to complete an international agreement banning the use, stockpiling, production and transfer of anti-personnel land-mines, and on other steps taken to implement bans, moratoriums or other restrictions on anti-personnel landmines and to submit such information to the Secretary-General by 15 April 1997.

NOTE

* Resolution reproduced from United Nations Website.

4

ICRC Position Paper No. 4

ICRC Briefing and Position Paper

December 1997

Landmines: Crucial Decisions in 1997

Decisions made in 1997 on how to build on the momentum for a ban on anti-personnel landmines will decide the fate of tens of thousands of innocent civilians in the coming years. Potential victims include children whose limbs may be torn apart by mines laid in 1997, families which will go hungry unless their land is cleared, and young men and women who have already lost limbs and whose hope for a brighter future is contingent upon access to rehabilitative care.

Whether, and for how long, the carnage caused by landmines continues will depend on the priority which governments, the media and the public give in 1997 to taking dramatic action to end this crisis in countries around the world.

In October 1996, 50 States met in Ottawa and pledged to work together, regionally and globally, for a total ban on anti-personnel mines and to increase by a significant amount the resources available for mine clearance and assistance to victims. This has provided a unique opportunity to adopt far-reaching measures in 1997, including a legal ban on these pernicious weapons.

At the UN General Assembly last December, 155 countries supported a call for a new treaty completely outlawing these weapons (resolution A/51/45 S). None voted against. The Canadian Minister of Foreign Affairs has invited States back to Ottawa in December 1997 to sign such a treaty. Several governments are planning to host conferences with a view to accelerating mine clearance, improving assistance for victims and negotiating a ban. Countries in Central America will begin implementing plans for the world's first regional zone free of anti-personnel mines.

In 1997, work to end the landmines crisis will shift away from negotiations aimed at reaching a consensus, which would tend to reflect the lowest common denominator, and focus instead on making rapid and clear progress under the leadership of concerned governments. The International Committee of the Red Cross (ICRC) believes that major breakthroughs are possible if vigorous efforts are pursued in the following four fields.

1. AWARENESS AND STIGMATIZATION

The appalling costs in terms of human suffering and the indiscriminate nature of landmines, particularly anti-personnel mines, need to be more widely known – especially in areas which have not yet been exposed to such information. Public abhorrence of these weapons must be such that combatants, whatever their knowledge of the law, will choose not to use them because mines, like poison gas, have become unacceptable to their own people.

The ICRC, in cooperation with National Red Cross and Red Crescent Societies, will continue its international public advertising campaign to increase awareness and stigmatization of these weapons. It will take part in new joint efforts with groups working in areas where information and concern about the landmines crisis are not yet widespread and it will encourage parallel efforts by journalists and the news media.

As a means of increasing awareness of the human costs, in February the ICRC will introduce on its World Wide Web site (http://www.icrc.org) a global database on landmine casualties based on information available from all sources. The database will include reported incidents by country, region, age, sex and status (civilian, soldier), where known. Drawing on its extensive field experience in medical and orthopaedic care for war casualties, the ICRC will publish, in the spring of 1997, a global study on the need for assistance to mine victims and what such assistance entails. It will also produce a short documentary film on the subject.

At the conclusion of a meeting held by the ICRC in Zagreb in October 1996 and attended by other members of the International Red Cross and Red Crescent Movement, National Societies reaffirmed their commitment to continue working for a comprehensive international ban on anti-personnel mines at national and regional levels. They agreed to work for heightened awareness of the problem in political and public fora and to expand their efforts so as to further promote the goals of the Ottawa Group. In pursuing these objectives, they pledged to encourage the adoption of national legislation, undertake bilateral and multilateral initiatives, build partnerships and increase their support for mine victims.

The ICRC urges National Societies, the news media and other organizations to ensure that crucial information about the abhorrent effects of landmines and the scale of the current crisis reaches into new circles and countries in 1997.

2. PROHIBITION

The ICRC believes that continued initiatives to prohibit and eliminate anti-personnel mines at the national, regional and global levels are essential. Efforts at each of these levels are complementary and mutually reinforcing.

a. National and regional initiatives

Ending the landmines crisis doesn't need to wait for a global solution. Mines in one country don't threaten people in distant parts of the globe. They most often threaten a country's own civilian population. States and regions can assume their humanitarian responsibilities to their own people by prohibiting the production, stockpiling, transfer and use of anti-personnel mines by their own armed forces. Countries affected by mines which unilaterally renounce the use of these weapons will strengthen their case for international assistance in clearance operations.

The ICRC will make available examples of national legislation banning anti-personnel mines and support initiatives by regional organizations to establish regional zones free of these weapons. Further dialogue on the military effectiveness of anti-personnel mines, which was called into question by more than 50 senior officers from 19 countries in a 1996 study commissioned by the ICRC, will be pursued through seminars for experts in southern Africa and Asia.

National Societies and other organizations are encouraged to pursue their efforts to end the use of anti-personnel mines, to eliminate them and to ban them through legislation in their own national and regional contexts.

b. A global ban in December 1997

With 50 States having committed themselves to a global ban as members of the Ottawa Group in October 1996 and 155 States having expressed their support for a treaty to that effect at the UN General Assembly the following month, prospects for the rapid prohibition of anti-personnel mines have never been better. But many major challenges remain and crucial decisions must still be made on the following issues:

i. The negotiating forum – The Canadian Minister of Foreign Affairs has invited States of the Ottawa Group, and others, to come to Ottawa in December 1997 to sign a treaty banning anti-personnel mines. The ICRC fully supports this initiative, which reasserts the right of States to enter into sovereign agreements that they consider to be in their own, and the international community's, best interests. If the treaty is adopted by all the members of the Ottawa Group, this would constitute broader acceptance than was the case only two years ago with the 1980 Convention on Certain Conventional Weapons (CCW), which currently governs mine use. A new international norm would thereby be established, which additional States could accept as they felt able.

It has been suggested that a new treaty should be negotiated in an arms control forum such as the Conference on Disarmament in Geneva, which can only make decisions by consensus and which includes States that feel they need to continue using anti-personnel mines. The ICRC believes that choosing such a forum would de-emphasize the humanitarian dimension of the problem and achieve no more than the modest results of the past three years of negotiations in the context of the CCW.

We urge National Societies and other organizations to encourage their governments to take an active part in the efforts of the Ottawa Group and to make an early public commitment to be among the original signatories of a treaty banning anti-personnel mines in December 1997.

ii. All mines designed to kill or injure people are anti-personnel mines – The ICRC strenuously opposed the inclusion in the recent revision of the CCW of language which defined an anti-personnel mine as one "primarily designed" to be exploded by the presence, proximity or contact of a person. The subjective judgement as to whether a mine is "primarily designed" to injure people or may instead be said to have another "primary" purpose introduces a dangerous ambiguity into the definition central to an effective ban on these weapons.

This issue is of paramount importance because of the increased use and rapid development of high-tech dual-purpose mines which can destroy, for example, either a vehicle or a person. Munitions other than mines which can be detonated in more than a way, including by a person, should be considered to be anti-personnel mines but may not be if an ambiguous definition is used. Although many States recognize the dangers of a vague definition, time pressure prevented the problem from being resolved in the CCW context. A future treaty containing such a definition could, in the long run, become ineffective on the technical level and unacceptable on the political level.

The ICRC encourages efforts to ensure that both a new ban treaty and national legislation contain a clear and unambiguous definition of anti-personnel mines which, contrary to the current definition in the CCW, does not include the word "primarily". Failure to do so would entail the risk of continued civilian suffering caused by weapons with new names but the same effects.

iii. A comprehensive ban – To be effective in ending the current crisis, a new treaty must comprehensively prohibit the **production, stockpiling, transfer and use** of anti-personnel mines and require their **destruction**. All of these activities are linked. Some have suggested an *à la carte* approach, in which States could agree to prohibit only one or two of the above activities but continue others (e.g., prohibit transfer or production but not use). The ICRC is concerned that such an approach risks being ineffective. Since most producing States have already enacted moratoria on the transfer of anti-personnel mines, making this provision optional would represent a step backwards. Likewise, continued production, stockpiling or transfer indicates the intention to use these weapons. Prolonged use will only aggravate the current crisis and add to the estimated 1,100 years and 33 billion US dollars it will cost to clear the 120 million mines already in the ground.

An effective treaty should immediately prohibit the use, production, stockpiling and transfer of anti-personnel landmines and provide a timetable for the destruction of existing stocks. The early entry into force of a comprehensive prohibition is the only effective global means of ending the landmines crisis and preventing further needless suffering and the waste of scarce resources for mine clearance.

iv. New and amended CCW protocols – Since not all States are prepared to renounce the use of anti-personnel mines in the near future, **the ICRC urges a maximum number to accept the CCW's amended Protocol II on landmines** so that it will enter into force as rapidly as possible. This would raise the minimum norms applicable when mines are used, by extending the Protocol to non-international armed conflicts, immediately prohibiting the transfer of non-detectable anti-personnel mines, increasing protection for humanitarian workers and other measures. Moreover, a new ban treaty may not govern the use of anti-vehicle mines, booby traps or other devices which are covered by the CCW.

The ICRC has also proposed that States, when accepting the new Protocol IV on blinding laser weapons, declare their intention to apply it "in all circumstances". Indeed, an understanding that the Protocol would apply beyond international conflicts was reached during the negotiation phase and reaffirmed by 135 States in a resolution of the 1995 International Conference of the Red Cross and Red Crescent but this is not reflected in the treaty text.

3. INCREASED ASSISTANCE TO MINE VICTIMS

The plight of victims, whose suffering has triggered the outpouring of international concern with landmines, is all too often overlooked in the attempt to find legal solutions and new ways of clearing mines. The work of the ICRC's 23 hospitals and rehabilitation centres in 10 countries, and of those run by governments and other humanitarian organizations, still only benefits a fraction of mine victims. Yet landmines continue to take a heavy toll, and only intensive and long-term care for individual victims can mitigate their horrible effects and reduce the suffering they inflict.

The ICRC urges a massive increase in resources for the care and rehabilitation of mine victims and is prepared to share its experience in this field within and beyond Red Cross and Red Crescent circles so as to support expanded assistance to these victims. As importantly, programmes to prevent casualties through "mine-awareness" education for affected communities need to be expanded and improved.

4. INCREASED RESOURCES FOR MINE CLEARANCE

The comprehensive approach to the landmines crisis to which the 50 States of the Ottawa Group have committed themselves includes legal prohibitions and significantly increased resources for both assistance to victims and mine clearance. This international commitment to increase resources must now be translated into national funding decisions.

In 1996 the UN Secretary-General increased his estimate of the resources needed to clear all existing mines, assuming no more are laid, from 33 billion to over 50 billion US dollars. Yet in 1996, funding for clearance is estimated to have been less than 150 million US dollars and some governments are finding it difficult to maintain their current level of commitment.

The ICRC calls on National Societies and other concerned bodies to help ensure that national funding commitments to mine-clearance activities, both through bilateral projects and through the UN Voluntary Trust Fund for mine clearance, are not only maintained but significantly increased in 1997.

* * *

In 1996 alone, 30 additional States called for the total prohibition of anti-pesornel mines, bringing to 53 the number of States supporting a ban. The same year, the UN General Assembly supported new negotiations for a global ban and 50 States formed the Ottawa Group to ensure that this would be achieved. To date, 27 governments have adopted policies prohibiting the use of these weapons and 14 have begun destroying their stockpiles – a very encouraging development.

The year 1997 will be crucial in ensuring that the momentum gathered by efforts to ban mines leads to a new legally binding norm, that mines are not only banned but existing ones eliminated (both from the ground and from stockpiles) and that victims receive the care they need. Increased efforts in this direction by the entire International Red Cross and Red Crescent Movement, and by many other organizations, will play a central role in determining whether these objectives are achieved.

5

Expert Meeting on the Convention for the Prohibition of Anti-personnel Mines

Vienna, Austria
12–14 February 1997

The Austrian government had prepared a draft treaty text at the end of 1996, which ultimately formed the basis for the anti-personnel mine ban Convention finally adopted at the Oslo Diplomatic Conference in September 1997. The Expert Meeting on the Convention for the Prohibition of Anti-personnel Mines was, however, convened to enable States to discuss the content of such a Convention in general terms, rather than to propose or to negotiate specific language. In addressing the meeting, the ICRC outlined what it believed to be the key elements in the treaty:

- a clear and unambiguous definition of an anti-personnel mine;
- a comprehensive prohibition on all anti-personnel mines, not the step-by-step approach favoured by certain governments;
- a verification mechanism, but not at the expense of a clear norm prohibiting anti-personnel mines;
- the importance of the ultimate – rather than the immediate – universality of the norm.

In the subsequently revised draft of the Convention, the ICRC was pleased to note that the definition of an anti-personnel mine no longer included the word 'primarily', even though anti-vehicle mines equipped with anti-handling devices were specifically excluded from the scope of the treaty.

Press Release

10 February 1997

FIRST MEETING TO DISCUSS A TREATY BANNING ANTI-PERSONNEL LANDMINES OPENS IN VIENNA ON 12 FEBRUARY 1997

Representatives of governments, the International Committee of the Red Cross (ICRC) and the United Nations will meet in Vienna next week to begin consultations on the text of an international agreement to ban the production, transfer, stockpiling and use of anti-personnel landmines. The meeting, to be hosted by Austria from 12–14 February, will consider a draft treaty prepared by the country's government. Discussions are expected to cover a number of issues, including the precise definition of an anti-personnel mine, the importance of an immediate and total prohibition on mines as against a phased approach and the need to establish a monitoring mechanism to ensure compliance with the treaty. The ICRC, which strongly supports a total ban on anti-personnel mines, will attend the meeting as an expert observer. The drafting process is intended to lead to the formal signature of a treaty in Ottawa in December 1997.

This initiative was taken in response to the disappointing results of the Review Conference of the 1980 United Nations Weapons Convention, which ended in Geneva in May 1996. Owing to the lack of a consensus, the Conference, which was convened specifically to address the problem of anti-personnel mines, was able to agree only on limited restrictions on the use of these weapons instead of an outright ban. The meeting next week in Vienna should build on the momentum created by the adoption in 1996 by the UN General Assembly of a resolution (51/45S) calling on States to complete "as soon as possible" negotiations for an effective, legally binding international agreement to ban the use, stockpiling, production and transfer of anti-personnel mines. This landmark resolution was supported by 155 States, with none opposing it.

The forthcoming meeting also reflects the efforts of the "Ottawa Group", which was set up in October 1996 and is composed of 50 pro-ban States. At the time, the Canadian Minister of Foreign Affairs invited all governments to return to Ottawa in December 1997 to sign an international treaty banning anti-personnel mines. To date, 53 States have publicly expressed their support for a global ban. Of these, 28 have already put an end to the use of anti-personnel mines by their own armed forces and 15 are destroying their stocks.

Further meetings to discuss the text of a new treaty are to be held in Vienna, Brussels and Oslo later this year. In parallel, the Geneva-based UN Conference on Disarmament is discussing proposals to put the negotiation of a ban on anti-personnel mines on its agenda. However, the consensus required to do so has not yet been achieved.

The ICRC will actively support the negotiation of an effective international agreement to outlaw the use of anti-personnel landmines, which each year kill or maim an estimated 24,000 people, mostly civilians in rural developing communities.

Statement of the International Committee of the Red Cross

12 February 1997

Sixteen months ago tomorrow the Review Conference of the 1980 Convention on Certain Conventional Weapons (CCW) ended its first session in Vienna, after nearly two years of preparation. At that time States were unable to agree upon provisions to mitigate the indiscriminate effects of anti-personnel mines. By the time agreement on a set of modest measures was reached last May nearly half of the States Parties to the CCW had concluded that only a total ban on anti-personnel mines would effectively end the immense human suffering they have caused. Two months ago the United Nations General Assembly, in resolution A51/45S, overwhelmingly reached a similar conclusion, namely, that a new treaty prohibiting the production, stockpiling, transfer and use of anti-personnel mines must be urgently negotiated and concluded "as soon as possible".

These sixteen months have been a period of intense diplomatic efforts and encouraging developments on the national and regional levels. Twenty-seven States have ended the use of anti-personnel mines, the Ottawa Group was established and Central America is on the way to becoming the first region to free itself of the scourge of these weapons. But for those living in mined areas sixteen months is a very long time. According to information available to the ICRC the level of casualties has not decreased; indeed indications are that our previous estimates were too low. At least 13,000 people are thought to have been killed by mines during these months; more than 20,000 have suffered unspeakable injuries. Only a fraction of these will ever receive adequate care.

The ICRC commends the government of Austria for hosting these consultations and for the new Austrian law totally prohibiting anti-personnel mines which took effect at the beginning of this year. The discussions you have convened in Vienna today are a sign of hope, the fruit of a new political environment and a reflection of what can be accomplished through political leadership. For the first time, governments from every region of the world have come together to discuss the contents of a treaty totally prohibiting anti-personnel mines. We are grateful to be associated with this process.

We welcome the Austrian draft convention on the prohibition of anti-personnel mines which will help focus discussions on the specific treaty arrangements which will be required. We look forward to discussing this text in detail in the coming days. However at the outset we would like to address some of what the ICRC considers to be key issues related to a future treaty.

1. DEFINITION OF AN "ANTI-PERSONNEL MINE"

In the course of negotiations in the CCW context a dangerous ambiguity was introduced into the definition of an anti-personnel mine. By defining this weapon as a mine which is "primarily designed" to be exploded by the presence, proximity or contact of a person the entire regime came to rely upon a subjective interpretation as to what the "primary design" of the munition in question is. The ICRC warned of this ambiguity as it was introduced in Vienna in 1995 and as negotiations were pursued in 1996. By the end of the CCW review process it was difficult to determine the origin or the precise purpose of this phrase and yet some considered it too late to reconsider definitions in the draft of an amended Protocol II.

As stated by President Sommaruga to the closing session of the CCW Review Conference, the ICRC considers the definition of an anti-personnel mine to be of fundamental importance. We would consider the inclusion of an ambiguous definition in a treaty attempting to outlaw anti-personnel mines to constitute a major weakness. The uncertainties engendered by such ambiguity could threaten the long-term viability of a future treaty.

The ICRC considers that any munition which is "designed" to be exploded by the presence, proximity or contact of a person must be considered an anti-personnel mine, regardless of what its "primary design" is said to be. This type of unambiguous definition is consistent with national legislation now in force in Austria, Belgium and the United States. We would be glad to make the relevant legislation available to delegations.

In order to spell out the implications of an ambiguous definition the ICRC will be distributing a conference room paper with an indicative list of mine technologies which could be said to have another "primary" purpose, even though they are in fact designed to explode upon contact with a person and are designed to kill or injure persons. These include (i) mines designed with both anti-personnel and anti-vehicle or anti-helicopter capabilities, (ii) anti-handling mechanisms on anti-tank mines and (iii) "runway denial" submunitions. We have noted several examples where mines once advertised as anti-personnel mines have recently been re-named "anti-vehicle" or "anti-armour" mines even though the technology is exactly the same.

Of particular concern to us are remotely delivered anti-tank mines equipped with anti-handling devices. These can be delivered on a massive scale, by the tens of thousands per hour, and over huge areas. Unlike anti-handling mechanisms attached to buried mines, which are unlikely to be disturbed by a footstep, these mines will explode and kill the person who treads on it. This technology is becoming smaller and less costly each and every year. Use of these weapons has the potential to recreate many of the humanitarian problems associated with traditional anti-personnel mines, with the exception that all victims will probably be killed.

Today we can identify some of the technologies which could undermine a regime based on an imprecise definition. Use of an ambiguous definition will only invite the design of new technologies to bypass a prohibition on anti-personnel mines and lead to a proliferation of such technologies. Such a result could quickly render ineffective a new treaty intended to end the suffering caused by these weapons.

2. A COMPREHENSIVE REGIME

To be effective in ending the current crisis, a new treaty must comprehensively prohibit the **production, stockpiling, transfer and use** of anti-personnel mines and require their **destruction**. All of these activities are linked. Some have suggested an initial agreement by which only one of the above activities would be prohibited but others would continue (e.g., prohibit transfer or production but not use). However civilians die each day not from the production, transfer or stockpiling of anti-personnel mines but from their use. It might prove difficult to describe the underlying principle of a treaty which would prohibit the production or transfer of a weapon the use of which is nonetheless permitted.

The ICRC recognizes that practical constraints may require the phased implementation of particular aspects of a comprehensive agreement, such as the destruction of stockpiles and the clearance of existing minefields. But we are firmly convinced that the first step in ending this crisis is the prohibition of use. Other approaches defy historical experience of how weapons prohibitions in international humanitarian law and in arms control have developed.

Efforts to prohibit the production, stockpiling and transfer of chemical and biological weapons, in the Conference on Disarmament and its predecessor bodies, were successful because they were underpinned for most of this century by well-established prohibitions on use contained in the 1925 Geneva Protocol. A political norm against use and public abhorrence have been important factors in the negotiation of nuclear arms control agreements, the end to nuclear testing and the 1977 prohibition on the use of environmental modification techniques as a method of warfare.

If there is to be a phased approach it would logically begin with an immediate prohibition on new deployments, production and transfers. A second phase, which should be as short as practical constraints permit, could provide for the phased destruction of existing stockpiles and the clearance and destruction of mines already deployed.

Prolonging the use of anti-personnel mines will only aggravate the current crisis, multiply the human suffering and add to the estimated 1,100 years and 33 billion US dollars it will cost to clear the 120 million mines already in the ground. It would also be inconsistent with UN General Assembly resolution A/51/45S which calls for a comprehensive agreement to prohibit production, transfer, stockpiling "and" (rather than "or") the use of anti-personnel mines.

3. COMPLIANCE MONITORING

Compliance monitoring will be an important element of a regime to end the use of anti-personnel mines. The best method would be for an independent mechanism to investigate credible reports of the use of this weapon following the entry into force of a new treaty.

It should be noted, however, that those most directly affected in most of the recent conflicts in which AP mines have been used are civilians. Verification of mine use in a given area has often occurred when civilians are injured or in the course of the routine work of humanitarian agencies. Therefore it is unlikely that any large-scale use of AP mines would go unnoticed or unreported.

While the ICRC supports the maximum possible verification of a future treaty we encourage States not to permit verification to stand in the way of an absolute prohibition on the use, production, stockpiling and transfer of this weapon. Indeed all of the humanitarian law norms prohibiting the use of specific weapons have been established without verification and yet they have been almost universally respected. Verification, together with prohibitions on production, stockpiling and transfer can reinforce such a norm but should not be allowed to override the importance of the norm itself.

4. UNIVERSALITY

Recently there has been great deal of discussion on the extent to which a future treaty must be universally adhered to at the outset. Universality is an important objective and the ICRC, in keeping with its mandate under international humanitarian law, has devoted a great deal of resources to its promotion in relation to existing agreements, including most recently the CCW and its new and amended Protocols.

None of the major instruments of humanitarian law has attracted universal adherence from the outset. Indeed in the case of Dum-Dum bullets two of the major powers of the day voted against a prohibition. And yet the vast majority of States have respected the norms created by these agreements.

The essential ingredients in achieving near universal adherence have been public abhorrence of the use of a weapon and the exercise of consistent political will to ensure that norms are adhered to and respected. These will be required regardless of the weapon and regardless of the forum in which an agreement is negotiated.

The ICRC is convinced that the "public conscience" of people throughout the world is revolted both by the indiscriminate nature of anti-personnel mines and by the horrific suffering they have caused. The question now before us is whether there is sufficient political will to establish an absolute prohibition on these weapons and to ensure respect for such a norm.

We look forward to working vigorously throughout this year with States committed to the achievement of a total ban on anti-personnel mines and to the signing of an historic treaty in December. The ICRC will do its utmost to ensure that the largest possible number of States adhere to this new international norm, both in December 1997 and in the years to come.

Sixteen months has been a long time for those living in mined areas. Ten more is long enough. But there is nothing inevitable about the loss of two thousand limbs and lives each and every month. Crucial decisions which your governments will make this year can stop this carnage.

Informal Information Paper on Dual-Use Munitions

12 February 1997

1. Mines which may be said to be "primarily" designed for use either against vehicles or persons

- The DNG Giant Shotgun is said to have a considerable potential as an "anti-personnel or anti-vehicle mine", according to Jane's Military Vehicles and Logistics (1994–1995, p. 170).

- The Valsella VS-DAFM 1 Directional Mine is said to be a fixed directional fragmentation anti-personnel mine in Jane's Military Vehicles and Logistics (1994–1995, p. 207) and a fixed directional anti-vehicle mine in Trends in Land Mine Warfare (Jane's Special Report, August 1995, p. 152). The mine can be coupled to two independent firing systems through two priming wells. The actuating load is said to be between 3–6 kg.

- The Valsella VS-DAFM 6 and VS-DAFM 7 Directional Anti-material Mines are said to be fixed directional fragmentation anti-material (anti-personnel and anti-armour) mines used primarily for the defence of tactically important areas such as airfields and landing zones in Jane's Military Vehicles and Logistics

(1994–1995, p. 207). In Jane's Trends in Land Mine Warfare (Jane's Special Report, August 1995, p. 153) they are fixed directional fragmentation anti-vehicle and anti-armour mines! The actuating load is also said to be between 3–6 kg.

• The Ambush Mine described in Trends in Land Mine Warfare (Jane's Special Report, August 1995, p. 154) and in Jane's Military Vehicles and Logistics (1994–1995, p. 221) is said to be an anti-vehicle or anti-personnel mine and to also have a potential role against low-flying helicopters. This mine is ready to be fired either by remote control or by a tripwire or similar static system.

2. Anti-handling devices on small scatterable A/T mines which equip them with A/P functions

• According to Trends in Land Mine Warfare (August 1995, p. 95) and to Jane's Military Vehicles and Logistics (1994–1995, p. 182) the PGMDM Scatterable Anti-tank Mine is operated by a single pressure or an accumulation of slight pressures, for example handling, and is electrically activated. The weight of the liquid explosive is between 1.4 kg and 2 kg.

• In Trends in Land Mine Warfare (August 1995, p. 108) and Jane's Military Vehicles and Logistics (1994–1995, p. 208) the Valsella VS-1.6 Anti-tank Scatter Mine and the BPD SB-81 Scatterable Anti-tank Mine are said to be available including an electronic model with anti-lift device. The weight of the main charge is between 1.85 kg and 2.0 kg.

3. Runway Denial Submunitions (HB876)

In conjunction with bomblets, which make holes in the runways, submunitions are used to impede the work of repair vehicles and personnel. These submunitions have a shape-charge and are fitted with an anti-tilt fuze. The weight of the main charge is between 1.5–2.0 kg and the casing is pre-fragmented.

6

Fourth International Non-governmental Organization Conference on Landmines: Toward a Mine-free Southern Africa

Maputo, Mozambique
25–28 February 1997
Hosted by the International Campaign to Ban Landmines (ICBL) and the Mozambique Campaign to Ban Landmines

Within the framework of an ICBL regional conference for Africa, the ICRC brought together twelve National Societies from southern Africa to discuss their role in moving the region forward towards a total ban on anti-personnel mines. The National Societies issued a strong declaration calling upon their governments – representing some of the world's most mine-affected countries – to support the Ottawa process, to prohibit anti-personnel mines at national level, and to work together to establish an anti-personnel mine-free zone in southern Africa.

Declaration of National Red Cross Societies of Southern Africa

Maputo, 28 February 1997

Immense long-term human suffering has been caused by the use of tens of millions of anti-personnel landmines in the countries of southern Africa over the past two decades. Mines have already killed or maimed tens of thousands of men, women and children in the region and inflicted deep and prolonged wounds on societies and economies. Even though peace and stability are returning to the region, millions of uncleared landmines remain a major obstacle to post-conflict development and reconstruction and will continue to claim innocent lives and limbs for years to come.

The National Red Cross Societies of southern Africa, together with the entire International Movement of the Red Cross and Red Crescent, have direct experience in caring for the innocent casualties of anti-personnel mines: our doctors, surgeons and volunteers have treated many of the victims and Red Cross physical rehabilitation centres have fitted artificial limbs to thousands of mine amputees. We shall continue to implement programmes of mine awareness to try to reduce the number of landmine incidents. But without a dedicated and comprehensive effort to tackle the root of the problem our best efforts can only mitigate the horrible effects of this weapon.

We are firmly convinced that the appalling human, social and economic impact of anti-personnel landmines far outweigh their limited military utility. We therefore call upon all countries of southern Africa:

(1) to participate actively in the Ottawa Group of pro-ban States;
(2) to support the early conclusion of an international treaty prohibiting the production, stockpiling, transfer and use of anti-personnel mines;
(3) to initiate the establishment of a zone totally free of anti-personnel landmines in southern Africa; and
(4) to adopt and implement national legal prohibitions of anti-personnel mines.

We would like to stress the urgent need for countries of the region to work together and in co-operation with the international community to strengthen programmes of humanitarian mine clearance, mine awareness and assistance to the victims of landmines. We support the initiatives of governments of the region to work together actively to prevent conflicts and to bring existing conflicts to an end through dialogue and other peaceful means.

Red Cross Society of Angola, Baphalali Swaziland Red Cross Society, Botswana Red Cross Society, Lesotho Red Cross Society, Madagascar Red Cross Society, Malawi Red Cross Society, Mozambique Red Cross Society, Namibia Red Cross Society, South African Red Cross Society, Tanzania Red Cross Society, Zambia Red Cross Society, Zimbabwe National Red Cross Society.

Press Release

Maputo, 27 February, 1997

SOUTHERN AFRICAN RED CROSS NATIONAL SOCIETIES CALL FOR REGIONAL ZONE FREE OF ANTI-PERSONNEL LANDMINES

Nine Red Cross Societies from the countries of southern Africa met in Maputo this week at the invitation of the ICRC, to devise a regional strategy in the context of the global campaign of the International Movement of the Red Cross and Red Crescent in support of an immediate ban on the use, production, stockpiling and transfer of anti-personnel landmines.

The National Red Cross Societies of the region recognize the immense long-term human suffering caused by the use of tens of millions of anti-personnel landmines in the countries of southern Africa over the past two decades. Even though peace and stability are returning to the region, millions of uncleared landmines remain a major obstacle to post-conflict development and reconstruction and will continue to claim innocent lives for years to come.

In a Declaration resulting from the meeting, twelve National Red Cross Societies urge all countries of southern Africa: to participate actively in the Ottawa Group of pro-ban States; to support the early conclusion of an international treaty prohibiting the production, stockpiling, transfer and use of anti-personnel mines; to initiate the establishment of a zone totally free of anti-personnel landmines in southern Africa and to adopt and implement national legal prohibitions of anti-personnel landmines.

The ICRC hopes that this Declaration will lead to the creation of a zone in southern Africa in which all landmines have been cleared and existing stockpiles destroyed, and will promote similar initiatives throughout the continent and globally. The creation of such a zone in southern Africa will require the active support in mine clearance and the destruction of mine stocks from the entire international community.

The ICRC and the National Red Cross Societies are participating in the Fourth International NGO Landmine Conference currently taking place in Maputo from 25 to 28 February.

(Red Cross National Societies signing Declaration: Angola, Botswana, Lesotho, Madagascar, Malawi, Mozambique, Namibia, Tanzania, South Africa, Swaziland, Zambia, Zimbabwe)

7

Seminar on Anti-Personnel Landmines and Their Humanitarian Implications

Budapest, Hungary

7 March 1997

Organized by the Council of Europe Parliamentary Assembly Migration, Refugees and Demography Committee in collaboration with the ICRC

The break-up of the former Soviet Union and the subsequent bloody disintegration of the former Yugoslavia led to widespread – and frequently indiscriminate – use of landmines in Europe for the first time since the end of the Second World War. Concerned by the impact of landmines on the displaced, the Migration, Refugees and Demography Committee of the Council of Europe's Parliamentary Assembly convened a meeting with the ICRC to discuss the problem. The result of the meeting was a clear recommendation for the total prohibition of anti-personnel mines. A report containing that recommendation was forwarded to the Parliamentary Assembly for consideration.

ICRC News

20 March 1997

PARLIAMENTARY ASSEMBLY OF THE COUNCIL OF EUROPE: APPEAL FOR A TOTAL BAN ON ANTI-PERSONNEL MINES

On 7 March the ICRC and the Council of Europe's Committee on Migration, Refugees and Demography held a joint seminar in Budapest on anti-personnel landmines, an issue on which the Parliamentary Assembly and the ICRC have been cooperating closely. The seminar addressed an 11-point appeal to Member States of the Council of Europe, urging them to take an active part in the effort to achieve a comprehensive ban on anti-personnel mines and to sign an international treaty to that effect at the Ottawa Conference scheduled for December 1997. The Committee also called for increased contributions to rehabilitation programmes for mine victims, mine-clearance operations and mine-awareness campaigns for populations at risk.

The States represented at the seminar included Austria, Belgium, Croatia, Cyprus, the Czech Republic, Estonia, Finland, France, Norway, Germany, Greece, Ireland, Italy, Poland, Portugal, Romania, the Russian Federation, Slovakia, Spain, Sweden, Switzerland, Turkey, Ukraine, and the United Kingdom. Armenia and Azerbaijan, which have special status within the Council, also sent representatives.

The discussions were sometimes charged with emotion. One participant told the tragic story of a close relative who was recently killed by a mine blast as he was working in his vineyard.

At present, 19 of the 40 member States of the Council of Europe support a total and immediate ban on anti-personnel mines. Ten have renounced the use of mines, and two have suspended their use.

The seminar's conclusions will be the subject of a report and a draft resolution, which will be submitted to a forthcoming session of the Parliamentary Assembly.

8

1997: The Year of a Treaty Banning Anti-personnel Mines?

Peter Herby, Adviser, Legal Division, ICRC

Published in the I*nternational Review of the Red Cross,* 317
March 1997

Following widespread disappointment with the modest amendments made in 1996 to Protocol II relating to landmines, of the 1980 Convention on Certain Conventional Weapons (CCW), hopes have risen that 1997 may see the adoption and signing of a new international treaty prohibiting the production, export, transfer and use of anti-personnel landmines. Although such a treaty might not attract universal adherence at the outset, it would nevertheless establish a significant international legal norm and represent a major advance towards the ICRC's goal of bringing the scourge of landmines to an end.

Recent diplomatic initiatives

On 3 May 1996, during the concluding session of the CCW Review Conference, the Canadian government announced that, given the limited progress that had been achieved over more than two years of negotiations, it would convene a meeting of like-minded States in Canada later in the year to discuss how to make further progress towards a total ban on anti-personnel mines. The Ottawa Conference, which took place over three days in October 1996, brought together 50 pro-ban States, representatives of the United Nations, the ICRC, and representatives of the International Campaign to Ban Landmines, a global coalition of non-governmental organizations (NGOs), to work out a strategy for the total prohibition of anti-personnel mines. An "Ottawa Group" of 50 States was formed around a political declaration calling for joint efforts to:

- prohibit and eliminate anti-personnel mines;
- significantly increase resources for mine clearance and victim assistance;
- progressively reduce or end their own use of anti-personnel mines;
- support a General Assembly resolution calling for a total ban;
- promote regional initiatives in favour of a ban.

In his closing address to the Conference, the Canadian Foreign Minister issued an invitation to all governments to return to Ottawa in December 1997 to sign a legally binding agreement banning anti-personnel landmines. Following the Ottawa Conference, the United States introduced a draft resolution at the United Nations General Assembly calling upon States to negotiate a new treaty totally prohibiting anti-personnel mines. In December 1996, 155 States voted in favour of the General Assembly resolution (A/51/45S), which urges States "to pursue vigorously an effective, legally binding international agreement to ban the use, stockpiling, production and transfer of anti-personnel landmines with a view to completing the negotiation as soon as possible". No votes were cast against this resolution and only 10 States abstained.

At the end of January 1997, the ICRC listed 53 countries which had publicly expressed their support for a global ban on the production, transfer, stockpiling and use of anti-personnel mines. Of these, 28 have either renounced or suspended use of anti-personnel mines by their own forces.

New initiatives towards a ban treaty are currently being pursued along two tracks. As part of the "Ottawa process", it is planned to hold a series of preparatory meetings of interested governments in 1997. The first of these meetings, an exchange of views on a draft treaty text, was held in Vienna from 12–14 February 1997. Representatives of 111 governments, the United Nations, the ICRC and the International Campaign to Ban Landmines participated in the meeting. A revised draft treaty text is being prepared on the basis of the discussions and will be the subject of detailed negotiations at the forthcoming meetings in Brussels (June) and Oslo (October) prior to its adoption and signature in Ottawa in December 1997. In parallel with the Ottawa process, the issue of a negotiated end to the scourge of landmines has recently been proposed as an agenda item for the United Nations Conference on Disarmament in Geneva.

The ICRC's latest position paper on landmines, entitled Landmines: Crucial decisions in 1997, offered strong support to the Ottawa process, without disregarding other important initiatives. The overriding objective must be the establishment of a binding international agreement to prohibit unequivocally the production, transfer, use and stockpiling of all anti-personnel mines. For such a prohibition to be truly meaningful, potential loopholes, such as the ambiguous definition of an anti-personnel mine introduced into amended Protocol II of the 1980 CCW, must be firmly closed. Even if such an agreement is initially endorsed by as few as the 38 States which were party to the 1980 CCW a few years ago, adherence as with other treaties, can and will increase over time, as has been the case with other treaties.

Campaigning for a ban, assistance to victims and mine clearance

In tandem with this process, the ICRC's public advocacy campaign launched in November 1995 continues to grow in strength. The campaign, which is run in cooperation with more than 45 National Red Cross and Red Crescent Societies, is designed to mobilize public opinion and foster political will, using the press, television and radio to stigmatize anti-personnel mines in the public conscience. An estimated 600 million people worldwide have already been reached by this message. The campaign also seeks to raise awareness of the need to increase assistance to mine victims and to strengthen mine-clearance programmes. A new series of powerful advertisements supporting the ICRC's message, *Landmines must be stopped*, are currently being finalized. These are intended for use in the international and national media during 1997; it is hoped that they will further reinforce the perception of anti-personnel mines as indiscriminate weapons which must be banned.

At the same time, the needs of the hundreds of thousands of mine victims are slowly being addressed. In early March 1997, an intergovernmental conference is being convened by the Japanese government in Tokyo to consider how to improve the assistance rendered to mine casualties. Despite the best efforts of the ICRC and other humanitarian organizations, too many mine victims are left to fend for themselves because of lack of resources, transport and trained medical specialists, and also because of security and access problems. The ICRC's Health Division has contributed to the Tokyo conference a detailed paper entitled *Assistance for victims of anti-personnel mines: Needs, constraints and strategies.* This document sets out the ICRC's approach to helping landmine victims and proposes practical action by the international community to increase the availability of, and access to, adequate assistance in terms of rehabilitation and care.

The Tokyo conference will also consider the pressing need to strengthen humanitarian mine-clearance programmes. Although the United Nations is developing a set of standards for mine-clearance operations, a major increase in resources is urgently needed to clear priority areas in dozens of severely mine-affected countries across Africa, the Americas, Asia and Europe. Mine clearance is expensive, but so is the cost – human, social and economic – of leaving tens of millions of uncleared landmines scattered over more than 70 countries. At the same time, resources must be committed to research on and development of low-cost but more effective mine-clearance technologies.

While continuing to encourage increased mine-clearance funding, the ICRC has highlighted the need for programmes to assist communities living with an existing mine threat. In order to alert civilians to the dangers of mines and to teach them how to reduce the risk of injury as they go about their daily lives, the ICRC and the National Red Cross and Red Crescent Societies have been conducting major mine-awareness programmes in at least 11 countries in Africa, Asia, Central America and Central and Eastern Europe (Afghanistan, Armenia, Azerbaijan, Bosnia-Herzegovina, Colombia, Croatia, Mozambique, Nicaragua, Somalia, Tajikistan and Yemen). Countless lives and limbs have already been saved in this way.

Regional initiatives

As a complement to global action, the ICRC continues to stress the value of regional and national initiatives. In this context the ICRC sponsored a seminar on "Anti-personnel mines, mine clearance and rehabilitation of victims in Central America" in Managua in May 1996. The two-day meeting, which brought together the governments of Costa Rica, El Salvador, Guatemala, Honduras, Mexico, Nicaragua and Panama as well as Red Cross Societies, United Nations agencies, donors and NGOs, drew up a broad strategy for the elimination of the mine threat in Central America. The seminar's recommendations called for a total international prohibition of anti-personnel mines backed by effective national legislation, strengthened mine-clearance programmes and improved access to rehabilitative care for mine victims, the ultimate objective being to establish a regional zone free of anti-personnel mines.

In June 1996, the General Assembly of the Organization of American States (OAS) passed a resolution calling for a global prohibition of anti-personnel mines and the establishment of an anti-personnel mine-free zone in the Americas. The OAS called upon Member States to institute national moratoria on the production, transfer and use of anti-personnel mines and to ratify the 1980 Convention, in particular amended Protocol II. The OAS also decided to create a registry of mine stocks, mine-clearance efforts and uncleared mines in the region. The OAS effort was soon bolstered by a "regional accord" of the Council of Foreign Ministers of Central America of 12 September 1996, in which Ministers committed their countries to establishing a regional mine-free zone and to enacting national laws to that effect.

Following the encouraging developments in the Americas, a number of initiatives are being launched in other regions in 1997. An international conference organized by the International Campaign to Ban Landmines in Maputo on 25–28 February 1997 brought together more than 450 participants from 60 countries, mostly African, to promote moves towards a mine-free southern Africa. During the conference, the ICRC convened a strategy meeting on landmines for National Red Cross Societies in the region. The meeting resulted in a forceful joint declaration by the Societies calling upon governments from the region among other things to support the Ottawa process and to implement national prohibitions on anti-personnel mines. In April 1997 military and political experts from southern Africa met at an ICRC seminar in Harare to consider global developments, regional approaches and possible ways forward. In May the Organization of African Unity will sponsor a continental conference in South Africa to address mine clearance, improved assistance to victims and political efforts to end the scourge of landmines in Africa.

In Europe the ICRC, jointly with the Refugee Commission of the Council of Europe, held a one-day seminar on anti-personnel mines in March. The gathering, for Council of Europe parliamentarians from Western Europe, the Russian Federation, the Balkans, the newly independent States and Central Europe, addressed the human, social and economic impact of mines, their military utility, and ongoing global, regional and national initiatives towards a ban. Plans are also under way to organize a consultation of Asian military and political analysts in Manila later in the year.

The agenda for 1997

The international, regional and national initiatives being supported by governments, regional organizations, NGOs and the ICRC throughout 1997 will keep the mines issue high on the international agenda, contribute to the stigmatization of anti-personnel mines and build up further momentum towards their elimination. Decisions made this year will decide the fate of tens of thousands of innocent civilians in the years to come. A new legal instrument prohibiting anti-personnel mines may crown the achievements of 1997. A legal agreement is in itself only one step towards ending the humanitarian emergency created by landmines. But it will stand as a beacon of hope, a signal that the international community has both the ability and the intention to end the "epidemic" of landmines which threatens the lives and livelihoods of so many individuals and communities around the world.

9

ICRC Regional Seminar for States of the Southern Africa Development Community (SADC)

Harare, Zimbabwe

21–23 April 1997

Organized by the ICRC in cooperation with the Organization of African Unity and the Republic of Zimbabwe

To further the developing momentum in the region towards a total ban on anti-personnel mines, the ICRC convened a meeting of defence and foreign affairs officials from the 12 SADC countries to discuss the military utility and humanitarian costs of anti-personnel mines. The findings of the military study commissioned by the ICRC and entitled *Anti-personnel Landmines: Friend or Foe?* were examined critically in light of the southern African experience. The seminar ended with an unequivocal call for an end to the landmine plague, and the participants unanimously endorsed the objective of a total prohibition of anti-personnel mines, including by their own governments.

Anti-personnel Mines: What Future for Southern Africa?

Final Declaration of Participants

(representing: Angola, Botswana, Lesotho, Malawi, Mauritius, Mozambique, Namibia, South Africa, Swaziland, Tanzania, Zambia, Zimbabwe)

Harare, 23 April 1997

Defence and foreign affairs officials from all 12 States of the Southern Africa Development Community gathered in Harare to consider the human, social and economic costs of anti-personnel mines in Southern Africa and throughout the world. Participants examined the military effectiveness of these weapons, the practical difficulties of their actual use and mine clearance. The seminar sought to develop recommendations for a common response in Southern Africa to the humanitarian crisis caused by landmines. The following statement was adopted by the participants.

Participants in the seminar "Anti-personnel Landmines: What Future for Southern Africa ?" agree that:

1. the global scourge of landmines, which kill and injure some 2,000 persons per month, most of whom are civilians, is unacceptable and must be stopped. Mines indiscriminately kill and maim civilians and combatants alike, and continue to do so long after conflicts end, which is a cause of grave concern;

2. since 1,100 years will be required, at the current rate, to clear the world of mines currently in the ground, parties to conflicts should not add to this burden through new deployments of anti-personnel mines;

3. those who have used and those who have supplied anti-personnel mines bear a joint responsibility to ensure the clearance of these weapons and the provision of adequate care to their victims;

4. the limited military utility of anti-personnel mines is far outweighed by the appalling humanitarian consequences of their use in actual conflicts;

5. resolutions of the United Nations General Assembly resolution (A/51/45S) calling for a new global treaty banning anti-personnel mines and of the Organisation of African Unity (CM/res.1593, LXII) calling for the total prohibition of these weapons should be urgently implemented throughout the continent;

6. the presence of an estimated 20 million mines in Southern Africa, one fifth of those in the world, represents a direct threat to the health, welfare and survival of many millions of people in the region; and

7. anti-personnel mines constitute a major obstacle to post-conflict reconstruction and development in Southern Africa.

Participants call upon States of the Southern Africa Development Community (SADC) to take the following steps, on the national, regional and global levels, towards ending the scourge of anti-personnel mines:

1. to launch an initiative, in the context of the SADC, for the establishment of a regional zone free of anti-personnel mines;

2. in this context to establish within the SADC Organ on Politics, Defence and Security a Sub-Committee on Landmines supported by a working group of experts to promote and coordinate as a matter of urgency:

a. planning of humanitarian mine clearance;
b. joint training of mine clearance personnel;
c. technological cooperation to facilitate more rapid and cost effective clearance operations in the region;

d. adoption of a SADC code of ethics and standards for humanitarian mine clearance (quality assurance);
e. mine awareness campaign programs;
f. support of national programs for victim assistance;
g. creation of a SADC data bank on landmine matters; and
h. funding for humanitarian mine clearance, victim assistance, rehabilitation programs and mine awareness;

3. to immediately end all new deployments of anti-personnel mines and to establish national prohibitions, such as those already adopted in the region, on their production, stockpiling, transfer and use;

4. for those States which are not yet Parties, to adhere to the 1980 United Nations Convention on Certain Conventional Weapons, including its Protocol II on landmines (as amended on 3 May 1996); for current States party to this Convention to adhere to its amended Protocol II at the earliest possible date to ensure its early entry into force;

5. to participate actively in the Conference of the Organisation of African Unity on "A Landmine Free Africa: The OAU and the Legacy of Anti-personnel Mines" to be held in Johannesburg from 19–21 May 1997 and to take into account there the results of this seminar;

6. to promote the strongest possible resolution on anti-personnel mines at the OAU Summit Meeting in Harare from 2–4 June 1997; and

7. to declare, at an early date, that they will participate actively in the Brussels Conference (24–26 June 1997) of the Ottawa Group of States supporting a global ban and officially to endorse there the conclusion of a new treaty comprehensively prohibiting the production, stockpiling, transfer and use of anti-personnel mines to be signed by the end of 1997.

Participants appeal to the international community, including governments, international agencies and non-governmental organisations to assist the Southern Africa region in becoming permanently free of the scourge of anti-personnel mines, in particular through the provision of technical, financial and other assistance in the clearance and destruction of mines, assistance to victims and mine awareness programs.

Participants express their thanks to the International Committee of the Red Cross for convening the seminar, and for its ongoing efforts on behalf of war victims in many of the countries of the region, to the Organisation of African Unity for its active leadership on the mines issue in Africa and to the Republic of Zimbabwe for its generous hospitality in hosting them in Harare.

10

International Meeting of Experts on the Possible Verification of a Comprehensive International Treaty Prohibiting Anti-Personnel Landmines

Bonn, Germany
24–25 April 1997
Organized by the German government

A number of States were concerned that the simple prohibition of anti-personnel mines, without a well-developed mechanism to ensure its respect, would not be effective. The German government, which had expressed a particular interest in the verification of any anti-personnel-mine ban convention, offered to host a meeting of interested States to discuss the subject. A total of 121 States, more than had attended the Vienna Expert Meeting, participated in the Bonn Verification Meeting, an indication not only of the importance attached to verification, but also of the growing momentum of the Ottawa process.

The ICRC reiterated its view that verification was important, but cautioned against the adoption of a mechanism purely designed to detect violations, reminding States that humanitarian law possessed a potent weapon against individuals in the form of compulsory universal jurisdiction, which was intended to ensure that violators would ultimately have no safe haven from prosecution. The ICRC again stressed the overriding importance of establishing a clear norm prohibiting anti-personnel mines since, historically, complete prohibitions on the use of specific weapons had been very largely respected.

The meeting as a whole was inconclusive, reflecting the divide between those who favoured more of a disarmament approach demanding extensive verification, and those who placed the emphasis on traditional humanitarian law mechanisms intended to ensure compliance with specific norms.

Statement of the International Committee of the Red Cross

24 April 1997

Mr Chairman,

The International Committee of the Red Cross commends the government of Germany for hosting this International Meeting of Experts and appreciates the opportunity to express its views on the possibility of including measures to verify and ensure compliance with a treaty totally prohibiting anti-personnel mines.

As we stated at the recent Experts Meeting held in Vienna from 12–14 February, while the ICRC supports the maximum possible verification of compliance with a future treaty we encourage States not to permit negotiation of such measures to stand in the way of an absolute prohibition on the use, production, stockpiling and transfer of this weapon. Indeed the humanitarian law norms prohibiting the use of specific weapons were initially established without verification and yet they have been almost universally respected. Verification, together with prohibitions on production, stockpiling and transfer, can reinforce such a norm but should not be allowed to override the importance of the norm itself.

Moreover, the ICRC believes that the monitoring of compliance with a total prohibition on anti-personnel landmines can be seen as a positive exercise, promoting the goal of the international community of eliminating anti-personnel landmines rather than operating primarily to detect possible violations. Given the high proportion of civilian casualties due to landmines, it is unlikely that any large-scale use of anti-personnel mines would go unnoticed or unreported. Accordingly, if States deem that it is not appropriate to incorporate an independent verification mechanism into the treaty at this stage, a series of reporting and transparency measures could help to build confidence and promote international assistance in the implementation of the treaty.

For example, an important component of the initiative of the Organization of American States (OAS) towards the establishment of a mine-free zone in the Americas requires the governments from the region to submit an annual report to the OAS Secretary General on the number and location of deployed anti-personnel mines, the number removed during the past year, plans for the removal and destruction of the remainder, and the number of civilian casualties caused by anti-personnel mines during the year. These reports will act not only as confidence-building measures but will also help to target assistance to priority areas. A similar regime could be developed to promote the implementation of the international treaty banning anti-personnel landmines. This could entail annual reporting to the Depositary on the implementation of treaty provisions and the types of cooperation and assistance a State needs or can offer in the areas of mine clearance and the destruction of stockpiles.

Nonetheless, some mechanisms for verification of compliance with international humanitarian law provisions have been developed and might be adapted for a treaty banning anti-personnel landmines. In particular while the 1925 Geneva Protocol did not provide for a verification mechanism, such a mechanism was elaborated in a 1989 report of the United Nations Secretary-General drawn up pursuant to General Assembly Resolution 42/37/C and endorsed by Resolution 45/117/C. This enables the Secretary-General, with the support of a qualified team of experts, to conduct on-site inspections into alleged violations. A verification mechanism of this kind, which seeks to avoid the creation of costly and cumbersome institutions, could play an important role in creating an atmosphere of mutual trust. However such confidence-building is clearly of greater importance in relation to weapons having a strategic significance than in relation to tactical weapons of limited utility such as anti-personnel landmines.

In seeking to prevent violations, international humanitarian law has generally focused on the conduct of the individual. While the production and use of certain high-technology weapons require a significant level of organization, compliance with most rules of humanitarian law ultimately depends on the individual. This is certainly true in relation to the use of small, low-technology weapons such as anti-personnel landmines. The production, transfer, acquisition and use of such weapons may readily be undertaken at the local level. It is therefore essential to ensure that individuals, whether combatant or civilian, are aware of the rules of international humanitarian law and are punished for serious violations of those rules.

Provisions on dissemination and training are to be found in the Geneva Conventions and the 1980 Convention on Certain Conventional Weapons. It is recommended that similar provisions are included in any treaty prohibiting the use of anti-personnel landmines. Beyond this, international humanitarian law has sought to stigmatize, deter and punish abuses through criminalizing individual violations. This approach has received increasing attention in recent years through renewed efforts to punish war crimes, notably at the international level. Prohibiting and punishing violations through the criminal law not only indicates the gravity with which the international community views such abuses, but also ensures that use can be made of existing national and international mechanisms.

Amended Protocol II to the Convention on Certain Conventional Weapons already seeks to criminalize certain violations in relation to landmines. However, international humanitarian law goes further by incorporating provisions to punish the worst violations on the basis of universal jurisdiction, seeking to ensure that there is no safe haven for those perpetrating such crimes. It would appear consistent with this approach to treat the most serious violations of the treaty under discussion in the same manner.

It is also important to penalize those responsible for the production, stockpiling and transfer. While such conduct should not be treated on the same basis as the most serious violations of humanitarian law, it would appear appropriate for criminal penalties to be imposed on such offenders. Some countries have already done so under existing national laws on production, stockpiling and transfer. Finally States should be encouraged to adopt all other regulatory measures to prevent and suppress violations.

Once again, Mr Chairman, it should be stressed that the International Committee of the Red Cross supports the most effective measures possible to promote and monitor compliance with a future treaty. However any compliance measures should emphasise the positive aspects of any new regime – by encouraging transparency, building confidence and promoting international cooperation. Such measures should draw, where appropriate, on the approaches already developed in the field of international humanitarian law and not delay the achievement of our principal objective – the establishment of a norm prohibiting the production, stockpiling, transfer and use of anti-personnel landmines.

Thank you.

INTERNATIONAL COMMITTEE OF THE RED CROSS

INFORMAL DRAFT

ARTICLE ON NATIONAL IMPLEMENTATION MEASURES

(1) Each State Party shall enact any legislation necessary to provide effective penal sanctions for persons who use, or order the use of, anti-personnel mines in violation of this Convention.

Each State Party shall search for all persons on its territory or under its jurisdiction or control, being nationals of a State party to this convention, alleged to have committed such acts, regardless of the place where such acts were committed, and shall either bring such persons before its own courts or hand such persons over for trial to another State Party.

[See Geneva Conventions 1949 I: Art. 49; II: Art. 50; III: Art. 129; IV: Art. 146. Additional Protocols 1977 l: Art. 85. Convention on Certain Conventional Weapons 1980: Amended Protocol II: Art. 14(2)]

(2) Each State Party shall enact any legislation necessary to provide effective penal sanctions for persons who develop, produce, acquire, stockpile, retain or transfer anti-personnel mines, or who order, assist, encourage or induce the commission of such acts, in violation of this Convention.

(3) Each State Party shall take all other appropriate legal, administrative or other measures necessary to prevent and suppress violations of this Convention.

[See Geneva Conventions 1949 I: Art. 49 para. 3; II: Art. 50 para. 3; III: Art. 129 para. 3; IV: Art. 146 para. 3. Additional Protocols 1977 I: Art. 80. Convention on Certain Conventional Weapons 1980, Amended Protocol II: Art. 14(1)].

(4) Each State Party shall disseminate as widely as possible the provisions of this Convention, and of any legislation or other measures adopted to implement this Convention, so that these provisions may become known to the armed forces and to the general civilian population.

[See Geneva Conventions 1949 I: Art. 47; II: Art. 48; III: Art. 127; IV: Art. 146 para. 3. Additional Protocols 1977 I: Art. 83; II: Art. 19. Convention on Certain Conventional Weapons 1980: Art. 6; Amended Protocol II: Art. 14(3)].

(5) Each State shall report to the Depositary, no later than six months after the entry into force of this convention in relation to that State, all measures adopted pursuant to paragraphs (1) to (4), and shall update this report annually.

[See Geneva Conventions 1949 I: Art. 48; II: Art. 49; III: Art. 128; IV: Art. 145. Additional Protocols 1977 I: Art. 84. Conventional Weapons Convention 1980, Amended Protocol II: 13(3)].

11

OAU Conference on a Landmine-free Africa: The OAU and the Legacy of Anti-personnel Mines

Johannesburg, South Africa
19–21 May 1997

In some ways, and with hindsight, the Johannesburg meeting may come to be seen as the watershed in the Ottawa process, as African governments sought to take responsibility for tackling the mines crisis in the region. If African participation in the Oslo Diplomatic Conference was both visible and highly effective, this must be put down, in part at least, to the momentum created beginning with the ICBL Conference in Maputo in February, increasing with the ICRC seminar in Harare in April, and culminating with the OAU Conference in Johannesburg. With the exception of one African government, all others were of a single mind, determined to ensure the total prohibition of anti-personnel mines which they saw as essential to stem the continuing proliferation of the weapon. This solidarity helped to ensure that the treaty ultimately adopted was clear and unequivocal.

The Provision of Assistance to Mine Victims

Dr Chris Giannou
Health Operations Division, International Committee of the Red Cross

19 May 1997

Algeria, Angola, Botswana, Burundi, Chad, Congo, Djibouti, Egypt, Eritrea, Ethiopia, Guinea Bissau, Liberia, Libya, Malawi, Mali, Mauritania . . . No: this is not a roll-call of the Member States of the Organization of African Unity. Morocco, Mozambique, Namibia, Nigeria, Rwanda, Senegal, Sierra Leone, Somalia, Sudan, Swaziland . . . This is a list of regions of the African continent which are or have been polluted to a varying extent by landmines. Tanzania, Tunisia, Uganda, Western Sahara, Zaire, Zambia, Zimbabwe. Many of these mines date back to World War II, others to the struggle for independence and the wars of decolonization, yet others to post-independence conflicts. It would be simpler to mention the handful of African countries not affected at some time in their history by the scourge of landmines.

The eras involved in these examples may well be different; the circumstances different; the politics different. The appalling humanitarian cost is the same, a legacy of lost and broken lives and limbs: soldiers and combatants, farmers and shepherds, women and children lose an arm or a leg; are physically, psychologically and socially scarred and mutilated. Entire societies, all too often already poverty-stricken, have become economic invalids struggling for rehabilitation and reconstruction. Communities remain impoverished and malnourished, fields and villages empty. The normal activities of life – work, travelling and play – become fraught with danger because of the presence of anti-personnel mines.

War is ugly. War wounds particularly so. But there is something especially horrendous about wounds due to anti-personnel mines. A colleague of mine, a surgeon working in an ICRC hospital on the Thai border of Cambodia, put it succinctly: "Whenever we were called in for an emergency case, we prayed that it would not be a mine injury, not another child or woman or peasant terribly mutilated". It was the war-hardened ICRC medical staff who prompted our organization to adopt its public stance against anti-personnel mines, a stance unprecedented since shortly after World War I and the ICRC position against poison gas as a weapon of war.

We have come to describe the present situation as a worldwide epidemic of anti-personnel mine injuries. This is a classic example of a pathology which is not simply biological in scope; like all epidemics, in their causes and consequences, it is a social, economic, health and political event which particularly targets the innocent, the weakest, and the least prepared.

Environmental prevention includes the political will to decide to ban anti-personnel mines, as well as marking mined areas and humanitarian mine clearance. Public education to change high-risk behaviour involves mine-awareness campaigns. Curative therapy for the victims extends from the evacuation of the wounded all the way to the discharge from hospital. Rehabilitation and social reintegration involves physiotherapy, the fitting of an artificial limb when necessary, and vocational training. Most important: it is anti-personnel mines which must be stigmatized and not the victims.

The needs are great, and the costs greater still. The task of clearing the world of the existing tens of millions of landmines is already daunting, and many more are being laid than are being cleared. But mine clearance is essential and the international community can undoubtedly do a great deal more to clear the scourge of mines from afflicted societies and communities. In addition, at a local level, mine awareness programmes – equipping individuals and communities with the information and skills they need to minimize the risk of injury – can help to reduce mine casualties while allowing people to get on with their daily lives but these injuries will still occur. Fields must still be cultivated, water drawn and firewood collected, even if that sometimes means entering a mined area.

Thus, for the foreseeable future, the global number of mine injured and mine amputees will continue to increase. The *ICRC database of mine incidents* has registered more than 9,000 victims since January 1995 in more than 40 countries. For African countries alone, 234 separate incidents (representing 36% of the total number of incidents in the database) were registered, causing 330 killed and 749 injured. It is impossible to know what proportion of all victims are registered; they are certain to be only a small minority; many will die in the fields and mountains before they can reach a health facility. Access to the injured thus remains an important obstacle to their receiving assistance. Mine injuries remain a large-scale, scattered and largely unattended problem. Accurate collection of data is therefore the first step in addressing the mines epidemic. Improving the availability and flow of reliable data is crucial not only to helping to meet the needs of the victims but also to determine priorities for mine clearance and community mine-awareness activities.

Survivors of mine incidents typically suffer from three patterns of injury. Those who step on blast landmines usually lose a foot or a leg and suffer severe injuries to the other leg, genitalia and arms. Survivors of explosions of fragmentation mines, on the other hand, receive wounds similar to those from any other fragmentation device; such wounds can affect any part of the body. The third pattern of injury occurs where mines are accidentally detonated, either by deminers or by children mistaking mines for toys. This pattern inevitably involves severe wounds to the hands and face.

The medical and social management of landmine injuries taxes the public health system in every respect and at every level. Once a person has been injured by a mine the first priority is to get him or her safely out of the minefield – the rescue effort puts the rescuers in danger – and to stop the bleeding. This "first aid" can usually be achieved by the application of a firm dressing, but a traumatic amputation may require some sort of tourniquet which must be applied as low as possible and released at regular intervals. Many limbs are lost or have to be amputated higher than otherwise necessary because tourniquets are applied too high on a limb and left on for more than six hours.

The earlier the wounded person reaches a competent medical facility the better. Transportation in many parts of rural Africa is inadequate at the best of times, during the rainy season it can be almost impossible. Again, many will succumb before reaching a clinic or a hospital.

Mine victims will often need an intravenous infusion. Early administration of simple antibiotics can help to prevent the onset of serious infection such as gangrene. In addition to antibiotics, measures must be taken to prevent tetanus. Active immunization is not universal in Africa, especially in rural areas, and the presence of landmines can, and, in Mozambique and Angola, effectively has, hampered the work of mobile vaccination teams.

As soon as the patient reaches a hospital certain routines must be followed. The patient must be registered and, if possible, the details of the incident recorded. This is often the most important and most overlooked step in data retrieval. The patient will need to be washed with simple soap and water, if his or her condition permits, especially if injured by a buried anti-personnel mine. Blood tests should be taken both to estimate the patient's haemoglobin value and to establish the blood group. Before going to the operating theatre, the patient should be assessed by the surgeon and the anaesthetist.

The surgical management of mine-injured patients can be a challenge to even the most competent surgeon; it is difficult and time-consuming. The blast of the mine tears through tissues, it burns and coagulates; it drives soil, grass, gravel, metal or plastic fragments of the mine casing, and pieces of shoe and shattered bone of the foot, up into the leg, penetrating between tissue layers and causing irregular contusions and contamination of the muscles. This renders the required level of surgical amputation higher than is apparent to the inexperienced eye. Wounds such as these are not seen in civilian practice and they do not correspond to any of the modern surgical specialities. The surgeon should therefore have a solid background in general surgery, The effects of anti-personnel mines are a relatively new subject in the medical literature. Until recently, the injuries inflicted by mines were considered to be the same as any other caused by conventional weapons. However, mines inflict a much more severe injury owing to the specific design of the weapon; the result is specific medical needs. Thus, the treatment of mine victims in ICRC hospitals has become a speciality in its own right. As a matter of policy ICRC surgeons work with a basic level of technology and follow a number of well recognized basic principles of wound surgery.

All war wounds are considered to be dirty and contaminated. Dead and contaminated tissue must be excised, leaving the wound open under a secure dressing. Closure of the wound skin edges with sutures takes place only after four or five days (this procedure is known as delayed primary closure). Large wounds may require skin grafts. A wound with a fracture is managed in the same manner and the fracture immobilized. A surgical amputation either of an entire limb that is beyond repair, or at a point above a traumatic amputation, should be carried out according to the same principles and closed after a delay. Certain amputation techniques, known as myoplastic amputations, are particularly appropriate for mine injuries. The ICRC has produced teaching material to pass on this experience to both military and civilian surgeons.

The severe nature of the injuries caused by mines and the lengthy time mine victims spend in hospital divert already limited resources from the needs of others. Moreover, the additional needs of mine victims may not be recognized by authorities or aid agencies because, compared with battle casualties, mine victims tend to arrive at a hospital in small numbers over a long period. These patients spend more time in hospital, require more operations, and greater quantities of blood transfusions. Blood is essential for the treatment of severe mine injuries. Those undergoing surgical amputation require the most. The ICRC has found that victims of blast mines require, on average, more than six times as much blood as those injured by bullets or fragments.

After any operation the patient must have at least 24 hours of close nursing supervision. Without this, surgery is dangerous. It is the nursing staff who are responsible for the efficient running of a hospital and who are most closely involved in patient care. To help other agencies, non-governmental organizations, and even Ministries of Health, the senior nurses who have worked in ICRC hospitals have written a book entitled *Hospitals for War-Wounded*, to be published before the end of 1997.

The physiotherapy requirements of mine victims are unique. In the initial phase, the injured limb must be kept moving by passive movement and isometric exercises (though without interfering with the wounds). Patients must not be allowed to languish in bed, instead they must be up and about on crutches as soon as possible. As soon as the wounds are closed, active physiotherapy should ensure a full range of movements and muscle strength, especially in the remainder of a limb above an amputation. A stiff knee, for example, may make the wearing of an artificial limb impossible.

After the surgical treatment and when all the wounds are healed and the swelling of the amputation stump has settled, the patient is ready to be fitted with an artificial limb. This is normally four to five weeks after completion of the surgical treatment. The fitting of an artificial limb is an essential part of the rehabilitation of a mine-injured amputee. In addition to ensuring mobility, it also constitutes the patient's first step in regaining his/her dignity, thereby facilitating the psychological recovery of the mine victim. There will be a life-long need for an artificial limb, and a need for regular replacement.

The ICRC follows a policy of "appropriate technology" in its limb-fitting centres. In many countries affected by mines, advanced modern technology may not be practicable or affordable. Thus, ICRC workshops now make artificial limbs out of polypropylene, a thermoformable plastic that is cheap, easy to repair or replace, and recyclable. In Colombia, for example, the introduction of polypropylene technology reduced the complete cost of an above-knee prosthesis from US$ 936 for the conventional method to US$ 390. The cost of physiotherapy, the prosthetic technician, psychologist and social worker remain constant in both cases; the difference in price is due entirely to the technique. Since 1979, the 45 ICRC rehabilitation projects in 22 countries have manufactured more than 100,000 artificial limbs for 80,000 amputees, as well as 140,000 pairs of crutches and 7,000 wheelchairs.

The ICRC puts great emphasis on the training of local technicians in the use of this technology; many mine amputees find employment in its limb-fitting centres. This ensures that the limb-fitting programme can continue once the ICRC withdraws. In recent years, partner organizations have been found to take over its programmes, the majority of them with some continued support from the ICRC. Most commonly, this is a governmental body in the country concerned, but some have been handed over to competent private foundations or to National Red Cross or Red Crescent Societies. Former ICRC projects include prosthetic centres in Chad, Eritrea, Mozambique, Sudan and Zimbabwe.

Ongoing programmes of the ICRC are to be found in Angola, and of the ICRC Special Fund for the Disabled in Cameroon, Ethiopia, Ghana, Kenya, Nigeria, Mali, Malawi, Rwanda, Somalia, Somaliland, Tanzania, Uganda, Zaire, Zambia and Zimbabwe. The centre in Addis Ababa has become an important venue for training prosthetic technicians from African, Asian and Latin American countries, and a new, special prosthetic kit has been developed there comprising ready-made polypropylene components for 100 artificial limbs at a material cost of US$ 21 apiece.

Of course, appropriate medical interventions are not cheap. Hospital care, artificial limbs, lodgings, logistics and equipment must all be taken into consideration. The global cost each year of providing surgical care and physical, psychological and social rehabilitation to thousands of mine victims amounts to tens of millions of US dollars. All agencies, however, are chronically short of funds, and no one agency can attend to the needs of all.

An effective strategy for the provision of assistance to victims of anti-personnel mines – to meet the needs of a maximum number of mine injured and mine amputees while allowing the rest of the population at high risk to feed, shelter and educate itself – entails a decision-making process based on an accurate and objective assessment of the needs, available resources, and likely constraints. Priorities must be established. Needs are great, money is scarce. The mine-injured are one category of the wounded; the wounded are one category of all people needing assistance. There are many constraints and limitations to working in conflict areas. Access to the wounded and the needy is imperative; security, in a war zone, for the wounded and for health professionals is essential. In the last 12 months, the ICRC has experienced security-related tragedies in Burundi and in Chechnya; so, too, have other non-governmental humanitarian organizations and UN human rights observers.

Only a small proportion of the immediate needs are met by the ICRC and other agencies because:

– there is an obvious lack of funds for the projects already underway;
– the specific constraints, imposed in any given country affected by the mine problem, may be insurmountable;
– a comprehensive and coordinated approach to the problem of victim assistance with a long-term view is lacking.

There are three reasonable and immediate goals for improved assistance for mine victims: first, that all mine victims should have access to adequate surgical care and rehabilitation; second, that the psychological and social needs of those injured and disabled by mines should be effectively addressed; and third, that the impact on the social and economic development of affected countries of the presence of mines should be measured. As a first step towards meeting these three goals, the ICRC proposes a number of initiatives:

1) To be optimally effective, any concrete assistance in mine-affected countries must follow an integrated approach of possible measures: preventive, curative, rehabilitative. The ICRC believes that the key to improving assistance, to bringing more and higher-quality assistance to mine victims in the long term, is the structured flow and analysis of information about the entire mine problem in any given situation.

In each region or country affected by mines, there should therefore be a Mines Information System, to exchange data amongst governments and local authorities, the ICRC and National Red Cross and Red Crescent Societies, United Nations agencies and the network of international and local non-governmental organizations involved. Information gathering, pertaining to every aspect of assistance and prevention, and sharing, can then be used to better coordinate humanitarian activities and to best define needs and operational priorities: mine-risk education, mine clearance, access to victims, proper surgical care of the injured, rehabilitation and social reintegration.

A mine-afflicted society will go through various phases, of differing operational priorities, as it comes to grips with the challenges created by the presence of anti-personnel mines, as well as those created by the conflict and post-conflict reconstruction and reconciliation. Improved coordination will help to better define the priorities at any particular time, and allow for the most efficient allocation of resources, foreign and local, to the various aspects of prevention and assistance.

2) On a regional basis, regular training sessions should be organized for civilian and military health workers on appropriate treatment and rehabilitation measures for the mine-injured.

3) Greater efforts should be made to assess and then address the psychological and social needs of mine victims, aspects that are far too easily forgotten or ignored in the maelstrom of misery that permeates a post-conflict society.

4) All organizations, governments and agencies concerned with the mines issue should mobilize to improve access to mine victims and, where possible, to provide logistical support to victims so that they can be treated and assisted at appropriate medical and rehabilitative centres.

The ICRC is present at this conference not only to emphasize its continuing work and commitment to assist mine victims but also to establish how the necessary information exchange can take place in practice. No single agency or organization can respond to the needs of the hundreds of thousands of mine amputees worldwide. But with better information, increased cooperation, greater resources, and the necessary political will, the provision and quality of assistance to mine victims will be greatly strengthened.

12

Debate on Landmines at the Angolan Parliament

Luanda, Angola

23 May 1997

Organized in collaboration with the ICRC and UNICEF

Despite Angola being one of the world's most mine-affected countries, the debate in the Angolan Parliament demonstrated that not everyone was aware of the sheer scale and extent of the mines problem in the country. The debate was critical in ensuring that all parties were aware of the acute need to tackle the mines problem and to make certain that no more mines were laid in the country.

Statement on International Political and Legal Efforts on Anti-personnel Landmines

Peter Herby, Legal Division, International Committee of the Red Cross, Geneva

23 May 1997

In 1918 the international community was horrified at the appalling effects of the widespread use of poison gas in the trenches and battlefields of Europe. The public and many political leaders were sickened at the sight of tens of thousands of soldiers gasping for breath, blinded, burned and vomiting blood as they returned from battle. Many thousands more died an agonizing death in the trenches. As a result, the use of chemical and biological weapons was stigmatized. Norms which had existed in ancient cultures and various religious traditions against the use of poison as a method of war were codified in the 1925 Geneva Protocol prohibiting gas warfare.

In the 1990s the same process of stigmatization is occurring in response to the horrendous effects of anti-personnel landmines. The distinguished Ministers who have spoken today have amply described the appalling human, economic and social costs of the use of this weapon in Angola. The experience of Angola, which the ICRC has witnessed first-hand through its delegates and medical staff in this country, should alone be sufficient to trigger an international response.

But your experience is being lived ten times over in other parts of Africa, such as Ethiopia, Mozambique, Somalia, Sudan and Zimbabwe. The same nightmare of anti-personnel mines is faced by communities in Afghanistan, Bosnia, Croatia, Cambodia, Guatemala, Nicaragua and the Chechnya region of Russia, to name only a few of the dozens of mine-affected regions of the world.

The ICRC, based on the experience of its medical personnel around the world, became convinced in the early 1990s that landmines had provoked a profound medical, human and social crisis in nearly all the situations where they had been used on a large scale. Indeed, in medical terms they had created an "epidemic" of injury, death and suffering. And, as with any epidemic, treating the victims alone will never be enough. As long as anti-personnel mines are used the epidemic will continue to spread far beyond the capacity of any country or even the entire international community to respond. As with other epidemics the only solution must be dramatic and comprehensive. This includes on the one hand **prevention,** through total prohibition of the weapon, community mine awareness and mine clearance, and on the other hand **treatment of victims** through improved emergency care, surgery, rehabilitation and reintegration into society. The ICRC, and now most of the international community, is convinced that a more modest response simply will not work.

The ICRC bases its position both on its medical expertise and existing norms of international humanitarian law – which we are mandated by States to promote and develop. On the medical side our war surgeons are faced with one of the most horrific injuries they ever see. The blast of an anti-personnel mine tears through tissues, it burns and coagulates, it drives soil, grass, gravel, fragments of the mine casing, and pieces of the shoe and shattered bone deep into the tissues of the leg, causing irregular wounds and contamination of the muscles. The medical resources required to treat such wounds are many times that required for other war casualties and require expertise not found in civilian medical practice. In many cases emergency care is simply not available and the victim dies a slow and painful death. An estimated 50% of victims die alone in the fields or mountains without ever reaching medical assistance.

The tragic fact of anti-personnel mines is that they are up to ten times more likely to kill or injure a civilian after a conflict than a soldier during hostilities. Even with poison gas the maiming ends when the fighting stops. But unlike other weapons landmines are indiscriminate both in their choice of target – once laid they cannot distinguish a civilian from a combatant, and in time – they will go on killing and maiming for decades, respecting neither cease-fires or peace agreements.

The use of weapons in combat is limited by several elements of customary international humanitarian law, which is binding on all States regardless of whether they have adhered to particular treaties. The use of weapons which are indiscriminate, that are unable to discriminate between a combatant and a civilian, is prohibited. The use of weapons which inflict more suffering and injury than is necessary to take a soldier out of combat is prohibited. And weapons which are abhorrent to the "dictates of the public conscience" may not be used.

In the case of anti-personnel mines both military doctrine and humanitarian law have tried to mitigate the inherently indiscriminate nature of anti-personnel mines by requiring that they be used only in areas from which civilians are excluded by fencing, warning signs and guards. A military study of some twenty-six conflicts since 1945, commissioned by the ICRC last year, concluded that in actual fact the implementation of the provisions of military doctrine and humanitarian law is extremely difficult, even for professional armies, and has rarely occurred, especially under combat conditions. The conclusions of this study, which have been distributed here today, have now been endorsed by fifty-five senior military commanders from nineteen countries. These commanders concluded that the limited military utility of anti-personnel mines is far outweighed by the appalling humanitarian consequences of their use and that, on this basis, they should be prohibited. The study demonstrated that, while the effects of anti-personnel mines on unprotected civilians were devastating, their effectiveness against combatants was limited and that they had not played a decisive role in any of the twenty-six conflicts reviewed.

The use of anti-personnel and other types of landmines is specifically governed by the 1980 Convention on Certain Conventional Weapons (CCW). As stated above the provisions of this Convention, aimed at protecting civilians, have been difficult to implement. In response to the global landmines crisis this treaty was somewhat strengthened in a process of negotiations between 1994 and 1996. Its provisions were extended to apply to internal armed conflicts. The responsibility for mine clearance was clearly assigned to those who use mines. The location of all mines are to be mapped and recorded. However, due to the need to adopt decisions by consensus, the provisions limiting the use of anti-personnel mines were, in the view of the ICRC, too weak and too complex. Some key provisions will not come into force until around 2007.

Although we encourage States to adhere to the amended version of the CCW Convention, in order to improve the absolute minimum norms applicable to States which continue to use mines, we are not convinced that these modest improvements will end the humanitarian crisis caused by landmines.

Fortunately a growing number of States have reached the same conclusion. By the end of the Review Conference of the Convention on Certain Conventional Weapons in May 1996 nearly half of the States Parties to the Convention had supported a total ban on anti-personnel mines. We welcome the compelling statement of the government of Angola to the closing session of the Review Conference endorsing the view that a "total ban on their production, export and use is the only effective solution to the humanitarian catastrophe they are causing all over the Third World". In December 1996 156 States voted for a United Nations General Assembly resolution (A51/45S) calling for the urgent negotiation of a new treaty comprehensively banning the production, stockpiling, international transfer and use of anti-personnel mines. No State voted against, although ten abstained.

An unprecedented international effort was launched by fifty pro-ban States in Ottawa, Canada last October to develop a comprehensive global response to the landmines crisis. States of the "Ottawa Group" which includes countries from all regions of the world, including Angola and seven of your African neighbours, committed themselves to working together for a global ban, ending new deployments of anti-personnel mines, promoting regional efforts, and supporting a massive increase in resources for mine clearance, mine awareness and victim assistance. The "Ottawa process" has led to an intensification of efforts at the global and regional levels to address the problems of mines clearance, the needs of victims and to promote the establishment of zones free of anti-personnel mines.

The "Ottawa process" has also accelerated efforts to achieve a legally binding international treaty banning anti-personnel mines. Canadian Foreign Minister Lloyd Axworthy has invited all States to come to Ottawa in December 1997 to sign such a treaty. An Austrian draft treaty was the basis for discussions among 111 States in Vienna in February 1997. A follow-up conference of the "Ottawa Group" will be convened in Brussels in June to launch formal negotiations on this treaty. These negotiations will be pursued in Oslo in September with a signing conference for Foreign Ministers to be held in Ottawa in December 1997. This agreement, even if not signed initially by all States, will signal the beginning of the end of anti-personnel mines as an acceptable means of warfare.

The Organization of African Unity has been a leading force in efforts to end the scourge of landmines. As early as June 1995 the Council of Ministers of the OAU called for the global prohibition of anti-personnel mines and increased international assistance for mine clearance and assistance to victims. In reaffirming this appeal in February 1996, OAU Ministers called on African regional organizations to launch initiatives for the prohibition and elimination of this weapon. Earlier this week the OAU hosted a first continental conference on landmines entitled "A Landmine Free Africa: The OAU and the Legacy of Anti-personnel Mines" in Johannesburg. This conference developed a comprehensive set of proposals for

African strategies for action in the fields of mine clearance, victim assistance and legal prohibitions.

States of the Southern African Development Community (SADC) have begun to respond to the OAU's call for regional action. At a conference in April sponsored in Harare by the ICRC in cooperation with the OAU and government of Zimbabwe, representatives of foreign and defence ministries of all twelve SADC countries agreed that the human costs of anti-personnel mines far outweigh their military utility and that all new deployments of this weapon should be stopped. Participants urged the leadership of SADC countries to launch an initiative for a regional zone free of anti-personnel mines and urgently to establish SADC programs to promote regional cooperation in mine clearance and victim assistance and to facilitate international assistance in these fields. A similar initiative among States of Central America, which is faced with a significant but smaller mine problem than the SADC region, has attracted international support to the extent that the region expects to be able to clear all existing mines by the year 2000.

Throughout the world States and their armed forces are realizing that they do not need to await an international consensus before taking measures to protect their own territory and populations from the devastating effects of landmines. Thirty States, including Mozambique, South Africa and Zimbabwe, have now renounced for themselves the use of anti-personnel mines. Sixteen, including South Africa, are in the process of destroying their stockpiles. Prohibitions on international transfers of anti-personnel mines have now been adopted by a large majority of mine-producing States. The parliaments of five States have adopted legislation making the production, stockpiling, transfer and use of anti-personnel mines illegal on the national level.

The ICRC considers the landmines crisis to represent a humanitarian emergency which should be addressed through dramatic national and international leadership. 1997 is a crucial year in which African States can play a leading role in freeing both themselves and the world from this scourge. Specific actions we have encouraged in the coming months include:

- The strongest possible resolution on anti-personnel mines at the OAU Summit in June;

- Active participation in the Brussels Conference of the "Ottawa Group" including national statements there of a country's readiness to join in negotiations for a comprehensive ban by the end of 1997;

- A decision at the upcoming summit of SADC States to establish a zone free of anti-personnel mines in the region and to create a SADC structure for regional cooperation in mine clearance, victim assistance and mine risk education programs; and

• National decisions to end the production, stockpiling, transfer and use of anti-personnel mines.

In conclusion, the ICRC would like to use this occasion to urge the newly united Parliament of Angola urgently to address the issue of anti-personnel mines as both a challenge and an opportunity. The enormity of the problem, as we have heard today, cannot be underestimated. Clearance and the rehabilitation of victims may take decades. But, as we have heard today, anti-personnel mines present a common threat to life and livelihoods of people throughout Angola. For the entire country they represent a common obstacle to development and reconstruction. The cooperation, commitment and mobilization of resources needed to address this common enemy can become, over time, a basis for reconstruction and reconciliation.

In accepting to take up the challenge of anti-personnel mines as a common national priority Angola will merit the full support of the international community. In joining your efforts to those of a growing number of States which have come together to eliminate this pernicious weapon from the world's arsenals you will provide a clear sign to your children and to the world that humanity is not powerless in the face of a particularly cruel technology. You will be helping to reinforce international humanitarian law norms through which humanity has attempted to protect itself from its most destructive impulses – even in times of war.

13

International Conference for a Global Ban on Anti-personnel Landmines

Brussels, Belgium

24–27 June 1997

Organized by the government of Belgium

The Brussels Conference, which was the official follow-up to the 1996 Ottawa Conference, was designed to ensure that those negotiating the treaty were all formally committed to achieving a total ban treaty. For this reason, governments that wanted a seat at the negotiating table at the forthcoming Diplomatic Conference in Oslo were asked to adhere to the 'Brussels Declaration'. By the end of the Conference, 97 governments had done so, prompting ICRC President Cornelio Sommaruga to remark that: 'The Brussels Conference has demonstrated that the momentum towards a ban of this pernicious weapon is now irreversible.'

In his address to one of the Conference's round-tables, President Sommaruga outlined what the ICRC considered to be the five key elements in the treaty to be concluded: (a) an unambiguous definition of an anti-personnel mine, (b) an absolute prohibition on production, transfer and use, valid from the entry into force of the treaty, (c) the destruction of stockpiles and emplaced mines in the shortest possible time, (d) positive provisions for technical assistance in implementation of the treaty and (e) penal sanctions to punish serious violations of the treaty. He added that the treaty should neither permit reservations nor allow for withdrawal while a party is engaged in armed conflict. It is pleasing to note that all these elements were ultimately included in the Ottawa treaty.

Humanitarian Aspects – an Integrated and Coordinated Approach

Statement of Cornelio Sommaruga, President, ICRC, Geneva

26 June 1997

It is a privilege to join you in Brussels on this solemn occasion on which States from around the globe will announce their intention to negotiate a comprehensive ban on anti-personnel mines by the end of 1997. Last October I had the pleasure of being present to offer the immediate support of the ICRC as Canadian Foreign Minister Lloyd Axworthy invited States to come back to Ottawa this year to sign a treaty banning this pernicious weapon. The "Ottawa process" is an unequivocal expression of the revulsion of public and statesmen alike at the "mass destruction in slow motion" we have witnessed through the widespread use of anti-personnel mines.

One-hundred and twenty three years ago Brussels was the site of the first attempt to codify customary international humanitarian law. Although the Brussels Declaration on the Laws and Customs of War of 1874 was not adopted as a treaty it was the basis of the subsequent Hague Conventions of 1899 and 1907 which reaffirmed prohibitions on the use of poison and of weapons causing unnecessary suffering.

The Brussels Conference of 1997 meets in the same spirit of limiting the horrors of war by banning excessively injurious weapons. We are here under particularly encouraging circumstances: the use of anti-personnel landmines has been stigmatized in large parts of the world, the UN General Assembly has voted overwhelmingly for a legally binding treaty to outlaw this weapon, twenty-nine States have prohibited its use and seventeen are destroying their stockpiles. The presence of more than half of the States of the world at this historic meeting to launch negotiations for a ban bodes well for its rapid negotiation in Oslo and the signing of a comprehensive treaty in Ottawa in December.

With most other prohibited weapons, including poison gas, the stream of victims ends shortly after use of the weapon is halted. The particularly pernicious nature of landmines means that the humanitarian work of this Conference will continue for decades even if not one more mine is sown from today. The Brussels Conference will achieve a major success in launching negotiations for a global ban. I take this opportunity to appeal to the conference to make an equally historic commitment to the less visible, more expensive and far more long-term humanitarian challenges of care and treatment for the victims, mine awareness and mine clearance. These are the tasks, beyond a ban, by which the commitment and effectiveness of this extraordinary global coalition of NGOs, governments, the United Nations and international agencies will be judged in time.

The ICRC's call, in 1994, for a total ban on anti-personnel mines was based on the experience of our medical staff around the world who had witnessed a profound medical, human and social crisis in nearly all of the situations where these weapons had been used. In medical terms they had created an "epidemic" of exceptionally severe injury, death and suffering. As with other epidemics the solution must be dramatic and comprehensive. This includes on the one hand

prevention, through a ban, community mine awareness and mine clearance, and on the other hand **assistance to victims** through dramatically improved access for landmine survivors to transport, emergency care, surgery, rehabilitation and reintegration into society.

The ICRC has stressed the necessity of an integrated approach, based on the structured flow and analysis of data, in which mine clearance, mine awareness and victim assistance efforts are coordinated and mutually reinforcing. At the Tokyo Conference in March we put forward the concept of a "mine information system" in affected areas through which all actors – governmental, NGO, deminers, the UN and the ICRC – can pool relevant information and work together to improve our effectiveness.

It is often medical personnel who first learn the location of mines through the tragic stories of their victims. Such knowledge is essential for prioritizing clearance operations. On the other hand, knowledge of the location of mined areas held by clearance teams is crucial to those engaged in community mine awareness. Mine awareness workers frequently have information from affected populations which can help target clearance efforts. Yet far too often this information is not consistently available to other actors at the national or even community levels, due to the absence of integrative communication structures, adequate resources and competition among humanitarian agencies. Mistaken policies are often adopted. The results are too often both tragic and avoidable.

Depending on circumstances such an information flow could be managed by a semi-governmental agency, a government body or an international organization. In some situations regular meetings of involved actors may suffice. What is essential here, as in other humanitarian emergencies, is that all parties must be prepared to contribute and cooperate to ensure an effective response to a situation of massive human suffering.

The ICRC is prepared to play a special role as regards victim assistance. To ensure the consistent and coherent provision of information we are preparing, with input from specialized NGOs and agencies, a standardized "mine incident" report form. We invite all our partners in this effort regularly to share with us reliable information on mine incidents worldwide.

The ICRC has been compiling, since 1995, a registry of landmine incidents from credible sources throughout the world. This register, which we will soon make available on the Internet, records incidents in which some 7,404 persons were killed or injured from landmine explosions in 1996 alone. These incidents are spread over 41 countries and territories in Africa, Asia, Europe, Latin America and the Middle East. As it is likely that these statistics record but a small fraction of the devastation of human lives which landmines are causing, they suggest that the estimates of 2,000 casualties per month may underestimate the scale of the problem. They also demonstrate, once again, that this epidemic knows no boundaries of time or geography and has taken root in virtually every situation where mines have been used.

Returning now to the humanitarian law side of this Conference I would like to stress that the ICRC welcomes the initiative being taken here not only as a means of creating a new international humanitarian law treaty but as a return to the practice, well established in classical international law, of developing and universalizing law through national initiatives rather than relying solely on consensus negotiations, which frequently can lead to a lowest-common-denominator result. Indeed the 1874 Brussels Conference and its successors in the Hague in 1899 and 1907 were the direct result of initiatives by successive Russian Czars which were taken up by leading statesmen of the era. Even recent documents such as the Geneva Conventions of 1949 and their Additional Protocols of 1977 were developed in negotiations which did not require consensus in their rules of procedure.

Experience in international humanitarian law has shown that clear and unambiguous norms, such as those being endorsed in this Conference, are more compelling, easier to promote and more readily implemented than complex and nuanced regimes. The result is that they are more easily universalized over time, even if all States cannot subscribe to a given rule at the outset. Examples include the absolute prohibitions on the use of exploding and expanding bullets and of chemical and biological weapons – all of which were far from universal at their adoption but have now become customary law. In this regard it is encouraging to note the participation of 111 and 121 States at recent meetings in Vienna and Bonn to discuss the text of a treaty banning anti-personnel mines, considerably more than participated in recent negotiations on the continued use of this weapon.

In an effort to ensure that the Ottawa treaty attracts the broadest possible adherence from the outset, particularly among mine-affected States, the ICRC hosted a regional seminar for the twelve countries of the Southern Africa Development Community in Harare in April 1997 and cooperated with the Organization of African Unity, the Parliamentary Assembly of the Council of Europe and the government of Turkmenistan in the preparation of regional conferences on landmines this year. The ICRC sponsored a regional seminar for States of Central America in Managua in May 1996 and will next month host a meeting for experts from Asian military and strategic studies institutes in Manila, in cooperation with the government of the Philippines. We are pleased to note that the regional conferences in Africa, Central America and Europe called for an immediate end to the use of anti-personnel mines at the national level and for the establishment of regional zones free of this weapon. They have also encouraged member States to sign the treaty in Ottawa in December.

We would like to commend Austria for having produced an excellent draft Convention which will form the basis for negotiations in Oslo and which already commands a high degree of acceptance. By consulting widely in the preparation of this draft the Austrian text has facilitated negotiations and thus saved both time and lives.

The ICRC would like to underline several elements which we consider crucial to the effectiveness of the future treaty and which, we are pleased to note, are contained in the current draft text. These are (a) an unambiguous definition of an anti-personnel mine, (b) an absolute prohibition on production, transfer and use from entry into force, (c) the destruction of stockpiles and emplaced mines in the shortest possible time, (d) positive provisions for technical assistance in implementation and (e) penal sanctions to punish serious violations of the treaty. The treaty should neither permit reservations nor allow for withdrawal while a party is engaged in armed conflict. We will be circulating detailed comments on the draft text in the coming weeks.

As we go forth from this Conference tomorrow let us never forget that the legal prohibition of anti-personnel mines must be firmly rooted in the stigmatization of this weapon as a weapon of war so that, whatever their knowledge of the law, combatants will choose not to use this arm because their consciences and communities will not tolerate it.

Let us also remember that mines, whether "dumb" or "smart", know nothing of the law and will terrorize generations of innocents until every last one has been removed and destroyed. While the debate over the legal status of anti-personnel landmines may soon be resolved, the humanitarian problem they have created has barely begun to be addressed. Being maimed by an illegal weapon will bring little solace to future victims. Today's child amputee with no hope of an artificial limb will find scant comfort in a total ban tomorrow. The legacy of this epidemic will demand massive resources and sustained commitment for decades. Let us commit ourselves today not only to Oslo and to Ottawa but also to the victims and to mobilizing within the international community and its institutions for the long road ahead.

Declaration of the Brussels Conference on Anti-personnel Landmines

The following States met in Brussels from June 24 to 27 1997 to pursue an enduring solution to the urgent humanitarian crisis caused by anti-personnel landmines. They are convinced that this solution must include the early conclusion of a comprehensive ban on anti-personnel landmines.

They recall the United Nations General Assembly resolution 51/45S, supported by 156 States, urged the vigorous pursuit of "an effective, legally binding international agreement to ban the use, stockpiling, production and transfer of anti-personnel landmines".

In that spirit they affirm that the essential elements of such an agreement should include:

– a comprehensive ban on the use, stockpiling, production and transfer of anti-personnel landmines,
– the destruction of stockpiled and removed anti-personnel landmines,
– international co-operation and assistance in the field of mine clearance in affected countries.

The following States:

encouraged by the work of the Brussels Conference;

encouraged further by numerous national and regional initiatives and measures taken to eliminate anti-personnel landmines;

encouraged by the attention given to this subject by the United Nations and by other fora;

encouraged, finally, by the active support of the International Committee of the Red Cross, the International Campaign to Ban Landmines and numerous other Non-governmental Organisations;

welcome the convening of a Diplomatic Conference by the Government of Norway in Oslo on 1 September 1997 to negotiate such an agreement;

also welcome the important work done by the Government of Austria on the text of a draft agreement which contains the essential elements identified above and decide to forward it to the Oslo Diplomatic Conference in order to be considered together with other relevant proposals which may be put forward there;

affirm their objective of concluding the negotiation and signing of such an agreement banning anti-personnel landmines before the end of 1997 in Ottawa;

invite all other States to join them in their efforts towards such an agreement.

Press Release

27 June 1997

BRUSSELS CONFERENCE SETS 95 STATES ON PATH TO ANTI-PERSONNEL MINE BAN

Cornelio Sommaruga, President of the International Committee of the Red Cross (ICRC), hailed the success of the Brussels International Conference for a Global Ban on Anti-Personnel Landmines, which this morning adopted a declaration in which 95 States committed themselves to conclude an anti-personnel landmine ban treaty by the end of 1997. "The Brussels Conference has demonstrated that the momentum towards a ban of this pernicious weapon is now irreversible," said the ICRC President.

In associating themselves with the Brussels Declaration, States affirm that the essential elements of the treaty should include: a comprehensive ban on the use, stockpiling, production and transfer of anti-personnel landmines; the destruction of stockpiled and removed anti-personnel landmines; and international cooperation and assistance in the field of mine clearance in affected countries.

Detailed negotiations on the wording of the new international humanitarian law treaty will take place at a Diplomatic Conference in Oslo in September. Only States which have signed the Brussels Declaration will be allowed to participate actively in these negotiations, although other States, the ICRC, UN agencies and the International Campaign to Ban Landmines will be able to attend as observers. It is expected that the final treaty will be signed in Ottawa in the first week of December.

In his address to the Conference, President Sommaruga described the Ottawa process as "an unequivocal expression of the revulsion of public and statesmen alike at the 'mass destruction in slow motion' that we have witnessed through the widespread use of anti-personnel mines." He urged the 155 States attending the Conference to commit themselves not only to a ban of anti-personnel mines in the short term but also to long-term programmes to meet the needs of mines victims and to clear existing mines.

14

The Third Austrian Draft of the Convention on the Prohibition of Anti-personnel Mines

May 1997

The third draft of an international treaty banning anti-personnel landmines was the starting point for the negotiations which took place in Oslo. The draft had been prepared by the Austrian government and integrated many of the proposals made by governments, the ICRC and international and non-governmental organizations during earlier sessions.

Comments of the
International Committee of the Red Cross
on the
Third Austrian Draft (13/5/97)
of the
Convention on the Prohibition of Anti-personnel Mines

General Comments – We commend and support the Austrian draft which has effectively integrated many of the proposals made in the Vienna and Bonn expert group meetings in a manner which should command broad support. In particular we welcome the unambiguous definition of an anti-personnel mine. The following comments reflect the views of the ICRC on the Austrian text. In preparing this document we have taken into account statements on the Austrian text made by States during the first two days of the Brussels Conference.

This document proposes specific provisions concerning mine victims and mine awareness – subjects which received considerable attention at the Brussels Conference. Proposals for new text on these subjects are contained under Article 1 (general obligations), a new Article 7 (labeled 6 bis) and under Article 8 (para. 3).

Article 1

The addition of the following paragraph under this article on "general obligations" would emphasize the comprehensive nature of the treaty as well as the positive benefits of adherence:

New para. 3 – *Each State Party undertakes to provide for the care and rehabilitation of landmine victims and for mine-awareness programs in accordance with the provisions of this Convention.*

Article 2

In our view this article reduces significantly the ambiguity created by the definition of an anti-personnel mine contained in Protocol II of the 1980 Convention on Certain Conventional Weapons (CCW, as amended on 3 May 1996) and accurately reflects the reason given by many States for the inclusion of the word "primarily" in the aforementioned Protocol.

We remain concerned about the humanitarian problems created by anti-vehicle mines, which will be exacerbated by the use of anti-handling mechanisms – especially on remotely delivered models. It is important that anti-vehicle mines are not designed in such a way that they can have the same effects as anti-personnel mines.

para. 1 – Further clarity would be achieved by avoiding the use of the opening phrase *"Anti-personnel mine" means a mine . . .* which is taken from the CCW. In that context the meaning of a mine is defined in a preceding paragraph. Two solutions may be considered. The CCW definition of a mine could be placed ahead of existing para. 1. Alternatively, existing para. 1 could be reformulated as follows:

"Anti-personnel mine" means a munition placed under, on or near the ground or other surface area and designed to be exploded by the presence, proximity or contact of a person and that will incapacitate, injure or kill one or more persons. This reformulation is more concise and does not include any element not currently in the CCW definition of a "mine" or the definition of an "anti-personnel" mine in the Austrian draft.

para. 5 – The definition of a "minefield" remains imprecise. Further clarification here is particularly important as different responsibilities under Articles 5 and 6 depend on whether a mine is located within or outside a minefield.

New para. 6 – We propose that the following definition of a "mined area", taken from Article 2 of the CCW, be added here and that this term be used under Article 6 to refer more precisely to mines located outside minefields. This definition would simplify Article 6 and make its provisions clearer.

"Mined area" is an area which is dangerous due to the presence of mines

Article 3

No comment

Article 4

No comment

Article 5

para. 1 – The 10-year time period for the destruction of mines in minefields should be maintained. Given the practical concerns which have been expressed we have proposed a mechanism for extension of this period under Article 9.

Article 6

We consider that this Article, in particular, is crucial for the protection of civilian populations and should be strengthened.

Title – Should read: *Destruction of anti-personnel mines in mined areas*

para. 1 – We propose removing the word *laid* from both the title (as above) and para. 1. Mines may accurately be said to be "laid" in minefields. However, mines may arrive in areas outside minefields either by laying, carelessness, remote-delivery, movement caused by rain or snow or by other means. Mined areas, which threaten civilians, are not solely those in which mines have been "laid".

In keeping with our proposal to use the term "mined areas" para. 1 would read: *Each State Party undertakes to destroy, as soon as possible, all anti-personnel mines in mined areas under its jurisdiction or control.*

para. 2 – Should read: *Each State Party shall make every effort to identify mined areas under its jurisdiction and control and to provide an immediate, effective and continuous warning to the civilian population until such time as the area is known to be clear of mines. Such warning shall include all feasible measures, including those contained in Article 5 (2) to ensure the effective exclusion of civilians from the area.*

This reformulation of the existing para. 2 makes clear that warning civilians of mined areas is a continuing obligation and indicates specific types of measures which will be necessary.

Proposed New Article 7 (6 bis)

Each State Party undertakes to provide, as far as it is able, for the care and rehabilitation of landmine victims and for mine-awareness programs in areas under its jurisdiction and control.

Article 7 (Existing)

new para. 3 bis – Should read: *Each State Party in a position to do so shall provide assistance for the care and rehabilitation of landmine victims and for mine-awareness programs.*

Article 8

No comment

Article 9

The ICRC considers arrangements to clarify issues related to implementation of the Convention to be desirable as they will build confidence in its effectiveness. However we feel strongly that disagreements over such measures should not be permitted to stand in the way of the establishment of the norms in Article 1 which are of fundamental importance.

To meet concerns expressed by some States about fulfillment of the obligations under Article 5, para. 1 (destruction of mines in minefields within 10 years) we suggest the following language be inserted into Article 9:

If a State Party believes that it will be unable to ensure the destruction of all anti-personnel mines referred to in Article 5 not later than 10 years after entry into force of this Convention for that State Party, it may submit a request to a Meeting of States Parties or a Review Conference of this Convention for an extension of the deadline for completing the destruction of such anti-personnel mines. Such a request must be made not later than nine years after entry into force of this Convention for the State Party.

The request shall contain:

(a) The duration of the proposed extension,
(b) A detailed explanation of the reasons for the proposed extension,
(c) A detailed analysis of the humanitarian and financial costs and benefits of destruction of such landmines during the proposed extension period, including an indication of the technical, financial and other resources available.

A decision on the request shall be taken by the next Meeting of States Parties or Review Conference. Any extension shall be the minimum necessary, but in no case shall the extension exceed 5 years, renewable upon the submission of a new request and its subsequent approval by the same procedure.

Problems associated with implementation of Article 6 para. 1 should be brought to the attention of a Meeting of States Parties or Review Conference. In doing so a State Party shall submit the detailed analysis described in para. c above and an estimate of the period in which it expects to fulfill the requirement to destroy anti-personnel mines in mined areas.

This extension mechanism, based on that contained in section C (paras. 24–28) of the Verification Annex of the 1993 Chemical Weapons Convention, would serve several purposes:

1. It would prevent an undermining of the principle, recently established in amended Protocol II of the CCW, that those who use mines are responsible for their clearance (This principle could be undermined by an indefinite extension);

2. It would provide a means for mine affected States which could not fulfill the obligations of Articles 5 or 6 to adhere to other provisions of the Convention and to draw attention to their specific problems in the field of clearance and destruction of anti-personnel mines.

3. It would provide for review of the obligation to remove and destroy all anti-personnel mines in light of mine-clearance technologies available during the proposed extension period.

Article 10

The ICRC intends to propose, during the Oslo negotiations, specific language to strengthen the provisions of this Article.

In our experience national implementation is facilitated by specific description, in the treaty itself, of the types of measures which States are expected to undertake. In the context of this Convention such measures would include:

(a) national penal sanctions for persons who develop, use, produce, acquire, stockpile, retain or transfer anti-personnel mines or who assist in the commission of such acts in violation of the Convention;
(b) a requirement to try or extradite nationals of States Parties alleged to have used, or to have ordered the use of, anti-personnel mines, regardless of the place of the offence;
(c) a requirement for Parties to undertake other legal and administrative measures necessary to prevent and suppress violations; and
(d) a provision on dissemination of the provisions of the Convention and related national law and administrative measures, so that these provisions may become known to the armed forces and to the general population.

Article 11

No comment

Article 12

In para. 4 we wonder if the reference to the UN, ICRC and other organizations might be strengthened to say *may attend these meetings as observers in accordance with the agreed Rules of Procedure*, with similar language being added to para. 3 of Article 13

Article 13

We suggest that a paragraph be added concerning decision making in Review Conferences. The rules on decision making should be the same as for the Oslo Conference itself, which are identical to those used for the adoption of the Additional Protocols of 1977 and similar to those used for the adoption of the Geneva Conventions of 1949.

Articles 14 and 15

No comment

Article 16

We consider that the threshold of 40 ratifications for entry into force is too high, particularly given that the threshold for the CCW Convention was only 20. Humanitarian law instruments such as the Geneva Conventions and Additional Protocols of 1977 entered into force upon the deposit of two ratifications. We would also favour entry into force for a particular State Party upon, rather than six months after, their deposit of the instrument of ratification, acceptance, approval or accession. As time delays for stockpiling and destruction of mines are already provided for, this would affect only development, use and transfer. Entry into force of the Convention is unlikely before 1999, at the earliest. Thus, avoiding further delay in ending the use of anti-personnel mines is of major importance.

Article 17

The ICRC is strongly opposed to permitting any reservations to the provisions of this Convention and supports the current text.

Article 18

We support the text as written. We consider that any withdrawal provision permitting the resumption of anti-personnel mine use during an armed conflict would go against the spirit and purposes of this Convention.

Articles 19 and 20

No comment

Austrian draft
13 May 1997

CONVENTION ON THE PROHIBITION OF THE USE, STOCKPILING, PRODUCTION AND TRANSFER OF ANTI-PERSONNEL MINES AND ON THEIR DESTRUCTION

The States Parties

Recalling that . . .

Have agreed as follows:

Article 1

General obligations

1. Each State Party undertakes never under any circumstances:

a) to use anti-personnel mines;

b) to develop, produce, otherwise acquire, stockpile, retain or transfer to anyone, directly or indirectly, anti-personnel mines;

c) to assist, encourage or induce, in any way, anyone to engage in any activity prohibited to a State Party under this Convention.

2. Each State Party undertakes to destroy all anti-personnel mines in accordance with the provisions of this Convention.

Article 2

Definitions

1. "Anti-personnel mines" means a mine designed to be exploded by the presence, proximity, or contact of a person and that will incapacitate, injure or kill one or more persons. Mines designed to be detonated by the presence, proximity or contact of a vehicle as opposed to a person, that are equipped with anti-handling devices, are not considered anti-personnel as a result of being so equipped.

2. "Mine" means a munition designed to be placed under, on or near the ground or other surface area and to be exploded by the presence, proximity or contact of a person or a vehicle.

3. "Anti-handling device" means a device intended to protect a mine and which is part of, linked to, attached to or placed under the mine and which activates when an attempt is made to tamper with the mine.

4. "Transfer" involves, in addition to the physical movement of anti-personnel mines into or from national territory, the transfer of title to and control over the mines, but does not involve the transfer of territory containing emplaced anti-personnel mines.

5. "Minefield" is a defined area in which mines have been emplaced.

Article 3

Exceptions

1. Notwithstanding the general obligations under Article 1, a State Party may retain or transfer a number of anti-personnel mines necessary for the development and teaching of mine detection, mine clearance, or mine destruction

techniques. The number of such mines shall not exceed that absolutely necessary for the above mentioned purposes.

2. The transfer of anti-personnel mines for the purpose of destruction is permitted.

Article 4

Destruction of stockpiled anti-personnel mines

Each State Party undertakes to destroy all stockpiled anti-personnel mines it owns or possesses, or that are under its jurisdiction or control, as soon as possible but not later than three years after the entry into force of this Convention for that State Party.

Article 5

Destruction of anti-personnel mines laid within minefields

1. Each State Party undertakes to destroy all anti-personnel mines laid within minefields under its jurisdiction or control, as soon as possible but not later that ten years after the entry into force of this Convention for that State Party.

2. Minefields where anti-personnel mines have been laid shall, until all anti-personnel mines contained therein have been destroyed, be perimeter-marked, monitored and protected by fencing or other means, to ensure the effective exclusion of civilians. The marking must be of a distinct and durable character and must at least be visible to a person who is about to enter the minefields.

Article 6

Destruction of anti-personnel mines laid in areas outside minefields

1. Each State Party undertakes to destroy, as soon as possible, all anti-personnel mines laid in areas under its jurisdiction or control outside minefields.

2. Each State Party shall make every effort to identify areas under its jurisdiction or control in which anti-personnel mines are known or suspected to be present and to provide an immediate and effective warning to the population.

Article 7

International cooperation and assistance

1. In fulfilling its obligations under this Convention each State Party has the right to seek and receive assistance from other States Parties to the extent possible.

2. Each State Party undertakes to facilitate and shall have the right to participate in the fullest possible exchange of equipment, material and scientific and technological information concerning the implementation of this Convention.

3. Each State Party in a position to do so shall provide assistance for mine clearance. Such assistance may be provided, inter alia, through the United Nations System, other international organizations or institutions, regional organizations, or on a bilateral basis, or by contributing to the United Nations Voluntary Trust Fund for Assistance in Mine Clearance.

4. Each State Party in a position to do so shall provide assistance for the destruction of stockpiled anti-personnel mines.

5. Each State Party undertakes to provide information to the database on mine clearance established within the United Nations System, especially information concerning various means and technologies of mine clearance, and list of experts, expert agencies or national points of contact on mine clearance.

Article 8

Transparency measures

1. Each State Party shall report to the Depositary not later than one year after the entry into force of this Convention for that State Party on:

    a) the national implementation measures referred to in Article 10;

    b) the types and quantities of all stockpiled anti-personnel mines owned or possessed by it, or under its jurisdiction or control;

    c) the location of all minefields under its jurisdiction or control containing anti-personnel mines;

    d) to the extent possible, the location of all areas outside minefields in which anti-personnel mines are known or suspected to be present;

    e) the types and quantities of all anti-personnel mines retained or transferred for the development and teaching of mine detection, mine clearance or mine destruction techniques, or transferred for the purpose of destruction, as well as the names and the institutions authorized by a State Party to retain or transfer anti-personnel mines, in accordance with Article 3;

    f) the status of programmes for the conversion or de-commissioning of anti-personnel mine production facilities; and

    g) the status of programmes for the destruction of anti-personnel mines in accordance with Articles 4, 5 and 6, including applicable safety and environmental standards to be observed.

2. The information provided in accordance with this Article shall be updated by the States Parties annually, covering the last calendar year, and reported to the Depositary not later than 1 March.

Article 9
Facilitation and clarification of compliance

1. The States Parties agree to consult and cooperate with each other regarding the implementation of the provisions of this Convention and to facilitate compliance by States Parties with their obligations under this Convention.

2. If one or more States Parties wish to clarify and seek to resolve questions relating to compliance with the provisions of this Convention by another State Party, it may submit, through the Depositary, a Request for Clarification of that matter to that State Party. Such a request shall be accompanied by all relevant information. A State Party that receives a Request for Clarification shall provide through the Depositary, within 30 days to the requesting State Party all information which would assist in clarifying this matter.

3. If the requesting State Party does not receive a response through the Depositary within the time period provided for in para. 2, or deems the response to the Request for Clarification to be unsatisfactory, it may submit through the Depositary, the matter to a Meeting of the State Parties. This submission shall include all relevant information. The Meeting of State Parties shall consider the matter at its next meeting.

4. The requesting State Party may propose to the Depositary to convene a Special Meeting of the States Parties to consider the matter. The Depositary shall thereupon communicate this proposal to the States Parties with a request that they indicate whether they favour a Special Meeting of the States Parties, for the purpose of considering the matter. In the event that within 30 days from the date of such communication, at least one third of the States Parties favour such a Special Meeting the Depositary shall convene this Special Meeting of the States Parties.

5. All States Parties shall cooperate fully with the Meeting of States Parties in the fulfilment of its review of the matter.

6. The Meeting of States Parties may request additional information from a State Party on matters under its consideration. If it concludes that it is necessary, the Meeting of States Parties may authorize a fact-finding mission to collect additional information on the spot or in other places directly related to the alleged compliance issue under the jurisdiction or control of the concerned State Party.

7. The fact-finding mission shall report, through the Depositary, to the States Parties and the Meeting of the States Parties the results of its findings, including any recommendations it considers appropriate in order to facilitate compliance with this Convention.

Article 10

National implementation measures

Each State Party shall take all appropriate legal, administrative and other measures, including the imposition of penal sanctions to prevent and suppress any activity prohibited to a State Party under this Convention undertaken by persons or on territory under its jurisdiction or control.

Article 11

Settlement of disputes

The States Parties undertake to consult each other and to cooperate with each other to resolve any problems that may arise with regard to the application and interpretation of this Convention. Each State Party may report any such problem to the Meeting of the States Parties.

Article 12

Meetings of the States Parties

1. The States Parties shall meet regularly in order to consider any matter with regard to the application or implementation of this Convention, including:

a) the operation and status of this Convention;

b) matters arising from the reports submitted under the provisions of this Convention;

c) international cooperation and assistance in accordance with Article 7;

d) the development of technologies to clear anti-personnel mines;

e) submissions of States Parties under Article 9.

2. The First Meeting of the States Parties shall be convened by the Depositary within one year after the entry into force of this Convention. The subsequent meetings shall be convened by the Depositary annually until the first Review Conference.

3. Under the conditions set out in Article 9, the Depositary shall convene a Special Meeting of the States Parties.

4. States not parties to this Convention, as well as the United Nations, other relevant international organizations or institutions, regional organizations, the International Committee of the Red Cross and relevant non-governmental organizations may be invited to attend these meetings as observers in accordance with the agreed Rules of Procedure.

5. The costs of the Meetings of the States Parties shall be borne by the States Parties and States not parties to this Convention participating in the meeting,

in accordance with the United Nations scale of assessment adjusted appropriately.

Article 13
Review Conferences

1. A Review Conference shall be convened by the Depositary five years after the entry into force of this Convention, and thereafter at five years' interval. All States Parties to this Convention shall be invited to the Review Conference.

2. The purpose of the Review Conference shall be to

a) review the operation and status of this Convention;

b) consider any proposal for amendments to this Convention;

c) consider the need for and the interval between further Meetings of the States Parties referred to in para. 2 of Article 12.

3. States not parties to this Convention, as well as the United Nations, other relevant international organizations or institutions, regional organizations, the International Committee of the Red Cross and relevant non-governmental organizations may be invited to attend the Review Conference as observers in accordance with the agreed Rules of Procedure.

4. The costs of the Review Conference shall be borne by the States Parties and States not parties to this Convention participating in the Review Conference; in accordance with the United Nations scale of assessment adjusted appropriately.

Article 14
Signature

This Convention shall be open to all States for signature before its entry into force.

Article 15
Ratification, acceptance, approval or accession

This convention is subject to ratification, acceptance or approval by the Signatories. Any State which has not signed this Convention before its entry into force may accede to it at any time.

Article 16
Entry into force

1. This Convention shall enter into force on the first day of the sixth month after the month in which the 40th instrument of ratification, acceptance or approval has been deposited.

2. For any State which deposits its instrument of ratification, acceptance, approval or accession after the date of the deposit of the 40th instrument of ratification, acceptance, approval or accession, this Convention shall enter into force six months after the date on which that State has deposited its instrument of ratification, acceptance, approval or accession.

Article 17

Reservations

The Articles of this Convention shall not be subject to reservations.

Article 18

Duration and withdrawal

1. This Convention shall be of unlimited duration.

2. Each State Party shall, in exercising its national sovereignty, have the right to withdraw from this Convention if it decides that extraordinary events, related to the subject-matter of this Convention, have jeopardized the supreme interests of its country. It shall give notice of such withdrawal one year in advance to all other States Parties, to the Depositary and to the United Nations Security Council. Such notice shall include a statement of the extraordinary events it regards as having jeopardized its supreme interests.

3. In case a State Party should withdraw from this Convention, the withdrawal shall only take effect one year after receipt of the instrument of withdrawal by the Depositary. If, however, on the expiry of that year the withdrawing State Party is engaged in an armed conflict the withdrawal shall not take effect before the end of the armed conflict.

4. The withdrawal of a State Party from this Convention shall not in any way affect the duty of States to continue fulfilling the obligations assumed under any relevant rules of international law.

Article 19

Depositary

The Secretary-General of the United Nations is hereby designated as the Depositary of this Convention and shall, inter alia,

a) Promptly inform all signatory and acceding States of the date of each signature, the date of deposit of each instrument of ratification, acceptance, approval or accession and the date of the entry into force of this Convention;

b) Transmit duly certified copies of this Convention to the Governments of all signatory and acceding States;

c) Register this Convention pursuant to Article 102 of the Charter of the United Nations;

d) Transmit to the States Parties not later than two months before the next Meeting of the States Parties the reports received under Article 8;

e) Transmit to the States Parties the submissions received under Article 9; and

f) Convene the Meetings of the States Parties referred to in Article 12 and the Review Conferences referred to in Article 13.

Article 20

Authentic texts

The original of this Convention, of which the Arabic, Chinese, English, French, Russian and Spanish texts are equally authentic, shall be deposited with the Secretary-General of the United Nations.

15

Anti-personnel Mines: What Future for Asia?

Regional Seminar for Asian Military and Strategic Studies Experts
Manila, The Philippines
20–23 July 1997
Sponsored by the ICRC in cooperation with the government of the Republic of the Philippines and the Philippine National Red Cross Society

Prior to the Manila Seminar, there had been little governmental discussion of the mines problem in Asia – even though a number of countries across the region were severely affected – and few Asian governments were prepared to support a total prohibition on anti-personnel mines. The Manila Seminar, which brought together regional military and strategic studies experts, was intended to consider the need for such a prohibition, balancing the military utility of landmines with their humanitarian costs. Discussions were extremely focused and, although several participants were unable fully to support the conclusions of the seminar, it was generally agreed that a total ban on anti-personnel mines was a necessary and justified objective.

Final Declaration of Participants to the Regional Seminar for Asian Military and Strategic Studies Experts

Manila, The Philippines
23 July 1997
(with corrected signatory list as at 24.07.97)

The undersigned Asian military and strategic studies analysts from 14 countries gathered in Manila to examine the experience of anti-personnel mine use in the region. The analysts discussed the military effectiveness of anti-personnel mines based on their actual combat performance in Asian and other conflicts. The military value of AP mines was considered in the context of the long-term human, social and economic costs incurred in many of the conflicts in which this weapon has been used. Particular attention was given to the difficulties and extremely high costs of post-conflict mine clearance.

The seminar sought to develop recommendations which will promote and broaden dialogue within Asian military and political circles on the question of anti-personnel mines. It is hoped that the work of the Manila seminar can contribute to the development of a common approach within the region to the humanitarian problems which anti-personnel mines have caused in Asia and globally. The following statement was adopted by participants acting in their personal capacities.

The undersigned participants in the regional seminar "Anti-personnel Landmines: What Future for Asia?" agree that:

1. **The global scourge of anti-personnel landmines, which kill and injure some 2,000 persons per month, most of whom are civilians, is unacceptable and must be stopped. These mines not only kill and maim combatants in an inhumane way, but also indiscriminately affect civilians and inflict on them enormous physical and psychological damage long after the conflict is over. This must be a grave and continuous concern of the international community;**

2. **In most conflicts, the appalling humanitarian consequences of the use of anti-personnel mines have far outweighed their military utility;**

3. The use of anti-personnel landmines in internal armed conflicts, either by State or non-State actors, should not be condoned;

4. The cases considered during the seminar, and the personal experience of participants, lead to some initial conclusions concerning **traditionally emplaced mines**:

• Establishing, monitoring and maintaining extensive border minefields is time-consuming, expensive and dangerous. In order to be effective they need to be under continuous observation and direct fire, which is not always possible. Because of these practical difficulties some armed forces have entirely refrained from using such minefields. Moreover, these minefields have not always proved successful in preventing infiltration.

• Under battlefield conditions the use, marking, mapping and removal of mines in accordance with classical military doctrine and international humanitarian law is extremely difficult, even for professional armed forces. History indicates that effective marking, mapping and removal of mines have rarely occurred [1]. The cost to forces using anti-personnel mines, in terms of casualties to one's own forces and civilians, the limitation of tactical flexibility and the loss of sympathy of the indigenous population is higher than has been generally acknowledged.

• Use in accordance with traditional military doctrine appears to have occurred infrequently and only when the following specific conditions were met:

– both parties to the conflict were disciplined professional armies with a high sense of responsibility and engaged in a short-lived international conflict,
– the tactical situations were fairly static,
– mines were not a major component of the conflict,
– forces possessed adequate time and resources to mark, monitor and maintain minefields in accordance with law and doctrine,
– mined areas were of sufficient economic or military value to ensure that mine clearance occurred,
– parties had sufficient resources to ensure clearance and it was carried out without delay, and
– the political will existed to limit strictly the use of mines and to clear them as indicated above;

5. Remotely delivered anti-personnel mines are not exclusively defensive weapons. They can easily be used in an offensive manner behind frontlines to prevent reinforcement and escape, and to saturate target areas.

Remotely delivered anti-personnel mines can cause vastly increased civilian casualties, even if such mines are designed to be self-destructing and self-deactivating, for the following reasons:

• they will be dangerous during their intended active life-time,
• the fencing and marking of such mines will be virtually impossible,
• in extended conflicts minefields may be re-laid many times,
• self-destructing and deactivating devices may be unreliable,
• inactive mines, as unexploded ordnance, can still be dangerous, and
• the mere presence of mined areas will produce fear, keeping civilians out of areas important for their livelihood;

6. Some barrier systems and other methods offer **more humane alternatives to anti-personnel mines** under certain circumstances. Additional alternatives should be pursued rather than further development of any new anti-personnel mine technologies. Developments which further increase, rather than reduce, the lethality of anti-personnel mines are to be deplored and are unnecessary;

7. Those who have used and those who have supplied anti-personnel mines bear a joint responsibility to ensure the clearance of these weapons and the provision of adequate care to their victims;

8. Improved mine clearance technologies for military, humanitarian and civilian agencies that are affordable and easy to use should be vigorously developed with a goal of making the use of anti-personnel mines progressively less useful;

9. Since resources are not currently available even to clear mines currently in the ground, any attempt to deploy additional anti-personnel mines is likely to impose an unacceptable level of cost to countries that are least able to bear it;

10. Countries in Asia, including Afghanistan, Cambodia, Laos and Vietnam are among those most affected by anti-personnel mines and similar remnants of war; and

11. Notwithstanding successive UN resolutions since 1994 calling for increased assistance by all States to mine-affected countries, the actual assistance rendered has fallen far short of the requirements.

The undersigned participants therefore call upon States of the Asian region to consider the following urgent measures:

1. The adoption of national prohibitions on the production, stockpiling, transfer and use of anti-personnel mines;

2. For those States which are not yet Parties, adherence to the 1980 United Nations Convention on Certain Conventional Weapons, including its Protocol II on landmines (as amended on 3 May 1996), and for current States party to this Convention that have not yet done so adherence to its amended Protocol II at the earliest possible date to ensure its early entry into force;

3. A substantial increase in assistance to mine-affected countries in the region, including Afghanistan, Cambodia, Laos and Vietnam. Such assistance might include provision of trained manpower, specialized equipment and funds to cope with the problems of landmines laid in those countries. The delivery of such assistance should be considered a purely humanitarian measure, should be free of political considerations, and should not be at the expense of other forms of humanitarian assistance;

4. The initiation, through all appropriate institutions, including the Asian Development Bank, of programs of regional cooperation in the fields of mine clearance, mine-risk education and victim assistance;

5. The rapid adoption of a regional agreement to prohibit remotely delivered anti-personnel landmines in Asia so as to prevent an escalation of mine warfare in the region and even higher levels of civilian casualties; and

6. Participation in upcoming negotiations aimed at the conclusion of a new treaty comprehensively prohibiting anti-personnel landmines by the end of 1997.

7. To build on and work towards the implementation of United Nations General Assembly Resolution 51/45S calling for the conclusion of a legally-binding agreement totally prohibiting anti-personnel landmines.

The undersigned participants appeal to the international community:

1. To pursue as a matter of urgency the prohibition and elimination of anti-personnel mines;

2. For those States which are not yet Parties, to adhere to the 1980 United Nations Convention on Certain Conventional Weapons, including its Protocol II on landmines (as amended on 3 May 1996), and for current States party to this Convention that have not yet done so to adhere to its amended Protocol II at the earliest possible date to ensure its early entry into force;

3. To recognize that the use of anti-personnel landmines in internal armed conflicts, either by State or non-State actors, should not be condoned;

4. To explore how non-State actors involved in internal armed conflicts can be encouraged to end the use of anti-personnel mines;

5. To assist mine-affected countries in Asia in ending the scourge of anti-personnel mines on their soil, in particular through the provision of technical, financial and other assistance in the clearance and destruction of mines, assistance to victims and mine awareness programs; and

6. To adopt a compassionate approach to the reunification of mine victims with family members living in mine-free countries.

Participants express their thanks to the International Committee of the Red Cross for convening the seminar, and for its ongoing efforts on behalf of war victims in many of the countries of the region and to the **Government of the Republic of the Philippines and Philippine National Red Cross for the generous hospitality they have provided in Manila.**

NOTE

1 According to a number of participants these requirements were successfully carried out in the India-Pakistan wars.

16

Diplomatic Conference on an International Total Ban on Anti-personnel Landmines

Oslo, Norway

1–18 September 1997

Organized by the government of Norway

The Diplomatic Conference opened in Oslo on 1 September 1997 with its work sharply focused in the international media by the tragic death of Diana, Princess of Wales, the day before its opening. The necessary political will clearly existed to ensure a successful outcome – indeed a good text (the Conference had before it the third Austrian draft treaty) was even strengthened in a number of key areas. An important factor, however, was the choice of the South African Ambassador, Jacob Selebi, to chair the Conference. His skill and dedication, supported in particular by the united voice of African governments and strong commitment of many others, ensured that compromises which would have substantially weakened the treaty were resisted. The Conference was able to adopt formally, on 18 September 1997, the Convention on the Prohibition of the Use, Stockpiling, Production and Transfer of Anti-Personnel Mines and on their Destruction.

Press Release

Geneva, 26 August 1997

ANTI-PERSONNEL MINES: CRUCIAL DECISIONS FOR OSLO CONFERENCE

On behalf of landmine victims worldwide, the International Committee of the Red Cross (ICRC) is renewing its appeal to States to conclude a treaty prohibiting anti-personnel landmines at the Oslo Diplomatic Conference due to open on 1 September. More than 100 pro-ban countries are expected to be represented at the Conference, which is scheduled to last three weeks. The ICRC urges these States to stand firm in their commitment to a total ban.

The ICRC maintains that a total and immediate prohibition on the use of all anti-personnel mines is essential for the effectiveness and credibility of the treaty. Such an unequivocal ban has been crucial to the success of similar controls on weapons imposed by humanitarian law, such as the Protocol on blinding laser weapons negotiated in September 1995, the Geneva Protocol on poison gas adopted in 1925 and the prohibition on expanding (Dum-Dum) bullets agreed in 1899.

The ICRC believes that the definition of an anti-personnel mine in the treaty should be clear and unambiguous, and should encompass all types of anti-personnel mines, whether used on their own or in conjunction with other weapons. In this regard, the ICRC supports the definition proposed by the government of Austria.

The ICRC has made a strong appeal for a "no reservations" clause to be maintained in the final version of the treaty, similar to the provision in the 1993 Chemical Weapons Convention. This will ensure that all States are subject to the same obligations and will thereby build confidence in the treaty's implementation.

Every year an estimated 24,000 people, mostly civilians, are killed or injured by anti-personnel mines and millions more lives are disrupted by the loss of fertile agricultural land and by problems of access to food and water. An unambiguous and legally binding treaty banning anti-personnel mines is central to efforts to put an end to this humanitarian emergency.

CONVENTION ON THE PROHIBITION OF THE USE, STOCKPILING, PRODUCTION AND TRANSFER OF ANTI-PERSONNEL MINES AND ON THEIR DESTRUCTION

Preamble

The States Parties,

Determined to put an end to the suffering and casualties caused by anti-personnel mines, that kill or maim hundreds of people every week, mostly innocent and defenceless civilians and especially children, obstruct economic development and reconstruction, inhibit the repatriation of refugees and internally displaced persons, and have other severe consequences for years after emplacement,

Believing it necessary to do their utmost to contribute in an efficient and coordinated manner to face the challenge of removing anti-personnel mines placed throughout the world, and to assure their destruction,

Wishing to do their utmost in providing assistance for the care and rehabilitation, including the social and economic reintegration of mine victims,

Recognizing that a total ban of anti-personnel mines would also be an important confidence-building measure,

Welcoming the adoption of the Protocol on Prohibitions or Restrictions on the Use of Mines, Booby-Traps and Other Devices, as amended on 3 May 1996, annexed to the Convention on Prohibitions or Restrictions on the Use of Certain Conventional Weapons Which May Be Deemed to Be Excessively Injurious or to Have Indiscriminate Effects, and calling for the early ratification of this Protocol by all States which have not yet done so,

Welcoming also United Nations General Assembly Resolution 51/45 S of 10 December 1996 urging all States to pursue vigorously an effective, legally-binding international agreement to ban the use, stockpiling, production and transfer of anti-personnel landmines,

Welcoming furthermore the measures taken over the past years, both unilaterally and multilaterally, aiming at prohibiting, restricting or suspending the use, stockpiling, production and transfer of anti-personnel mines,

Stressing the role of public conscience in furthering the principles of humanity as evidenced by the call for a total ban of anti-personnel mines and recognizing the efforts to that end undertaken by the International Red Cross and Red Crescent Movement, the International Campaign to Ban Landmines and numerous other non-governmental organizations around the world,

Recalling the Ottawa Declaration of 5 October 1996 and the Brussels Declaration of 27 June 1997 urging the international community to negotiate an international and legally binding agreement prohibiting the use, stockpiling, production and transfer of anti-personnel mines,

Emphasizing the desirability of attracting the adherence of all States to this Convention, and determined to work strenuously towards the promotion of its universalization in all relevant fora including, inter alia, the United Nations, the Conference on Disarmament, regional organizations, and groupings, and review conferences of the Convention on Prohibitions or Restrictions on the Use of Certain Conventional Weapons Which May Be Deemed to Be Excessively Injurious or to Have Indiscriminate Effects,

Basing themselves on the principle of international humanitarian law that the right of the parties to an armed conflict to choose methods or means of warfare is not unlimited, on the principle that prohibits the employment in armed conflicts of weapons, projectiles and materials and methods of warfare of a nature to cause superfluous injury or unnecessary suffering and on the principle that a distinction must be made between civilians and combatants,

Have agreed as follows:

Article 1 – General obligations

1. Each State Party undertakes never under any circumstances:

a) To use anti-personnel mines;

b) To develop, produce, otherwise acquire, stockpile, retain or transfer to anyone, directly or indirectly, anti-personnel mines;

c) To assist, encourage or induce, in any way, anyone to engage in any activity prohibited to a State Party under this Convention.

2. Each State Party undertakes to destroy or ensure the destruction of all anti-personnel mines in accordance with the provisions of this Convention.

Article 2 – Definitions

1. "Anti-personnel mine" means a mine designed to be exploded by the presence, proximity or contact of a person and that will incapacitate, injure or kill one or more persons. Mines designed to be detonated by the presence, proximity or contact of a vehicle as opposed to a person, that are equipped with anti-handling devices, are not considered anti-personnel mines as a result of being so equipped.

2. "Mine" means a munition designed to be placed under, on or near the ground or other surface area and to be exploded by the presence, proximity or contact of a person or a vehicle.

3. "Anti-handling device" means a device intended to protect a mine and which is part of, linked to, attached to or placed under the mine and which activates when an attempt is made to tamper with or otherwise intentionally disturb the mine.

4. "Transfer" involves, in addition to the physical movement of anti-personnel mines into or from national territory, the transfer of title to and control over the mines, but does not involve the transfer of territory containing emplaced anti-personnel mines.

5. "Mined area" means an area which is dangerous due to the presence or suspected presence of mines.

Article 3 – Exceptions

1. Notwithstanding the general obligations under Article 1, the retention or transfer of a number of anti-personnel mines for the development of and training in mine detection, mine clearance, or mine destruction techniques is permitted.

The amount of such mines shall not exceed the minimum number absolutely necessary for the above-mentioned purposes.

2. The transfer of anti-personnel mines for the purpose of destruction is permitted.

Article 4 – Destruction of stockpiled anti-personnel mines

Except as provided for in Article 3, each State Party undertakes to destroy or ensure the destruction of all stockpiled anti-personnel mines it owns or possesses, or that are under its jurisdiction or control, as soon as possible but not later than four years after the entry into force of this Convention for that State Party.

Article 5 – Destruction of anti-personnel mines in mined areas

1. Each State Party undertakes to destroy or ensure the destruction of all anti-personnel mines in mined areas under its jurisdiction or control, as soon as possible but not later than ten years after the entry into force of this Convention for that State Party.

2. Each State Party shall make every effort to identify all areas under its jurisdiction or control in which anti-personnel mines are known or suspected to be emplaced and shall ensure as soon as possible that all anti-personnel mines in mined areas under its jurisdiction or control are perimeter-marked, monitored and protected by fencing or other means, to ensure the effective exclusion of civilians, until all anti-personnel mines contained therein have been destroyed. The marking shall at least be to the standards set out in the Protocol on Prohibitions or Restrictions on the Use of Mines, Booby-Traps and Other Devices, as amended on 3 May 1996, annexed to the Convention on Prohibitions or Restrictions on the Use of Certain Conventional Weapons Which May Be Deemed to Be Excessively Injurious or to Have Indiscriminate Effects.

3. If a State Party believes that it will be unable to destroy or ensure the destruction of all anti-personnel mines referred to in paragraph 1 within that time period, it may submit a request to a Meeting of the States Parties or a Review Conference for an extension of the deadline for completing the destruction of such anti-personnel mines, for a period of up to ten years.

4. Each request shall contain:

a) The duration of the proposed extension;

b) A detailed explanation of the reasons for the proposed extension, including:

   (i) The preparation and status of work conducted under national demining programs;

(ii) The financial and technical means available to the State Party for the destruction of all the anti-personnel mines; and

(iii) Circumstances which impede the ability of the State Party to destroy all the anti-personnel mines in mined areas;

c) The humanitarian, social, economic, and environmental implications of the extension; and

d) Any other information relevant to the request for the proposed extension.

5. The Meeting of the States Parties or the Review Conference shall, taking into consideration the factors contained in paragraph 4, assess the request and decide by a majority of votes of States Parties present and voting whether to grant the request for an extension period.

6. Such an extension may be renewed upon the submission of a new request in accordance with paragraphs 3, 4 and 5 of this Article. In requesting a further extension period a State Party shall submit relevant additional information on what has been undertaken in the previous extension period pursuant to this Article.

Article 6 – International cooperation and assistance

1. In fulfilling its obligations under this Convention each State Party has the right to seek and receive assistance, where feasible, from other States Parties to the extent possible.

2. Each State Party undertakes to facilitate and shall have the right to participate in the fullest possible exchange of equipment, material and scientific and technological information concerning the implementation of this Convention. The States Parties shall not impose undue restrictions on the provision of mine clearance equipment and related technological information for humanitarian purposes.

3. Each State Party in a position to do so shall provide assistance for the care and rehabilitation, and social and economic reintegration, of mine victims and for mine awareness programs. Such assistance may be provided, inter alia, through the United Nations system, international, regional or national organizations or institutions, the International Committee of the Red Cross, national Red Cross and Red Crescent societies and their International Federation, non-governmental organizations, or on a bilateral basis.

4. Each State Party in a position to do so shall provide assistance for mine clearance and related activities. Such assistance may be provided, inter alia, through the United Nations system, international or regional organizations or institutions, non-governmental organizations or institutions, or on a bilateral basis, or

by contributing to the United Nations Voluntary Trust Fund for Assistance in Mine Clearance, or other regional funds that deal with demining.

5. Each State Party in a position to do so shall provide assistance for the destruction of stockpiled anti-personnel mines.

6. Each State Party undertakes to provide information to the database on mine clearance established within the United Nations system, especially information concerning various means and technologies of mine clearance, and lists of experts, expert agencies or national points of contact on mine clearance.

7. States Parties may request the United Nations, regional organizations, other States Parties or other competent intergovernmental or non-governmental fora to assist its authorities in the elaboration of a national demining program to determine, inter alia:

a) The extent and scope of the anti-personnel mine problem;

b) The financial, technological and human resources that are required for the implementation of the program;

c) The estimated number of years necessary to destroy all anti-personnel mines in mined areas under the jurisdiction or control of the concerned State Party;

d) Mine awareness activities to reduce the incidence of mine-related injuries or deaths;

e) Assistance to mine victims;

f) The relationship between the Government of the concerned State Party and the relevant governmental, inter-governmental or non-governmental entities that will work in the implementation of the program.

8. Each State Party giving and receiving assistance under the provisions of this Article shall cooperate with a view to ensuring the full and prompt implementation of agreed assistance programs.

Article 7 – Transparency measures

1. Each State Party shall report to the Secretary-General of the United Nations as soon as practicable, and in any event not later than 180 days after the entry into force of this Convention for that State Party on:

a) The national implementation measures referred to in Article 9;

b) The total of all stockpiled anti-personnel mines owned or possessed by it, or under its jurisdiction or control, to include a breakdown of the type, quantity and, if possible, lot numbers of each type of anti-personnel mine stockpiled;

c) To the extent possible, the location of all mined areas that contain, or are suspected to contain, anti-personnel mines under its jurisdiction or control, to include as much detail as possible regarding the type and quantity of each type of anti-personnel mine in each mined area and when they were emplaced;

d) The types, quantities and, if possible, lot numbers of all anti-personnel mines retained or transferred for the development of and training in mine detection, mine clearance or mine destruction techniques, or transferred for the purpose of destruction, as well as the institutions authorized by a State Party to retain or transfer anti-personnel mines, in accordance with Article 3;

e) The status of programs for the conversion or de-commissioning of anti-personnel mine production facilities;

f) The status of programs for the destruction of anti-personnel mines in accordance with Articles 4 and 5, including details of the methods which will be used in destruction, the location of all destruction sites and the applicable safety and environmental standards to be observed;

g) The types and quantities of all anti-personnel mines destroyed after the entry into force of this Convention for that State Party, to include a breakdown of the quantity of each type of anti-personnel mine destroyed, in accordance with Articles 4 and 5, respectively, along with, if possible, the lot numbers of each type of anti-personnel mine in the case of destruction in accordance with Article 4;

h) The technical characteristics of each type of anti-personnel mine produced, to the extent known, and those currently owned or possessed by a State Party, giving, where reasonably possible, such categories of information as may facilitate identification and clearance of anti-personnel mines; at a minimum, this information shall include the dimensions, fusing, explosive content, metallic content, colour photographs and other information which may facilitate mine clearance;

and

i) The measures taken to provide an immediate and effective warning to the population in relation to all areas identified under paragraph 2 of Article 5.

2. The information provided in accordance with this Article shall be updated by the States Parties annually, covering the last calendar year, and reported to the Secretary-General of the United Nations not later than 30 April of each year.

3. The Secretary-General of the United Nations shall transmit all such reports received to the States Parties.

Article 8 – Facilitation and clarification of compliance

1. The States Parties agree to consult and cooperate with each other regarding the implementation of the provisions of this Convention, and to work together in a spirit of cooperation to facilitate compliance by States Parties with their obligations under this Convention.

2. If one or more States Parties wish to clarify and seek to resolve questions relating to compliance with the provisions of this Convention by another State Party, it may submit, through the Secretary-General of the United Nations, a Request for Clarification of that matter to that State Party. Such a request shall be accompanied by all appropriate information. Each State Party shall refrain from unfounded Requests for Clarification, care being taken to avoid abuse. A State Party that receives a Request for Clarification shall provide, through the Secretary-General of the United Nations, within 28 days to the requesting State Party all information which would assist in clarifying this matter.

3. If the requesting State Party does not receive a response through the Secretary-General of the United Nations within that time period, or deems the response to the Request for Clarification to be unsatisfactory, it may submit the matter through the Secretary-General of the United Nations to the next Meeting of the States Parties. The Secretary-General of the United Nations shall transmit the submission, accompanied by all appropriate information pertaining to the Request for Clarification, to all States Parties. All such information shall be presented to the requested State Party which shall have the right to respond.

4. Pending the convening of any meeting of the States Parties, any of the States Parties concerned may request the Secretary-General of the United Nations to exercise his or her good offices to facilitate the clarification requested.

5. The requesting State Party may propose through the Secretary-General of the United Nations the convening of a Special Meeting of the States Parties to consider the matter. The Secretary-General of the United Nations shall thereupon communicate this proposal and all information submitted by the States Parties concerned, to all States Parties with a request that they indicate whether they favour a Special Meeting of the States Parties, for the purpose of considering the matter. In the event that within 14 days from the date of such communication, at least one-third of the States Parties favours such a Special Meeting, the Secretary-General of the United Nations shall convene this Special Meeting of the States Parties within a further 14 days. A quorum for this Meeting shall consist of a majority of States Parties.

6. The Meeting of the States Parties or the Special Meeting of the States Parties, as the case may be, shall first determine whether to consider the matter further,

taking into account all information submitted by the States Parties concerned. The Meeting of the States Parties or the Special Meeting of the States Parties shall make every effort to reach a decision by consensus. If despite all efforts to that end no agreement has been reached, it shall take this decision by a majority of States Parties present and voting.

7. All States Parties shall cooperate fully with the Meeting of the States Parties or the Special Meeting of the States Parties in the fulfilment of its review of the matter, including any fact-finding missions that are authorized in accordance with paragraph 8.

8. If further clarification is required, the Meeting of the States Parties or the Special Meeting of the States Parties shall authorize a fact-finding mission and decide on its mandate by a majority of States Parties present and voting. At any time the requested State Party may invite a fact-finding mission to its territory. Such a mission shall take place without a decision by a Meeting of the States Parties or a Special Meeting of the States Parties to authorize such a mission. The mission, consisting of up to 9 experts, designated and approved in accordance with paragraphs 9 and 10, may collect additional information on the spot or in other places directly related to the alleged compliance issue under the jurisdiction or control of the requested State Party.

9. The Secretary-General of the United Nations shall prepare and update a list of the names, nationalities and other relevant data of qualified experts provided by States Parties and communicate it to all States Parties. Any expert included on this list shall be regarded as designated for all fact-finding missions unless a State Party declares its non-acceptance in writing. In the event of non-acceptance, the expert shall not participate in fact-finding missions on the territory or any other place under the jurisdiction or control of the objecting State Party, if the non-acceptance was declared prior to the appointment of the expert to such missions.

10. Upon receiving a request from the Meeting of the States Parties or a Special Meeting of the States Parties, the Secretary-General of the United Nations shall, after consultations with the requested State Party, appoint the members of the mission, including its leader. Nationals of States Parties requesting the fact-finding mission or directly affected by it shall not be appointed to the mission. The members of the fact-finding mission shall enjoy privileges and immunities under Article VI of the Convention on the Privileges and Immunities of the United Nations, adopted on 13 February 1946.

11. Upon at least 72 hours notice, the members of the fact-finding mission shall arrive in the territory of the requested State Party at the earliest opportunity. The requested State Party shall take the necessary administrative measures to receive,

transport and accommodate the mission, and shall be responsible for ensuring the security of the mission to the maximum extent possible while they are on territory under its control.

12. Without prejudice to the sovereignty of the requested State Party, the fact-finding mission may bring into the territory of the requested State Party the necessary equipment which shall be used exclusively for gathering information on the alleged compliance issue. Prior to its arrival, the mission will advise the requested State Party of the equipment that it intends to utilize in the course of its fact-finding mission.

13. The requested State Party shall make all efforts to ensure that the fact-finding mission is given the opportunity to speak with all relevant persons who may be able to provide information related to the alleged compliance issue.

14. The requested State Party shall grant access for the fact-finding mission to all areas and installations under its control where facts relevant to the compliance issue could be expected to be collected. This shall be subject to any arrangements that the requested State Party considers necessary for:

a) The protection of sensitive equipment, information and areas;

b) The protection of any constitutional obligations the requested State Party may have with regard to proprietary rights, searches and seizures, or other constitutional rights; or

c) The physical protection and safety of the members of the fact-finding mission.

In the event that the requested State Party makes such arrangements, it shall make every reasonable effort to demonstrate through alternative means its compliance with this Convention.

15. The fact-finding mission may remain in the territory of the State Party concerned for no more than 14 days, and at any particular site no more than 7 days, unless otherwise agreed.

16. All information provided in confidence and not related to the subject matter of the fact-finding mission shall be treated on a confidential basis.

17. The fact-finding mission shall report, through the Secretary-General of the United Nations, to the Meeting of the States Parties or the Special Meeting of the States Parties the results of its findings.

18. The Meeting of the States Parties or the Special Meeting of the States Parties shall consider all relevant information, including the report submitted by the

fact-finding mission, and may request the requested State Party to take measures to address the compliance issue within a specified period of time. The requested State Party shall report on all measures taken in response to this request.

19. The Meeting of the States Parties or the Special Meeting of the States Parties may suggest to the States Parties concerned ways and means to further clarify or resolve the matter under consideration, including the initiation of appropriate procedures in conformity with international law. In circumstances where the issue at hand is determined to be due to circumstances beyond the control of the requested State Party, the Meeting of the States Parties or the Special Meeting of the States Parties may recommend appropriate measures, including the use of cooperative measures referred to in Article 6.

20. The Meeting of the States Parties or the Special Meeting of the States Parties shall make every effort to reach its decisions referred to in paragraphs 18 and 19 by consensus, otherwise by a two-thirds majority of States Parties present and voting.

Article 9 – National implementation measures

Each State Party shall take all appropriate legal, administrative and other measures, including the imposition of penal sanctions, to prevent and suppress any activity prohibited to a State Party under this Convention undertaken by persons or on territory under its jurisdiction or control.

Article 10 – Settlement of disputes

1. The States Parties shall consult and cooperate with each other to settle any dispute that may arise with regard to the application or the interpretation of this Convention. Each State Party may bring any such dispute before the Meeting of the States Parties.

2. The Meeting of the States Parties may contribute to the settlement of the dispute by whatever means it deems appropriate, including offering its good offices, calling upon the States parties to a dispute to start the settlement procedure of their choice and recommending a time-limit for any agreed procedure.

3. This Article is without prejudice to the provisions of this Convention on facilitation and clarification of compliance.

Article 11 – Meetings of the States Parties

1. The States Parties shall meet regularly in order to consider any matter with regard to the application or implementation of this Convention, including:

a) The operation and status of this Convention;

b) Matters arising from the reports submitted under the provisions of this Convention;

c) International cooperation and assistance in accordance with Article 6;

d) The development of technologies to clear anti-personnel mines;

e) Submissions of States Parties under Article 8; and

f) Decisions relating to submissions of States Parties as provided for in Article 5.

2. The First Meeting of the States Parties shall be convened by the Secretary-General of the United Nations within one year after the entry into force of this Convention. The subsequent meetings shall be convened by the Secretary-General of the United Nations annually until the first Review Conference.

3. Under the conditions set out in Article 8, the Secretary-General of the United Nations shall convene a Special Meeting of the States Parties.

4. States not parties to this Convention, as well as the United Nations, other relevant international organizations or institutions, regional organizations, the International Committee of the Red Cross and relevant non-governmental organizations may be invited to attend these meetings as observers in accordance with the agreed Rules of Procedure.

Article 12 – Review Conferences

1. A Review Conference shall be convened by the Secretary-General of the United Nations five years after the entry into force of this Convention. Further Review Conferences shall be convened by the Secretary-General of the United Nations if so requested by one or more States Parties, provided that the interval between Review Conferences shall in no case be less than five years. All States Parties to this Convention shall be invited to each Review Conference.

2. The purpose of the Review Conference shall be:

a) To review the operation and status of this Convention;

b) To consider the need for and the interval between further Meetings of the States Parties referred to in paragraph 2 of Article 11;

c) To take decisions on submissions of States Parties as provided for in Article 5; and

d) To adopt, if necessary, in its final report conclusions related to the implementation of this Convention.

3. States not parties to this Convention, as well as the United Nations, other relevant international organizations or institutions, regional organizations, the International Committee of the Red Cross and relevant non-governmental organizations may be invited to attend each Review Conference as observers in accordance with the agreed Rules of Procedure.

Article 13 – Amendments

1. At any time after the entry into force of this Convention any State Party may propose amendments to this Convention. Any proposal for an amendment shall be communicated to the Depositary, who shall circulate it to all States Parties and shall seek their views on whether an Amendment Conference should be convened to consider the proposal. If a majority of the States Parties notify the Depositary no later than 30 days after its circulation that they support further consideration of the proposal, the Depositary shall convene an Amendment Conference to which all States Parties shall be invited.

2. States not parties to this Convention, as well as the United Nations, other relevant international organizations or institutions, regional organizations, the International Committee of the Red Cross and relevant non-governmental organizations may be invited to attend each Amendment Conference as observers in accordance with the agreed Rules of Procedure.

3. The Amendment Conference shall be held immediately following a Meeting of the States Parties or a Review Conference unless a majority of the States Parties request that it be held earlier.

4. Any amendment to this Convention shall be adopted by a majority of two-thirds of the States Parties present and voting at the Amendment Conference. The Depositary shall communicate any amendment so adopted to the States Parties.

5. An amendment to this Convention shall enter into force for all States Parties to this Convention which have accepted it, upon the deposit with the Depositary of instruments of acceptance by a majority of States Parties. Thereafter it shall enter into force for any remaining State Party on the date of deposit of its instrument of acceptance.

Article 14 – Costs

1. The costs of the Meetings of the States Parties, the Special Meetings of the States Parties, the Review Conferences and the Amendment Conferences shall be borne by the States Parties and States not parties to this Convention participating therein, in accordance with the United Nations scale of assessment adjusted appropriately.

2. The costs incurred by the Secretary-General of the United Nations under Articles 7 and 8 and the costs of any fact-finding mission shall be borne by the States Parties in accordance with the United Nations scale of assessment adjusted appropriately.

Article 15 – Signature

This Convention, done at Oslo, Norway, on 18 September 1997, shall be open for signature at Ottawa, Canada, by all States from 3 December 1997 until 4 December 1997, and at the United Nations Headquarters in New York from 5 December 1997 until its entry into force.

Article 16 – Ratification, acceptance, approval or accession

1. This Convention is subject to ratification, acceptance or approval of the Signatories.

2. It shall be open for accession by any State which has not signed the Convention.

3. The instruments of ratification, acceptance, approval or accession shall be deposited with the Depositary.

Article 17 – Entry into force

1. This Convention shall enter into force on the first day of the sixth month after the month in which the 40th instrument of ratification, acceptance, approval or accession has been deposited.

2. For any State which deposits its instrument of ratification, acceptance, approval or accession after the date of the deposit of the 40th instrument of ratification, acceptance, approval or accession, this Convention shall enter into force on the first day of the sixth month after the date on which that State has deposited its instrument of ratification, acceptance, approval or accession.

Article 18 – Provisional application

Any State may at the time of its ratification, acceptance, approval or accession, declare that it will apply provisionally paragraph 1 of Article 1 of this Convention pending its entry into force.

Article 19 – Reservations

The Articles of this Convention shall not be subject to reservations.

Article 20 – Duration and withdrawal

1. This Convention shall be of unlimited duration.

2. Each State Party shall, in exercising its national sovereignty, have the right to withdraw from this Convention. It shall give notice of such withdrawal to all other States Parties, to the Depositary and to the United Nations Security Council. Such instrument of withdrawal shall include a full explanation of the reasons motivating this withdrawal.

3. Such withdrawal shall only take effect six months after the receipt of the instrument of withdrawal by the Depositary. If, however, on the expiry of that six-month period, the withdrawing State Party is engaged in an armed conflict, the withdrawal shall not take effect before the end of the armed conflict.

4. The withdrawal of a State Party from this Convention shall not in any way affect the duty of States to continue fulfilling the obligations assumed under any relevant rules of international law.

Article 21 – Depositary

The Secretary-General of the United Nations is hereby designated as the Depositary of this Convention.

Article 22 – Authentic texts

The original of this Convention, of which the Arabic, Chinese, English, French, Russian and Spanish texts are equally authentic, shall be deposited with the Secretary-General of the United Nations.

Press Release

17 September 1997

OSLO CONFERENCE: LANDMINES BANNED

The International Committee of the Red Cross (ICRC) and the International Federation of Red Cross and Red Crescent Societies wholeheartedly welcome and endorse the new treaty of international humanitarian law forever banning anti-personnel mines, which was concluded earlier today at a Diplomatic Conference in Oslo. "This is a victory for humanity. It shows that it is possible, with determination and perseverance, to make significant improvements in international humanitarian law", said Louise Doswald-Beck, head of the ICRC delegation to the Conference.

The treaty contains an unambiguous prohibition on the use, production, stockpiling and transfer of anti-personnel mines. It requires the destruction of stockpiled mines within four years and of mines in the ground within ten years.

Since the ICRC launched its appeal for a total ban on anti-personnel mines in February 1994, the entire International Red Cross and Red Crescent Movement has worked strenuously to raise public awareness of the problem and to encourage diplomatic and military circles to strive for this historic breakthrough. "The Red Cross and Red Crescent Movement will do its utmost to achieve universal ratification of the treaty and to ensure its earliest possible entry into force so as to end once and for all the horrendous suffering caused by anti-personnel mines," added Louise Doswald-Beck.

The treaty will be formally adopted by the Oslo Diplomatic Conference on Thursday and opened for signature in Ottawa, Canada, on 3–4 December. The vast majority of the 106 States that officially launched the process in Brussels in June 1997 are expected to sign the treaty, which will enter into force once 40 signatory States have ratified it. In the meantime, States are nonetheless expected to respect the core obligations it lays down.

Summary of the Convention

The Convention on the Prohibition of the Use, Stockpiling, Production and Transfer of Anti-personnel Mines and on their Destruction ("The Ottawa treaty") is part of the international response to the widespread suffering caused by anti-personnel mines. The Convention is based on customary rules of international humanitarian law applicable to all States. These rules prohibit the use of weapons which by their nature do not discriminate between civilians and combatants or cause unnecessary suffering or superfluous injury. The Convention will be open for signature in Ottawa from 3 to 4 December 1997.

Why a ban on anti-personnel mines?

Anti-personnel mines cannot distinguish between soldiers and civilians and usually kill or severely mutilate their victims. Relatively cheap, small and easy to use, they have proliferated by the tens of millions, inflicting untold suffering and wreaking social and economic havoc in dozens of countries throughout the world. Because it is far easier to lay than to remove them, it has been difficult or impossible to use this weapon in accordance with the rules of international humanitarian law in most of the conflicts where they have been employed. In December 1996, 157 States supported a United Nations General Assembly resolution calling for the urgent negotiation of a legally binding ban on these weapons.

What are the basic obligations contained in the Ottawa treaty?

States adhering to this treaty must never under any circumstances use, develop, produce, stockpile or transfer anti-personnel mines or help anyone else to do so.

They must also destroy existing anti-personnel mines, whether in stockpiles or in the ground, within a fixed time period. A small number of these mines may be retained for the sole purpose of developing mine clearance and destruction techniques and training people in their use.

Which mines are affected by this treaty?

Anti-personnel mines are designed to be placed on or near the ground and detonated by the presence, proximity or contact of a person. It was the understanding of the negotiators that "improvised" devices produced by adapting other munitions to function as anti-personnel mines are also banned by the treaty. The Ottawa treaty prohibits anti-personnel mines only. It does not affect (a) anti-tank or anti-vehicle mines (regulated by the 1980 UN Convention on Certain Conventional Weapons and the general rules of international humanitarian law); (b) "anti-handling devices" attached to an anti-vehicle mine to prevent its removal or (c) "command-detonated" munitions which can only be triggered manually by a combatant and cannot be detonated simply by the presence, proximity or contact of a person.

When and how will existing anti-personnel mines be destroyed?

Stockpiled anti-personnel mines must be destroyed within four years of entry into force of the Convention for a particular State. As for mines in the ground, whether in minefields or elsewhere, they must be destroyed within 10 years after entry into force. Pending such destruction every effort must be made to identify mined areas and to have them marked, monitored and protected by fencing or other means to ensure the exclusion of civilians. If a State cannot complete the destruction of emplaced mines within 10 years it may request a meeting of the States Parties to extend the deadline and to assist it in fulfilling this obligation.

How will the treaty help mine victims?

The treaty is a comprehensive response to the landmines crisis. Not only are States Parties prohibited from using anti-personnel mines, but those able to do so agree to provide assistance for mine clearance, mine-awareness programs and the care and rehabilitation of mine victims. Mine-affected States have a right to seek and receive such assistance directly from other parties to the treaty and through the United Nations, regional or national organizations, components of the International Red Cross and Red Crescent Movement or non-governmental organizations. These cooperative aspects of the Convention will play as great a role as the ban it imposes in providing an effective international response to the suffering caused by these weapons.

How will compliance with the treaty be monitored?

The Ottawa treaty includes a variety of measures designed to promote confidence that its provisions are being respected and to deal with suspected violations. States are required to report annually to the United Nations Secretary-General on all stockpiled anti-personnel mines, mined areas, mines retained for training purposes, destruction of mines and measures taken to prevent civilians from entering mined areas. To facilitate mine clearance, States must also provide detailed technical information about mines they have produced in the past.

If there are concerns about a State's compliance with the treaty, clarification may be sought through the UN Secretary-General and if necessary a meeting of States Parties may be held. This meeting can decide to send an obligatory fact-finding mission to relevant territory of the State concerned for up to 14 days. On the basis of the mission's report the meeting of States Parties may propose corrective actions or legal measures in accordance with the UN Charter.

What must a country do to implement the Ottawa treaty?

Until its entry into force, States may sign the treaty in Ottawa or at UN Headquarters in New York and then express their consent to be bound by transmitting a ratification (or similar) instrument to the treaty depositary, the UN Secretary General. States which have signed the treaty must not take any action which would undermine its purpose. The treaty enters into force six months after the fortieth State has expressed its consent to be bound. Before or after entry into force States may adhere directly to the treaty, without signing, by transmitting an instrument of accession to the Depositary.

The treaty also requires governments to take national legal and administrative measures, including the imposition of penal sanctions, to ensure respect for its provisions by persons or on territory under their jurisdiction or control. This may involve the adoption of criminal legislation. It may also require administrative instructions to armed forces and changes in military planning.

Can a State ban anti-personnel mines and adhere to other agreements which permit mine use?

Earlier rules regulating the use of anti-personnel mines are contained in Protocol II to the 1980 UN Convention on Certain Conventional Weapons. This regulates the use of all types of mines and similar devices, including mines used against tanks. States which adhere to the Ottawa treaty are advised to adhere also to the 1980 Convention. This Convention's three other Protocols provide important rules on the use or prohibition of other arms, such as blinding laser weapons. Adherence to the 1980 Convention will enable a State to invoke its provisions, such as those requiring a party which uses mines to remove them at the end of hostilities, in any conflict with a State which has adhered to the 1980 Convention but not to the Ottawa treaty.

17

A Global Ban on Landmines: Treaty Signing Conference and Mine Action Forum

Ottawa, Canada

2–4 December 1997

Organized by the government of Canada

Fourteen months after Foreign Minister Axworthy's audacious challenge to the world's governments, representatives of 121 States queued up to sign the Convention on the Prohibition of the Use, Stockpiling, Production and Transfer of Anti-personnel Mines and on Their Destruction, and three of these – Canada, Ireland and Mauritius – also ratified the Convention. In addition to the signing ceremony, participating governments pledged a total of more than US$ 500 million for mine action programmes world-wide.

Statement of Cornelio Sommaruga, President, ICRC

3 December 1997

We celebrate today a victory for humanity; for the cause of humanitarian values in the face of cruelty and indifference.

This historic movement against the horrors of anti-personnel mines began as an expression of human compassion on the part of medical and other humanitarian workers in mine-affected countries. It grew as their compelling testimony and images of the appalling effects of this weapon were transmitted by a myriad of non-governmental organizations and international agencies. It became unstoppable as the public conscience began to view this weapon as an abomination. An absolute ban on anti-personnel mines was transformed from an "idealistic dream" into the Ottawa treaty as diplomats, political leaders and generals allowed themselves to move beyond "business as usual" in the world of international negotiations and respond to the suffering this weapon inflicts.

The International Committee of the Red Cross, and the entire International Movement of the Red Cross and Red Crescent on behalf of which I speak, pay tribute to those whose untiring efforts have brought us to this solemn moment in which the Convention on the Prohibition of Anti-personnel Mines and on their Destruction is signed by distinguished leaders from around the world. In particular we commend the International Campaign to Ban Landmines, Foreign Minister Lloyd Axworthy and the government of Canada, the United Nations Secretary-General and his predecessor, the many governments which took unilateral action to stop the use of anti-personnel mines, the fifty States which committed themselves to the Ottawa process in this room 14 months ago, and each State which stands here today, often after difficult internal deliberations, ready to sign the new Convention.

This extraordinary coalition of civil society, international institutions and governments has proven that humanity is not powerless in the face of its worst instincts or the destructive uses of modern technology. But this experience, far from being a reason for self-satisfaction, should be the basis of sober reflection. In demonstrating that it is indeed possible to respond resolutely to the trauma and suffering of humanity we assume a solemn responsibility to ask why this happens so seldom? Why is war waged with one's whole mind and heart and soul, while struggles for humanity are too often waged, if at all, with only half of our being?

The Ottawa treaty is historic not only due to the process through which it was created but also because of its content. For the first time a weapon which has been in widespread use by armed forces throughout the world is being withdrawn from arsenals due to its appalling human, economic and social costs. And for the first time the use, development, production, stockpiling and transfer of a weapon are being prohibited in one decisive step. This reflects an important insight with implications for the future development of international humanitarian law. The Ottawa treaty recognizes that outlawing the use of a weapon while permitting its continued production, possession and transfer is not enough. Had this insight prevailed when poison gas was banned in 1925 fifty to seventy years of uncertainty about the retaliatory use of biological and chemical weapons would have been avoided.

This week the Ottawa treaty becomes the common heritage of those who have given it birth and those who will put their signature to it. Together we assume the responsibility for the long-term task of assuring its early entry into force, universalization and implementation. The ICRC is committed to continuing its work in all regions of the world to promote acceptance of the treaty.

The ICRC is painfully aware that a ban on anti-personnel mines, which will save lives and limbs, will nonetheless do nothing to improve the plight of the hundreds of thousands of existing victims whose needs have barely begun to be addressed. The ban itself will provide little comfort to some two thousand people whose lives will be forever shattered this month by mines currently in the ground. In some respects banning anti-personnel mines was the easiest of our challenges. Mine clearance and the provision of adequate assistance to victims will require enduring engagement and will certainly be more costly. The ICRC alone has already spent close to 15 million US dollars in 1997 on assistance programmes related to mine incidents. I appeal to each of the Heads of State, Heads of Government, Ministers and organizations here present to personally ensure that this day marks the beginning of the mobilization of the international community and its institutions for the long road ahead.

For its part the ICRC is committed to continuing its surgical and rehabilitative services for all war victims, including mine victims, and intends to increase its efforts to reinforce the capabilities of National Red Cross and Red Crescent Societies to provide medical and rehabilitative care in mine-affected countries.

The tragic plight of mine victims has been one of the driving forces behind efforts for a ban and to improve mine clearance. Yet relatively little attention has been given, at the international level, to coordinating and improving the delivery of care to mine victims. This situation must begin to change in 1998.

If we fail to learn from our mistakes we are doomed to repeat them. In the coming decades the potential for development of particularly heinous and indiscriminate arms threatens to outpace the ability of humanity to respond. For this reason States bear a weighty responsibility, enshrined in the Geneva Law, to determine, before deployment, whether weapons under development are covered by the general prohibitions of weapons which are inherently indiscriminate or of a nature to cause superfluous injury or unnecessary suffering.

The landmines issue is but part of a phenomena of great concern to the ICRC: the virtually unrestricted flow of vast quantities of weapons, particularly small arms, throughout the world and their consistent use in violation of the basic norms of international humanitarian law. As this trend continues efforts to teach respect for these norms are being overwhelmed by the flow of weapons. We have learned with landmines that it is both easier and faster to distribute arms than to teach humanitarian law principles to those who possess them. I call on States, as I did two years ago at the opening of the Vienna Review Conference, to address the issue of arms availability as a matter of pressing international concern.

Let us celebrate today a victory for humanity. But in doing so let us recognize that the real victory for humanity lies ahead. It will be celebrated on that day when we must no longer pour our best efforts into picking up the debris of war and mending the human wounds of a destructive technology out of control; it will be the day when humanity's wisdom, respect for basic humanitarian norms and instinct for self-protection converge to prevent such horrors from occurring.

Press Release

5 December 1997

RED CROSS / RED CRESCENT MOVEMENT WELCOMES SIGNING OF OTTAWA TREATY BY 121 STATES

The International Committee of the Red Cross (ICRC) and the International Federation of Red Cross and Red Crescent Societies enthusiastically welcomed the signing by 121 States in Ottawa, Canada, of the Convention on the Prohibition of the Use, Stockpiling, Production and Transfer of Anti-personnel Mines and on their Destruction. The President of the ICRC, Cornelio Sommaruga, declared that he was "absolutely delighted" by the outcome. "We can now be optimistic that, if the momentum is maintained, adherence to the Convention could soon be universal and an end to the scourge of landmines will be in sight. We will do all that we can to make this a reality."

The Convention will formally enter into force once 40 signatories have ratified it. States party to the treaty will then have four years to destroy existing stockpiles and ten years to clear all anti-personnel mines from the ground. Yet until the treaty is fully implemented, thousands of people will continue to be killed or maimed by these weapons. "We must put the spotlight back on the victims. Ensuring lifelong assistance to mine victims and their communities is going to be a major challenge facing the international community for decades to come", said Astrid Noklebye Heiberg, President of the International Federation of Red Cross and Red Crescent Societies. The Ottawa treaty specifically encourages States to provide increased support for mine victims through bodies such as the International Red Cross and Red Crescent Movement. A meeting involving the ICRC, the International Federation and interested National Societies will be held in Phnom Penh, Cambodia, in February 1998 with a view to planning future victim-assistance strategies.

The Ottawa Conference also adopted an Agenda for Action for the coming months and years, which includes plans announced by the ICRC to convene international conferences on the coordination of assistance to mine victims and on mine-awareness activities in 1998, and to hold regional seminars in Central and Eastern Europe and East and South Asia to promote adherence to the treaty.

18

ICRC Position Paper No. 5
Anti-personnel Mines:
Agenda 1998 – From Prohibition to Elimination and Adequate Care for the Victims

ICRC Briefing and Position Paper

January 1998

With the opening for signature of the Convention on the Prohibition of the Use, Stockpiling, Production and Transfer of Anti-Personnel Mines and on their Destruction (the Convention) in Ottawa, Canada, from 3–4 December 1997, the international community is moving one step closer towards a comprehensive response to the scourge of anti-personnel mines. This new instrument of international humanitarian law represents a landmark in efforts to end the landmines epidemic and is a testament to the efforts of the International Red Cross and Red Crescent Movement, the International Campaign to Ban Landmines and governments which have joined the "Ottawa process".

Yet major challenges remain before the effects of this Convention are felt in fields, villages and conflict zones around the world. The Ottawa treaty represents a prescription for ending the mines epidemic. But the cure will require sustained, and costly, long-term efforts. The International Red Cross and Red Crescent Movement will need to work vigorously to: (a) promote universal adherence to and full implementation of the Convention; (b) strengthen its efforts to provide adequate care for mine victims; and (c) encourage comprehensive humanitarian mine clearance in affected countries. For many years to come, the Movement will continue to have an essential role to play in addressing the aftermath of the landmines epidemic.

In 1997, the tide turned against anti-personnel mines as they became stigmatized in international public opinion as an unacceptable weapon, the use of which could no longer be justified. With determined action in 1998 and beyond, the human suffering caused by landmines will slowly begin to diminish.

A. THE INTERNATIONAL BAN OF ANTI-PERSONNEL MINES

1. The Convention on the Prohibition of Anti-Personnel Mines and on their Destruction

The Convention on the Prohibition of the Use, Stockpiling, Production, and Transfer of Anti-Personnel Mines and on their Destruction was adopted by 89 States on 18 September 1997 at the Oslo Diplomatic Conference on an International Total Ban on Anti-Personnel Landmines. For the first time in a humanitarian law treaty, the development, production, stockpiling, transfer and use of a weapon are totally prohibited. States party to the Convention will be required to destroy stockpiled anti-personnel mines within four years of its entry into force and to clear emplaced anti-personnel mines within ten years of its entry into force. No reservations to the Convention are permitted and effective withdrawal is not allowed during an ongoing armed conflict.

As at 1 January 1998, a total of 123 governments had signed the Convention and three of these – Canada, Ireland and Mauritius – have deposited their instrument of ratification with the depositary, the UN Secretary-General. The Convention will enter into force six months after 40 signatories have ratified it. In accordance with Article 18 of the Convention it is possible for States to declare, at the time of ratification, that they will provisionally apply, pending the Convention's entry into force, its core prohibitions of the development, production, stockpiling, transfer and use of anti-personnel mines. "Ratification kits", including a short summary of the Convention for the public and parliamentarians and guidelines for State adherence and implementation, are available from the ICRC in seven languages (Arabic, Chinese, English, French, Portuguese, Russian and Spanish).

The ICRC urges the National Red Cross and Red Crescent Societies of signatory States to work actively for early national ratification of the Convention and to encourage governments to declare, at the time of ratification, that they will provisionally apply its core prohibitions.

Although the humanitarian objective of a total ban on anti-personnel mines appears to enjoy near-universal support, it is clear that a number of States are not yet ready to adhere to the Convention. One of the major objectives of the international effort against landmines in 1998 must be to promote its rapid universalization. To this end, the ICRC plans in 1998 to convene a number of seminars of experts, particularly in Asia and Central and Eastern Europe, to encourage adherence to the Convention and to expand discussion of comprehensive solutions to the mines problem.

The ICRC urges the National Red Cross and Red Crescent Societies of non-signatory States to encourage strongly their respective governments to respect the new norm created by the Convention and to adhere to it officially at an early date.

2. The 1980 Convention on Conventional Weapons

Although early ratification of the Ottawa Convention is the priority for 1998, adherence to the 1980 Convention on Conventional Weapons and its annexed Protocols must also be actively promoted. Of particular importance are Protocol II as amended on 3 May 1996, on landmines, booby-traps and other devices, and Protocol IV on blinding laser weapons. Protocol II as amended, although allowing continued use of certain anti-personnel mines, restricts the use of anti-vehicle/anti-tank mines and directional fragmentation munitions (when command-detonated by remote control), and prohibits the use of any mines that are specifically designed to detonate as a result of the use of electro-magnetic mine detectors. In addition, at the end of an armed conflict, States party to Protocol II as amended are obliged to remove all mines laid by them.

By 8 January 1998, 18 States had adhered to Protocol IV and 14 to Protocol II as amended. Both instruments will enter into force as binding international law six months after 20 States have notified the UN Secretary-General of their consent to be bound.

The ICRC urges National Societies to encourage their governments, if they have not yet done so, to adhere to the 1980 Convention on Conventional Weapons and all annexed Protocols, particularly Protocol II as amended and Protocol IV.

B. VICTIM PREVENTION AND ASSISTANCE

The Ottawa Convention requires States Parties to provide assistance for the care, rehabilitation and socio-economic reintegration of mine victims and for mine-awareness programmes. In doing so, it encourages the channelling of such assistance through the ICRC, National Red Cross and Red Crescent Societies and their International Federation. The ICRC plans to reinforce its work on behalf of mine victims in the context of its overall assistance to war-wounded. It is also prepared to work with National Societies to strengthen their capacity to respond to the needs of mine victims.

The ICRC urges National Societies to expand their efforts on behalf of mine victims and to encourage their governments to increase funding for the care of mine victims and other war casualties.

1. Addressing the needs of victims

The ICRC will continue its support to health facilities treating patients injured by mines in Afghanistan, Angola, Azerbaijan, Cambodia, Ethiopia, Georgia, Iraq, Somalia, Sudan, Tajikistan, Uganda and Zimbabwe. This will involve direct assistance to first-aid posts, hospitals, surgical programmes, and prosthetic and orthotic workshop projects. The ICRC will continue to train surgeons in the appropriate surgical treatment of mine injuries. In 1998, projects will include a series of war surgery seminars in various countries.

While the international response has focused on relieving the physical suffering of mine victims, relatively little attention has been given to their psychological and socio-economic needs. Responding to these needs is an essential part of restoring mine victims to whole and productive lives. The ICRC, with the assistance of National Societies, intends to begin examining the psycho-social needs of mine victims with a view to identifying the best methods of intervention for vocational training and social reintegration according to local customs and cultural restraints.

National Societies and NGOs interested in contributing substantively to the study of the psychological and socio-economic needs of disabled mine victims are urged to contact the ICRC's Medical Division.

2. Coordination of assistance

In mine-affected countries, the ICRC will promote the concept of a "Mines Information System" with the participation of governments, non-governmental organizations, UN agencies, the ICRC and National Red Cross and Red Crescent Societies. This approach will include (a) standardized data collection, (b) the promotion of humanitarian criteria for setting operational priorities in mine-infested areas and (c) cooperation and data exchange among those engaged in mine clearance, victim assistance, mine awareness and governmental decision-making.

3. Mine Awareness

Mine awareness can save precious lives and limbs before mines are cleared. The ICRC, in cooperation with interested National Societies, will further develop mine-awareness programmes in affected countries, based on detailed evaluations of existing projects and the development of guidelines for future projects. As part of this process the ICRC convened, in early 1998, an international conference on mine awareness in Sarajevo, Bosnia-Herzegovina.

National Societies with a special interest in mine-awareness work should contact the ICRC (Division for Promotion of IHL) to coordinate the use and adaptation of existing materials, exchange of experiences and development of new programmes.

C. MINE CLEARANCE

Although the International Red Cross and Red Crescent Movement is not involved directly in mine clearance, it advocates greatly increased resources for humanitarian mine clearance. Moreover, whenever a member of the Movement is engaged in field work in a situation where demining activities are under way, it should encourage the competent authorities and mine-clearance organizations to conduct demining operations in accordance with humanitarian criteria (available from the ICRC).

National Societies are requested to encourage their governments to increase support significantly for mine-clearance operations. Furthermore, National Societies should encourage governments, as well as academic and military researchers, to develop more rapid and less expensive mine-clearance techniques. So far research work has focused on the development of the weapons which have created this humanitarian problem; today greater efforts should be devoted to coping with the consequences of their use.

19

Regional Conference on Landmines

Budapest, Hungary

26–28 March 1998

Hosted by the government of Hungary in cooperation with the ICRC and the International Campaign to Ban Landmines

The Budapest Conference provided an opportunity for States from central and eastern Europe to come together to discuss the mines problem in the region and beyond. Although most of the countries of central Europe had signed the Ottawa treaty, adherence in eastern Europe was limited, and there were hopes that a number of States could be encouraged to look again at the military need for anti-personnel mines.

Accordingly, under the umbrella of the regional conference, the ICRC convened a seminar for defence and foreign affairs officials from the region, on the military utility and humanitarian costs of anti-personnel mines. Participants were asked to consider the actual effectiveness of landmines in combat compared with their long-term effects and to discuss alternatives to anti-personnel mines, for example through an evolution in military doctrine. A strong final declaration was adopted by the seminar, although participants from Belarus and the Russian Federation were unable fully to support it.

Final Declaration of Participants

ICRC seminar on the humanitarian impact and military utility of anti-personnel mines

Budapest, Hungary

27–28 March 1998

(Participants from ministries of foreign affairs and defence of Albania, Belarus*, Bosnia and Herzegovina, Bulgaria, Croatia, the Czech Republic, Estonia, the Federal Republic of Yugoslavia, FYR of Macedonia, Hungary, Latvia, Lithuania, Moldova, Poland, Romania, the Russian Federation*, Slovakia, Slovenia, and Ukraine)

Defence and foreign affairs officials from 19 European States gathered in Budapest, Hungary, to examine the experience of anti-personnel mine use in the region. Participants discussed the military effectiveness of anti-personnel mines based on their actual combat performance in European and other conflicts. The military value of anti-personnel mines was considered in the context of the long-term human, social and economic costs incurred in many of the conflicts in which this weapon has been used. Particular attention was given to the difficulties and extremely high costs of post-conflict mine clearance.

The seminar sought to develop recommendations which will promote and broaden dialogue within European military and political circles on the question of anti-personnel mines. It is hoped that the work of the Budapest seminar can contribute to the development of an effective response to the humanitarian problems which anti-personnel mines have caused in Europe and throughout the world. The following statement was adopted by participants acting in an individual capacity.

* Participants from Belarus and the Russian Federation, while associating with the humanitarian concerns expressed in this declaration, cannot fully associate with all of the conclusions contained herein.

Participants in the ICRC regional seminar on the humanitarian impact and military utility of anti-personnel mines agree that:

1. The global scourge of anti-personnel landmines, which kill and injure some 2,000 persons per month, most of whom are civilians, is unacceptable and must be stopped. These mines not only kill and maim combatants in an inhumane way, but also indiscriminately affect civilians and inflict on them enormous physical and psychological damage long after the conflict is over. This must be a grave and continuous concern of the international community;

2. The appalling humanitarian consequences of the use of anti-personnel mines far outweigh their military utility;

3. The cases considered during the seminar, and the personal experience of participants, lead to some initial conclusions concerning traditionally emplaced mines:

Establishing, monitoring and maintaining extensive border minefields is time-consuming, expensive and dangerous, and in order to be effective they need to be under continuous observation and direct fire. Because of these practical difficulties some armed forces have entirely refrained from using such minefields. Moreover, these minefields have not always proved successful in preventing infiltration.

Under battlefield conditions the use, marking, mapping and removal of mines in accordance with classical military doctrine and international humanitarian law is extremely difficult, even for professional armed forces. History indicates that effective marking, mapping and removal of mines have rarely occurred. The cost to forces using anti-personnel mines, in terms of casualties to one's own forces and civilians, the limitation of tactical flexibility and the loss of sympathy of the indigenous population is higher than has been generally acknowledged.

4. Remotely delivered anti-personnel mines can cause extensive civilian casualties, even if such mines are designed to be self-destructing and self-deactivating, for the following reasons:

- they will be dangerous during their intended active life-time,
- the fencing and marking of such mines will be virtually impossible,
- in prolonged conflicts the same territory may be re-laid with remotely delivered mines many times,
- self-destructing and deactivating devices are not 100 per cent reliable,
- inactive mines, as unexploded ordnance, can still be dangerous, and
- the mere presence of mined areas will produce fear, keeping civilians out of areas important for their livelihood:

5. Some alternatives to anti-personnel mines already exist. Additional alternatives should be encouraged rather than further development of any new anti-personnel mine technologies.

Participants agree that European States should consider taking the following steps, on the national, regional and global levels, towards ending the scourge of anti-personnel mines:

1. for those States which are not yet signatories, to sign the Convention on the Prohibition of the Use, Stockpiling, Production and Transfer of Anti-personnel Mines and on their Destruction (the Ottawa treaty) and for those signatory States which are not yet parties, to ratify and implement the Convention as soon as possible;

2. for those States not yet able to adhere to the Ottawa treaty, to initiate plans for other alternatives to anti-personnel mines and to end all new deployments of these weapons;

3. for those States which have not yet done so, to announce permanent bans on the transfer of all types of anti-personnel mines;

4. for those States which are not yet Parties, to adhere to the 1980 United Nations Convention on Certain Conventional Weapons, including its Protocol II on

landmines (as amended on 3 May 1996); for current States party to this Convention to adhere to its amended Protocol II at the earliest possible date to ensure its early entry into force.

Participants appeal to the international community, including governments, international agencies and non-governmental organizations to continue to assist, where necessary, mine-affected countries and regions in Europe to become permanently free of the scourge of anti-personnel mines, in particular through the provision of technical, financial and other assistance in the clearance and destruction of mines, assistance to victims and mine awareness programs.

Participants express their thanks to the International Committee of the Red Cross for convening the seminar, and to the **Hungarian government, the Hungarian National Assembly and the City of Budapest** for their generous hospitality in hosting them in Budapest.

ICRC NEWS

1 April 1998

HUNGARY: ONE MORE STEP AWAY FROM LANDMINES

A three-day regional conference on anti-personnel landmines hosted by the Hungarian government in Budapest ended on 28 March with a declaration by ICRC President Cornelio Sommaruga that mines were "weapons of the past, not of the future". The conference, which comprised parallel two-day seminars sponsored by the ICRC and the International Campaign to Ban Landmines respectively, brought together representatives of governments and non-governmental organizations from 19 Central and Eastern European countries.

The ICRC seminar, attended by representatives of the region's ministries of defence and foreign affairs, focused on the human cost and military utility of anti-personnel mines. The participants heard powerful evidence from outside military experts that, on the basis of their actual use, the military effectiveness of anti-personnel mines was extremely questionable, especially for the protection of long, unguarded borders. It was pointed out that alternatives already existed to fulfil the functions of these weapons in military doctrine.

The participants in the seminar adopted a very strong final declaration stating that the human cost of anti-personnel mines far outweighed their limited military utility and urging early adherence to the Ottawa treaty. The declaration was fully endorsed by all the participants in their personal capacity except for those from Belarus and the Russian Federation, who expressed support for its humanitarian objectives but were not able to agree with all its conclusions. Ten governments in the region have already signed the Ottawa treaty, and during the conference Hungarian President Arpad Goncz signed his country's instrument of ratification.

20

Implementing the Ottawa Treaty: Questions and Answers

1 June 1998

The 126 States which have signed the Convention on the Prohibition on the Use, Stockpiling, Production and Transfer of Anti-personnel Mines and on their Destruction (commonly referred to as the Ottawa treaty) have taken an important step towards resolving the humanitarian crisis caused by these weapons. While the treaty is a remarkable achievement, much remains to be done if the global landmines epidemic is to be successfully addressed. The treaty has been endorsed by a large number of governments, but signing it only reflects the political commitment of the signatory State to be bound by it in the future. Every such State must now take action at the national level for the treaty to become *legally* binding upon it. The treaty will not come into effect until 40 States have taken such action, and at that time it will apply to them only. In addition, in order to reduce the threat that anti-personnel mines represent for civilians, States must ensure that the obligations contained in the treaty are fully implemented. Putting an end to the scourge of landmines is a process which will require concerted and long-term action.

Q. How does a State become legally bound by the Ottawa treaty?

A. Generally, a State becomes bound by a treaty by ratifying it (in some instances this is referred to as acceptance or approval). Ratification is normally a two-step process. First, a State must follow its own national procedures for adhering to international agreements. In many cases this means submitting the treaty to the national parliament or national assembly for approval. Once these national requirements are met, the government must then notify the United Nations Secretary-General, who is the depositary of the treaty. This is done by depositing an "instrument of ratification" with the United Nations in New York. The ICRC has prepared kits providing additional information on the ratification process and model instruments of ratification which States are welcome to use.

The Ottawa treaty will become binding on the first 40 consenting States six months after the 40th instrument has been deposited with the United Nations. After that time, each additional State will become bound six months after its instrument is deposited. At that point the State is considered to be a party to the treaty or a "State Party". The treaty invites, and the ICRC encourages, States to declare that they will provisionally observe its basic obligations even during this six-month waiting period (Article 18).

Q. What "national measures" must a State Party take to implement the treaty?

A. The Ottawa treaty prohibits the use, stockpiling, production, development and transfer of anti-personnel mines. To ensure that these controls become a reality in practice, a State Party will have to take a number of regulatory steps at the national level. Specifically, it must adopt appropriate legal, administrative and other measures to prevent and punish any prohibited activity by persons or on territory under its jurisdiction or control. Depending on the national law or procedures, this may require the adoption of specific criminal legislation.

In addition, implementing the treaty's provisions will almost always entail the adoption of administrative measures to make sure that the necessary changes in military doctrine and operating procedures are made. Organizations and corporations involved in the development, production, sale, and transfer of anti-personnel mines must also be notified to ensure their compliance. The relevant ministries will also have to review any export licences involving the sale of anti-personnel mines.

States are required to take additional measures to ensure the destruction of stockpiles and the clearing of mined areas (see below).

The ICRC's Legal Division is available to provide guidance on national legislation, and can supply examples of instruments already adopted by certain States. Copies of such legislation will soon be available on the ICRC Website.

Q. What must a State Party do with the anti-personnel mines it currently possesses?

A. Under the terms of the Ottawa treaty, a State must destroy all anti-personnel mines in its possession or under its jurisdiction or control within four years of the treaty becoming binding on it (Article 4). It may, however, retain a small number of mines – generally accepted to mean no more than several thousand – for training in mine-clearance techniques. A State Party will have to draw up plans for the collection, transport and destruction of its anti-personnel mine stockpiles.

As a party to the treaty, each State has the right to seek assistance in the destruction of its stockpiles. This includes the exchange of equipment, material and scientific and technological information. Conversely, every State Party in a position to do so is obliged to provide assistance to other Parties for the destruction of their stocks.

Q. What does a State Party have to do about mines already in the ground?

A. A State Party is required to take a number of steps to ensure that anti-personnel mines no longer pose a threat to the civilian population. First it must, as soon as possible, identify all mined areas under its control which are known or suspected to contain anti-personnel mines. These areas must be marked and civilians have to be kept out of the area by fencing or other means. Secondly, within ten years of its becoming bound by the treaty, all mined areas under its jurisdiction or control must be cleared of anti-personnel mines, irrespective of how they came to be there or who was responsible for laying them. Land believed to contain anti-personnel and anti-tank or anti-vehicle mines must be surveyed, and at least the anti-personnel mines must be removed and/or destroyed. In States affected by large-scale mine contamination, a plan of action will be needed. Requests for extensions of the ten-year deadline can be submitted at meetings of States Parties. These meetings provide a forum for the latter to report on what has been accomplished, indicate where additional work is needed and seek any assistance they may require.

As with the destruction of stockpiles, each State Party has the right to seek international assistance in fulfilling these obligations and to participate in the fullest possible exchange of equipment, material and scientific and technological information. No State Party may impose undue restrictions on the provision of mine-clearance equipment and related technological information intended for humanitarian purposes. Indeed, each State Party in a position to do so is obliged to provide assistance for mine clearance and related activities such as mine-awareness campaigns.

Q. What about programmes to help landmine victims?

A. People who survive landmine explosions usually face a dismal future. In addition to the immediate medical treatment and physical rehabilitation they require (if any is available), they are likely to suffer psychological trauma and have difficulty in finding employment.

Under the terms of the Ottawa treaty, each State Party in a position to do so has a duty to provide assistance for the care and rehabilitation of mine victims, including social and economic integration. This assistance may be provided through the United Nations System, international, regional or national organizations and institutions, the ICRC, National Red Cross and Red Crescent Societies and their International Federation, non-governmental organizations, or on a bilateral basis. As with the obligations discussed above, a State Party has the right to seek such assistance from others in its efforts to care for its mine victims and help them surmount the difficulties that lie before them.

Q. Does a State Party have to report on its activities?

A. The Ottawa treaty does not establish an independent monitoring mechanism. In an effort to promote transparency, however, it requires each State Party to send an annual report to the Secretary-General of the United Nations which must include the following information:

- the total number and the types of anti-personnel mines it has stockpiled;
- the progress made in its mine-destruction programmes, including the total number and the types of mines destroyed;
- the total number and the types of mines kept for training purposes;
- the technical characteristics of each type of mine it has produced in the past;
- the location of all mined areas under its jurisdiction or control; details of the type, quantity and age of the mines laid there (to the extent known); and steps taken to warn the civilian population;
- the national measures, such as legislation or administrative regulations, taken to prevent and suppress violations of the treaty;
- the status of programmes for the conversion or decommissioning of anti-personnel mine production facilities.*

The first report must be submitted no later than 180 days after the treaty becomes binding on the State in question.

In a further effort to promote transparency, the treaty also provides for annual meetings of States Parties and for a review conference to be held every five years.

Q. What should be the role of National Red Cross and Red Crescent Societies in promoting the entry into force and implementation of the Ottawa treaty?

A. National Societies can play an important role in ensuring that the Ottawa treaty comes into force at an early date and that its provisions are implemented and respected at the national level. They can contact their governments to discuss progress made and the timetable for ratification. They can also work with their governments to promote the adoption of effective national legislation or other administrative measures so as to ensure that the aims of the treaty are achieved.

In mine-affected countries, the National Societies can help determine the extent of the problem. With ICRC and Federation support, National Societies can carry out surveys to identify the number, location and needs of mine victims and, on the basis of that information, propose programmes to meet those needs. They can also work with the government to help make sure that mine clearance is conducted in accordance with humanitarian, as opposed to political or economic, priorities.

National Societies in countries not affected by mines can promote and monitor mine-clearance and victim assistance programmes supported by their governments. They can also establish bilateral victim assistance projects with other National Societies or provide support for programmes run by the ICRC or the Federation.

21

An International Ban on Anti-personnel Mines: History and Negotiation of the "Ottawa Treaty"

Stuart Maslen, Legal Adviser, and Peter Herby, Coordinator,
Mines-Arms Unit, Legal Division

Published in the *International Review of the Red Cross*, 325
December 1998

The background to the Ottawa process

As the First Review Conference of the 1980 Convention on Conventional Weapons (1) (CCW) closed in Geneva on 3 May 1996, there was widespread dismay at the failure of the States Parties to reach consensus on effective ways to combat the global scourge of landmines. The CCW Protocol II as amended on 3 May 1996 (2) (Protocol II as amended) introduced a number of changes that were widely welcomed, but it fell far short of totally prohibiting these weapons, a move already supported by more than 40 States. Keen to sustain the international momentum that might otherwise have slackened, the Canadian delegation announced that Canada would host a meeting of pro-ban States later in the year to develop a strategy to move the international community towards a global ban on anti-personnel mines.

That meeting, the "International strategy conference: Towards a global ban on anti-personnel mines" (usually referred to as "the 1996 or the first Ottawa Conference"), was held in the Canadian capital from 3 to 5 October 1996; it set the scene for what would become known as the "Ottawa process" – a fast-track negotiation of a convention banning anti-personnel mines. At the closing session of the Conference the host country's Foreign Minister, Lloyd Axworthy, ended his address with an appeal to all governments to return to Ottawa before the end of 1997 to sign such a treaty. This bold initiative was immediately supported by ICRC President Cornelio Sommaruga, who was attending the Conference, by the UN Secretary-General and by the International Campaign to Ban Landmines (ICBL), but it came as a surprise to the governments taking part, and it was by no means certain that it would be successful. Indeed, at that stage only about 50 governments had publicly declared their support for a comprehensive, worldwide ban on anti-personnel mines, (3) and Protocol II as amended was widely considered to be the most stringent international agreement possible in the prevailing climate.

Yet only 14 months later, at the Treaty Signing Forum held in Ottawa in December 1997, representatives of 121 governments queued up to sign the *Convention on the Prohibition of the Use, Stockpiling, Production and Transfer of Anti-Personnel Mines and on their Destruction,* (4) and three of them – Canada, Ireland and Mauritius also deposited their instruments of ratification. As at the end of April 1998, there were a total of 124 signatories and 11 States had already ratified the Convention, which will enter into force six months after 40 States have formally adhered to it. The present article aims to discuss this remarkable achievement in the context of the negotiation of the treaty and its various provisions. It also considers some of the implications of the process and its successful outcome for the future development of international humanitarian law. It does not, however, purport to serve as a commentary of the treaty *per se,* a task which is best left to a later date.

The negotiation of the treaty

Only a few weeks after the challenge issued by Canada's Foreign Minister, the Austrian government circulated a first draft of a treaty banning anti-personnel mines. The text was clearly inspired by the negotiations in the First Review Conference of the CCW and by disarmament law, especially the Convention on the Prohibition of the Development, Production, Stockpiling and Use of Chemical Weapons and on their Destruction, adopted by the UN General Assembly on 30 November 1992. It contained clear prohibitions of the development, production, stockpiling, transfer and use of anti-personnel mines – though it maintained the ambiguous definition of anti-personnel mines contained in Protocol II as amended (5) – CD and required destruction of stockpiles within one year and clearance of emplaced anti-personnel mines within five years. In December 1996, the UN General Assembly adopted its landmark resolution 51/45 S (157 votes in favour, 10 abstentions and none against) in which States were urged to "pursue vigorously an effective, legally binding international agreement to ban the use, stockpiling, production and transfer of anti-personnel landmines with a view to completing the negotiation as soon as possible."

Thus began the process of negotiating a treaty providing for a comprehensive ban on anti-personnel mines, the target date for its adoption and signature being the end of 1997. A core group of committed governments from different geographical regions began meeting informally to discuss how to push the Ottawa process forward. The Austrian government, which had prepared a draft text in late 1996, hosted a meeting in Vienna from 12 to 14 February 1997 to enable States to exchange views on the content of such a treaty. The Expert Meeting on the Text of a Convention to Ban Anti-personnel Mines (the Vienna Expert Meeting), which was attended by representatives of 111 governments, heard the

ICRC outline what it considered to be the key issues at stake. First, the ICRC emphasized the crucial importance of having an unambiguous definition of anti-personnel mines. Second, if a prohibition was to be effective, the new treaty should comprehensively ban the production, stockpiling, transfer and use of anti-personnel mines and require their destruction. A phased approach should begin with an immediate prohibition on new deployments, production and transfers; a second phase, which should be as short as practical constraints allow, could provide for the destruction of existing stockpiles and the clearance and destruction of mines already deployed.

Third, the ICRC noted that compliance monitoring would be an important element of a regime put in place to end the use of anti-personnel mines, and suggested that the best method would be for an independent mechanism to investigate credible reports of their use following the entry into force of the new treaty. But while advocating the maximum possible degree of verification, the ICRC specifically encouraged States not to allow this question to stand in the way of the basic norm prohibiting anti-personnel mines. It reminded States that earlier norms of humanitarian law prohibiting the use of specific weapons had been enacted without provisions on verification, but this did not prevent them from being very largely respected.

Fourth, the ICRC addressed the issue of universality. Universal application of legal norms is an important objective and the ICRC, in keeping with its mandate under the Geneva Conventions, has devoted a great deal of time and effort to the promotion of existing agreements. It is a fact, however, that no major instrument of humanitarian law has attracted universal adherence from the outset. Indeed, a number of States took decades to ratify the 1925 Geneva Gas Protocol, (6) and in 1899 two of the major powers of the day voted against a prohibition of dum-dum bullets. And yet the vast majority of States have observed the norms laid down by these agreements.

Following the Vienna Expert Meeting, and taking into account the comments made by participating governments, the Austrian government revised its original text and on 14 March 1997 issued its second draft of the *Convention on the Prohibition of the Use, Stockpiling, Production and Transfer of Anti-Personnel Mines and on their Destruction.* The title, and much of the text, would remain the same in the treaty ultimately adopted.

The discussions at the Vienna Expert Meeting had made it clear that the question of verification would give rise to considerable debate, for some governments favoured a humanitarian law approach (i.e., one entailing a minimum of verification) while others sought a complex verification regime similar to those enshrined in negotiated disarmament agreements. Given its interest in the issue of verification, the German government offered to host a meeting devoted exclusively to this question.

The International Expert Meeting on possible Verification Measures to Ban Anti-Personnel Landmines (the Bonn Expert Meeting) was held on 24 and 25 April 1997. To stimulate debate on the topic, Germany had drafted an "Option Paper for a Possible Verification Scheme for a Convention to Ban Anti-personnel Landmines". A total of 121 countries were represented at the meeting, and views were again divided between States which felt that detailed verification was essential to ensure that any agreement was effective and those which followed an approach similar to that of the ICRC, arguing that the proposed agreement was essentially humanitarian in character and stressing the overriding importance of a clear norm prohibiting anti-personnel mines.

From 24 to 27 June 1997, the Belgian government hosted the International Conference for a Global Ban on Anti-Personnel Mines (the Brussels Conference), the official follow-up to the 1996 Ottawa Conference. Its primary task was to adopt a declaration forwarding the latest (third) Austrian draft text for negotiation and adoption to the Diplomatic Conference being convened in Oslo, Norway, in September 1997. Out of the 156 States attending the Brussels Conference, a total of 97 signed the "Brussels Declaration", as it came to be known, which affirmed that the essential elements of a treaty to ban anti-personnel mines were:

- a comprehensive ban on the use, stockpiling, production and transfer of anti-personnel mines;
- the destruction of all stockpiled and cleared anti-personnel mines;
- international cooperation and assistance in the area of mine clearance in affected countries.

Notable by their absence from the declaration were references to the importance of international support for assistance to mine victims and to an absolute duty to clear emplaced mines (the duty being only to destroy "cleared" anti-personnel mines).

In addition to forwarding the Austrian draft text to the Oslo Diplomatic Conference, States supporting the Brussels Declaration also reaffirmed the goal set by the Canadian Foreign Minister of signing the treaty in Ottawa before the end of 1997. The outcome of the meeting was hailed by the ICRC as a major step forward. In the words of its President, "[t]he Brussels Conference has demonstrated that the momentum towards a ban on this pernicious weapon is now irreversible."

The Diplomatic Conference on an International Total Ban on Anti-Personnel Land Mines (the Oslo Diplomatic Conference), convened by Norway, opened in

Oslo on 1 September 1997. It was scheduled to last a maximum of three weeks. The host country had already announced the draft Rules of Procedure at the end of the Brussels Conference. Modelled on those used in the negotiation of the 1977 Protocols additional to the Geneva Conventions of 12 August 1949, they provided for resolution of issues by a two-thirds majority vote in cases where consensus proved impossible. Only those States which had formally supported the Brussels Declaration were officially recognized as participants in the Oslo Diplomatic Conference and therefore entitled to vote. All other States present were officially classed as observers, together with the United Nations, the ICRC, the International Federation of Red Cross and Red Crescent Societies, and the ICBL.

The success of the Oslo Diplomatic Conference can be attributed to many factors: the creation and sustenance of the necessary political will, and extensive media attention following the tragic death of Diana, Princess of Wales, are of obvious significance. But the crucial role played by the Chairman of the Conference, Ambassador Jakob Selebi of South Africa, should not be forgotten. With skill and determination he drove the process forward to a successful conclusion without the need for the full three-week negotiation period. His contribution to the favourable outcome of the Ottawa process should be duly recognized.

Key issues in the Ottawa negotiations

The scope of the treaty

The first Austrian draft text contained an article on the proposed scope of the treaty to the effect that the Convention would apply "in all circumstances, including armed conflict and times of peace." The issue of whether such a provision was needed in a treaty entirely prohibiting a weapon was debated at the Vienna Expert Meeting. As a result of those discussions the scope article was dropped from the second Austrian draft, for it was deemed superfluous in an agreement in which States Parties undertook "never under any circumstances" to develop, produce, stockpile, transfer or use anti-personnel mines.

The lack of such a provision, however, entailed the absence of a specific reference to the application of the treaty to all parties to a conflict. This dashed the hopes of a number of countries, particularly Colombia, (7) and the ICBL that the treaty would expressly regulate the behaviour of all protagonists in a conflict, and not only that of States. However, on signing the Ottawa treaty Colombia stated its understanding that while it had no effect on the legal status of the various parties involved, the Convention applied to all warring parties which are subjects of international humanitarian law (i.e., under Article 3 common to the 1949 Geneva Conventions and their Additional Protocol II of 1977). This understanding was not contested.

In its Article 9 the Ottawa treaty sets out the duty of each State Party to take all appropriate legal, administrative and other measures at the national level to prevent and suppress violations of the Convention. This means that the production, stockpiling, transfer or use of anti-personnel mines by individuals under the jurisdiction or control of States Parties, including members of insurgent forces, are covered, at least in theory. Moreover, one of the preambular paragraphs states that the agreement by the States Parties is based on "the principle of international humanitarian law that the right of the parties to an armed conflict to choose methods or means of warfare is not unlimited, on the principle that prohibits the employment in armed conflicts of weapons, projectiles and materials and methods of warfare of a nature to cause superfluous injury or unnecessary suffering and on the principle that a distinction must be made between civilians and combatants." (8) These principles are elements of customary international law which apply to all parties to any conflict.

Definitions

(a) Mine

Although it had been suggested that "improvised explosive devices" were regulated as mines under Protocol II as amended, a number of States were keen to ensure that they were encompassed by the Ottawa treaty. The definition of a mine given in Protocol II as amended was therefore supplemented by the expression "designed to be" inserted immediately after the word "munition", so that the new definition read: "a munition designed to be placed under, on or near the ground or other surface area and exploded by the presence, proximity or contact of a person or vehicle." (9) As a result of an Australian initiative, in the course of the Oslo negotiations on definitions it was accepted that the Convention also prohibited explosive devices improvised or adapted to serve as anti-personnel mines (i.e., if a device functions as an anti-personnel mine, it is to be considered as such).

(b) Anti-personnel mine

The definition of anti-personnel mines was inevitably one of the most highly charged issues. The definition given in Protocol II as amended was unnecessarily ambiguous, a fact recognized even by some of those States which were in favour of keeping it in the new treaty. At the Vienna Expert Meeting opinions appeared fairly evenly divided as to whether the wording contained in the first Austrian draft should be maintained or amended. The ICRC and the ICBL, for their part, made a strong call for the word "primarily" to be removed from the definition.

In the months following the Vienna Expert Meeting, however, more and more States – particularly those most active in the Ottawa process – espoused the view that the definition of anti-personnel mines had to be clarified. Accordingly, in its second draft text the Austrian government removed the term "primarily" from the definition and included an exemption for anti-vehicle mines equipped with anti-handling devices. (10) The phrase on anti-handling devices mirrored the text of an interpretative statement made by Germany and more than 20 other delegations at the adoption of Protocol II as amended concerning the meaning of the word "primarily".

At the Brussels Conference a number of States, such as Ethiopia, commented favourably on the removal of the word "primarily" from the text. Others, such as Sweden and the United Kingdom, concerned by an apparent discrepancy with the definition contained in Protocol II as amended, indicated that they might wish to discuss the definition further.

Indeed, at the Oslo Diplomatic Conference there was considerable debate on the issue. The United States, for instance, sought to include an exception for anti-personnel mines contained within mixed weapons systems, (11) but this was not acceptable to other delegations. Likewise, a proposal by Australia to exempt mines whose effects could be limited exclusively to combatants (12) was not approved, in part at least because no such mines were believed to exist or to be under development. Finally, Norway's proposal to clarify the exemption for anti-vehicle mines, "including those equipped with anti-personnel mines", (13) was not retained. The definition of anti-personnel mines in the Ottawa treaty thus covers all anti-personnel mines, including tripwire-activated directional fragmentation devices, but excludes anti-vehicle mines equipped with anti-handling devices. The ICRC will continue to monitor the development of anti-vehicle mines to ensure that they are not capable of detonation under the weight of a person.

(c) Anti-handling device

The second Austrian draft included a definition identical to that contained in Protocol II as amended. (14) At the Oslo Diplomatic Conference, following a proposal by the United Kingdom, (15) language was added to the effect that in addition to activation upon tampering with the mine, a device that detonated the mine when an attempt was made to "otherwise intentionally disturb [it]" would also be considered as an anti-handling device. A number of States invoked the so-called "doctrine of the innocent act", which holds that a mechanism so sensitive that merely touching it would cause the mine to detonate should not be considered an anti-handling device. Such a mine would therefore fall within the definition of an anti-personnel mine. If it did not, innocent passers-by would be exposed to excessive risk of injury; this would betray the very aim of the treaty, namely preventing the indiscriminate effects of anti-personnel mines. The ICRC intends to monitor developments in this field closely to ensure that both the spirit and the letter of this provision are fully observed.

(d) Transfer

The definition of transfer given in the Ottawa treaty is an exact replica of that contained in Protocol II as amended. This in turn was based on the definition of "transfer" used since 1993 for the UN Register of Conventional Arms. It was introduced into the second Austrian draft and adopted in Oslo without major changes (the word "anti-personnel" was added to the Protocol II definition). Thus, under Article 2, para. 4, of the Ottawa treaty "'[t]ransfer' involves, in addition to the physical movement of anti-personnel mines into or from national territory, the transfer of title to and control over the mines, but does not involve the transfer of territory containing emplaced anti-personnel mines." There is, however, some disagreement as to whether the first two possibilities are cumulative or alternative. This issue has some relevance to the discussions within NATO as to the possibility of transit of United States anti-personnel mines through NATO countries.

Core prohibitions

(a) The prohibition on the use of anti-personnel mines

The prohibition on the use of anti-personnel mines was central to the success of the treaty and represented its primary object and purpose. Without an absolute prohibition on use, the other prohibitions could not be absolute, either. The first Austrian draft had stated that "it is prohibited to use anti-personnel mines as they are deemed to be excessively injurious and to have indiscriminate effects." After some negative comments regarding the implication that States had in fact been violating international law by using anti-personnel mines, in the second draft the prohibition became simply an undertaking "never under any circumstances to use anti-personnel mines". This remained the text of the provision as it was finally adopted despite the wish of a number of countries to include exceptions or transition periods. At the Brussels Conference, for example, one State called for an exception to the prohibition on use "in exceptional circumstances". In Oslo, the same State proposed a "temporary arrangement", whereby a State "under exceptional circumstances for its national security, may resort to the use of anti-personnel mines in accordance with the international laws of armed conflict." This proposal was roundly rejected by the other participating States which felt that any such exception would undermine the entire treaty.

(b) The prohibition on the development of anti-personnel mines

The prohibition on the development of anti-personnel mines (16) represents the first time that such a provision has been included in a humanitarian law treaty.

States will have to take great care in their military research and development programmes to ensure that the rule is fully respected. Special attention will have to be paid to dual-use technologies and anti-handling devices which could cause an anti-tank mine to function as an anti-personnel mine.

(c) The prohibition on "inciting" a violation

It is also prohibited to "assist, encourage or induce, in any way, anyone to engage in any activity prohibited to a State Party under this Convention." (17) This provision covers, for example, the granting of licences to manufacture anti-personnel mines. It is also relevant to the situation of some NATO countries. On ratifying the treaty, Canada entered an understanding that mere participation by Canadian soldiers in a United Nations operation involving a State not party to the Convention and the use of mines by such a State would not constitute assistance within the meaning of the treaty. This would also apply to the transit of anti-personnel mines owned by a State not party to the Convention across the territory of a State Party.

(d) Proposed exceptions to the general obligations

At the Oslo Diplomatic Conference, with the situation in Korea uppermost in its mind, the United States proposed that an exception be made to the general treaty obligations with respect to "activities in support of a United Nations command or its successor, by a State Party participating in that command, where a military armistice agreement had been concluded by a United Nations command." (18) This issue was discussed at great length, both inside and outside the conference hall, but ultimately it was decided that such an exception was not acceptable. The possibility of a general transition period under which any State could defer implementation of the Convention for a set period was also considered. However, the United States did not feel that the provision as it stood was sufficient to ensure its early adherence to the treaty, and the issue was dropped.

The destruction of stockpiled anti-personnel mines

The original Austrian draft text had called for stockpiled anti-personnel mines to be destroyed within one year, though it offered the possibility of an additional one-year deferral period. (19) Following remarks by a number of States that the allotted time was unrealistically short, the second draft raised the limit to three years, but removed the possibility of deferral. At the Oslo Diplomatic Conference the limit was again raised, this time to four years, at which point final agreement was secured. The United States had suggested that an exception be made for mines not owned or possessed by a State Party though present on its territory, (20) but this proposal was rejected. Finally, Article 6, para. 5, of the Ottawa treaty specifically requires States Parties in a position to do so to provide assistance in the destruction of stockpiled anti-personnel mines. A good many States had emphasized the importance of international cooperation and assistance in fulfilling the obligation within what is, undoubtedly, a fairly short time period.

Exception for training in demining: the first Austrian draft contained a proposed exception to the prohibition on acquiring or retaining anti-personnel mines "if they are exclusively used for the development and teaching of mine detection, mine clearance, or mine destruction techniques and if the responsible institutions, the amount and the types are registered with the Depositary." A suggestion by Spain at the Brussels Conference that an exception be made to the prohibition on production in order that stocks of mines for training be replenished was not retained. At the Brussels Conference, Italy had called for a more specific limit on the number of mines needed for training in detection and clearance. This proved impossible to achieve during the negotiations in Oslo, so that the provision finally read: "[t]he amount of such mines shall not exceed the minimum number absolutely necessary" for such purposes. (21) At the adoption of the treaty a number of States declared that they considered one to two thousand anti-personnel mines to be sufficient for training purposes. (22) This was not contested. Under Article 7 ("Transparency measures"), the types, quantities and, if possible, lot numbers of all anti-personnel mines retained for training purposes must be reported annually to the UN Secretary-General.

Clearance and destruction of emplaced anti-personnel mines

The original Austrian draft required that emplaced anti-personnel mines be destroyed within five years, with the possibility of an additional two-year period for States requesting an extension. At the subsequent Vienna Expert Meeting it was rightly pointed out that the ability of a State to remove and destroy anti-personnel mines depended on its technical capacity and resources, and that in the case of States with massive problems this process could take decades. It was clear that the treaty needed to allow such States to adhere without fear that they might find themselves in breach of its provisions because of the continued presence of uncleared mines. It was even suggested that emplaced mines should be fenced and marked but that their removal and destruction should not be subject to a strict timetable.

Taking account of those concerns, the second Austrian draft differentiated between the destruction of anti-personnel mines laid within minefields (deemed to be a "defined area in which mines have been emplaced") and those laid in areas outside minefields. The former were to be marked in accordance with international standards and then destroyed within 10 years of the treaty's entry into force, whereas in the case of the latter the obligation was simply to destroy the mines, without a fixed deadline, and to give an immediate and effective warning to the population of the danger posed by their presence.

A number of States pointed out that a 10-year fixed period for the clearance of minefields was unrealistic. In addition, at the Brussels Conference the United Kingdom, referring to the problems of clearing "undetectable" plastic mines in the Falkland Islands, proposed that an exception be made in the case of land of marginal economic value where the risk to the civilian population was minimal. The ICRC, however, asked States not to extend the time period and reminded them that the recently adopted obligation in Protocol II as amended to clear mines at the end of hostilities should not be weakened. The ICRC insisted, as did a number of others, that an open-ended commitment to clear all anti-personnel mines "as soon as possible" was inadequate, as it risked detracting from some of the undoubted urgency that would be contained in a specified time-period. The shortest possible time frame was desirable from a humanitarian point of view; on the other hand, a number of countries were so severely contaminated by mines that too short a period would be unrealistic and might deter their adherence to the treaty.

For this reason, the ICRC proposed that the obligation to clear anti-personnel mines laid in minefields within 10 years should be maintained, but that States Parties needing more time could apply for an extension period. (23) In this way progress in mine clearance and the need for greater international assistance and support could be objectively assessed. Given the difficulty of distinguishing "minefields" from "mined areas", at the Oslo Diplomatic Conference the obligation to demine within 10 years of the treaty's entry into force for each State was expanded to include all emplaced anti-personnel mines, though with the possibility for severely mine-affected States to be granted extension periods of up to 10 years at a time. As a result of this agreement the definition of a minefield was removed and only that of a mined area remained; this was deemed to be "an area which is dangerous due to the presence or suspected presence of mines." (24)

The final issue of importance to note is that the obligation to demine covers all territory under the jurisdiction *or* control of a State Party. (25) This means that the obligation extends to a situation where an insurgent or separatist armed force controls a certain portion of territory within the boundaries of a State. Of course, when deciding whether to grant an extension period, States Parties can be expected to sympathize with a fellow State unable to clear emplaced mines because it does not have physical control of all of its territory.

International cooperation and assistance

It was evident that where both the clearance of emplaced anti-personnel mines and the destruction of stockpiled mines were concerned, international cooperation and assistance would play an essential role in ensuring early adherence to the treaty and its successful implementation on the ground. (26) At the Vienna Expert Meeting, the ICRC pointed out that the treaty's technical demands would be entirely different from those set out in Protocol II as amended, which envisages the use of new types of mines. Therefore, the text relating to technical assistance could not be identical to that painstakingly put together in the CCW negotiations.

Assistance to mine victims

Similarly, international support would be a crucial element with regard to the need to provide long-term care and assistance for mine victims. In its formal comments on the third Austrian draft, (27) the ICRC called for the inclusion of a provision requiring each State Party in a position to do so to provide assistance for the care and rehabilitation of landmine victims and for mine-awareness programmes. A further proposal that States Parties accept a duty under Article 1 to assist mine victims was, however, not retained. The relevant provision incorporated in the final text of the treaty expressly mentioned the possibility of channelling assistance to mine victims through relevant non-governmental organizations (NGOs), the United Nations and components of the International Red Cross and Red Crescent Movement. (28) Following a proposal by the ICBL, the provision also set out an obligation to provide international assistance for the social and economic reintegration of survivors of mine explosions. (29) As ICBL representatives rightly noted during the negotiations in Oslo, comprehensive assistance to mine victims demands more than surgical care and physical rehabilitation.

Promoting compliance and implementation

As already mentioned, the issue of verification of compliance with the treaty was the subject of very detailed debate. At the Vienna Expert Meeting, differences arose between States which thought that little or no verification was necessary to oversee what was essentially a humanitarian treaty, and those which felt very strongly about the security implications of a ban and therefore pushed for comprehensive verification procedures similar to those provided for in earlier disarmament agreements. One country, Saint Lucia, speaking on behalf of the Organization of American States, pointed out that countries in Central America were eager to comply with the treaty and to cooperate fully where transparency and exchange of information were concerned in order to obtain outside assistance in eliminating their anti-personnel mines. This positive approach, it suggested, was preferable to attempts to "catch" treaty violators through traditional verification mechanisms.

In the same spirit, at the subsequent Bonn Expert Meeting one government proposed the creation of an implementation or fact-finding commission which would work on a cooperative basis. While no consensus on any approach emerged from the discussions, the Chairman's summary mentioned the need to ensure that any mechanism was cost-effective and practical, stressed the value of information exchange and raised the possibility of providing for fact-finding missions. The Chairman declared that a possible adoption of the approach used in arms control agreements would require further discussion and emphasized the reference to an "effective" ban in UN General Assembly resolution 51/45S.

Discussions continued in June 1997 at the Brussels Conference, during which a number of countries, including Norway, Sweden and Switzerland, asserted that since this was primarily a humanitarian law treaty, detailed verification was not essential, and might even dissuade adherence. Ecuador claimed that the prohibitions on production and export were needed to ensure that the treaty would be respected. Others, notably Australia, called for a disarmament treaty that would attract universal adherence, suggesting that the Ottawa treaty might be a "permanent partial solution". Uruguay called for a "balanced formula" including effective verification provisions so that the treaty did not become merely an expression of good intentions.

Several "transparency measures" had already been included in the first Austrian draft. By the time the definitive text was drawn up, the list of items to be reported had grown considerably. Under Article 7 of the Ottawa treaty, each State Party must provide the Depositary with a detailed report on many compliance-related matters within 180 days of the treaty's entry into force. Provision is also made for fact-finding missions to clarify any doubts as to the compliance with the treaty by a State Party. (30) For obvious reasons, this was a difficult provision to negotiate and it is easily the longest article in the treaty. Less controversially, it was agreed that States Parties would meet annually following the treaty's entry into force up until the date of the first review conference, scheduled to take place five years after the treaty comes into force. Amendments may, however, be proposed at any time after the treaty becomes legally binding. The relatively simple procedures regarding the facilitation and clarification of compliance are likely to be complemented by a "citizen-based monitoring mechanism", the details of which are to be elaborated during the course of 1998.

National implementation

Taking up a proposal made by the United States during the negotiations leading to the adoption of Protocol II as amended, the first Austrian draft text had incorporated a provision for compulsory universal jurisdiction for wilful acts committed during armed conflict and causing death or serious injury. In the second draft, however, the proposed provision had been considerably watered down to a duty only to take "all appropriate legal, administrative and other measures, including the imposition of penal sanctions, to prevent and suppress any activity prohibited to a State Party (. . .) undertaken by persons or on territory under its jurisdiction or control." (31) Accordingly, at the Bonn Expert Meeting the ICRC circulated an informal proposal on detailed national implementation measures. At the Oslo Diplomatic Conference, Switzerland introduced a proposal for compulsory jurisdiction over any national of a State Party who has used anti-personnel mines or ordered them to be used. Unfortunately, and perhaps as a consequence of the largely disarmament background of many of the negotiators, this proposal was not retained. The provision finally adopted is largely that contained in the second Austrian draft, and therefore normally – though not always – requires the adoption of national legislation.

Reservations

An article prohibiting reservations to the provisions of the treaty, similar to the one found in the 1992 Chemical Weapons Convention, was already included in the first Austrian draft. It remained unchanged throughout the period of negotiations, although at the Oslo Diplomatic Conference several States tried to weaken it by inserting an exception for periods of armed conflict or to have it deleted altogether. (32)

Entry into force

The first Austrian draft had proposed that the treaty should enter into force six months after the deposit of the fortieth instrument of ratification. Some States felt that the number was too high; the ICRC, for its part, reminded the participants at the Oslo Diplomatic Conference that the 1949 Geneva Conventions and their 1977 Additional Protocols had entered into force after the deposit of just two ratifications. Indeed, from a humanitarian viewpoint the application of the new treaty even by a limited number of States would provide significant benefits and encourage others to follow. However, some other States felt that the security implications of forgoing anti-personnel mines were significant and that even more than 40 ratifications should be required.

Ultimately, a compromise agreement was reached, stipulating that 40 ratifications would be needed for the treaty to come into force. (33) As at the time of writing, 11 States – Belize, Canada, the Holy See, Hungary, Ireland, Mauritius, Niue, San Marino, Switzerland, Trinidad and Tobago, and Turkmenistan – had deposited their instruments of ratification with the Depositary, the UN Secretary-General. It is hoped that the number of 40 ratifications may be reached well before the end of 1998, so that the treaty could come into force in early 1999.

Following a suggestion by Belgium, an opportunity was given to the first 40 States to declare, at the time of ratification, that they would provisionally apply the treaty's core provisions, set out in Article 1, para. 1 (i.e., the prohibitions on use, development, production and transfer) until the entry into force of the treaty as a whole. (34) As at the time of writing, two States – Mauritius and Switzerland – had taken advantage of this possibility.

Withdrawal

The first Austrian draft text had provided for withdrawal from the treaty on 90 days' notice if a State decided that "extraordinary events [had] (. . .) jeopardized [its] supreme interests." At the Vienna Expert Meeting a number of different approaches to this question were already emerging. Some States were of the opinion that no right of withdrawal should be permitted in view of the danger that it might be used when a country was engaged in an armed conflict, which was precisely when compliance with the treaty's provisions was most important. Mexico made a proposal along the lines of the withdrawal clause contained in 1977 Additional Protocol I whereby withdrawal was possible but would not be effective during an ongoing armed conflict. (35) Other States proposed a straightforward withdrawal clause similar to that contained in the Austrian draft text. At the Oslo Diplomatic Conference there was extensive negotiation on this issue. Finally, agreement was reached on a provision that allows effective withdrawal six months after receipt of the relevant instrument by the Depositary. If, however, at the end of that six-month period the withdrawing State Party is engaged in an armed conflict, the withdrawal will not take effect before the end of the armed conflict. (36)

The implications of the Ottawa process for international humanitarian law

The success of the Ottawa process marks a welcome return to the traditional approach to the development of international humanitarian law whereby treaties are adopted without a consensus rule. Indeed, recent negotiations in the domain of international humanitarian law, such as those on the 1980 CCW, were governed by the practice of consensus. The formal CCW review process demonstrated the limitations of this method where landmines were concerned. With this in mind, it may well be necessary to review the practice of adopting agreements by consensus for future CCW negotiations (a Second Review Conference is scheduled for 2001).

In addition, the continuing importance of the CCW should be stressed. It is the only framework, based on international humanitarian law, for the specific regulation of existing conventional weapons and for responding to the emergence of new weapons. The regulation of the employment of anti-personnel mines by Protocol II as amended should be seen as the absolute minimum norm for States which continue to use them. Protocol II as amended also remains the only international instrument specifically governing the use and transfer of anti-vehicle mines and establishes the rule that those who use mines of whatever type are responsible for their removal at the end of hostilities. This could prove to be an important protection, even for Parties to the Ottawa treaty, in situations of conflict with a State not party to it.

For the ICRC, and indeed for the International Red Cross and Red Crescent Movement as a whole, the mines campaign has provided an example of how successful advocacy in the interests of war victims can be carried out in the post-Cold War environment. Dozens of National Red Cross and Red Crescent Societies, including many new to campaigning, have felt empowered to advocate on behalf of mine casualties present and future. The campaign has been conducted without compromising the fundamental principle of neutrality, which prohibits components of the Movement from taking sides in a conflict or favouring particular political parties or groups. This principle is intended to ensure that all victims of war receive protection and assistance – it is therefore a means to an end, not an end in itself.

The Ottawa process has also shown that civil society has a crucial role to play in strengthening international law. The complementary role played by key governments, the ICBL and the ICRC augurs well for the future development of international humanitarian law. In contrast to the ICRC, the ICBL has been able to criticize the positions of specific governments directly and publicly. On the other hand, the ICRC's special status as an international organization and its network of professional military officers working with armed forces on humanitarian law issues give it access to governmental and military circles, which NGOs often do not have.

The global mobilization that has been necessary to achieve an international legal norm prohibiting anti-personnel mines has also clearly shown that the international community must seek a more preventive approach to the control or prohibition of weapons which go against international humanitarian law, and a more dynamic method of developing that law. The Ottawa process has raised awareness among the general public of the limits that must be placed on the conduct of warfare. As a result, higher expectations will be placed on the behaviour of States. At present the ICRC is examining, together with medical experts, possible objective medical criteria to determine whether the effects on health of a given weapon are of a nature to cause superfluous injury or unnecessary suffering. (37)

Despite its success, the Ottawa process has most clearly demonstrated the need for a more preventive approach to arms issues under international humanitarian law. One must ask whether appalling levels of civilian death and injury have to be reached before the use of each new weapon which may violate international humanitarian law is either regulated or prohibited, as the case may be. Far more systematic analysis and informed debate is needed before any new weapon is deployed. The recent agreement to prohibit, in advance, the use and transfer of blinding laser weapons is a basis for hope. (38) Given the rapid development of new technologies, the protection provided by humanitarian law will be of crucial importance in making sure that humankind is the beneficiary, and not the victim, of technical advances which have profound implications on the waging of war.

NOTES

1 United Nations Convention on Prohibitions or Restrictions on the Use of Certain Conventional Weapons Which May be Deemed to be Excessively Injurious or to Have Indiscriminate Effects, of 10 October 1980.

2 Protocol on Prohibitions or Restrictions on the Use of Mines, Booby-Traps and Other Devices as amended on 3 May 1996 (Protocol II as amended on 3 May 1996), annexed to the CCW, *supra* note 1.

3 Fifty States were full participants at the first Ottawa Conference: Angola, Australia, Austria, Belgium, Bolivia, Bosnia and Herzegovina, Burkina Faso, Cambodia, Cameroon, Canada, Colombia, Croatia, Denmark, Ethiopia, Finland, France, Gabon, Germany, Greece, Guatemala, Guinea, Honduras, Hungary, Iceland, Iran, Ireland, Italy, Japan, Luxembourg, Mexico, Mozambique, the Netherlands, New Zealand, Nicaragua, Norway, Peru, the Philippines, Poland, Portugal, Slovakia, Slovenia, South Africa, Spain, Sweden, Switzerland, Trinidad and Tobago, the United Kingdom, the United States, Uruguay, and Zimbabwe. A further 24 countries – Albania, Argentina, Armenia, the Bahamas, Benin, Bulgaria, Brazil, Brunei Darussalam, Chile, Cuba, the Czech Republic, Egypt, the Federal Republic of Yugoslavia, the Holy See, India, Israel, Malaysia, Morocco, Pakistan, the Republic of Korea, Romania, the Russian Federation, Rwanda, and Ukraine – attended as official observers.

4 Convention on the Prohibition of the Use, Stockpiling, Production and Transfer of Anti-Personnel Mines and on their Destruction, of 18 September 1997, reprinted in the *International Review of the Red Cross* (*IRRC*), No. 320, September–October 1997, pp. 563–578.

5 Article 2, para. 3, of Protocol II as amended defines an anti-personnel mine as one "primarily designed to be exploded by the presence, proximity or contact of a person". The use of the phrase "primarily designed" was strongly opposed by the ICRC which feared its abuse in cases where a munition which was clearly an anti-personnel mine could be claimed to have another "primary" purpose.

6 Geneva Protocol of 17 June 1925 for the Prohibition of the Use in War of Asphyxiating, Poisonous or other Gases and of Bacteriological Methods of Warfare.

7 See APL/CW.46 of 3 September 1997.

8 11th preambular paragraph.

9 Article 2, para. 2.

10 "Mines designed to be detonated by the presence, proximity or contact of a vehicle as opposed to a person, that are equipped with anti-handling devices, are not considered anti-personnel mines as a result of being so equipped."
11 See APL/CW.9 of 1 September 1997.
12 See APL/CW.2 of 1 September 1997.
13 See APL/CW.4 of 1 September 1997.
14 "Anti-handling device means a device intended to protect a mine and which is part of, linked to, attached to or placed under the mine and which activates when an attempt is made to tamper with the mine." See Protocol II as amended, Art. 2, para. 14.
15 See APL/CW.32 of 2 September 1997.
16 See Art. 1, para. 1(b).
17 Art. 1, para. 1 (c).
18 See APL/CW.8 of 1 September 1997.
19 Art. 5, first Austrian draft treaty text.
20 See APL/CW.10 of 1 September 1997.
21 Art. 3, para. 1, Ottawa treaty.
22 Canada stated it would retain around 1,500, the Netherlands 2,000, and Germany "thousands, not tens of thousands". Belgium supported this interpretation.
23 *Comments of the International Committee of the Red Cross on the Third Austrian Draft (13/5/97) of the Convention on the Prohibition of Anti-personnel Mines*, Informal Working Paper for the Oslo Negotiations, September 1997. A similar extension period is contained in section C (paras. 24–28) of the Verification Annex of the 1992 Chemical Weapons Convention.
24 Art. 2, para. 5.
25 Art. 5, para. 1.
26 Art. 6.
27 *Supra*, note 23.
28 Article 6, para. 3.
29 *Idem.*
30 See Art. 8.
31 See Art. 10, Second Austrian Draft.
32 Art. 19.
33 Art. 17.
34 Art. 18.
35 Protocol Additional to the Geneva Conventions of 12 August 1949, and relating to the protection of victims of international armed conflicts (Protocol I), Article 99.
36 Art. 20.
37 See Coupland, R. M. (ed.), *The SIrUS Project, Towards a determination of which weapons cause "superfluous injury or unnecessary suffering"*, ICRC, Geneva, 1997.
38 Louise Doswald-Beck, "New Protocol on Blinding Laser Weapons", *IRRC*, No. 312, May–June 1996, pp. 272–299.

22

The Entry into Force of the Ottawa Treaty

1 March 1999

On 1 March 1999, the Ottawa treaty entered into force and became binding international law. While 123 States had signed the treaty in Ottawa, Canada, in December 1997, their signature was essentially a political commitment to ratify the treaty at a later date. To enter into force and become a legally binding instrument, forty States had to notify the Secretary-General of the United Nations formally that they were ready to be bound by the treaty. Burkina Faso became the fortieth State to submit an instrument of ratification and in accordance with its provisions, the Ottawa treaty became binding upon these 40 States six months later. This marked the fastest entry into force of a multilateral arms-related treaty and established a new international norm governing anti-personnel mines. Events marking the treaty's entry into force were held throughout the world. At the United Nations in Geneva, States marked the occasion with a ceremony and met to prepare the first meeting of States Parties.

Statement of Eric Roethlisberger
Vice-President, International Committee of the Red Cross

Ceremony on the
entry into force of the treaty banning anti-personnel mines

Palais des Nations, Geneva

1 March 1999

When the torrential rains of Hurricane Mitch finally stopped pounding the Matagalpa region in northern Nicaragua four months ago, Degliz Lopez and his friend Celestin Murilto rode their bicycles to the Esquirin river to view the damage. The river had swelled to eight times its normal width; a well-known farm was gone; the bridge washed away. But the extensive debris on the riverbank fascinated both Degliz and Celestin who went down to explore. Returning up the embankment, Degliz heard the explosion which hurled him through the air. When he regained consciousness he had shrapnel in his face and eye and Celestin was dead. Those trying to rescue him couldn't approach – he was laying amidst a scattering of mines deposited by the floodwaters. Eventually, a long enough pole was found to pull him to the roadside. Celestin left behind four children. Degliz is thankful he is alive and that his little daughter is not an orphan.

In the name of the hundreds of thousands of victims and of the millions who live each day in fear of becoming victims, the ICRC whole-heartedly welcomes the entry into force today of the Convention on the Prohibition of the Use, Stockpiling, Production, and Transfer of Anti-personnel Mines and on their Destruction. This Convention, which will enter into force more rapidly than any previous multilateral arms-related treaty, represents a comprehensive response to the landmine crisis by the 134 States which have now signed or acceded to it. It represents the norm by which all efforts to address this humanitarian tragedy will be judged.

The incident in Nicaragua reminds us of the urgency of the task before us – to ensure rapid universalisation and implementation of the Ottawa treaty. Until this is achieved the story of Degliz and Celestin will be repeated, in various forms, some 2,000 times per month, in all regions of the world. This treaty is the only adequate response to a major humanitarian emergency; each day its implementation is delayed the cost in human lives and limbs is high. As we learned with Hurricane Mitch, even the bold efforts of the countries of Central America to clear the region of mines by the year 2000 can be defied and set back many years by an act of nature. Progress is vital; time is precious.

Landmines are not aware that a treaty has been signed; nor that it has entered into force. While the treaty may be the <u>prescription</u> for ending the landmines epidemic it is the difficult, dangerous and costly work in individual fields and communities which is the cure. Organizing these efforts for the long-term is the fundamental challenge faced by us all – States, NGOs and international agencies alike. But the task of organizing for the mid-to-long term does not come easily for governments, funding bodies and agencies. Admittedly, the outline of structures which will support increased efforts in the fields of victim assistance, of mine clearance and of mine awareness are beginning to become apparent. But the results of too many of our meetings, conferences and seminars are not yet to be seen on the ground.

Looking back over 1998, it becomes clear as never before what a daunting challenge we have taken up in the Ottawa treaty. Since it was signed the ICRC has tried to do its part in promoting the treaty through the hosting of and substantial support for regional conferences in Budapest, Moscow, Dhaka, Mexico City and Beirut as well as a host of events of a national character. Our delegations and National Societies of the Red Cross and Red Crescent have used the ICRC "ratification kits" as tools to encourage rapid ratification of the Convention and the early adoption of appropriate laws for national implementation. Surgical assistance and physical rehabilitation programs have been provided or supported by the ICRC in some 22 mine-affected countries; two new rehabilitation centres opened in Uganda and the Democratic Republic of Congo. Mine awareness work, often in cooperation with National Societies, is continuing in five countries and is being considered in several others. We are preparing, with our partners in the International Movement of the Red Cross and Red Crescent, a comprehensive, long-term landmines strategy for the Movement – to be adopted in October of this year by the Movement's Council of Delegates.

We are also pleased to present to this audience the ICRC's new traveling exhibition on the Ottawa treaty, now available in both English and Spanish; you are encouraged to make use of it for upcoming national and international events.

In just over two months States Parties will gather in Maputo to begin laying the groundwork for implementation in the years ahead. In the view of the ICRC, some of the key issues to be considered are:

How to bring about an effective exchange of information on treaty implementation among States, agencies and NGOs? What mechanisms are called for to guide such efforts and to avoid needless duplication of conferences, databases and other activities?

How to assure long-term funding – to provide confidence to agencies, NGOs and government health services which are willing to address the lifetime needs of mine and other war victims, but which hesitate to do so for fear of being left with unsustainable programs in a few years time?

How to respond collectively to the new use of landmines in such places as Angola and, on a lesser scale, Kosovo? This is a particular concern when such use involves a signatory State.

The ICRC today commends the 65 States which have ratified or acceded to the Ottawa treaty; we appeal to the additional 69 signatory States to do so as a matter of urgency. As the Ottawa process grew from a core of 50 States in October 1996 to a group of 134 States in 1999, we are confident that the core of States Parties will continue to increase until the absolute norms contained in this landmark treaty of international humanitarian law are universal.

With commitment and creativity the daunting task of implementation which officially begins today can and must be met. The first faltering steps, which began in 1998, will be steadied in Maputo; our treaty can be fully on its feet by the end of the year. We have begun together today the long march towards a world free of anti-personnel mines. In taking up this challenge we can relieve the burden of many millions of our fellow human beings who either have paid the price for these perverse weapons or live with their silent menace each day. When the job is done the rains which fall on the earth will no longer spread the seeds of death but will nourish, unhindered, the seeds of life.

Press Release

1 March 1999

OTTAWA TREATY BECOMES LAW

Geneva (ICRC/Federation) – On behalf of the hundreds of thousands of mine victims and the millions who live each day in fear of those weapons, the International Committee of the Red Cross (ICRC) and the International Federation of Red Cross and Red Crescent Societies whole-heartedly welcome the entry into force of the Convention on the Prohibition of the Use, Stockpiling, Production, and Transfer of Anti-personnel Mines and on their Destruction (the Ottawa treaty).

The treaty, which has become law more quickly than any previous multilateral arms-related agreement, represents a comprehensive response to the landmine crisis on the part of the 134 States that have now signed or acceded to it. The Red Cross and Red Crescent Movement congratulates the 65 States that have become bound by the Ottawa treaty and appeals to the other 69 signatory States to do likewise as a matter of urgency. "The treaty represents the standard by which all efforts to deal with this humanitarian tragedy will be judged", said ICRC Vice-President Eric Roethlisberger at a ceremony held at the United Nations' Geneva headquarters to mark the event. Mr Roethlisberger drew attention to the daunting challenges that lie ahead for States, international agencies and non-governmental organizations in ensuring that the treaty becomes binding worldwide and fully implemented in mine-affected communities. He committed the ICRC to doing its share in this regard.

"The task before us – to ensure rapid universalization and implementation of the Ottawa treaty – is a matter of high priority for National Red Cross and Red Crescent Societies", said Ms Astrid N. Heiberg, President of the International Federation. "Those Societies continue to play a key role by advising their governments on national legislation needed to ensure swift implementation of the treaty's provisions and by keeping the plight of mine victims in the public eye."

The ICRC is currently running 25 limb-fitting and rehabilitation programmes in 13 countries (Afghanistan, Angola, Azerbaijan, Cambodia, the Democratic Republic of the Congo, Georgia, Iraq, Kenya, Rwanda, Sri Lanka, Sudan, Tajikistan and Uganda). Twenty-four ICRC projects in 12 other countries have now been handed over to local or international NGO control, though many continue to receive financial and technical support from the ICRC. In a number of countries, the National Red Cross and Red Crescent Societies, supported by their International Federation, care for mine-injured people through health, rehabilitation and social welfare programmes.

Between 1979 and the end of 1998, the ICRC manufactured over 130,000 artificial limbs, over 175,000 pairs of crutches and close to 9,000 wheelchairs. In 1998 alone, the ICRC manufactured over 11,500 prostheses; of these, more than 6,500 were for mine victims. During the same year it produced over 17,200 pairs of crutches and more than 700 wheelchairs.

In addition, the ICRC and National Societies are conducting mine-awareness programmes in several countries in order to reduce the number of incidents in mine-affected areas.

23

The First Meeting of States Parties to the Ottawa Treaty

Maputo, Mozambique
3–7 May 1999

With the historic entry into force of the Ottawa treaty in March 1999, there was considerable support to convene a meeting of States Parties and begin immediate consultations on converting the treaty's obligations into action on the ground. The First Meeting of States Parties (FMSP), held only two months after entry into force, focused on the implementation of the Ottawa treaty and adopted several important mechanisms to facilitate this. Firstly, the meeting adopted the 'Maputo Declaration', which reaffirms the commitment of States Parties to the total elimination of anti-personnel mines, thus pledging themselves to universalize the treaty and eradicate the human suffering the weapons cause. In very clear terms, the Declaration condemns the continued use of mines and calls upon all States to adhere to and implement the Ottawa treaty and intensify their efforts to rid the world of these weapons and help mine victims and mine-affected communities.

To assist States in fulfilling their obligation under Article 7 of the treaty, whereby States Parties must submit annual reports on mine-related issues to the UN Secretary-General, the FMSP approved a reporting format to ensure that such reports were standardized and comprehensive. States Parties also established a structure to continue dialogue and consultation on mine-action issues between the annual meetings of States Parties. Five Standing Committees of Experts (SCEs) were created to ensure that the goals of the treaty were accomplished in an efficient and coordinated manner. SCEs were created to deal with issues related to mine clearance, victim assistance, stockpile destruction, technologies for mine action and general status and operation of the treaty.

The FMSP was attended by representatives of 108 States and a large number of international, regional and non-governmental organizations. Of those States participating, 59 had ratified the treaty and 36 were signatories. Importantly, representatives from a number of non-signatory States also attended as observers. Most notable was the participation of officials from China, Finland, Sri Lanka and Turkey.

Statement of Eric Roethlisberger
Vice-President, International Committee of the Red Cross

First Meeting of States Parties to the Ottawa treaty

Maputo, Mozambique

4 May 1999

It is most fitting that the international community has come to Mozambique for this – the first meeting of States Parties to the Convention on the Prohibition of the Use, Stockpiling, Production and Transfer of Anti-personnel Mines and on their Destruction. Mozambique is a particularly appropriate setting because this country and this region know all too well the horrific consequences of anti-personnel mines. Southern Africa has long been regarded as one of the world's most heavily mined regions. Here in Mozambique, demining teams are at work in the countryside; mine awareness is being taught in the classrooms and rehabilitation clinics are producing and fitting artificial limbs. While much remains to be done before this country is free from the curse of these horrific weapons, coming here reminds us of both our accomplishments and of the daunting challenge of doing away once and for all with anti-personnel mines.

It was only two months ago that the international community marked the entry into force of the Ottawa treaty. This instrument, which became international law more rapidly than any previous multilateral arms related convention, is a remarkable achievement. The Ottawa treaty represents the <u>comprehensive</u> framework for ending the man-made epidemic of landmine injuries. Its universal acceptance and full implementation are imperative if that goal is to be achieved.

We now move into a new phase of the Ottawa process: that of implementing the convention. This a formidable task and will demand much of all of us. Even in the best of circumstances, ending the scourge of anti-personnel mines will be a slow and difficult process. Given the need for effective coordination, the ICRC firmly supports the proposal to establish an intersessional process, within the framework of the convention, through which issues of mine action will be addressed at working level.

Since the signing of the Ottawa treaty in December 1997, the ICRC has tried to do its part to encourage universalization and implementation while continuing its field work to alleviate the suffering of those in mine-affected countries. The ICRC has promoted the Ottawa treaty by hosting and providing substantial support for regional conferences such as those held last year in Budapest, Moscow, Dhaka, Mexico City and Beirut. Later this year, the ICRC will be organizing a meeting on anti-personnel mines for States of south Asia which will be hosted by the Government of Sri Lanka. To promote greater understanding of the Ottawa treaty, the ICRC has produced a travelling exhibition explaining its provisions which is available in English, Spanish and Arabic. We are happy to invite you to view the English version which is currently on display in the tent at the Polana Hotel. We have also produced a short video outlining the treaty's scope and obligations for military personnel, demining experts, parliamentarians and others involved in the Convention's implementation.

In the field, the ICRC continues to carry out mine awareness programmes and provide assistance to mine victims and other victims of war. It currently conducts mine awareness in Azerbaijan, Bosnia and Herzegovina, Croatia, Georgia and in the region of Nagorni Karabach. Surgical assistance and physical rehabilitation programmes are being provided or supported by the ICRC in 22 mine affected countries. Two new rehabilitation centres have opened in Uganda and the Democratic Republic of the Congo. In 1998 the ICRC provided and fitted prostheses to over 7,000 mine victims. The ICRC has also been cooperating with the World Health Organization to develop an integrated public health approach to the treatment of mine victims.

The ICRC would like to take this opportunity to draw the attention of States Parties to a potential threat to the purpose of this Convention. As States prohibit the use of anti-personnel mines by their armed forces, many are likely to employ anti-vehicle mines with anti-handling devices as an alternative. The ICRC is concerned that increased use of <u>certain</u> anti-handling devices will endanger civilian populations. Of specific concern are devices which will trigger the mine's detonation through the innocent passage of a person over or near the mine or through inadvertent or accidental contact with the mine itself. This threat is particularly serious with regard to remotely delivered mines which lie on the ground. In such cases the anti-handling device can cause the anti-vehicle mine to function as an anti-personnel mine. A range of experts consulted by the ICRC believe that anti-handling devices can be designed in such a manner so as to limit the danger to innocent civilians. The ICRC calls upon all States to examine the technical characteristics of existing and proposed anti-handling devices and to ensure that they are designed so as to minimize the risks of detonation through inadvertent or accidental contact.

Finally, the ICRC would like to voice concern about the reports of new use of landmines in some countries. There is clearly a need for a collective response from States Parties on this issue. This concern is particularly acute when such use involves a signatory State. The ICRC urges the conference to send a clear message that anti-personnel mines are no longer an acceptable weapon of warfare and remind any signatory State using them that such use is contrary to the spirit and purpose of the Ottawa treaty.

During the sessions of this conference being held later this week, the ICRC will actively contribute and provide practical materials on victim assistance and national legislation. These papers are intended to assist States with their implementation efforts. The ICRC will also be distributing an informal paper on the issue of anti-handling devices. This paper will explain in further detail our concerns about the increased use of such devices and identify some of the systems which States should consider examining.

With the entry into force of the Ottawa treaty the focus now turns to the practical work of eliminating anti-personnel mines. The results of this meeting will lay the ground for the treaty's implementation in the years to come. It was the plight of mine victims and people living in mine-affected communities which led most of the world to reject anti-personnel mines as a weapon of war. But in communities throughout the world people await tangible results in the form of cleared fields, unhindered travel, physical rehabilitation and socio-economic re-integration. Progress in achieving their goals and preventing the laying of new mines will be the scale against which the success of our actions will be measured.

ICRC Information Paper

First Meeting of States Parties to the Ottawa treaty

Implementing the Ottawa treaty:

National Legislation

The Ottawa treaty provides a comprehensive framework for solving the anti-personnel (AP) mine problem. At its core are prohibitions on the use, production, development, stockpiling and transfer of AP mines (see article 1 of the treaty). To ensure that these obligations are enforced, States Parties will have to take a number of steps at the national level. The treaty itself requires that they take,

> *"all appropriate legal, administrative and other measures, including the imposition of penal sanctions, to prevent and suppress any activity prohibited to a State party under this Convention undertaken by person or on territory under its jurisdiction or control." (art. 9).*

For most States this will entail the adoption of criminal legislation. It is also likely to require other regulatory measures such as changes in military doctrine and procedures, the notification of organizations and corporations involved in AP mine production and sale and instructions to the relevant ministries to rescind export licences which may involve AP mines.

In drafting national legislation to implement the treaty's prohibitions certain issues need to be addressed so as to ensure consistency between national law and the treaty's language and purpose. This paper identifies some of the key elements which States should consider when preparing such legislation.

Definitions

The Ottawa treaty contains a clear definition of an anti-personnel mine and removes the ambiguities found in earlier instruments. For this reason it is essential that national laws use the definition found in the Ottawa treaty and define an AP mine as "a mine designed to be exploded by the presence, proximity or contact of a person and that will incapacitate, injure or kill one or more persons." This will prevent discrepancies between the national law and the treaty and prevent undesirable loopholes. In their national laws some States have substituted the word "mine" with "munition" but the essence of the definition remains the same.

Other treaty definitions may also be incorporated into national legislation. These include the definition of "mine", "anti-handling device", "transfer" and "mined areas".

Comprehensive Prohibitions

National laws must cover the broad range of activity forbidden by the treaty. This includes prohibitions on the use, development, production, stockpiling and transfer of AP mines. In their national laws many States have detailed the elements understood to comprise the "transfer" of AP mines and have specifically banned their import, export, sale and transit in addition to the activities mentioned above.

Criminal sanctions

Effectively preventing and suppressing breaches will require penal sanctions. States have generally sought to punish violations with terms of imprisonment and/or a fine. Prison sentences can be as long as 14 years and fines range up to hundreds of thousands of dollars.

Component parts

The treaty itself does not explicitly refer to the component parts of AP mines. Nonetheless, knowingly manufacturing, transferring or selling components intended to be used to assemble AP mines would violate the prohibition on assisting in the production of AP mines. The national laws passed by a number of States classify AP mine components as prohibited objects.

Exclusions

The Ottawa treaty permits AP mines to be retained or transferred for training in mine detection, mine clearance and mine destruction techniques and this exception is often included in national laws. The number of mines kept shall not exceed the absolute number necessary and has been generally understood to mean no more than several thousand units.

Other regulatory measures

In addition to the prevention and punishment of violations through national legislation, States must ensure that other treaty obligations are implemented at the national level. These include:

- the drawing up of plans for the collection, transport and destruction of AP mines stockpiles;
- development and implementation of plans for mine field marking and mine clearance;
- the provision of resources for mine awareness and victim assistance programs.

Implementing such actions may not require the promulgation of national legislation but might be accomplished through other regulatory measures.

Attached is a copy of the national law passed by the government of Austria (annex 1).[1] This law was one of the first containing a comprehensive ban on AP mines and provides a good example of legislation incorporating most of the above elements. A list of the States which have passed similiar legislation and which have been supplied to the ICRC is found in annex 2 (additions or corrections are welcomed). The ICRC's Legal Division is available to provide guidance on national legislation and can furnish additional examples of the national legislation adopted by other governments.

International Committee of the Red Cross
Geneva, April 1999

Annex 1

Unofficial translation by the Austrian Red Cross

FEDERAL LAW GAZETTE

OF THE AUSTRIAN REPUBLIC

Year 1997 Published on 10. January 1997
Part I

13. Federal Law on the Prohibition of Anti-Personnel Mines

The Parliament has decided as follows:

Definitions

§ l. In this Federal Law:

1. "Anti-Personnel Mine" means a munition designed to be placed under, on or near the ground or another surface area and to be detonated or exploded by the presence, proximity or contact of persons;

2. "Anti-Detection-Device" means a device designed to explode or detonate an anti-personnel mine through the use of a mine-detecting device.

Prohibitions

§ 2. The manufacture, acquisition, sale, procurement, import, export, transit, use and possession of anti-personnel mines or anti-detection devices shall be prohibited.

Restrictions

§ 3. (1) Mines foreseen exclusively for training purposes within the Austrian Federal Army or the Mine Clearance or Disarming Service are not subject to the prohibition laid down in § 2.

(2) The import, possession and storage anti-personnel mines for immediate dismantling or other destruction shall be excluded from the prohibition laid down in § 2.

Destruction of existing stocks

§ 4. Existing stocks of anti-personnel mines or anti-detection devices under prohibition of § 2 shall be recorded to the Federal Ministry of the Interior within one month and destroyed by the Federal Ministry of the Interior not latter than one year after this Federal Law has entered into force by reimbursement of costs.

Penalty

§ 5. Whoever takes, and be it negligently only, to contravene the prohibition laid down in § 2 of this Federal Law shall, if the offense is not subject to a more severe sanction under other Federal Laws, be sentenced to imprisonment for up to two years or a fine equivalent to up to 360 per diem rates.

Confiscation and forfeiture

§ 6. (1) Anti-personnel mines or anti-detection devices and components thereof which are object to an act punishable under § 5 shall be confiscated by the court.

(2) Machines and facilities for the manufacture of items subject to the prohibition laid down in § 2 may be forfeited by the court. It shall be secured at the owner's expense that they are no longer in a state to be used in contravention of the prohibition laid down in § 2.

(3) Means used to transport items subject to the prohibition laid down in §2 may be forfeited by the court.

(4) Items under forfeiture according to paras. 2 and 3 shall become property of the public authority. Items confiscated under para. 1 shall became property of the public authority and must be handed over to the Federal Ministry of the Interior for destruction in accordance with § 4.

Execution

§ 7. Entrusted with the execution of this Federal Law are:

1. With reference to § 3 para. 1 the Federal Minister of the Interior and the Federal Minister of Defence,
2. with reference to §§ 5 and 6 the Federal Minister of Justice and
3. with reference to the other provisions the Federal Minister of the Interior.

Entry into force

§ 8. This Federal Law shall enter into force on 1 January 1997.

Annex 2

National legislation

List of States

Below is a list of States which have passed legislation enacting comprehensive prohibitions on anti-personnel mines into their national law and which have sent copies of such legislation to the ICRC. The ICRC welcomes any additional information on the measures taken by other States.

Australia
Austria
Belgium
Canada
France
Germany
Guatemala
Hungary
Italy
New Zealand
Spain
Switzerland
United Kingdom

NOTE

1 Unofficial translation from the Austrian law provided by the Austrian Red Cross.

ICRC Information Paper

First Meeting of States Parties to the Ottawa treaty

Anti-Vehicle Mines Equipped with Anti-Handling Devices

For the purpose of this paper "anti-vehicle mines" means all landmines other than anti-personnel mines

The Ottawa process and the swift entry into force of the Ottawa treaty reflect the international community's commitment to addressing the anti-personnel (AP) mine problem. One of the primary concerns about these weapons is their effect upon innocent civilians especially those moving through an area of conflict or trying to rebuild their lives following the end of hostilities. Through the Ottawa treaty, States have established a total ban on AP mines as well as obligations to destroy AP mine stocks, clear mined areas, and provide resources for mine awareness and victim assistance programs. This comprehensive approach is designed to alleviate the terrible consequences of AP mines and ensure that such a humanitarian tragedy never occurs again.

Under the terms of the Ottawa treaty, *anti-vehicle mines* equipped with anti-handling devices are ***not*** considered to be AP mines and therefore are ***not*** prohibited. As States Parties have renounced the use of AP mines, many armed forces are likely to employ anti-vehicle mines with anti-handling devices as an alternative. The ICRC is concerned that increased use of certain anti-handling devices will endanger civilian populations. Of specific concern are devices which will trigger the mine's detonation through the innocent passage of a person over or near the mine or through inadvertent or accidental contact with the mine itself. This threat is particularly serious with regard to remotely delivered mines which lie on the surface. In such cases the anti-handling device can cause the anti-vehicle mine to function as an AP mine and thus seriously threaten civilian populations.

Experts consulted by the ICRC confirm that anti-handling devices can be designed to minimize the likelihood of detonation resulting from innocent or inadvertent contact. The ICRC urges all States to ensure that anti-vehicle mines they use are so designed so as to avoid creating a new humanitarian problem through the use of anti-vehicle mines with overly sensitive anti-handling mechanisms.

This paper identifies a number of surface-laid anti-vehicle mines fixed with anti-handling devices which should be examined to determine if they are so sensitive so as to detonate through accidental or inadvertent contact. This list is not intended to be exhaustive and States are encouraged to also examine all types of anti-vehicle mines equipped with such devices.

++++

International Committee of the Red Cross
Geneva, April 1999

Surface-laid anti-vehicle mines with anti-handling devices
(includes mines which are mechanically-laid or remotely-delivered)

Name of the AV mine	*AH device*	*Source*
Adrushy	The AH device operates when the mine is tilted more than 20°.	Jane's Mines and Mine Clearance (1997–1998, p. 410)
BLU-91/B GATOR	This mine has a mercury tilt switch, which initiates the mine when a person kicks it, handles it or moves it. Once armed, the mine is dangerous to move.	Jane's Mines and Mine Clearance (1997–1998, p. 269)

Name of the AV mine	*AH device*	*Source*
MI AC DISP F1	The magnetic fuse will initiate the mine if moved.	MineFacts CD-ROM (US Dep. of State, US Dep. of Defense)
MN-111	The magnetic fuse will initiate the mine by any attempt to move or disturb the mine.	MineFacts CD-ROM (US Dep. of State, US Dep. of Defense); Jane's Mines and Mine Clearance (1997–1998, p. 431)
MN-121	The magnetic fuse will initiate the mine by any attempt to move or disturb the mine.	MineFacts CD-ROM (US Dep. of State, US Dep. of Defense); Jane's Mines and Mine Clearance (1997–1998, p. 432)
PTM-3	The magnetic fuse will initiate the mine if moved.	MineFacts CD-ROM (US Dep. of State, US Dep. of Defense)
PYRKAL	The magnetic fuse will initiate the mine if moved.	MineFacts CD-ROM (US Dep. of State, US Dep. of Defense)
SATM	The magnetic fuse will initiate the mine if moved.	MineFacts CD-ROM (US Dep. of State, US Dep. of Defense)
SB-81/AR	electronic anti-disturbance	MineFacts CD-ROM (US Dep. of State, US Dep. of Defense); Jane's Mines and Mine Clearance (1997–1998, p. 135)

Surface-laid anti-vehicle mines with anti-handling devices (*cont.*)
(includes mines which are mechanically-laid or remotely-delivered)

Name of the AV mine	*AH device*	*Source*
TM-72	If the MVN-72 magnetic influence fuse is used, it is possible that this mine may be initiated by movement.	MineFacts CD-ROM (US Dep. of State, US Dep. of Defense); Jane's Mines and Mine Clearance (1997–1998, p. 227)
TMD-1	The magnetic fuse will initiate the mine if moved.	MineFacts CD-ROM (US Dep. of State, US Dep. of Defense)
VS-SATM1	The magnetic fuse will initiate the mine if moved.	MineFacts CD-ROM (US Dep. of State, US Dep. of Defense)
AT2	The AH device prevents the mine to be lifted.	Jane's Mines and Mine Clearance (1997–1998, p. 408)
TMRP-6	This mine can be equipped with an anti-lift device.	MineFacts CD-ROM (US Dep. of State, US Dep. of Defense); Jane's Mines and Mine Clearance (1997–1998, p. 357)
Intelligent Horizontal Mine	This mine is available with an integrated anti-handling mechansism.	MineFacts CD-ROM (US Dep. of State, US Dep. of Defense)
VS-AT4–EL	An electronic anti-lift device can be fitted into the detonator plug on the bottom of the mine.	MineFacts CD-ROM (US Dep. of State, US Dep. of Defense); Jane's Mines and Mine Clearance (1997–1998, p. 421)

Name of the AV mine	*AH device*	*Source*
VS-1.6; VS-2.2; VS-3.6; VS-6.0; VS-9.0	An electronic AH device can be fitted into the detonator plug on the bottom of the mine.	MineFacts CD-ROM (US Dep. of State, US Dep. of Defense); Jane's Mines and Mine Clearance (1997–1998, p. 149)
Type 84	Yes, with magnetic fuse	MineFacts CD-ROM (US Dep. of State, US Dep. of Defense)

Maputo Declaration

Adopted at the First Meeting of States Parties to the Ottawa Treaty

Maputo, Mozambique

7 May 1999

1. We, the States Parties to the Convention on the Prohibition of the Use, Stockpiling, Production and Transfer of Anti-Personnel Mines and on Their Destruction, together with signatory States, are gathered in Maputo, Mozambique, joined by international organizations and institutions and non-governmental organizations, to reaffirm our unwavering commitment to the total eradication of an insidious instrument of war and terror: anti-personnel mines.

2. Even now, at the end of the century, anti-personnel mines continue to maim and kill countless innocent people each day; force families to flee their lands and children to abandon their schools and playgrounds; and prevent long-suffering refugees and displaced persons from returning to rebuild their homes and their lives. The real or suspected presence of antipersonnel mines continues to deny access to much-needed resources and services and cripples normal social and economic development.

3. We raise our serious concern at the continued use of anti-personnel mines in areas of instability around the world. Such acts are contrary to the aims of the

Convention; they exacerbate tensions, undermine confidence and impede diplomatic efforts to find peaceful solutions to conflicts.

4. Therefore, even as we celebrate this First Meeting of the States Parties two months after the rapid entry-into-force of the Convention, we recognize that the enduring value of this unique international instrument rests in fully realizing the obligations and the promise contained within the Convention
– to ensure no new use;
– to eradicate stocks;
– to cease development, production and transfers;
– to clear mined areas and thus free the land from its deadly bondage;
– to assist the victims to reclaim their lives and to prevent new victims.

5. We believe these to be common tasks for humanity and therefore call on governments and people everywhere to join us in this effort.

6. To those who continue to use, develop, produce, otherwise acquire, stockpile, retain and transfer these weapons: cease now, and join us in this task.

7. To those who can offer technical and financial assistance to meet the enormous challenges of humanitarian mine action: intensify your efforts and help build the capacity of mine-affected countries themselves to increasingly take on these tasks.

8. To those who can offer assistance: help with the physical and psycho-social treatment and social and economic reintegration of mine victims, support mine awareness education programmes, and help those States in need to meet treaty obligations to demine and to destroy stockpiles, thus facilitating the widest possible adherence to the Convention.

9. To those that have not yet joined this community of States Parties: accede quickly to the Convention. To those who have signed: ratify. If ratification will take more time: provisionally apply the terms of the Convention while you put in place the necessary domestic legislation.

10. To the international community: promulgate, implement and universalize the Convention, the new international standard and norm of behaviour it is establishing.

11. In this spirit, we voice our outrage at the unabated use of anti-personnel mines in conflicts around the world. To those few signatories who continue to use these weapons, this is a violation of the object and purpose of the Convention that you solemnly signed. We call upon you to respect and implement your commitments.

12. Know that, as a community dedicated to seeing an end to the use of anti-personnel mines, our assistance and cooperation will flow primarily to those who have forsworn the use of these weapons forever through adherence to and implementation of the Convention.

13. Driven by the sad reality that the people of the world will continue to suffer the consequences of the use of anti-personnel mines for many years to come, we believe it crucial that we use this First Meeting of the States Parties to ensure that we make continued, measurable progress in our future efforts to eradicate anti-personnel mines and to alleviate the humanitarian crisis caused by them.

14. We recognize that anti-personnel mines represent a major public health threat. The plight of mine victims has revealed the inadequacy of assistance for victims in the countries most affected. Such assistance must be integrated into broader public health and socioeconomic strategies to ensure not simply short-term care for victims, but special attention to the serious long-term needs for social and economic reintegration. Mine victims must be permitted to realize, with dignity, their place within their families and their societies. These issues must be accorded the highest political importance and practical commitment by States Parties and all those in the international community who care about this issue.

15. To this end, we commit ourselves to mobilise resources and energies to universalize the Convention, alleviate and eventually eradicate the human suffering caused by anti-personnel mines, including by striving to meet the goal of "zero victims".

16. For these purposes, we, the States Parties, will implement an intersessional work programme to take us steadily forward to the next Meeting of the States Parties, which will take place in Geneva from 11 to 15 September 2000. This will enable us to focus and advance our mine action efforts and to measure progress made in achieving our objectives. This work will be based on our tradition of inclusivity, partnership, dialogue, openness and practical cooperation. In this regard, we invite all interested governments, international organizations and institutions and non-governmental organizations to join us in this task.

17. Our work programme will draw together experts, building on the discussions held here in Maputo, to address the key thematic issues of:

- the general status and operation of the Convention;
- mine clearance;
- victim assistance and mine awareness;
- stockpile destruction; and,
- technologies for mine action.

This intersessional work will, *inter alia,* assist us in developing, with the United Nations, a global picture of priorities consistent with the obligations and time-frames contained within the Convention, including with regard to international cooperation and assistance. It will also take into account important work done at the international, regional and sub-regional levels.

18. The work of our experts will begin just four months from now, in Geneva. We appreciate and accept the offer of the Geneva International Centre for Humanitarian Demining to support our efforts. Our work will complement and reinforce the important mine action activities being undertaken by mine-affected States working in partnership with other States, international and regional organizations, non-governmental organizations and the private sector – also recognizing the United Nations system as an important actor in global mine action efforts.

19. Meeting here in one of the most mine-affected continents on earth and in a country which has experienced the ravages wreaked by these weapons on the Mozambican people and the social fabric of the nation, we focus our minds and strengthen our conviction on the need to make the killing fields of anti-personnel mines that have terrorized, maimed and killed people, destroyed lives and hope for too long, a relic of the past.

We are determined to succeed in our common task.

We are determined to work in partnership to this end.

We are determined to apply the principle of international humanitarian law, enunciated in the final preambular paragraph of the Convention itself that "... the right of the parties to an armed conflict to choose methods or means of warfare is not unlimited..."

This is our firm pledge to future generations.

24

Regional Conference on Landmines

Zagreb, Croatia

27–29 June 1999

Organized by the government of Croatia

This regional conference was organized by the government of Croatia as a follow-up to the meeting hosted by the government of Hungary in March 1998. Its purpose was to examine the current state of affairs in the regional efforts to eliminate landmines and to promote the ratification and implementation of the Ottawa treaty. The meeting brought together representatives from 33 countries from within and outside central and eastern Europe to discuss issues related to mine awareness, mine victim assistance, mine clearance, the military utility of anti-personnel mines and the status of ratification and implementation of the Ottawa treaty.

Address by Cornelio Sommaruga
President of the International Committee of the Red Cross

Zagreb Regional Conference on Landmines

28 June 1999

Eliminating Anti-personnel Landmines: Continuing the process

It is an honour and a pleasure to address the Zagreb Regional Conference on Landmines. I thank the Government of Croatia, the Croatian Red Cross and the Croatian Mine Action Centre for inviting the International Committee of the Red Cross to take part in this event and for inviting me to address this important meeting. By hosting this meeting, Croatia is helping to keep the spotlight on the plague of landmines which this region knows all too well. But we will also focus on the comprehensive solution to this problem contained in the Ottawa treaty and the crucial efforts being made in the region to implement its provisions.

Much has occurred since the countries of this region last came together to discuss landmines. As many of you will recall, in March 1998, only 3 months after the signing of the Convention on the Prohibition on the Use, Stockpiling, Production and Transfer of Anti-personnel Mines, the Government of Hungary took an important initiative in hosting the Budapest Regional Conference on Anti-personnel Landmines. That meeting was a significant step because it recognized the importance of regional co-operation in addressing the landmine problem and helped further the dialogue on this issue between governments and civil society. By reconvening governments, non-governmental organizations, international agencies and other interested parties in Zagreb this week, the Government of Croatia continues that process, provides an opportunity to examine what has been accomplished thus far in the region and focuses our attention on what remains to be done before the scourge of landmines will be forever removed from this part of the world.

Since Budapest, the international community has moved with unusual speed. On 1 March 1999, the Ottawa treaty entered into force becoming international law more rapidly than any previous multilateral arms related convention. Today, 82 States have now ratified the treaty. Another 53 States are committed to its object and purpose as a result of having signed it. Thus, over two thirds of the world's governments have now committed themselves to ending the use of these pernicious weapons, destroying their stocks of anti-personnel mines, and acting to address the plight of mine victims and mine-affected communities. Further indications of this trend are the decision of Ukraine to sign the treaty and begin the destruction of its stockpiles, the decision of Russia to end its production of some types of anti-personnel blast mines and the recent bilateral agreement between Turkey and Bulgaria not to use anti-personnel mines under any circumstances and to remove and destroy all such mines along their common border.

The Ottawa treaty represents the <u>comprehensive</u> framework for ending the man-made epidemic of death, injury and suffering caused by anti-personnel mines. Yet, its objectives will only be achieved when the treaty is <u>fully implemented</u> and <u>unversalized</u>. States which have ratified the treaty must begin to take measures to ensure its complete implementation. This includes, among other things, the adoption of national legislation to prevent and punish violations, institution of changes in military doctrine and procedures, the promulgation of measures to ensure the destruction of stockpiles and the identification and clearing of mined areas. States in a position to do so must also take steps to provide for the care and rehabilitation of landmine victims. But, governments do not need to wait until they have formally ratified the treaty to begin to undertake such action. Signatories in particular, are encouraged to begin the process of implementation and by doing so hasten the complete elimination of these weapons. Even non-signatories have done, and can do, much to contribute to solving the problem on the ground.

Yet, even in light of these developments there are States which still consider the anti-personnel mine to be a legitimate weapon of warfare. I would say to them today that you rely upon a weapon the military value of which has been questioned and the use of which has been stigmatized by world opinion. Many of the 135 countries which have adhered to the Ottawa treaty used these devices extensively in the past. Many are former producers. But they have nonetheless all come to agree that any utility the anti-personnel mine may possess is clearly outweighed by its appalling human costs. The use of these weapons by any State in any circumstances offends the public conscience and basic principles of humanity.

Let us recall that the anti-personnel mine is unable to discriminate between civilian and combatant. This is also a weapon which, by design, inflicts some of the most horrific wounds seen by war surgeons. The explosion from a buried anti-personnel mine forces earth, grass and pieces of the mine casing and victim's shoe into the wounded areas. It causes injuries so severe, that the victim will have to undergo multiple operations, prolonged rehabilitation and be left with a permanent and severe disability. Coupled with these are the social, psychological and economic implications of being an amputee. Should not such injuries, even when inflicted on soldiers, be considered to exceed that which is needed to take a soldier "out of combat" – which is the only legitimate purpose of anti-personnel weapons? I again urge countries which have not yet done so to join the Ottawa treaty as a matter of urgency.

I would also encourage all States to ratify amended Protocol II to the 1980 United Nations Convention on Conventional Weapons. This instrument, regulating landmines, booby traps and other similar devices is particularly important, even for States Parties to the Ottawa treaty, because it governs the use of anti-vehicle mines. These devices are not covered by the Ottawa treaty but nonetheless have had severe impact on civilian populations in many regions of the world. Amended Protocol II entered into force on 3 December 1998 and thus far has been ratified by 37 governments.

Many of the countries represented here know well the horrible consequences associated with the use of landmines. This area, in particular, is a testament to the full range of human, social and economic costs of these weapons. In this sub-region scores of men, women and children have been killed and maimed by these devices. The presence of landmines has slowed the pace of post-war reconstruction, hindered the return of refugees and displaced people and blocked the cultivation of valuable farmland. Today, in parts of Bosnia and Herzegovina and

Croatia landmines continue to attack civilian populations nearly 4 years after the end of the fighting. One cannot help but feel a sense of outrage about the senseless deaths, injury and suffering landmines are inflicting here and in other parts of the world. It is time for this man-made plague to come to an end. It is my hope that this common belief and the desire to contribute to the efforts have brought us together here today.

This conference will learn of and pay tribute to the many efforts already underway to alleviate the suffering that these weapons have wreaked and prevent further destruction. Demining teams are at work in the countryside, mine awareness is being taught in the classrooms, rehabilitation clinics are producing and fitting prostheses and mine survivor organizations are active in several countries. In Bosnia and Herzegovina reports indicate that mine casualty levels have dropped significantly due to these courageous and painstaking efforts. Yet, much remains to be done before one will walk in these lands without fear. Vast areas remain mine-affected and too many victims do not have access to the full range of rehabilitation they require. Our task in the coming 2 days is to determine how, through regional co-operation and the support of the international community, the objectives of the Ottawa treaty can become realities on the ground sooner rather than later.

For its part, the International Committee of the Red Cross has been actively involved in this region in protecting civilian populations from the effects of mines. Since early 1996, the ICRC has conducted extensive mine awareness training in both Croatia and Bosnia and Herzegovina in close co-operation with the respective Red Cross Societies. In both countries the ICRC employs a community based approach to involve local communities in mine awareness activities. In Croatia, the ICRC works very closely with the Croatian Red Cross which together have trained 150 volunteer mine awareness instructors. Thus far, over 140,000 men, women and children have been taught about the dangers of landmines and unexploded ordnance. In Bosnia and Herzegovina, 122 volunteers from approximately 100 municipalities across the country have been instructed in conducting mine awareness training. The ICRC is currently working with other organizations to ensure that refugees returning to Kosovo are provided with basic information about the dangers of mines prior to their return. The ICRC is also working with a number of governments to assist them in their ratification and implementation process for both the Ottawa treaty and the 1980 Convention on Conventional weapons.

In closing, let us recall that it was in this region that Europe learned the bitter lessons of landmines firsthand. It is this experience which helped convince most European States to renounce the use of anti-personnel mines. Let us make this

Zagreb meeting the moment when governments and civil society in the region pay tribute to those who have already lost lives and limbs by redoubling efforts to save others from this plight, to assist those in need and to ensure that no country will ever use anti-personnel mines again.

Chairman's Summary
Zagreb Regional Conference on Landmines

27–29 June 1999

The Second Regional Conference on Antipersonnel Landmines was held in Zagreb, Croatia, 27–29 June 1999. Participants from 33 Countries, as well as 14 International Organizations and 50 nongovernmental organizations attended the Conference. At the beginning of the Conference Mrs. Jadranka Kosor, Deputy President of the Croatian State Parliament welcomed the Participants. The Conference was opened by Dr. Mate Grani, Deputy Prime Minister and Minister of Foreign Affairs of Croatia. Cornelio Sommaruga, President of The ICRC, Mrs. Jody Williams, ICBL Ambassador and the recipient of the 1997 Nobel Peace Prize and Ozren Unec, Croatian Campaign to Ban Landmines also delivered opening statements.

Following the opening session a general debate was held and 19 delegates took part in the discussion. Vladimir Drobnjak, Croatian Assistant Minister of Foreign Affairs, chaired the Conference. The work of the Conference proceeded in the form of Panels, where following topics were discussed: The Status and Implementation of the Ottawa Treaty, Mine Awareness, Victim Assistance, Rethinking Military Doctrine, Demining and Stockpile Destruction.

On the first day of the Conference, on the Military Field in Slunj, Croatian Army destroyed 3434 antipersonnel landmines. This has marked the beginning of Croatia's fulfillment of Article 4 of the Ottawa Convention. Thus the Republic of Croatia has set an example to all other countries in the region to follow her footsteps.

During the discussion the Participants reaffirmed the importance of universality of the Convention on the Prohibition of the Use, Stockpiling, Production and Transfer of Antipersonnel Mines and their Destruction and called for the full support to implementation of the Convention. They have welcomed the fact that 82 countries so far have ratified the Convention, hence demonstrating a strong determination of the overwhelming majority of international community to eliminate this weapon.

With the aim of contributing to the universality of the Ottawa Convention, the Participants called upon all countries which have not yet done so to ratify or accede to the Convention. In that context several delegations have stated that the ratification process in their countries has already begun. That fact was welcomed with appreciation and further support to the implementation of the Convention was pledged.

The Participants also agreed that continued cooperation at both regional and global level is a fundamental framework for ensuring the implementation of the Convention, and affirmed their readiness to continue with such cooperation.

Attention was specially drawn to the recent Kosovo crisis, where antipersonnel landmines were recently implanted. The problem of antipersonnel landmines in the other countries in the region, notably Croatia and Bosnia and Herzegovina, which are among those that are mostly affected both in the region and the world, was also emphasized.

The international community was urged to help the mine affected countries in the region through necessary means, both financial and technical, in order to facilitate, speed up and make safer the process of mine clearance.

Participants to the Regional Conference welcomed a notable cooperation between Governments and non-governmental organizations. Strenuous efforts undertaken by the NGOs, as well as their valuable initiatives and influence contribute to the greater dynamics of the whole process. The Participants also expressed their hope that the cooperation between Governments and civil society would intensify throughout the region.

The Participants agreed that antipersonnel landmines cause enormous damage and casualties in the aftermath of a military conflict, thus primarily endangering civilian populations, and causing serious obstacles to the post-conflict rebuilding. In that context, the Participants attached great importance to the issue of economic reconstruction in the contaminated areas where landmines pose a serious obstacle to the economic and social development. They stressed the need for their rapid elimination as an important precondition for the revitalization and normalization of life in those areas.

The participants also expressed their concern regarding the danger that landmines pose to the return of refugees and displaced persons, slowing down the process, and called for their elimination in order to allow unimpeded and safe return to the affected areas.

The importance of rehabilitation of landmine victims was specially emphasized. The Participants called upon all the countries and international agencies to cooperate and participate with mine affected countries in the creation and implementation of the programmes and methods for each country, bearing in mind its own characteristics aiming to provide all sort of necessary assistance to the victims. The role of international agencies and NGOs is absolutely crucial in this segment.

On the issue of mine-awareness it was emphasized that it should not only be a method of educating people how to avoid and cope with the problem of land-mines, but also a preventive mechanism which develops a long term general awareness aimed at total ban of this appalling weapon and its eventual elimination.

Conference provided an opportunity to assess again the limited military usefulness of landmines in military conflicts compared to their long-term socio-economic impact on civilian populations. The need for alternative means and techniques which would replace antipersonnel landmines in the military doctrine, was especially emphasized.

The Participants expressed their gratitude to the Republic of Croatia for organizing the Regional Conference on Landmines in 1999, as a follow up to the First Regional Conference on Antipersonnel Landmines, held in Budapest, 1998.

It is noted with pleasure that the Republic of Slovenia has accepted to host the Third Regional Conference next year. This will provide an opportunity to strengthen the cooperation between countries in the region, as well as to asses the ongoing achievements and developments connected to the implementation of the Convention and the process in general.

25

South Asia Regional Seminar on Landmines

Wadduwa, Sri Lanka
18–20 August 1999
Organized by the ICRC

The South Asia Regional Seminar on Landmines marked the first time that senior officials from the region gathered to discuss the landmines issue. Participants included representatives from the ministries of foreign affairs and defence of Bangladesh, Bhutan, India, Nepal, Pakistan and Sri Lanka. Discussions were held on a wide range of topics, including the historical use of mines in South Asia, the use of mines and improvised explosive devices by non-State actors, the military utility of and alternatives to anti-personnel mines and the developing norms on the control and prohibition of the weapon. At the time of the meeting few States in the region were parties to the Ottawa treaty or the 1980 Convention on Certain Conventional Weapons. Yet all participants recognized the humanitarian consequences associated with anti-personnel mines and the need to respond positively to the international movement towards the weapon's elimination. As several countries in the region were reluctant to renounce immediate use of anti-personnel mines, preferring a more gradual approach, the meeting also examined incremental measures which could be taken at a national or a regional level to further their elimination. Many of these proposals were favourably received and participants agreed to pursue their consideration in their respective capitals.

South Asia Regional Seminar on Landmines

Wadduwa, Sri Lanka

18–20 August 1999

Summary Report

The South Asia Regional Seminar on Landmines was held 18 to 20 August in Wadduwa, Sri Lanka. Participants included senior officials from ministries of foreign affairs and ministries of defence of Bangladesh, Bhutan, India, Nepal, Pakistan and Sri Lanka as well as international and non-governmental organizations, the International Committee of the Red Cross and independent experts from within and outside of the region.

This was the first time that governments in the region gathered to discuss the issue of landmines and regional and national responses to this global problem. Participants were welcomed to the conference by Mr. D. Serasinghe, Director-General for Political Affairs, Ministry of Foreign Affairs, Sri Lanka and Max Hadorn, the head of ICRC's delegation in Sri Lanka.

Informal discussion addressed the following issues:

- the historical use of anti-personnel mines in South Asia and regional security;
- the rapidly developing norms on the control and prohibition of these weapons;
- the evolution of military doctrine and alternatives embraced by States which have chosen to renounce the use of anti-personnel mines.
- the use of landmines and improvised explosive devices by non-State actors.

All participants agreed that the global scourge of anti-personnel mines, which kill and injure some 2,000 persons per month including a significant number of civilian victims in certain countries in South Asia, is unacceptable and must be stopped. Because these mines kill and maim long after the conflict is over, it was recognized that addressing this problem should be a priority for the international community and the South Asia region.

Participants recognized the commitment of the vast majority of States to the elimination of anti-personnel mines and the need for the region to play a constructive role in this process.

Discussion revealed that South Asia includes several states which are prepared to adhere to the Ottawa treaty banning anti-personnel mines in the relatively near future and others which see their elimination as a step by step process requiring the identification of alternatives, adjustments to military practices and the reduction of tensions in the region.

Ending the use of anti-personnel mines and similar devices by non-State actors was seen as an important element in the elimination of these weapons in the region. Participants highlighted existing legal obligations prohibiting the transfer of landmines to non-State entities[1] and welcomed the efforts of non-governmental organizations to engage such groups in dialogue on this subject.

Participants appreciated the presentations on alternative technologies and tactics which would reduce the need for anti-personnel mines. It was felt that the active pursuit of such alternatives in the region should be encouraged.

The central task of this meeting was to identify incremental steps which could be taken nationally and/or regionally to further the goal of the eventual elimination of anti-personnel mines. In particular the following proposals were discussed.

1. Non-use of mines on certain common borders.

States not yet able to adhere to a ban on anti-personnel mines and neighbors with whom those States have friendly relations could elaborate bilateral agreements on the non-use of anti-personnel mines on common borders. This could strengthen bilateral ties, facilitate adherence to the Ottawa treaty as well as amended Protocol II of the 1980 Convention on Certain Conventional Weapons (CCW) by a number of States in the region and promote a positive climate for further cooperative measures on mine-related issues in South Asia. Such agreements would undoubtedly be warmly welcomed by the international community. Most participants felt that such agreements would provide a valuable sign of the region's willingness to move towards a ban on anti-personnel mines and should be actively encouraged.

2. Regionalising existing national or legal bans on transfers of anti-personnel mines.

Parties to amended Protocol II of the CCW, have undertaken a legal commitment not to transfer any landmines to non-State entities. All regional States have national policies prohibiting international transfers of anti-personnel mines. Conversion of these commitments into binding national law, as has occurred in Pakistan, was considered to be an important step.

Representatives of some States expressed the desire for channels of communication, consultation and information exchange to help identify the sources of illegal flows of mines into their territory. Regional agreements or mechanisms to structure such exchanges could reinforce existing transfer bans and provide confidence in their implementation. It was noted that an open ended mine forum for discussion of transfer, mine clearance and other issues has been found to be useful by countries of the Southern Africa Development Community.

These proposals were generally welcomed by the participants.

3. Agreement to prohibit the introduction of remotely-delivered mines into the region.

The introduction of remotely delivered mines into the region could increase the likelihood of massive use of landmines in a future conflict and significantly increase civilian casualty levels. India has previously proposed a global ban on these weapons. It was felt that both bilateral and multi-lateral agreements on the non-use of such systems and an international prohibition on their use should be encouraged.

4. An increased regional contribution to global efforts in the fields of mine clearance and victim assistance.

Some South Asian countries have considerable expertise in mine clearance and have contributed to international efforts in this field. There was general agreement that this expertise could be made more widely available if adequate resources from multi-lateral institutions could be identified.

All participants agreed that cooperation in the field of assistance for mine victims could be an important focus of cooperation in the South Asian Association for Regional Cooperation (SAARC).

5. Adherence by additional States to existing international conventions on landmines

a. The Ottawa Treaty banning anti-personnel mines

Discussion indicated that most countries in the region support the eventual elimination of anti-personnel mines and appropriate steps towards this goal. As indicated above, several States anticipate an early ratification of the Ottawa Treaty whereas others favor a gradual approach.

b. Protocol II to the 1980 Convention on Certain Conventional Weapons.

The recently amended version of this protocol strengthens prohibitions on the use of anti-vehicle mines, improvised explosive devices and booby traps. Adherence to this instrument is considered consistent with adherence to the Ottawa Treaty and is an important element in addressing the broader landmine problem. The CCW also contains important restrictions and prohibitions on other inhumane weapons such as blinding laser weapons and incendiary weapons. Participants from several states which have not yet adhered to this instrument anticipated that their governments would be able to give favorable consideration to adherence in the not too distant future.

6. Limiting landmine use by non-State actors

The use of mines by non-State actors was an important preoccupation of most participants. In addition to strengthened controls on transfers, as mentioned above, concerted political efforts to persuade non-State actors to end the use of anti-personnel mines were considered essential. Non-governmental organizations, religious bodies and international organizations were recognized as having an important role to play in these efforts.

Participants expressed thanks to the Government of Sri Lanka and the ICRC for providing an important opportunity for regional dialogue and for their hospitality. They expressed a willingness to pursue consideration of the proposals indicated above upon their return to their respective capitals.

NOTE

1 Specifically, Article 8(b) of amended Protocol II to the 1980 Convention on Certain Conventional Weapons (CCW).

INDEX